Optofluidics

Biophotonics Series

Series Editors: Israel Gannot and Joseph Neev

Artificial Tactile Sensing in Biomedical Engineering by Siamak Najarian, Javad Dargahi, and Ali Abouei Mehrizi

Biomedical Applications of Light Scattering, edited by Adam Wax and Vadim Backman

Optofluidics: Fundamentals, Devices, and Applications, edited by Yeshaiahu Fainman, Luke P. Lee, Demetri Psaltis, and Changhuei Yang

Organic Electronics in Sensors and Biotechnology, edited by Ruth Shinar and Joseph Shinar

Optofluidics

Fundamentals, Devices, and Applications

Yeshaiahu Fainman

Luke P. Lee

Demetri Psaltis

Changhuei Yang

McGraw Hill

New York Chicago San Francisco
Lisbon London Madrid Mexico City
Milan New Delhi San Juan
Seoul Singapore Sydney Toronto

The McGraw·Hill Companies

Cataloging-in-Publication Data is on file with the Library of Congress

1 2 3 4 5 6 7 8 9 0 DOC/DOC 0 1 5 4 3 2 1 0 9

ISBN 978-0-07-160156-6
MHID 0-07-160156-2

Sponsoring Editor
Taisuke Soda

Editing Supervisor
Stephen M. Smith

Production Supervisor
Pamela A. Pelton

Project Manager
Ekta Dixit, International Typesetting and Composition

Copy Editor
Mohinder Bhatnagar, International Typesetting and Composition

Proofreader
Ojen Yumnam, International Typesetting and Composition

Indexer
Robert Swanson

Art Director, Cover
Jeff Weeks

Composition
International Typesetting and Composition

Printed and bound by RR Donnelley.

Contents

Contributors

Andrea Armani *Mork Family Department of Chemical Engineering and Materials Science, University of Southern California, Los Angeles, California* (Chap. 13)

Sung Hwan Cho *Materials Science and Engineering Program, Jacobs School of Engineering, University of California at San Diego* (Chap. 9)

Su Eun Chung *Department of Electrical Engineering, Seoul National University, Seoul, Republic of Korea* (Chap. 17)

Tomáš Čižmár *SUPA, School of Physics and Astronomy, University of St Andrews, North Haugh, Fife, Scotland* (Chap. 15)

Xiquan Cui *Department of Electrical Engineering and Bioengineering, California Institute of Technology, Pasadena, California* (Chap. 11)

Kishan Dholakia *SUPA, School of Physics and Astronomy, University of St Andrews, North Haugh, Fife, Scotland* (Chap. 15)

Peter Domachuk *CUDOS, School of Physics, University of Sydney, Sydney, Australia* (Chap. 7)

Benjamin J. Eggleton *Institute of Photonics and Optical Science (IPOS), Centre for Ultrahigh-Bandwidth Devices for Optical Systems (CUDOS), School of Physics, University of Sydney, Sydney, Australia* (Chaps. 7, 18)

David Erickson *Sibley School of Mechanical and Aerospace Engineering, Cornell University, Ithaca, New York* (Chap. 5)

Yeshaiahu Fainman *Department of Electrical Engineering, University of California, San Diego, California* (Chaps. 8, 14)

Jessica Godin *Electrical and Computer Engineering, Jacobs School of Engineering, University of California at San Diego* (Chap. 9)

Christian Karnutsch *Institute of Photonics and Optical Science (IPOS), Centre for Ultrahigh-Bandwidth Devices for Optical Systems (CUDOS), School of Physics, University of Sydney, Sydney, Australia* (Chap. 18)

Shin-Hyun Kim *National Creative Research Initiative Center for Integrated Optofluidic Systems and Department of Chemical and Biomolecular Engineering, KAIST, Daejeon, Republic of Korea* (Chap. 6)

Anders Kristensen *Department of Micro and Nanotechnology, Technical University of Denmark* (CHAP. 10)

B. Kuhlmey *CUDOS, School of Physics, University of Sydney, Sydney, Australia* (CHAP. 7)

Sunghoon Kwon *Department of Electrical Engineering, Seoul National University, Seoul, Republic of Korea* (CHAP. 17)

Luke P. Lee *Department of Bioengineering, University of California—Berkeley* (CHAP. 19)

Seung Ah Lee *Department of Electrical Engineering, Seoul National University, Seoul, Republic of Korea* (CHAP. 17)

Seung-Kon Lee *National Creative Research Initiative Center for Integrated Optofluidic Systems and Department of Chemical and Biomolecular Engineering, KAIST, Daejeon, Republic of Korea* (CHAP. 6)

Uriel Levy *Department of Applied Physics, The Benin School of Engineering and Computer Science, The Hebrew University of Jerusalem, Jerusalem, Israel* (CHAP. 4)

G. Logan Liu *Department of Electrical and Computer Engineering, University of Illinois at Urbana-Champaign* (CHAP. 19)

Yu-Hwa Lo *Electrical and Computer Engineering, Jacobs School of Engineering, University of California at San Diego* (CHAP. 9)

Christelle Monat *Institute of Photonics and Optical Science (IPOS), Centre for Ultrahigh-Bandwidth Devices for Optical Systems (CUDOS), School of Physics, University of Sydney, Sydney, Australia* (CHAP. 18)

N. Asger Mortensen *Department of Photonics Engineering, Technical University of Denmark* (CHAP. 10)

Lin Pang *Department of Electrical Engineering, University of California at San Diego* (CHAP. 14)

Shuo Pang *Department of Electrical Engineering, California Institute of Technology, Pasadena, California* (CHAP. 1)

Wook Park *Department of Electrical Engineering, Seoul National University, Seoul, Republic of Korea* (CHAP. 17)

Joanna Ptasinski *Department of Electrical Engineering, University of California at San Diego* (CHAP. 14)

Wen Qiao *Electrical and Computer Engineering, Jacobs School of Engineering, University of California at San Diego, and 3 State Key Laboratory, National Engineering Research Center (NERC) for Optical Instruments, Zhejiang University, Hangzhou, People's Republic of China* (CHAP. 9)

Dominik G. Rabus *Baskin School of Engineering, University of California, Santa Cruz* (CHAPS. 12, 16)

Boris Slutsky *Department of Electrical Engineering, University of California at San Diego* (Chap. 14)

P. Steinvurzel *CUDOS, School of Physics, University of Sydney, Sydney, Australia* (Chap. 7)

Sindy K. Y. Tang *Department of Chemistry and Chemical Biology, Harvard University, Cambridge, Massachusetts* (Chaps. 2, 3)

Snjezana Tomljenovic-Hanic *Institute of Photonics and Optical Science (IPOS), Centre for Ultrahigh-Bandwidth Devices for Optical Systems (CUDOS), School of Physics, University of Sydney, Sydney, Australia* (Chap. 18)

Frank S. Tsai *Electrical and Computer Engineering, Jacobs School of Engineering, University of California at San Diego* (Chap. 9)

George M. Whitesides *Department of Chemistry and Chemical Biology, Harvard University, Cambridge, Massachusetts* (Chaps. 2, 3)

Allen H. J. Yang *Chemical and Biomolecular Engineering, Cornell University, Ithaca, New York* (Chap. 5)

Changhuei Yang *Department of Electrical Engineering, California Institute of Technology, Pasadena, California* (Chaps. 1, 11)

Seung-Man Yang *National Creative Research Initiative Center for Integrated Optofluidic Systems and Department of Chemical and Biomolecular Engineering, KAIST, Daejeon, Republic of Korea* (Chap. 6)

Steve Zamek *Department of Electrical Engineering, University of California, San Diego, California* (Chap. 8)

About the Editors

Yeshaiahu Fainman is Cymer Professor in the Department of Electrical and Computer Engineering at the University of California San Diego. He is a Fellow of the Optical Society of America (OSA), the Institute of Electrical and Electronics Engineers (IEEE), and the Society of Photo-Optical Instrumentation Engineers (SPIE).

Luke P. Lee is the Lloyd Distinguished Professor of Bioengineering at the University of California Berkeley. He is also the Director of the Biomolecular Nanotechnology Center and Co-Director of the Berkeley Sensor and Actuator Center at UC Berkeley. He is the leader of Berkeley's BioPOETS (Biologically-Inspired Photonics-Optofluidics-Electronics Technology and Science) group.

Demetri Psaltis is a Professor of Optics and the Dean of Engineering at Ecole Polytechnique Fédérale de Lausanne. He is a Fellow of the Optical Society of America (OSA), the Institute of Electrical and Electronics Engineers (IEEE), and the Society of Photo-Optical Instrumentation Engineers (SPIE).

Changhuei Yang is an Assistant Professor of Electrical Engineering and Bioengineering at the California Institute of Technology. He was named one of the top 20 scientists under 40 in *Discover* magazine's list of Best Brains in Science, 2008.

CHAPTER 1

Introduction

Changhuei Yang and Shuo Pang

Department of Electrical Engineering, California Institute of Technology, Pasadena, California

1-1 Introduction

The term "optofluidics" was coined in 2003 to define an emergent research field that focuses on combining microfluidic and optical technology [1–3]. In the space of 5 years, this terminology has become widely adopted and applied as a categorical descriptor for a large number of research directions. The input of the term "optofluidic" in Google yields around 24,000 webpage results.

It is certainly true that some of the research projects that have adopted "optofluidics" as part of their descriptor could have evolved independently. After all, we can find examples of research projects that combine fluidics and optics that predated the genesis of "optofluidics"—the electrowetting lens (Chap. 9) is an excellent example. However, the large number and wide variety of "optofluidic"-themed research projects that have cropped up over the past 5 years indicate that the definition of optofluidics as a field is causally linked to the proliferation of at least a few such projects. Once the seed idea of combining the advantages of microfluidics and optics was formally defined, it did not take long for the concept to take hold in the minds of researchers and germinate prolifically. The optofluidic microscope (Chap. 11) and optofluidic lasers (Chap. 10) are some of the projects for which causal links between the birth of the term "optofluidics" and the initiation of the projects can be traced.

This leads to the question: "What exactly is optofluidics?" In the next subsection, we will address this question. We will then briefly examine the advantages of optics and microfluidics and discuss briefly some of the ways these two disciplines can combine to generate optofluidic technologies with unique capabilities.

1-2 What Is Optofluidics? A Historical Perspective

Remarkably, the definition of optofluidics has evolved significantly over the few years that it has been in existence. The term "optofluidics" first appeared in the name of a University Research Center funded by the Defense Advanced Research Projects Agency (DARPA) in 2003. The charter of the center was to "develop adaptive optical circuits by integrating optical and fluidic devices." This optics-centric definition points to an interesting aspect of this field's origin—optics researchers were trying to incorporate microfluidic technologies into their research to create novel optical devices. It was recognized from the start that microfluidic technologies can enable changeable and reconfigurable optical devices (see Chaps. 2, 3, and 4 for some examples).

It quickly became apparent that microfluidics can bring other advantages to bear. In Ref. [1], several other aspects of fluidics were identified as key advantages for optofluidics: "the ability to change the optical property of the fluid medium within a device by simply replacing one fluid with another; the optically smooth interface between two immiscible fluids; and the ability of flowing streams of miscible fluids to create gradients in optical properties by diffusion." The focus of optofluidics on the creation of novel optical devices remained.

A review paper in 2007 [2] marked the shift to a more symmetric definition in which the advantages of optofluidic technologies were discussed as beneficial to both the optics and the microfluidics fields.

In the present context, an appropriate description of optofluidics would be to broadly define it as the combination of optics and microfluidics in the same platform to leverage specific advantages of these two disciplines.

1-3 Fluidic Advantages

There are numerous advantages associated with fluid media that optofluidic researchers have utilized. In this section, we shall look at some of these features.

1-3-1 Immiscible Fluid-Fluid Interfaces Are Smooth

It has long been recognized that the optical smoothness of fluid interfaces can be a useful and cost-effective way to create optical surfaces. Due to surface tension, an immiscible fluid-fluid interface is uniform and smooth. Liquid telescope mirrors that are created by spinning large dishes of mercury work on this principle [4]. On a much smaller scale, most optofluidic lens projects, likewise, make use of this principle (Chap. 9). It is worth noting that the meniscus between two immiscible fluids of equal density in a column is perfectly spherical—a curvature profile that is used in most commercially available lenses.

It is also interesting to note that the usefulness of this advantage extends beyond devices that have dynamically controllable fluidic surfaces; this advantage also enables low-cost and easy fabrication of optical components. For example, the toroid optical resonators discussed in Chap. 13 are able to achieve their high optical quality factor through the melting and solidification of the resonators' rims to create smooth optical tracks.

1-3-2 Diffusion Can Create Controllable Blend of Optical Properties

Miscible liquids and their interfaces can also be of significant use in the optofluidic context [1].The solid-based structures failed to provide the property that can be created by the diffusion across the interface of two liquids. Specifically, the diffusion process can create a concentration and refractive-index gradient which is smooth and controllable. The controllability and flexibility by which this gradient can be adjusted through flow parameters, fluid choices, and the device structures enable the creation of novel optical interconects. For example, an optical splitter and wavelength filter based on the selective mixing of two fluid jets in a third fluidic medium has been demonstrated (Chap. 3). Unlike a conventional beamsplitter, the split ratio of the optofluidic beamsplitter can be dynamically tuned for any given wavelength.

1-3-3 Fluid Can Be an Excellent Transport Medium

It is relatively easy to input, move, and manipulate fluid in an optofluidic device. Pressure differential is a common and convenient means. Electrokinetic approaches provide another set of flow control mechanisms (see Chap. 2 for more information). Over the past few years, several optical approaches for manipulating fluid have also been developed (Chaps. 5, 8, and 19). The optofluidic microscope (OFM) (Chap. 11) capitalizes on this advantage by using microfluidics as the means for sample input and microfluidic flow as the scanning mechanism during image acquisition. The optofluidic maskless lithography approach (Chap. 17) is another excellent example of an optofluidic technology that makes good use of fluid transport.

The easy transport of fluids benefits the field of optofluidics in three other ways. First, we can use the change of fluid media in an optofluidic device as a way to alter the properties of the device—thus, allowing us to create adaptable devices (see Chaps. 7, 12, and 18 for some excellent examples). Some of the properties that can be altered this way include refractive indices, spectral absorption coefficients, and scattering coefficients. The optofluidic lasers (Chap. 10), for example, depend on the switching of laser dye medium as a way to accomplish wavelength tuning.

Easy fluid transport is also useful for "renewing" optofluidic devices—an advantage that solid devices do not possess. Specifically, as and when the fluid media in an optofluidic device deteriorates, we can easily infuse the device with fresh fluid replacements. This advantage is very useful for optofluidic lasers as the lasing media in such devices need to be replaced when the dyes are bleached.

Finally, easy fluid transport enables the intriguing possibility of on-chip chemical analysis and synthesis by providing an easy means for inputs and transport. See Chap. 16 for a discussion on this topic.

1-3-4 Fluid Can Be an Excellent Buoyancy Mediator

The density of fluid media ranges widely—mercury has a density of 13.6 g/cm^3 while pentane has a density of 0.63 g/cm^3. By mixing two miscible fluids, we can create fluid with arbitrary intermediate density values.

The buoyancy of fluid facilitates manipulation of small objects that are suspended in a suitable fluid medium. Optical tweezer technology (Chaps. 5 and 15) provides an excellent illustration of this advantage. Optical tweezing force is relatively weak in comparison with gravitational pull. It is only by neutralizing the impact of gravitational pull by suspending objects in fluid that we can manipulate these objects by optical tweezing. The assembly of colloidal photonics crystal (see Chap. 6) is another good example of an application where neutralizing gravitational pull by using fluid is important.

1-4 Optical Advantages

Optics brings a complementary (and sometimes, overlapping) set of advantages to optofluidics. In this section, we shall look at some of these features.

1-4-1 Numerous High-Sensitivity Optical Sensing Techniques Exist

The range of light-matter interaction mechanisms is remarkably broad; to name a few of these mechanisms—fluorescence, phosphorescence, Raman scattering, polarization, elastic scattering, refraction, second harmonic generation, and stimulated emission. These mechanisms form the basis of optical sensing methods that are broadly used for chemical and biological sensing, because of their fast response and high specificity and sensitivity that are ideal for sensing applications.

For example, fluorescence and Raman scattering are commonly used tools to probe the dynamics of biological processes.

1-4-2 Light Localization Can Occur at Biologically Interesting Scale

We can focus light to a spot of a few hundred nanometers with conventional optics with relative ease. This is a fairly unique property of light in the EM spectrum. The long wavelengths of RF, microwaves, and even terahertz wave preclude focusing at such scale. X-ray does not suffer from such a limitation, but focusing X-ray requires relatively elaborate schemes. Unlike the X-ray, optical waves are nonionizing EM waves, which will not impose health hazards, and therefore are more favorable for bio applications.

The scale of a few hundred nanometers is biologically interesting as organelles are typically of that size. A microscope with such resolution can provide good imaging of cells. Microfluidics is a good match at this scale as well because this is a scale size at which fluidic controls are still possible.

By using optical near-fields, it is also possible to achieve even better length-scale or proximity sensitivity. The resonance-based biosensors described in Chaps. 12 and 13 are good examples of optofluidic devices that take advantage of this.

1-4-3 Light Can Manipulate Fluids and Objects Suspended in Fluids

Despite the fact that the force that light can directly exert is relatively weak, the extent of that force can be significant when it is exerted on small objects. Optical tweezers (Chap. 15) is a growing research field that capitalizes on this force to manipulate objects. Recently, there has been significant progress made in the use of waveguides to exert related types of controls (see Chap. 5).

Beyond direct force exertions (through momentum transfer), there are other more subtle ways in which light can be used to manipulate and move fluids and/or objects in fluids. The use of optically induced heating and fluid vaporization as a means to manipulate fluid is a new development that shows significant advantages for optofluidics (Chap. 19).

1-5 Future

Optofluidics is a rapidly growing field. The permutations of optics and microfluidics combinations are numerous and exciting to explore; we can reasonably expect this field to continue its rapid growth over the next decade. Optofluidics have brought about new and potentially better ways to build or use established optical structures and devices. Some of the growth directions in recent years have also been remarkably unanticipated. For example, the optofluidic maskless lithography technique (see Chap. 17) is unique and elegant in its implementation and applications.

We believe that the field of optofluidics will continue to surprise us with its new and unique devices and techniques.

References

1. Psaltis, D., R. S. Quake, and C. Yang, "Developing optofluidic technology through the fusion of microfluidics and optics," *Nature*, 2006, **442**: p. 381.
2. Monat, C., P. Domachuk, and B. J. Eggleton, "Integrated optofluidics: A new river of light," *Nat Photon*, 2007, **1**(2): pp. 106–114.
3. Horowitz, V. R., D. D. Awshalom, and S. Pennathur, "Optofluidics: Field or technique?" *Lan on a Chip*, 2008, **8**: pp. 1856–1863.
4. Borra, E. F., "The liquid-mirror telescope as a viable astronomical tool," *Journal of the Royal Astronomical Society of Canada*, 1982, **76**: pp. 245–256.

CHAPTER 2

Basic Microfluidic and Soft Lithographic Techniques

Sindy K. Y. Tang and George M. Whitesides

Department of Chemistry and Chemical Biology, Harvard University, Cambridge, Massachusetts

2-1 Introduction

Most optofluidic devices share a similar fluidic platform. The design, fabrication, and operation of the fluidic systems in these devices are based on those developed for microfluidics used in biochemical analysis. This chapter describes the basic ideas of microfluidics. We first summarize the materials most commonly used in fabricating microfluidic systems and the techniques developed for fabricating them. We then describe the characteristics of flow in these systems and illustrate the principle of operation of some important microfluidic components.

We focus our discussion on the use of polydimethylsiloxane (PDMS) for fabricating microfluidic systems. PDMS has been the most widely used material in the research and development of microfluidics. PDMS is an optically transparent elastomer whose stiffness can be controlled from very soft (easily deformed by finger pressure) to much stiffer. The fabrication of systems of microchannels in PDMS is particularly straightforward. The use of PDMS as a material allows rapid prototyping of devices, and facilitates the demonstration and the testing of new concepts. The physical and chemical properties of PDMS also make possible the fabrication of devices with a useful range of functions, ranging from molecular analysis to frequency-tunable lasing.

2-2 Historical Background

Microfluidic systems have the properties required for applications in a wide range of areas: molecular analysis, biodefense, molecular biology, microelectronics, clinical diagnostics, and drug development [1]. There are many benefits resulting from the miniaturization of devices for use in these areas, including decreased cost in manufacture, use, and disposal; decreased time of analysis; reduced consumption of reagents and analytes; reduced production of potentially harmful by-products; increased separation efficiency; decreased weight and volume; and increased portability [1]. The growth of molecular biology has stimulated the development of systems for analysis of biomolecules, DNA, and proteins. The first microfluidic device was a miniaturized gas chromatography (GC) system developed by Terry et al. [2] at Stanford University in the 1970s. The laboratories of Manz [3–5], Harrison [6–10], Ramsey [11–15], and Mathies [16–18] were among the first to develop microfluidic systems to analyze aqueous solutions. The technology used to fabricate these early systems—photolithography and etching in silicon and glass—was derived from microelectronics, as these technologies were available and highly developed. These materials and techniques are expensive and time-consuming, however, they require access to specialized facilities. They are therefore only marginally useful in research requiring rapid evaluation of prototypes. Their major advantage—chemical inertness—is so far required only in the still-undeveloped area of organic synthesis.

2-3 Materials for Fabricating Microfluidic Devices

Most research in microfluidic systems is now carried out in PDMS and other polymers. Fabrication in polymers is easier, more flexible, and much less expensive than in silicon or glass. It also avoids other problems of hard materials (e.g., formation of sharp shards on breakage) and enables certain components (e.g., pneumatic valves) that cannot be fabricated in rigid materials. In the following sections, we will focus on the use of PDMS for the development of microfluidic systems. PDMS has several attractive properties that make it suitable as a material for rapid prototyping of microfluidic devices capable of supporting a wide range of applications. Table 2-1 summarizes some of these properties and consequences.

2-3-1 Mechanical Properties of PDMS

PDMS is elastomeric. It has tunable Young's modulus, typically around 750 kPa [19]. It deforms easily, conforms to surfaces, and

Property	Characteristics	Consequence
Optical	Transparent; UV cutoff, 240 nm	Optical detection from 240–1100 nm
Electrical	Insulating; breakdown voltage, 2×10^7 V/m	Allows embedded circuits; intentional breakdown to open connections
Mechanical	Elastomeric; tunable Young's modulus, typical value of ~750 kPa	Conforms to surfaces; allows actuation by reversible deformation; facilitates release from molds
Thermal	Insulating; thermal conductivity, 0.2 W/(m·K); coefficient of thermal expansion, 310 μm/(m·°C); stable up to ~300°C	Can be used to insulate heated solutions; does not allow dissipation of resistive heating from electrophoretic separation
Interfacial	Low surface free energy ~ 20 erg/cm^2	Replicas release easily from molds; can be reversibly sealed to materials; not wetted by water unless oxidized to Si-OH presenting surface
Permeability	Low permeability to liquid water; permeable to gases and nonpolar organic solvents	Contains aqueous solutions in channels; allows gas transport through the bulk material; incompatible with many organic solvents
Reactivity	Inert; can be oxidized by exposure to a plasma	Unreactive toward most reagents; surface can be etched; can be modified to be hydrophilic and also reactive toward silanes
Toxicity	Nontoxic	Can be implanted in vivo; supports mammalian cell growth

(Adapted from J. C. McDonald and G. M. Whitesides, "Poly(dimethylsiloxane) as a material for fabricating microfluidic devices," *Acc. Chem. Res.*, 35, (2002), 491–499.)

TABLE 2-1 Physical and Chemical Properties of PDMS

releases from features of a mold—including undercut features—without damaging them or itself. Microfluidic channels and other features on the micron scale can therefore be reproduced with high fidelity in PDMS by replica molding. Using composite stamps composed of two layers—a stiff layer supported by a flexible layer of PDMS, replication of features below 1 nm has been demonstrated [20,21]. Because

PDMS is elastomeric, it is possible to form optical components whose dimensions can be tuned mechanically. Stretching or compressing a surface-relief grating or Fresnel lens made of PDMS, for example, changes the periodicity of the grooves on the grating or the lens, and the respective diffraction pattern generated or the focal properties of the lens [22,23].

2-3-2 Surface Chemistry of PDMS

The surface of PDMS is hydrophobic as it contains repeating units of –O-Si$(CH_3)_2$–groups. By exposing it to oxygen or air plasma, this surface can be made hydrophilic. Exposure to plasma introduces silanol (Si–OH) groups, and destroys methyl groups (Si–CH_3). Plasma-oxidized PDMS can be wetted by aqueous, polar solvents, and eutectic gallium-indium, a liquid metal alloy. On standing, a hydrophilic, oxidized PDMS surface becomes hydrophobic, as the surface reconstructs and as non-crosslinked components of the prepolymer bloom to the surface. It is possible to keep PDMS that has been plasma-treated hydrophilic indefinitely by keeping the surfaces in contact with water or polar organic solvents.

The silanol groups on the surface of PDMS allow it to react with a wide range of silanes (Si–R) that are terminated with important functional groups (i.e., R = NH_2, COOH, SH). By using different functional groups, it is possible to adjust the surface of PDMS to be hydrophilic or hydrophobic, or to introduce other reactive groups. Grafting a poly(ethylene glycol)di-(triethoxy)silane onto an oxidized PDMS surface makes the surface hydrophilic permanently, and reduces non-specific adsorption of proteins. Silanizing oxidized PDMS with an amino-terminated silane (aminopropyltriethoxysilate) provides a reactive surface for a bifunctional cross-linker for protein attachment [24]. These modified polar surfaces can, however, become hydrophobic again through blooming of mobile, nonpolar siloxanes. Application of a sol-gel coating may be more protective, but has not been extensively developed [25].

Irreversible Sealing

It is simpler to seal channels made in PDMS than channels that are made in glass, silicon, or thermoplastics, as high temperatures, pressures, and voltages are not required. For example, sealing glass to glass or silicon to silicon requires high temperatures (~600°C for glass; >800°C for silicon) and/or voltages (500–1500 V for anodic bonding of glass). Sealing of channels in PDMS can be performed in ambient laboratory conditions. By exposing the surface of PDMS and the surface of the substrate to an air- or oxygen-based plasma, PDMS channels can be sealed irreversibly to PDMS, glass, silicon, polystyrene, polyethylene, or silicon nitride [24]. Plasma oxidization produces silanol groups on PDMS, and –OH-containing functional groups on the other materials. When the surfaces are brought into contact, the polar groups form covalent –O-Si-O-bonds with oxidized PDMS; the channel is therefore sealed irreversibly. It should be noted that the two surfaces must be brought into contact quickly (< 1 min) after oxidation, because the surface of the oxidized PDMS reconstructs in air. Empirical evidence shows that sealing works best when the samples and chamber are clean, the samples are dry, the surfaces are smooth (on the micron scale), and the oxidized surfaces are not mechanically stressed. Heating a weak seal at 70°C can sometimes improve the strength of the seal [19]. Another way to seal two pieces of PDMS irreversibly involves adding an excess of the monomer to one surface and an excess of the curing agent to the other. When the two surfaces are cured together, an irreversible seal that is indistinguishable from the bulk properties of PDMS forms [24].

Reversible Sealing

Another advantage of PDMS over glass, silicon, and hard plastics is that it makes reversible conformal contact (van der Waals contact) to smooth surfaces. PDMS devices can therefore be demountable, and resealing can occur multiple times without degradation in the PDMS.

Microfluidic devices that are demountable can be used to pattern surfaces with proteins, cells, and other biomolecules using fluid flow [24]. Our group [26] and others [27] have performed binding assays using a demountable device. Antibodies were first patterned on a glass substrate by flowing a solution of antibody through a set of parallel channels. The PDMS device was then peeled off from the glass substrate, rinsed, and placed perpendicular to the first set of channels. Solutions containing antigens were then introduced through the channels. Antibody-antigen complexes were subsequently detected at the crossings of stripes of antibodies and the channels.

PDMS channels can also seal reversibly to silicone (or cellophane) adhesive tapes [19]. To make a mechanically stable support, double-sided tape—with one side applied to a flat plastic or glass slab—is a valuable component. Polymeric adhesive tapes are convenient because they are mechanically flexible, and they form a stronger (but still reversible) bond than that between PDMS and other flat surfaces. They also allow nonsealing functional layers such as filter papers and membranes to be incorporated into the microfluidic system [26].

Compatibility with Solvents

PDMS is compatible with water, and most polar organic solvents (such as methanol and glycerol); it swells, however, in nonpolar organic solvents (such as pentane and chloroform) [28], and will absorb nonpolar solutes from aqueous solutions. To reduce the absorption of small molecules and the swelling by nonpolar organic solvents, one can modify PDMS with silica particles [29], or coat the surface with a glass-like layer using sol-gel chemistry [25] (Fig. 2-1).

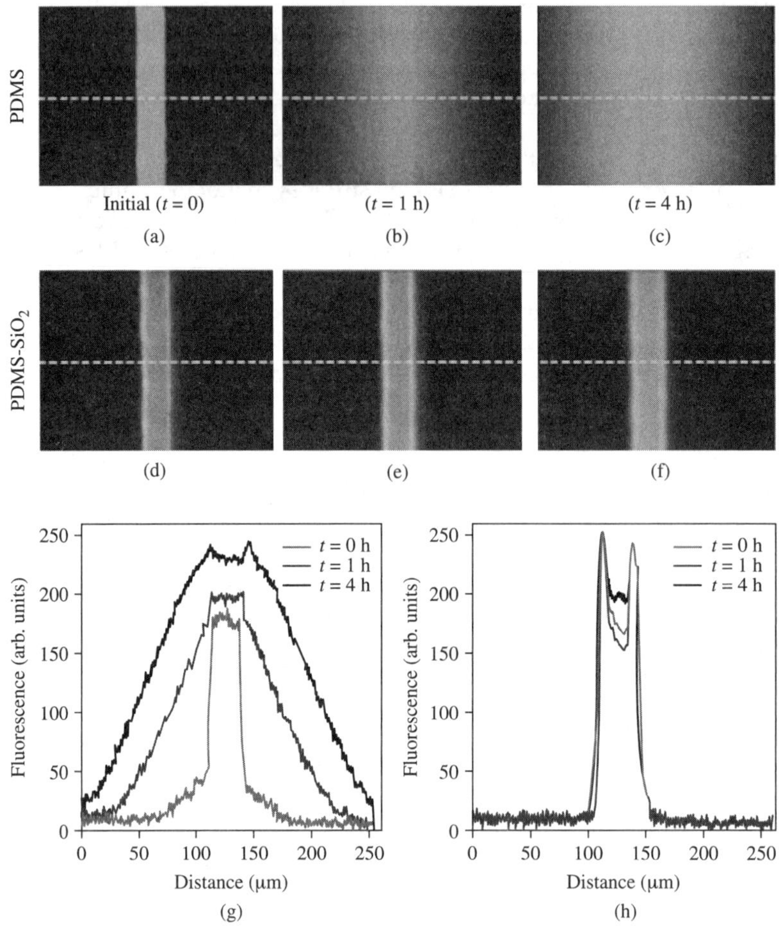

FIGURE 2-1 Images of PDMS (a-c) and PDMS-SiO_2 (d-f) devices are shown. The channels on these devices are filled with 10-μM rhodamine B in a 10-mM (pH 9.5) sodium borate solution. The images were acquired over a 4-h period. Fluorescent profiles of the PDMS and PDMS-SiO_2 channels are also shown in (g) and (h), respectively. These profiles were taken along the white dotted line in images (a-f). (*Adapted with permission from G. T. Roman, T. Hlaus, K. J. Bass, T. G. Seelhammer, and C. T. Culbertson, "Sol-gel modified poly(dimethylsiloxane) microfluidic devices with high electroosmotic mobilities and hydrophilic channel wall characteristics," Anal. Chem., 77, (2005), 1414–1422.Copyright 2005 American Chemical Society.*)

Toxicity

PDMS is nontoxic to proteins and cells. It is permeable to oxygen and carbon dioxide, but only slowly permeable to water. It is therefore suitable for biological studies: for example, mammalian cells can be cultured on it directly [30].

2-3-3 Optical Properties of PDMS

PDMS is optically transparent from 240 to 1100 nm [19], and has a refractive index around 1.41. It has negligible birefringence. It is therefore possible to enclose optofluidic components in PDMS, and couple light through PDMS, with minimal loss due to absorption. Commercially available PDMS—Silgard 184—does, unfortunately, contain nanoparticles of silica that introduce unwanted scattering of light. In the devices we and others have fabricated, the thickness of PDMS for enclosure of microfluidic components is limited (usually < 1 cm), and thus scattering due to passage of light through PDMS does not cause significant loss during the coupling of light into and out of the devices.

We have not identified a polymer that lacks these scatterers, and still possesses the other desirable qualities of PDMS. The Norland optical adhesives (photocurable polyurethanes), for example, contain no scattering particles, but they are not soft, and cannot be processed the same way as PDMS. This need for an elastomeric polymer with high optical transparency and easy sealability presents an opportunity for future research in material science. To summarize, PDMS has attractive features that make it useful for a wide range of applications in laboratory, and for prototyping in research, though it may not be the ultimate material used in large-scale manufacturing. Other polymers used for fabricating microfluidic systems include h-PDMS, photocurable perfluoropolyethers (PFPE), cyclic olefin copolymer (a thermoplastic polymer), thermoset polyester, polymethylmethacrylate, polycarbonate, and polyurethanes [31]. Each material has its own advantages and disadvantages; depending on the application, one material may be more suitable than the other. For example, PFPEs, a class of fluoropolymers that are liquids at room temperature, are chemically resistant (like Teflon). They are compatible with organic solvents such as toluene and dichloromethane (both of which swell PDMS). The fabrication process for channels in PFPE involves procedures that are more complicated than with PDMS, however. There is no simple procedure for adhesive-free contact sealing, and these polymers are much more expensive.

2-4 Fabrication of Microfluidic Systems in PDMS

Systems in PDMS are typically fabricated using techniques in soft lithography [19]. Soft lithography involves the replication of a topographically defined (typically in photoresist) structure on a master in a soft elastomer. The process can be carried out in ambient laboratory conditions. Replication can also be repeated multiple times. Soft lithography therefore enables rapid, simple, and inexpensive fabrication processes.

The details of fabrication using soft lithography can be found elsewhere [19]. Here we provide a summary of the processes involved. The microfluidic channels are designed in a CAD program and printed onto a high-resolution transparency (~5000 dpi) (or, somewhat less conveniently and more expensively, converted into a conventional chrome mask). This transparency is used as a photomask in 1:1 contact photolithography (usually using SU-8 or PMMA as photoresist) to produce a master. This master consists of a positive bas-relief of photoresist on a silicon wafer, and serves as a mold for PDMS. Liquid PDMS prepolymer is poured over the master and cured for 1 h at 70°C. The PDMS replica is then peeled from the master and sealed (following plasma oxidation of the interfaces involved) to a flat PDMS, glass, or silicon surface to form the microfluidic channels. The overall process takes approximately 24 h. Figure 2-2 shows a schematic diagram of the procedures involved.

2-5 Characteristics of Flow in Microchannels

A basic understanding of fluid dynamics in microsystems is useful in the design and development of microfluidic devices. This section summarizes a few characteristics of flow in microchannels that are important in common microfluidic components. Comprehensive reviews on the physics of fluids in microfluidic systems can be found elsewhere [32–34].

In general, as the physical length scale of the system decreases, gravity becomes less important. Surface forces (surface tension, electrical, van der Waals, and surface roughness) become dominant [33]. Most microfluidic devices are in the micro- or nanoscale range, and the relative importance of forces typically follows this order: interfacial force >> viscous forces > gravitation ~ inertial force > buoyancy [35]. Most microfluidic devices involve the use of miscible liquids only. Interfacial tension is therefore usually negligible. Viscous forces dominate, and as a result, the flow is primarily laminar without turbulence; mixing occurs by diffusion only [32]. We will describe laminar flow and diffusion in more details in the following section.

2-5-1 Laminar Flow

Flow in microchannels is commonly characterized by the Reynolds number, Re. The Reynolds number describes the tendency of fluid to develop turbulence. It represents the relative importance of inertia to viscous dissipation ($\mathrm{Re} = vl\rho/\mu$, where v is the average flow speed, l is the characteristic length scale of the channel, ρ is the density of the fluid and μ is the dynamic viscosity) [32]. For Re much less than 2000, viscous forces dominate, and the flow is laminar. As Re increases above 2000, the flow becomes dominated by inertial forces, which tend to produce instability leading to turbulence.

Since the length scale of microfluidic systems is small (< 500 μm typically), the flow of fluids in microchannels takes place in the regime

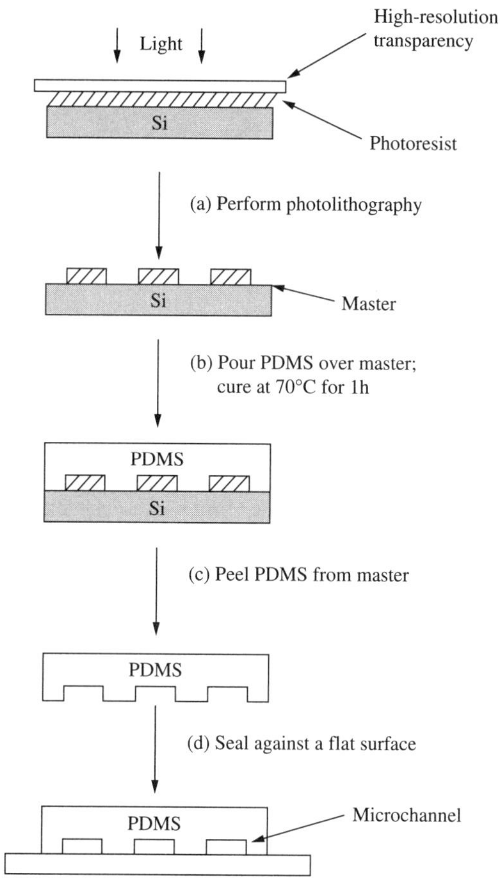

FIGURE 2-2 Scheme describing rapid prototyping of microfluidic systems. A system of channels is designed in a CAD program. A commercial printer uses the CAD file to produce a high-resolution transparency (~5000 dpi). (a) This transparency is used as a photomask in contact photolithography to produce a master. A master consists of a positive relief of photoresist on a silicon wafer and serves as a mold for PDMS. (b) Liquid PDMS prepolymer is poured over the master and cured for 1 h at 70°C. (c) The PDMS replica is peeled from the master. (d) The replica is sealed to a flat surface to enclose the channels. The overall process takes ~24 h. (*Adapted with permission from J. C. McDonald and G. M. Whitesides, "Poly(dimethylsiloxane) as a material for fabricating microfluidic devices," Acc. Chem. Res., 35, (2002), 491–499. Copyright 2002 American Chemical Society.*)

where the Reynolds number is low (typical Re < 10). Viscous forces dominate, and the flow is laminar. The liquids can be treated as laminae (layers) of uniform thickness; their boundaries remain fixed as the liquid moves between them; the only mixing of the streams occurs by diffusion across the liquid-liquid interface [36]. Figure 2-3 shows an

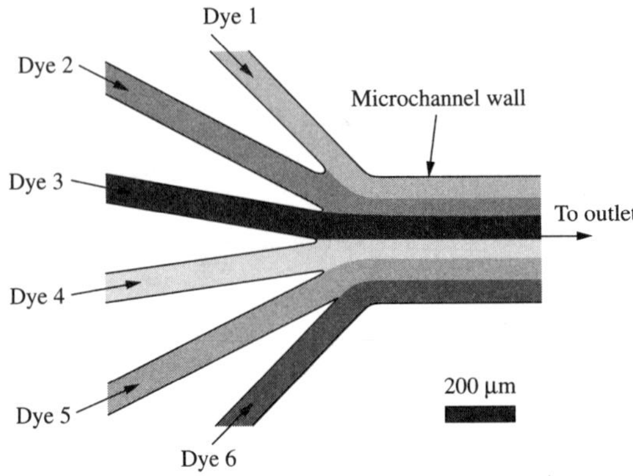

FIGURE 2-3 Optical micrograph of laminar flow of six streams of different dye solutions in a microchannel flowing from left to right. The height of the channel is 100 μm. (*Adapted with permission from D. B. Weibel, M. Kruithof, S. Potenta, S. K. Sia, A. Lee, and G. M. Whitesides, "Torque-actuated valves for microfluidics," Anal. Chem., 77, (2005), 4726–4733. Copyright 2005 American Chemical Society.*)

example of laminar flow of six different streams of solutions of dye flowing (from left to right) into a single channel. There is no turbulent mixing, and the interface between these laminar streams remains parallel and distinct. Note that this interface is at dynamic steady state: a continuous flow of liquids is necessary to maintain this interface. As a result, manipulating the conditions of flow can reconfigure the shape and position of this interface dynamically. This feature is attractive for some applications in optics, and has been exploited for constructing optical devices; more details can be found in Chap. 3.

2-5-2 Diffusion

Although there is no turbulence, mixing in laminar streams still occurs due to diffusion. There is a transverse (in the y direction in Fig. 2-4) concentration gradient across laminar streams that contain different concentrations of solutes (or other properties, such as temperature). Transverse diffusion of solutes broadens the laminar interface, flattens the concentration gradient across the streams, and homogenizes the liquids. Figure 2-4 shows this idea. To visualize the spatial extent of transverse diffusive mixing, two nonfluorescent chemical species (carried separately by the two flowing solution streams) are used. The product of these two species is fluorescent, and can therefore be imaged with a confocal microscope.

The Péclet number ($\mathrm{Pe} \equiv vw/D$) compares the typical time scale for diffusive transport to that for convective transport (for channel width w, velocity of fluid v, and diffusivity D) [32]. For solute ions with typical diffusivity $D \sim 2 \times 10^3\ \mu\mathrm{m}^2\mathrm{s}^{-1}$ flowing through a channel

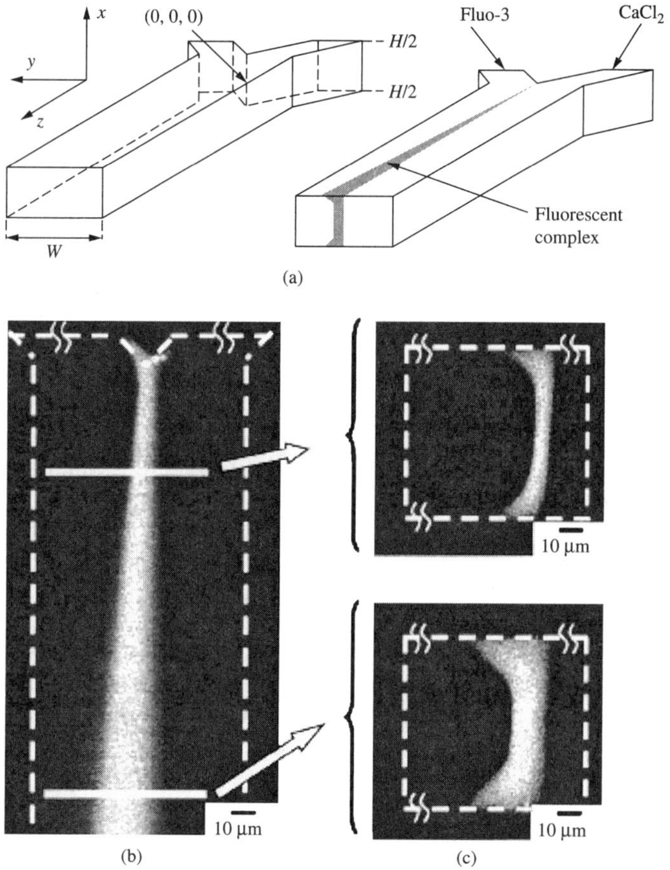

FIGURE 2-4 (a) Schematic drawing of the experiment used to generate fluorescence in the microchannel, and the coordinate system employed. The dashed lines indicate the internal edges of the channel. *H* is the height of the channel. The origin of the coordinate system is located in the middle of the vertical edge shared by both channels in the Y junction. (b) and (c) Data obtained for the fluo-3/Ca^{2+} system by confocal fluorescence microscopy in: (b) A 100 × 250 μm² *y-z* slice obtained 12 μm from the top of the channel. (c) Corresponding 90 × 90 μm² cross sections in the *x-y* plane of the channel obtained 50 μm (top) and 200 μm (bottom) from the Y junction (*Reprinted with permission from R. F. Ismagilov, A. D. Stroock, P. J. A. Kenis, G. Whitesides, and H. A. Stone, "Experimental and theoretical scaling laws for transverse diffusive broadening in two-phase laminar flows in microchannels," Appl. Phys. Lett., 76, (2000), 2376–2378. Copyright (2000), American Institute of Physics.*)

with width $w = 100$ μm at velocity $v = 100$ μm/s, it would take a time of $t \sim w^2/D = 5$ s for the ions to diffuse across the entire width of the channel. During this time, the stripe will have moved a distance $Z \sim vw^2/D = 500$ μm down the channel. The number of channel widths required for complete mixing would be of order $\mathrm{Pe} \equiv Z/w = vw/D = 5$. It is possible to increase the time before complete mixing occurs [or to decrease the spatial extent of transverse diffusive broadening for a given channel length (in the z-direction)] by applying a higher rate of flow, as long as the Reynolds number is still small enough for the flow to remain laminar, or by using liquids with higher viscosities and thereby lowering diffusivity.

For larger species with lower diffusivities, pure diffusive mixing can be slow. For example, small proteins ($D \sim 40\ \mu\mathrm{m}^2\mathrm{s}^{-1}$) flowing through a 100-μm channel at 100 μm/s would require approximately 4 min to mix completely. This time scale can be undesirably long for some biochemical applications. To enhance mixing, special channel designs have been developed. We will discuss various forms of on-chip mixers in the next section.

Note that the extents of diffusive mixing in the middle of the channel and close to the top wall (ceiling) and bottom wall (floor) of the channel are different. The cross-sectional profile (in the xy plane) of the laminar interface is not entirely vertical to the ceiling/floor of the channel (Fig. 2-4). At steady state, near the ceiling and the floor of the channel, the extent of transverse diffusive mixing across the liquid-liquid interface scales as the one-third power of the axial distance (in the z direction) along the channel [37]. Near the middle of the channel, the extent of mixing scales is the one-half power of the axial distance, and is therefore smaller than that close to the ceiling/floor at the same position (z) down the channel. As a result, the cross-sectional profile of the laminar interface becomes curved.

2-6 Components Fabricated in PDMS

This section describes examples of microfluidic components, which are the building blocks of more complex, multifunctional microfluidic systems with applications in polymerase chain reaction (PCR), protein crystallization, lab-on-a-chip, and other micro total analytical systems (μTAS). These examples illustrate the general methods to manipulate fluids in microchannels, and the basic design rules of microfluidic devices.

2-6-1 Inlets, Outlets, and Connecters

To introduce and recover liquids from microchannels made in PDMS, polyethylene tubing can be inserted into holes bore in PDMS that are slightly too small, so the PDMS must stretch to fit. This fitting provides a waterproof seal, and prevents leaking of liquids at this PDMS-tubing interface [19]. Syringes are usually used to provide pressure or vacuum, and thus to drive the flow of fluids in the channels. The polyethylene tubing also conforms to syringe needles. This ability allows for syringes (and syringe pumps) to be coupled easily to microfluidic channels.

2-6-2 Valves and Pumps

Several groups have used the elasticity of PDMS in the actuation of valves and pumps [19]. The valves operate by applying a force that pinches a fluidic channel closed at a precise location. Compression of the channels can be introduced in various ways, including: fluid pressure [38,39], torque actuation from embedded machine screws [40] or solenoids [41], expansion of a hydrogel [42], magnetic actuation [43], or the thermal response of shape-memory alloys [44]. Takayama et al. have also used the pins of a piezoelectric Braille display as valves in microfluidic systems [45].

Quake valves are perhaps the most commonly used microfluidic valves in elastomeric devices. The Quake valve is a three-layer microfluidic structure, consisting of a flow channel in one layer separated by a thin elastomeric membrane from a (usually perpendicular) control channel in the layer above. The application of pressurized air to the control channel closes the flow channel. These valves are compatible with soft lithography, and can be used in parallel at high densities because of their small footprint. Their fabrication and operation are complicated, however, and require costly and bulky off-chip infrastructure (computer-controlled pneumatic actuators, gas distribution system, etc.). These valves are sometimes overkill for simple microfluidic applications that require only one, or a small number, of valves.

TWIST and solenoid valves developed by our group are simpler to construct and operate, and are suitable for situations that require only small number of valves [40,41]. To construct a TWIST valve, a small machine screw is introduced directly above a microfluidic channel in a PDMS device. Rotation of the screw results in downward motion of the screw and compression of the underlying channel, and therefore the closing of the channel. To construct a solenoid valve, a cylindrical, push-type solenoid is placed directly on top of a channel. To focus the force from the solenoid onto a small area, a small bead is inserted between the armature of the solenoid and the top of the channel. Applying a voltage to the solenoid actuates the valve.

Recently Hulme et al. showed that it is possible to fabricate these valves [pneumatic (Quake-like), screw (TWIST-like), and solenoid valves] en masse, ahead of time, and then positioned and embedded in microfluidic devices as needed [41] (Fig. 2-5). These valves are suitable for systems in which they are needed only in small numbers, and in which fabrication of an integrated system is not required. Since the valves are prefabricated using a standardized procedure, uniform operation of the valves is possible. The disadvantage of this type of valves is the need for component-level assembly and a relatively large footprint for each valve.

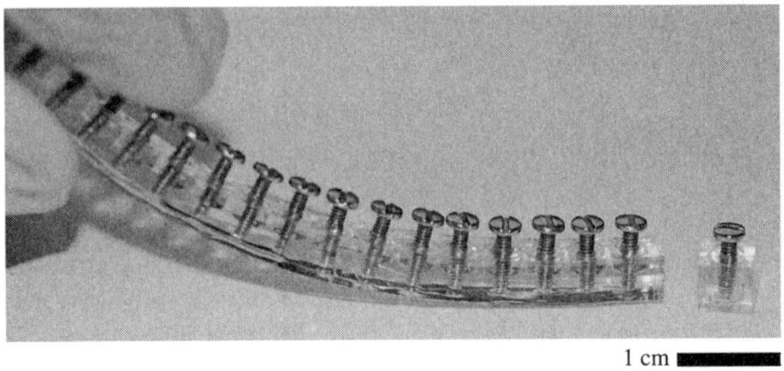

(a)

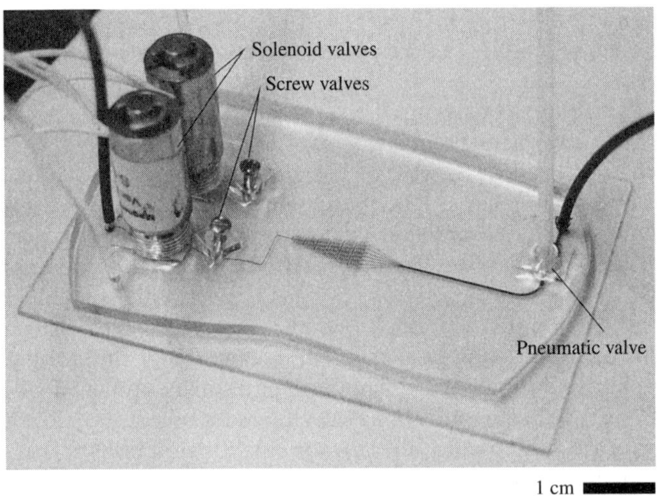

(b)

FIGURE 2-5 (a) A photograph of a strip of prefabricated screw valves. A single valve has been separated from the strip using a razor blade. (b) A photograph of a microfluidic gradient generator containing two embedded solenoid valves, two embedded screw valves, and one embedded pneumatic valve. (*S. E. Hulme, S. S. S., and W. G. M., "Incorporation of prefabricated screw, pneumatic, and solenoid valves into microfluidic devices," Lab Chip, submitted. Reproduced by permission of the Royal Society of Chemistry.*)

2-6-3 Mixers

Mixing of fluids in microchannels is important for many biological and chemical applications. Mixing in simple microchannels can be slow, as discussed in the preceding section. Mixers are therefore essential in enhancing mixing efficiency and in homogenizing reagents rapidly. All mixing ultimately occurs due to molecular diffusion. The basic idea behind mixers is reducing the distance over which mixing must occur [32].

A wide variety of mixers have been developed. They can be broadly classified as active (involving input of external energy) or passive (making use of the fluid dynamics in specific geometry of the channel in the absence of external forces). Passive mixers are usually easier to fabricate than active mixers, and are more suitable for applications involving sensitive species as they do not impose electrical, mechanical, or thermal agitation [46].

One of the passive mixers developed involves a staggered herringbone structure to generate chaotic advection in a microchannel [47] (Fig. 2-6). This mixer uses asymmetric grooves on the floor of the channel (the "staggered herringbone" design) to generate a transverse component to the flow when an axial pressure gradient is applied. Because of this transverse component, the fluid elements are stretched and folded into one another; this process increases the contact area between the flowing streams and facilitates mixing by diffusion. Channels with the staggered herringbone design thus have a higher efficiency of mixing laminar streams of fluid than channels with smooth walls.

Another type of passive mixer involves the use of serpentine channels [42,46]. Fluids flowing through curved channels experience both inertial forces and centrifugal forces. Under suitable conditions, these effects establish a radial pressure gradient whose magnitude can

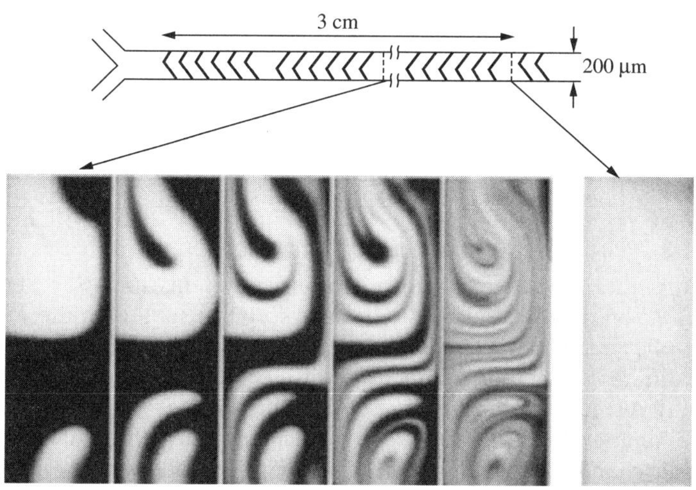

Figure 2-6 Continuous-flow staggered herringbone mixer, in which grooved channel walls drive alternating, asymmetric helical secondary flows that chaotically stir the fluid. Each cycle cuts the distance between stripes in half, so that the distance between stripes decreases exponentially with the number of cycles. Diffusive mixing occurs when the tracer can diffuse from one stripe to the next before another cycle has occurred, giving a mixing time that depends logarithmically on Pe. Thus the channel cross section is rapidly mixed. (*From A. D. Stroock, S. K. W. Dertinger, A. Ajdari, I. Mezit, H. A. Stone, and G. M. Whitesides, "Chaotic mixer for microchannels," Science, 295, (2002), 647–651. Reprinted with permission from AAAS.*)

become sufficient to generate a transverse flow ("Dean flow") [32] across the streams. This transverse flow increases the contact area between the streams, and enables more efficient mixing of the liquids.

Active mixers have also been developed for enhancing mixing: rotary mixers, where solutions to be mixed are actively pumped peristaltically in a circulating loop [48]; mixers based on electrowetting [49], nonlinear electrokinetic effects [50,51], and acoustic streaming [52]. These systems are usually complicated to fabricate; however, recently, a simple, portable, hand-powered mixer has been developed that exploits the introduction and movement of bubbles in microchannels to mix the continuous fluids [53].

2-6-4 Diluters for Generating Concentration Gradients in Microchannels

Gradients in the concentration of solutions are important in many biological and chemical processes, such as chemotaxis and nerve growth cone guidance. Various forms of diffusion-based dilution microfluidic devices have been developed to generate concentration gradients. The general design consists of two inlets, one for the reagent to be diluted, and the other for the diluting agent or buffer, leading into a network of multistep fluid-dividers [54] (Fig. 2-7). Mixers are usually incorporated to ensure the complete mixing of the reagent and the buffer. The ratio of fluidic resistance in the branches determines the ratio of volumetric flow of the reagent and the buffer in each branch, which in turn determines the output concentration. The fluidic resistance can be increased by increasing the length of the channel, or by decreasing the cross-sectional area of the channel. Different schemes have been developed to generate linear and logarithmic gradients [54–62].

2-6-5 Local Heaters and Electromagnets

Incorporation of metals into microfluidic systems for applications such as on-chip heating and magnetic sorting usually require more complicated procedures as the materials and the fabrication processes are different from those of microfluidic channels, which are polymer-based. A simple method—*microsolidics*—has been developed to fabricate complex metallic structures by injecting liquid solder into microfluidic channels, and allowing the solder to cool and solidify [63,64]. The general procedure consists of five steps (Fig. 2-8a):

1. Fabrication of microfluidic channels in PDMS.
2. Plasma oxidation and silanization of the inside surfaces of the microchannels with 3-mercaptopropyltrimethoxysilane (0.1 M solution in acetonitrile). This reduces the surface free energy of the channel surface, and allows the solders (such as In100, or 100% Indium) to wet the channel wall.
3. Injection of molten solder into the channels by applying a vacuum to draw metal into the channels.

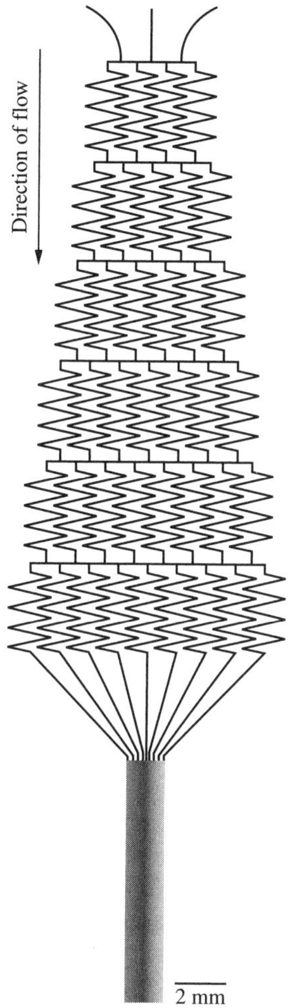

Figure 2-7 Photograph showing a microfluidic device we used for generating gradients of different dyes in solution. The three incoming channels (top part of the photograph) were connected to syringes via tubings (not visible). After combining the streams into a single, wide channel (bottom of the photograph), a gradient was formed across the channel, perpendicular to the direction of flow. (*Adapted with permission from S. K. W. Dertinger, D. T. Chiu, N. L. Jeon, and G. M. Whitesides, "Generation of gradients having complex shapes using microfluidic networks," Anal. Chem., 73, (2001), 1240–1246. Copyright 2001 American Chemical Society.*)

4. Cooling the channels to form solid metal microstructures.
5. Deforming the solder-filled system of channels into nonplanar structures (if desired).

Next, we will describe two components fabricated using this method.

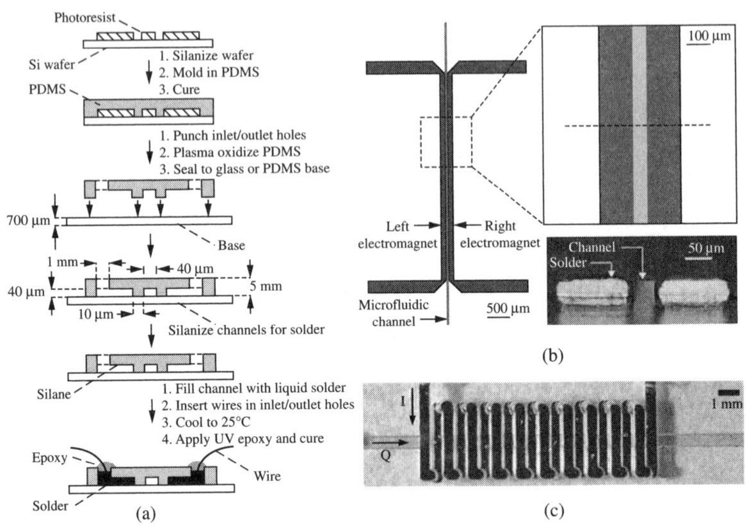

FIGURE 2-8 (a) A schematic diagram depicting the fabrication of electromagnets in PDMS. The completed device has three microfluidic channels: two outer channels filled with solder (length = 1.5 cm, width = 120 μm, height = 40 μm), and a central channel for fluids (length = 1.5 cm, width = 40 μm, height = 40 μm). (b) Electromagnet: photographs of the three channels as viewed from above at low magnification (left), high magnification (upper right) and the cross-section of the three channels (lower right). The photograph of the cross-section was obtained by sectioning the channels with a razor blade (shown as the dashed line in the upper right image); the dark line in the left electromagnet is the result of imperfect sectioning; the light areas at the bottom of the image are reflections of the metal. In the photograph at low magnification, lines were drawn to outline the location of the microfluidic channel. (c) Microheater: photograph of a solder coil positioned axisymmetrically around a central microfluidic channel; the channel was filled with dyed deionized water to make it easier to see; the exterior walls of the central microfluidic channel were highlighted for clarity. (*A. C. Siegel, D. A. Bruzewicz, D. B. Weibel, and G. M. Whitesides, "Microsolidics: fabrication of three-dimensional metallic microstructures in poly(dimethylsiloxane)," Adv. Mater., 19, (2007), 727–733. A. C. Siegel, S. S. Shevkoplyas, D. B. Weibel, D. A. Bruzewicz, A. W. Martinez, and G. M. Whitesides, "Cofabrication of electromagnets and microfluidic systems in poly(dimethylsiloxane)," Angew. Chem., Int. Ed., 45, (2006), 6877–6882. Copyright Wiley-VCH Verlag Gmbh & Co. KGaA. Reproduced with permission.*)

Electromagnet

The channels for electromagnets [64] were fabricated in close proximity (~10-μm separation), and in the same plane as the channels used to transport fluid (Fig. 2-8b). By passing electrical current through the wires, magnetic fields up to 2.8 mT and field gradients up to 40 Tm^{-1} have been generated inside microfluidic channels. This electromagnet has been demonstrated to be capable of sorting super-paramagnetic beads.

Microheater

Figure 2-8c shows a photograph of a coil microheater [63] fabricated by the injection of solder into a PDMS channel. The device consists of a solder coil (In100, height = 80 μm, width = 800 μm, length = 12 cm) wrapped around a central microfluidic channel (height = 80 μm, width = 800 μm, length = 3 cm). This device was fabricated using a procedure similar to that used to fabricate a "basket-weave" microstructure: three layers of PDMS containing microfluidic channels were aligned, bonded together, and mounted to a glass slide to form a multilayer network of microfluidic channels. The network was composed of two channels: a central microfluidic channel and a "coil channel" that passed through all three microfluidic layers to surround the central channel. Solder was injected into the coil channel and cooled to form the microheater.

To characterize the microheater, electrical currents (I = 0–600 mA, at 100 mA intervals) were applied through the wire while deionized water flowed through the central channel (flow rate, Q = 100 μL/min). As the current passing through the solder coil increased, the temperature of the fluid passing through the microfluidic channel increased up to 40°C as a result of Joule heating.

Microsolidics simplifies the incorporation of metals into microfluidic channels, but it also has several limitations. This method can only be used with metals and alloys with a low melting point (generally < 300°C) and affinity for the surface of the channel wall. These low-melting-point solders are usually more expensive than commonly used solders, and some (those containing Pb or Cd) are not biocompatible. In addition, the wire must be fabricated as a loop; this method cannot be used to fill "dead-end" channels. Lastly, it is currently difficult to use this process to fabricate wires with cross-sectional dimensions less than 10 μm.

2-6-6 Bubble and Droplet Generator

We have focused primarily on miscible systems so far. The use of immiscible fluids for the formation of emulsions and foams in microfluidic systems is also interesting, and has undergone rapid development in recent years. The controlled formation of microscale, individual fluid segments allow compartmentalized biochemical reactions and analyses using small volumes of reagents. It has also been shown that droplet and bubble-based microfluidics can perform simple Boolean logic functions [65,66].

There are several ways to generate droplets and bubbles in microfluidic systems; details are reviewed elsewhere [67]. Here we describe two common methods that depend on the geometry of the channel to control the generation of droplets and bubbles: the flow-focusing device and the T-junction.

Flow-Focusing Device

Figure 2-9a and 2-9b illustrates the flow-focusing device [68–70]. Gas and liquid meet upstream from the orifice at the junction of the three inlet channels. The pressure drop along the axis of the device forces the tip of the gas stream into the orifice. Here the thread breaks and

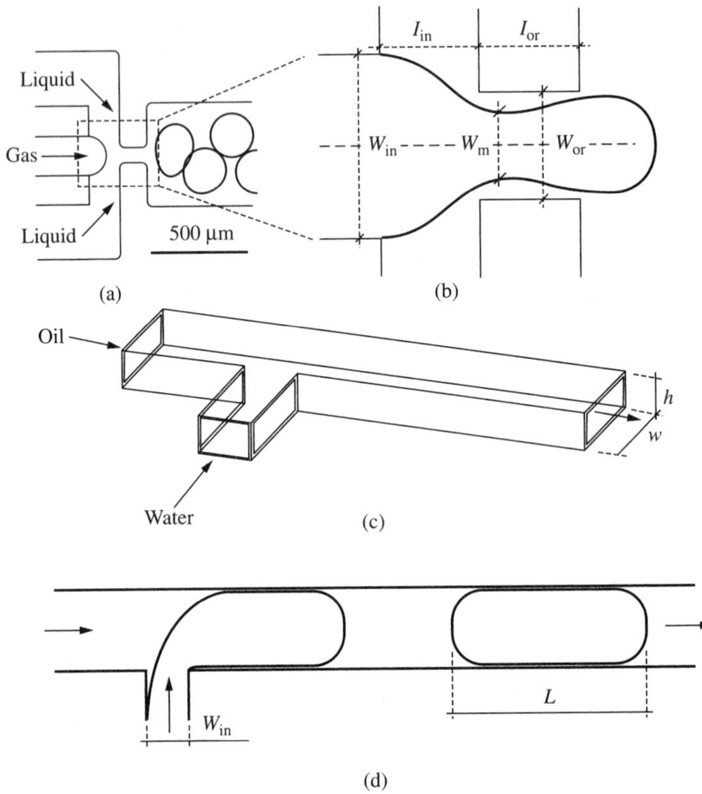

Figure 2-9 (a) Optical micrograph of the flow-focusing device (top view). The channels are fabricated in PDMS and have a uniform height *h*. (b) Schematics of the orifice region and the gas-liquid interface. The shaded areas correspond to PDMS walls and the long dashed line marks the center line and axis of mirror symmetry of the channel. (c) A schematic illustration of the microfluidic T-junction composed of rectangular channels. The channels are planar and have uniform height *h*. (d) A top view of the same schematic in a two dimensional representation. Flow along the main channel proceeds from left to right. The continuous fluid flows along the main channel of width *w*, and the fluid that will be dispersed flows via the orthogonal inlet of width w_{in}. (*Reprinted with permission from P. Garstecki, H. A. Stone, and G. M. Whitesides, "Mechanism for flow-rate controlled breakup in confined geometries: a route to monodisperse emulsions," Phys. Rev. Lett., 94, (2005), 164501/164501–164501/164504. P. Garstecki, M. J. Fuerstman, H. A. Stone, and G. M. Whitesides, "Formation of droplets and bubbles in a microfluidic T-junction-scaling and mechanism of break-up," Lab Chip, 6, (2006), 437–446. Copyright (2005, 2006) by the American Physical Society.*)

periodically releases bubbles into the outlet channel. Over a wide range of parameters, this system produces almost ideally monodispersed bubbles. The frequency of bubble formation can be greater than 10 kHz. The volume of the bubbles and the volume fraction of the dispersed phase can be varied independently by adjusting the pressure applied to the gas stream, and the rate of flow of the liquid. The same device can be used to generate liquid droplets in another immiscible liquid.

T-junction

Figure 2-9c and 2-9d illustrates the geometry of a T-junction [71,72]. Two channels merge at a right angle. The main channel carries the continuous fluid (oil here) and the orthogonal channel supplies the fluid that will be dispersed (water here). As the dispersed phase penetrates into the main channel, shear forces generated by the continuous phase and the subsequent pressure gradient cause the tip of the dispersed phase to elongate into the main channel until the neck connecting the inlet channel with the droplet breaks. The disconnected liquid plug flows downstream in the main channel, while the tip of the stream of the dispersed phase retracts to the end of the inlet and the process repeats. The viscosity of the fluids, the interfacial tension, volumetric rates of flow of the two phases, and the geometry of junction determine the size of the droplets or gas bubbles formed.

2-6-7 Optical Components

Because PDMS is soft and deformable, it is possible to form optical components whose physical dimensions can be tuned mechanically or thermally. These components can be prepared by molding PDMS elastomers into the desired shapes. Tunable lenses and mirrors, diffraction gratings, interferometric sensors, and distributed feedback lasers have been fabricated out of PDMS [22,23,73–76]. Some of these devices will be described in detail in later chapters.

2-7 Conclusions

We have illustrated the basic design and construction of some important microfluidic components. Methods for the manipulation of fluids in these microfluidic systems can be used to incorporate multiple functions on the same chip, and to develop more complex optofluidic systems.

The fabrication of microfluidic components in PDMS is easier and more flexible than in silicon or glass. The use of PDMS as a material reduces the time, complexity, and cost of prototyping. Its influence on costs of manufactured systems remains to be established, but polymers are, in general, less expensive than ceramics as materials.

Some of the properties of PDMS may be disadvantageous in certain situations. For example, PDMS is incompatible with many organic solvents; it has therefore been applied primarily to aqueous solutions. When working with biological samples, nonspecific adsorption may occur. The presence of nanoparticles of silica in commercial PDMS causes undesired scattering of light. Methods to control the surface chemistry of PDMS are being actively developed to overcome these problems, however, and to expand the range of properties of PDMS-based systems.

References

1. G. M. Whitesides, "The origins and the future of microfluidics," *Nature*, 442, (2006), 368–373.
2. S. C. Terry, J. H. Jerman, and J. B. Angell, "A gas chromatographic air analyzer fabricated on a silicon wafer," *IEEE Trans. Electron Devices*, ed. 26, (1979), 1880–1886.
3. D. J. Harrison, K. Fluri, K. Seiler, Z. Fan, C. S. Effenhauser, and A. Manz, "Micromachining a miniaturized capillary electrophoresis-based chemical analysis system on a chip," *Science*, 261, (1993), 895–897.
4. M. U. Kopp, A. J. De Mello, and A. Manz, "Chemical amplification: continuous-flow PCR on a chip," *Science*, 280, (1998), 1046–1048.
5. A. Manz and H. Becker, "Parallel capillaries for high throughput in electrophoretic separations and electroosmotic drug discovery systems," *Transducers 97, Int. Conf. Solid-State Sens. Actuators*, 2, (1997), 915–918.
6. N. Chiem and D. J. Harrison, "Microchip-based capillary electrophoresis for immunoassays: analysis of monoclonal antibodies and theophylline," *Anal. Chem.*, 69, (1997), 373–378.
7. K. Fluri, G. Fitzpatrick, N. Chiem, and D. J. Harrison, "Integrated capillary electrophoresis devices with an efficient postcolumn reactor in planar quartz and glass chips," *Anal. Chem.*, 68, (1996), 4285–4290.
8. P. C. H. Li and D. J. Harrison, "Transport, manipulation, and reaction of biological cells on-chip using electrokinetic effects," *Anal. Chem.*, 69, (1997), 1564–1568.
9. Z. Liang, N. Chiem, G. Ocvirk, T. Tang, K. Fluri, and D. J. Harrison, "Microfabrication of a planar absorbance and fluorescence cell for integrated capillary electrophoresis devices," *Anal. Chem.*, 68, (1996), 1040–1046.
10. H. Salimi-Moosavi, T. Tang, and D. J. Harrison, "Electroosmotic pumping of organic solvents and reagents in microfabricated reactor chips," *J. Am. Chem. Soc.*, 119, (1997), 8716–8717.
11. A. G. Hadd, D. E. Raymond, J. W. Halliwell, S. C. Jacobson, and J. M. Ramsey, "Microchip device for performing enzyme assays," *Anal. Chem.*, 69, (1997), 3407–3412.
12. S. C. Jacobson, C. T. Culbertson, J. E. Daler, and J. M. Ramsey, "Microchip structures for submillisecond electrophoresis," *Anal. Chem.*, 70, (1998), 3476–3480.
13. J. Khandurina, S. C. Jacobson, L. C. Waters, R. S. Foote, and J. M. Ramsey, "Microfabricated porous membrane structure for sample concentration and electrophoretic analysis," *Anal. Chem.*, 71, (1999), 1815–1819.
14. J. P. Kutter, S. C. Jacobson, N. Matsubara, and J. M. Ramsey, "Solvent-programmed microchip open-channel electrochromatography," *Anal. Chem.*, 70, (1998), 3291–3297.
15. L. C. Waters, S. C. Jacobson, N. Kroutchinina, J. Khandurina, R. S. Foote, and J. M. Ramsey, "Multiple sample pcr amplification and electrophoretic analysis on a microchip," *Anal. Chem.*, 70, (1998), 5172–5176.
16. S. Liu, Y. Shi, W. W. Ja, and R. A. Mathies, "Optimization of high-speed DNA sequencing on microfabricated capillary electrophoresis channels," *Anal. Chem.*, 71, (1999), 566–573.
17. P. C. Simpson, D. Roach, A. T. Woolley, T. Thorsen, R. Johnston, G. F. Sensabaugh, and R. A. Mathies, "High-throughput genetic analysis using microfabricated 96-sample capillary array electrophoresis microplates," *Proc. Natl. Acad. Sci.*, 95, (1998), 2256–2261.
18. A. T. Woolley and R. A. Mathies, "Ultra-high-speed DNA sequencing using capillary electrophoresis chips," *Anal. Chem.*, 67, (1995), 3676–3680.
19. J. C. McDonald and G. M. Whitesides, "Poly(dimethylsiloxane) as a material for fabricating microfluidic devices," *Acc. Chem. Res.*, 35, (2002), 491–499.

20. B. D. Gates and G. M. Whitesides, "Replication of vertical features smaller than 2 nm by soft lithography," *J. Am. Chem. Soc.*, 125, (2003), 14986–14987.
21. Q. Xu, B. T. Mayers, M. Lahav, D. V. Vezenov, and G. M. Whitesides, "Approaching zero: using fractured crystals in metrology for replica molding," *J. Am. Chem. Soc.*, 127, (2005), 854–855.
22. B. Grzybowski, D. Qin, R. Haag, and G. M. Whitesides, "Elastomeric optical elements with deformable surface topographies: applications to force measurements, tunable light transmission and light focusing," *Sens. Actuators, A*, A86, (2000), 81–85.
23. J. L. Wilbur, R. J. Jackman, G. M. Whitesides, E. Chang, L. Lee, and M. Prentiss, "Elastomeric optics," *Chem. Mater.*, 8, (1996), 1380–1385.
24. J. M. K. Ng, I. Gitlin, A. D. Stroock, and G. M. Whitesides, "Components for integrated poly(dimethylsiloxane) microfluidic systems," *Electrophoresis*, 23, (2002), 3461–3473.
25. A. R. Abate, D. Lee, T. Do, C. Holtze, and D. A. Weitz, "Glass coating for PDMS microfluidic channels by sol-gel methods," *Lab Chip*, 8, (2008), 516–518.
26. J. C. McDonald, M. L. Chabinyc, S. J. Metallo, J. R. Anderson, A. D. Stroock, and G. M. Whitesides, "Prototyping of microfluidic devices in poly(dimethylsiloxane) using solid-object printing," *Anal. Chem.*, 74, (2002), 1537–1545.
27. A. Bernard, B. Michel, and E. Delamarche, "Micromosaic immunoassays," *Anal. Chem.*, 73, (2001), 8–12.
28. J. N. Lee, C. Park, and G. M. Whitesides, "Solvent compatibility of poly(dimethylsiloxane)-based microfluidic devices," *Anal. Chem.*, 75, (2003), 6544–6554.
29. G. T. Roman, T. Hlaus, K. J. Bass, T. G. Seelhammer, and C. T. Culbertson, "Sol-gel modified poly(dimethylsiloxane) microfluidic devices with high electroosmotic mobilities and hydrophilic channel wall characteristics," *Anal. Chem.*, 77, (2005), 1414–1422.
30. J. N. Lee, X. Jiang, D. Ryan, and G. M. Whitesides, "Compatibility of mammalian cells on surfaces of poly(dimethylsiloxane)," *Langmuir*, 20, (2004), 11684–11691.
31. R. Mukhopadhyay, "When PDMS isn't the best," *Anal. Chem.*, 79, (2007), 3248–3253.
32. T. M. Squires and S. R. Quake, "Microfluidics: fluid physics at the nanoliter scale," *Rev. Mod. Phys.*, 77, (2005), 977–1026.
33. H. A. Stone and S. Kim, "Microfluidics: basic issues, applications, and challenges," *AIChE J.*, 47, (2001), 1250–1254.
34. H. A. Stone, A. D. Stroock, and A. Ajdari, "Engineering flows in small devices: microfluidics toward a lab-on-a-chip," *Annu. Rev. Fluid Mech.*, 36, (2004), 381–411.
35. L. Shui, J. C. T. Eijkel and A. van den Berg, "Multiphase flow in micro- and nanochannels," *Sens. Actuators, B*, B121, (2007), 263–276.
36. T. E. Faber, *Fluid Dynamics for Physicists*, Cambridge University Press, New York, 1995.
37. R. F. Ismagilov, A. D. Stroock, P. J. A. Kenis, G. Whitesides, and H. A. Stone, "Experimental and theoretical scaling laws for transverse diffusive broadening in two-phase laminar flows in microchannels," *Appl. Phys. Lett.*, 76, (2000), 2376–2378.
38. V. Studer, G. Hang, A. Pandolfi, M. Ortiz, W. F. Anderson, and S. R. Quake, "Scaling properties of a low-actuation pressure microfluidic valve," *J. Appl. Phys.*, 95, (2004), 393–398.
39. M. A. Unger, H.-P. Chou, T. Thorsen, A. Scherer, and S. R. Quake, "Monolithic microfabricated valves and pumps by multilayer soft lithography," *Science*, 288, (2000), 113–116.
40. D. B. Weibel, M. Kruithof, S. Potenta, S. K. Sia, A. Lee, and G. M. Whitesides, "Torque-actuated valves for microfluidics," *Anal. Chem.*, 77, (2005), 4726–4733.
41. S. E. Hulme, S. S. S., and W. G. M., "Incorporation of prefabricated screw, pneumatic, and solenoid valves into microfluidic devices," *Lab Chip, submitted*.
42. D. J. Beebe, J. S. Moore, J. M. Bauer, Q. Yu, R. H. Liu, C. Devadoss, and B.-H. Jo, "Functional hydrogel structures for autonomous flow control inside microfluidic channels," *Nature*, 404, (2000), 588–590.

43. W. C. Jackson, H. D. Tran, M. J. O'Brien, E. Rabinovich, and G. P. Lopez, "Rapid prototyping of active microfluidic components based on magnetically modified elastomeric materials," *J. Vac. Sci. Technol., B*, 19, (2001), 596–599.
44. M. Kohl, D. Dittmann, E. Quandt, and B. Winzek, "Thin film shape memory microvalves with adjustable operation temperature," *Sens. Actuators, A*, A83, (2000), 214–219.
45. N. Futai, W. Gu, J. W. Song, and S. Takayama, "Handheld recirculation system and customized media for microfluidic cell culture," *Lab Chip*, 6, (2006), 149–154.
46. A. P. Sudarsan and V. M. Ugaz, "Multivortex micromixing," *Proc. Natl. Acad. Sci. U.S.A.*, 103, (2006), 7228–7233.
47. A. D. Stroock, S. K. W. Dertinger, A. Ajdari, I. Mezit, H. A. Stone, and G. M. Whitesides, "Chaotic mixer for microchannels," *Science*, 295, (2002), 647–651.
48. H.-P. Chou, M. A. Unger, and R. Quake Stephen, "A microfabricated rotary pump," *Biomed. Microdevices*, 3, (2001), 323–330.
49. P. Paik, V. K. Pamula, M. G. Pollack, and R. B. Fair, "Electrowetting-based droplet mixers for microfluidic systems," *Lab Chip*, 3, (2003), 28–33.
50. M. Z. Bazant and T. M. Squires, "Induced-charge electrokinetic phenomena: theory and microfluidic applications," *Phys. Rev. Lett.*, 92, (2004), 066101/066101–066101/066104.
51. P. Takhistov, K. Duginova, and H.-C. Chang, "Electrokinetic mixing vortices due to electrolyte depletion at microchannel junctions," *J. Colloid Interface Sci.*, 263, (2003), 133–143.
52. Z. Yang, S. Matsumoto, H. Goto, M. Matsumoto, and R. Maeda, "Ultrasonic micromixer for microfluid systems," *Sens. Actuators, A*, A93, (2001), 266–272.
53. P. Garstecki, M. J. Fuerstman, M. A. Fischbach, S. K. Sia, and G. M. Whitesides, "Mixing with bubbles: a practical technology for use with portable microfluidic devices," *Lab Chip*, 6, (2006), 207–212.
54. N. L. Jeon, S. K. W. Dertinger, D. T. Chiu, I. S. Choi, A. D. Stroock, and G. M. Whitesides, "Generation of solution and surface gradients using microfluidic systems," *Langmuir*, 16, (2000), 8311–8316.
55. H. Bang, S. H. Lim, Y. K. Lee, S. Chung, C. Chung, D.-C. Han, and J. K. Chang, "Serial dilution microchip for cytotoxicity test," *J. Micromech. Microeng.*, 14, (2004), 1165–1170.
56. K. Campbell and A. Groisman, "Generation of complex concentration profiles in microchannels in a logarithmically small number of steps," *Lab Chip*, 7, (2007), 264–272.
57. J. K. Chang, H. Bang, S. J. Park, S. Chung, C. Chung, and D. C. Han, "Fabrication of the PDMS microchip for serially diluting sample with buffer," *Microsyst. Technol.*, 9, (2003), 555–558.
58. S. K. W. Dertinger, D. T. Chiu, N. L. Jeon, and G. M. Whitesides, "Generation of gradients having complex shapes using microfluidic networks," *Anal. Chem.*, 73, (2001), 1240–1246.
59. C. Kim, K. Lee, J. H. Kim, K. S. Shin, K.-J. Lee, T. S. Kim, and J. Y. Kang, "A serial dilution microfluidic device using a ladder network generating logarithmic or linear concentrations," *Lab Chip*, 8, (2008), 473–479.
60. C. Neils, Z. Tyree, B. Finlayson, and A. Folch, "Combinatorial mixing of microfluidic streams," *Lab Chip*, 4, (2004), 342–350.
61. G. M. Walker, N. Monteiro-Riviere, J. Rouse, and A. T. O'Neill, "A linear dilution microfluidic device for cytotoxicity assays," *Lab Chip*, 7, (2007), 226–232.
62. M. Yamada, T. Hirano, M. Yasuda, and M. Seki, "A microfluidic flow distributor generating stepwise concentrations for high-throughput biochemical processing," *Lab Chip*, 6, (2006), 179–184.
63. A. C. Siegel, D. A. Bruzewicz, D. B. Weibel, and G. M. Whitesides, "Microsolidics: fabrication of three-dimensional metallic microstructures in poly(dimethylsilo xane),"*Adv. Mater.*, 19, (2007), 727–733.
64. A. C. Siegel, S. S. Shevkoplyas, D. B. Weibel, D. A. Bruzewicz, A. W. Martinez, and G. M. Whitesides, "Cofabrication of electromagnets and microfluidic systems in poly(dimethylsiloxane)," *Angew. Chem., Int. Ed.*, 45, (2006), 6877–6882.

65. M. J. Fuerstman, P. Garstecki, and G. M. Whitesides, "Coding/Decoding and reversibility of droplet trains in microfluidic networks," *Science*, 315, (2007), 828–832.
66. M. Prakash and N. Gershenfeld, "Microfluidic bubble logic," *Science*, 315, (2007), 832–835.
67. S.-Y. Teh, R. Lin, L.-H. Hung, and A. P. Lee, "Droplet microfluidics," *Lab Chip*, 8, (2008), 198–220.
68. P. Garstecki, I. Gitlin, W. DiLuzio, G. M. Whitesides, E. Kumacheva, and H. A. Stone, "Formation of monodisperse bubbles in a microfluidic flow-focusing device," *Appl. Phys. Lett.*, 85, (2004), 2649–2651.
69. P. Garstecki, H. A. Stone, and G. M. Whitesides, "Mechanism for flow-rate controlled breakup in confined geometries: a route to monodisperse emulsions," *Phys. Rev. Lett.*, 94, (2005), 164501/164501–164501/164504.
70. P. Garstecki and G. M. Whitesides, "Flowing crystals: nonequilibrium structure of foam," *Phys. Rev. Lett.*, 97, (2006), 024503/024501–024503/024504.
71. P. Garstecki, M. J. Fuerstman, H. A. Stone, and G. M. Whitesides, "Formation of droplets and bubbles in a microfluidic T-junction-scaling and mechanism of break-up," *Lab Chip*, 6, (2006), 437–446.
72. T. Thorsen, R. W. Roberts, F. H. Arnold, and S. R. Quake, "Dynamic pattern formation in a vesicle-generating microfluidic device," *Phys. Rev. Lett.*, 86, (2001), 4163–4166.
73. B. A. Grzybowski, S. T. Brittain, and G. M. Whitesides, "Thermally actuated interferometric sensors based on the thermal expansion of transparent elastomeric media," *Rev. Sci. Instrum.*, 70, (1999), 2031–2037.
74. Z. Li, Z. Zhang, A. Scherer, and D. Psaltis, "Mechanically tunable optofluidic distributed feedback dye laser," *Opt. Exp.*, 14, (2006), 10494–10499.
75. J. A. Rogers, R. J. Jackman, O. J. A. Schueller, and G. M. Whitesides, "Elastomeric diffraction gratings as photothermal detectors," *Appl. Opt.*, 35, (1996), 6641–6647.
76. J. A. Rogers, O. J. A. Schueller, C. Marzolin, and G. M. Whitesides, "Wave-front engineering by use of transparent elastomeric optical elements," *Appl. Opt.*, 36, (1997), 5792–5795.

CHAPTER 3

Optical Components Based on Dynamic Liquid-Liquid Interfaces

Sindy K. Y. Tang and George M. Whitesides

Department of Chemistry and Chemical Biology, Harvard University, Cambridge, Massachusetts

3-1 Introduction

This chapter describes optical components based on dynamic liquid-liquid (L^2) interfaces between liquids with different optical properties (such as index of refraction) in microfluidic systems. Devices with optical interfaces formed by liquids have characteristics that are quite different from solid-gas and solid-liquid systems commonly used in conventional optics.

L^2 systems have four attractive characteristics:

1. It is simple to reconfigure the properties and functions of L^2 systems in real time by adjusting the compositions of the liquids, and their rates of flow.
2. Unlike their solid-state counterparts, polishing or high-precision microfabrication is not necessary to obtain smooth optical interfaces for L^2 devices: the L^2 interfaces are intrinsically smooth as a result of laminar flow that is characteristic of microfluidic systems.

3. It is straightforward to obtain a graded profile of refractive index in L^2 systems by taking advantage of diffusion between miscible liquids possessing different refractive indices.
4. Since the L^2 devices are formed inside a microfluidic channel, the manipulation of the liquids used for optics in microchannels is the same as that of liquids used for other purposes (separations, reagent storage, sample preparation, etc.). It is thus possible to design and cofabricate the channels for the optical parts of integrated system, and for other parts simultaneously. This feature facilitates integration and prealignment of L^2 devices to the relevant components on the same microfluidic platform.

This chapter has two objectives:

1. To discuss the basic construction of L^2 devices, and the characteristics of dynamic L^2 interfaces formed between laminar streams in microchannels
2. To give examples of optofluidic devices—L^2 waveguides, L^2 lenses, L^2 light sources, and bubble diffraction grating—to demonstrate the design and operation of these devices.

3-2 Basic Design and Construction of Liquid-Liquid Devices

Typically, L^2 devices consist of multiple streams of liquids possessing different optical properties (such as refractive indices) coflowing in a single microchannel. Figure 3-1 shows a representative design of an L^2 device. It consists of multiple inlets for different liquids to flow into a main channel. Depending on the application, this main channel can have different geometries (a straight channel of uniform width is shown in Fig. 3-1). To form and maintain the L^2 interface, liquids are

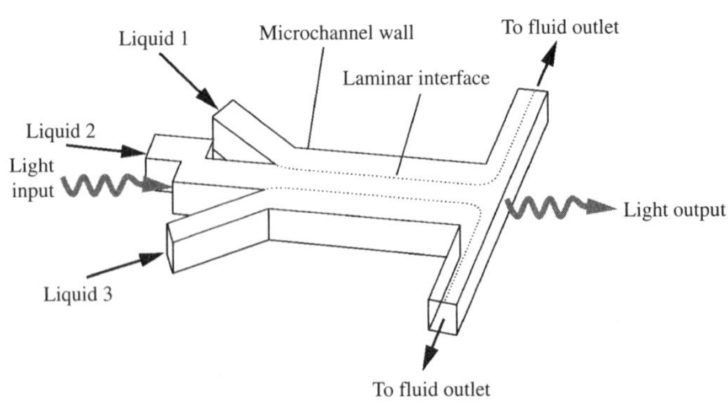

FIGURE 3-1 Schematic representation of the typical design of an L^2 device.

injected continuously into the channel. The rate of flow is sufficiently small such that the flow is laminar.

To couple light into and out of the L^2 devices, external lenses can be used to focus light from an off-chip light source into the microchannel across the polydimethylsiloxane (PDMS) wall. Alternatively, light can be coupled into an optical fiber, which is then inserted into the PDMS device through appropriate ports. The use of fibers facilitates optical alignment between external light sources, or detectors, and the microfluidic channel, and allows substantial flexibility in system design. It is therefore a common way of introducing light into L^2 devices.

Ports for insertion of optical fiber (Fig. 3-2) are often included in the design of L^2 devices [1]; they are fabricated at the same time as the rest of the microchannels. Light introduced through these inserted fibers is in the same plane of the microchannels. The port for the optical fiber is usually left sealed in the PDMS during the fabrication of the device; this port is opened later by cutting the back part of the

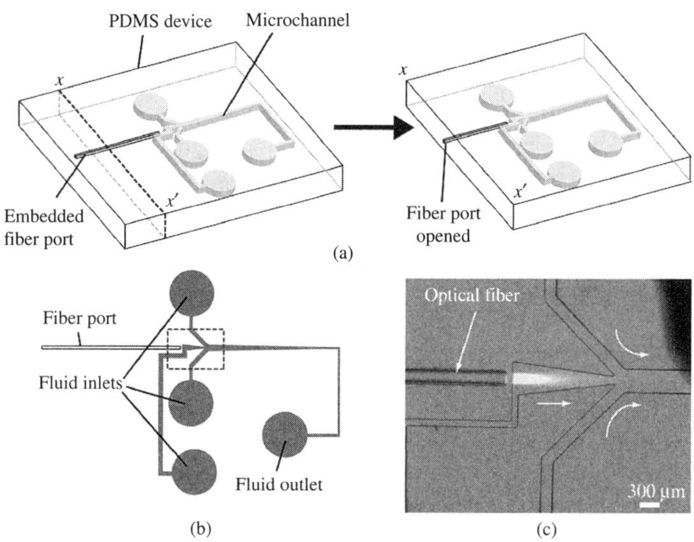

FIGURE 3-2 (a) Diagram of the sealed channel. The dotted line (*x*-*x*') depicts a typical location for cutting the sealed channel to expose the inlet for the optical fiber. (b) Top-down view of the schematic diagram of the microfluidic channel. (c) Optical micrograph of the inlet portion of the channel inside the dashed lines in (b) after the insertion of an optical fiber. The light from the optical fiber is from a fiber-coupled deuterium lamp. The channel is filled with a solution of fluorescein (1 mM). The bright area to the right of the fiber is the fluorescence of the fluorescein, and it shows the path of the light from the fiber into the fluid-filled channel. The small arrows depict the direction of the flow of the guiding-liquid and cladding-liquid streams. [(*D. J. Wolfe, R. S. Conroy, P. Garstecki, B. T. Mayers, M. A. Fischback, K. E. Paul, M. Prentiss, and G. M. Whitesides, "Dynamic control of liquid core/liquid-cladding optical waveguides," Proc. Natl. Acad. Sci. U.S.A., 101, (2004), 12434–12438.* (*Copyright 2004*) *National Academy of Sciences, U.S.A*).]

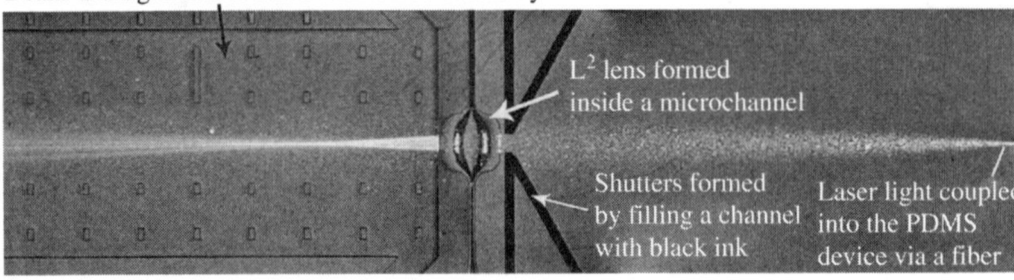

Figure 3-3 Bright-field image of beam-tracing chamber showing the optical path behind the L^2 lens. The laser beam from the fiber is visible in front of the aperture because PDMS contains nanoparticles of silica that scatter light. The focused beam in the beam-tracing chamber is visualized by the fluorescence of a rhodamine dye filling the chamber. (S. K. Y. Tang, C. A. Stan, and G. M. Whitesides, "Dynamically reconfigurable liquid-core liquid-cladding lens in a microfluidic channel," *Lab Chip*, 8, (2008), 395–401. *Reproduced by permission of the Royal Society of Chemistry.*)

sealed channels with a razor blade (Fig. 3-2a; x-x'). This cut opens a channel at the edge of the PDMS that has the dimensions of the fiber (width × height ~ 100 μm × 100 μm). The open channel accommodating the optical fiber ends at a distance from the fluidic channel, and is isolated from the fluids. Depending on the application, this distance varies from a few 10s of microns (for L^2 waveguides) to a few millimeters (for L^2 lens). The optical fiber is then manually inserted into this open channel. Index-matching liquids can be applied to fill any air gap between PDMS and the optical fiber. The center of the fiber channel is collinear with the center of the microfluidic channel.

To visualize the propagation of light inside the PDMS device, one can introduce fluorescent dyes in a chamber fabricated in the optical path [2]. This "beam-tracing" chamber is used for characterization of the focal distance and the quality of the focused beam of the L^2 lens, for example (Fig. 3-3). The solution of dye fluoresces only in regions where there was optical illumination. The concentration of the dye solution should be sufficiently low such that the incident light could propagate through the beam-tracing chamber without being significantly attenuated or absorbed. To avoid photobleaching of the dye during the experiment, the intensity of the incident light should also be sufficiently low; alternatively, new dye solution can be injected continuously to replace the photobleached dyes.

3-3 Index of Refraction of Common Liquids

Contrast of refractive index in liquids can be provided in several ways, including

1. Different liquids: A wide range of common liquids are transparent in the visible region of the spectrum, and have refractive indices ranging from 1.28 to 1.75 [3]. Table 3-1 lists the refractive indices of some common solvents.

Name	Refractive index n_d	NFPA*	Compatibility with PDMS†
Methylene iodide	1.749	H3; F1; R0	No
1,2-Diiodobenzene	1.718	H1; F1; R1	(not available)
1-Iodonaphthalene	1.701	H2; F1; R0	(not available)
Benzothiazole	1.642	H2; F0; R0	Yes
Carbon disulfide	1.627	H3; F4; R0	(not available)
1-Methylnaphthalene	1.618	H1; F2; R1	(not available)
Bromoform	1.596	H3; F0; R0	No
Thiophenol	1.588	H4; F2; R0	Yes
Nitrobenzene	1.550	H3; F2; R1	Yes
Benzaldehyde	1.544	H2; F2; R0	Yes
Benzyl alcohol	1.538	H1; F1; R0	Yes
Thiophene	1.526	H2; F3; R0	(not available)
Dimethyl sulfoxide	1.479	H1; F1; R0	Yes
Formamide	1.447	H2; F1; R0	Yes
N,N-Dimethylacetamide	1.438	H2; F2; R0	Yes
Ethylene glycol	1.429	H1; F1; R0	Yes
Propylene carbonate	1.422	H1; F1; R0	Yes
Ethanol	1.359	H0; F3; R0	Yes
Water	1.333	H0; F0; R0	Yes
Methanol	1.328	H1; F3; R0	Yes
2,2,2-Trifluoroethanol	1.300	H2; F3; R0	Yes

*The National Fire Protection Agency (NFPA) hazard identification system: H, F, R represents health, flammability, and reactivity, respectively. The number ratings range from 0 to 4. A rating of 0 represents essentially no hazard; a rating of 4 indicates extreme danger. http://www.ssrl.slac.stanford.edu/safety/nfpa_hazard_class.html.

† We define a liquid to be compatible with PDMS if the swelling ratio (defined as D/D_0, where D is the length of the solid PDMS in the solvent and D_0 is the length of the dry, solid PDMS) is less than 1.1 after a solid piece of PDMS is placed in a solvent for 24 h. Details regarding individual solvent's compatibility with PDMS can be found in Ref. 4.

(S. K. Y. Tang, C. A. Stan, and G. M. Whitesides, "Dynamically Reconfigurable Liquid-Core Liquid-Cladding Lens in a Microfluidic Channel," *Lab Chip*, 8, (2008), 395–401. Reproduced by permission of the Royal Society of Chemistry.)

TABLE 3-1 Refractive Index, Hazard Codes, and Compatibility with PDMS of Selected Liquids

2. Different concentrations of solutes in aqueous or organic solutions. Table 3-2 [5] shows the variation of refractive index with concentration of aqueous solutions of calcium chloride and sucrose, respectively.
3. A single homogeneous liquid with a difference in temperature. The thermo-optical coefficient (the change in refractive index with change in temperature) is typically around 1×10^{-4}°C^{-1} for common liquids between 10 and 80°C [6]. Water, for example, has an index $n_d = 1.3325$ at 10°C and an index $n_d = 1.3235$ at 75°C.
4. A suspension of dielectric nanoparticles, for example, a suspension of polystyrene particles in water (diameter = 160 nm, volume fraction = 0.25%) [7].
5. Liquids with refractive indices modulated by external forces such as electric fields (e.g., liquid crystals), and magnetic fields (e.g., magnetic fluid such as water-based Fe_3O_4 is reported to have an increase in refractive index $\Delta n = 0.005$ when the magnetic field increased from 25 to 300 Oe [8].

The choice of liquids to provide the contrast in refractive index depends on the application. The refractive index of organic solvents can go up to 1.75, but they may be toxic and incompatible with PDMS. The use of a single homogeneous liquid at different temperatures

Calcium Chloride Solution		Sucrose Solution	
% by Mass	**Refractive Index**	**% by Mass**	**Refractive Index**
1.00	1.3354	1.00	1.3344
5.00	1.3451	10.00	1.3478
10.00	1.3575	20.00	1.3639
15.00	1.3704	30.00	1.3812
20.00	1.3839	40.00	1.3999
26.00	1.4008	50.00	1.4201
30.00	1.4124	60.00	1.4419
34.00	1.4242	70.00	1.4654
38.00	1.4361	80.00	1.4906
40.00	1.4420	84.00	1.5010

TABLE 3-2 Refractive Index of Aqueous Solutions of Calcium Chloride and Sucrose at Different Concentrations

simplifies recycling, and facilitates closed-loop operation. Thermal diffusivity in liquids is typically two orders of magnitude higher than mass diffusivity of solute ions [6], however. A much higher rate of flow is therefore necessary to maintain the contrast in refractive index across the L^2 interface.

3-4 Dynamic Liquid-Liquid Interfaces in Microfluidic Systems

The interface between laminar streams in microfluidic systems is at dynamic steady state: continuous flow is required to maintain the interface between the streams. The use of this dynamic interface as part of an optical component has advantages and disadvantages, as discussed next.

3-4-1 L^2 Interfaces Are Reconfigurable in Real Time

Liquids can be replaced and/or replenished continuously in L^2 systems. This capability for replacement allows injection of liquids with different properties (e.g., index of refraction, absorption, and fluorescence) to tune the optical output of the system in real time. The ability to replenish liquids makes photobleaching and related phenomena relatively unimportant, since the component that is bleached is replaced continuously. This latter feature is especially important for the operation of microfluidic dye lasers—without a continuous replacement of solutions of dye, the lasing action would stop in a few seconds when the dye is photobleached. The disadvantage here is the need for constant supply of liquids. Microfluidic systems allow economical use of solutions and reagents, however; the consumption of fluids is therefore limited.

Another way to reconfigure the L^2 interface is by manipulating the flow conditions. The L^2 interface is deformable: it is possible to change the position or the shape of the liquid-liquid interface, and therefore the path of light inside the optofluidic devices by changing rates of flow (and other properties such as viscosity) of the fluids. Changing the relative volumetric rates of flow between the streams of liquids changes the position or the shape of the L^2 interface. The L^2 lens, for example, can take up shapes varying from biconvex to plano-convex to meniscus simply by changing the relative rates of flow between the core and cladding streams.

The switching time of liquids in microchannels is on the order of seconds. This time scale is limited by the time required for mass transport of liquids in the microfluidic system. This value is much longer than that in conventional optical systems. Nevertheless, the liquid-liquid system should meet the demands of applications that do not require fast switching, such as optical sensing and bioassays.

3-4-2 L² Interfaces Are Smooth

Unlike their solid-state counterparts, polishing or high-precision fabrication is not necessary to obtain smooth optical surfaces in L² devices. Because of their small length scale, L² devices operate in the low Reynolds number regime, and the flow is laminar (i.e., nonturbulent). Fluid flows at low Reynolds number generate an intrinsically optically smooth interface between streams of liquids. Small irregularities in the solid walls of the channels (having roughness of r) do not propagate into the liquid interfaces, as long as the width of the flowing streams is larger than $2r$ [9]. Figure 3-4 shows that the walls of the PDMS microfluidic channel are relatively rough (there is obvious roughness with dimensions > 5 μm). The L² interface, as viewed in this image, is still smooth. The generation of optically smooth interface in this rough channel is possible due to laminar flow of the streams of liquids. When the roughness is less than 5% of the total width of the channel, its effect is negligible on the interfaces between streams. It implies that it is possible to use low-precision fabrication to make the microfluidic channels, and still produce high-quality optical fluidic interfaces.

By introducing a liquid with refractive index matched to that of PDMS ($n_d = 1.41$) to "line" the channel, it is possible to reduce losses due to scattering of light that passes through the side wall of the channel. In the case of the L² lens, for example, the use of a mixture of 73.5% ethylene glycol ($n_d = 1.43$) and 26.5% ethanol ($n_d = 1.36$) (effective index $n_{deff} = 1.41$) as the cladding liquid reduced undesired scattering of light across the PDMS-liquid interface, and improved the quality of the focused beam (Fig. 3-10b and c). Other mixtures of liquids or solutions of different salt concentrations should also work.

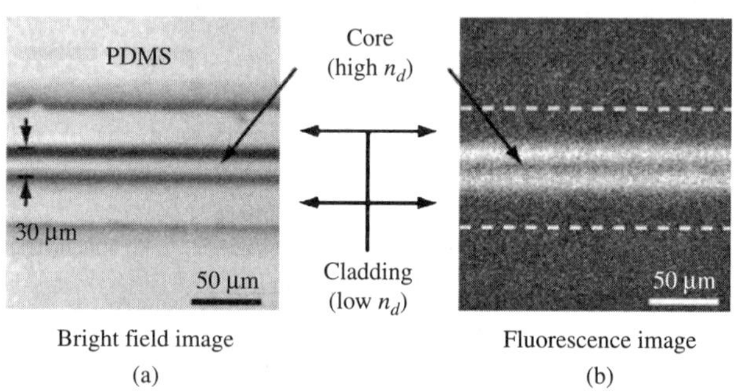

FIGURE 3-4 (a) Optical micrograph of the L² waveguide. The core fluid was dyed to aid visualization. (b) Fluorescence micrograph of the same region of the channel as in a. The visible fluorescence signal has been produced by excitation with a broadband deuterium, fiber-coupled light source leaking into the evanescent field from the core of the waveguide. The dotted lines indicate the location of the walls of the microchannel. [(*D. J. Wolfe, R. S. Conroy, P. Garstecki, B. T. Mayers, M. A. Fischback, K. E. Paul, M. Prentiss, and G. M. Whitesides, "Dynamic control of liquid core/liquid-cladding optical waveguides," Proc. Natl. Acad. Sci, U.S.A, 101, (2004), 12434–12438.* (*Copyright 2004*) *National Academy of Sciences, U.S.A*).]

3-4-3 L^2 Interface between Miscible Liquids Is Diffuse

The L^2 interface between miscible liquids is diffuse—it is a gradient of chemical/physical composition and refractive indices. Diffusion of molecules or ions between different liquids broadens the interface between the streams. This diffusion creates a graded profile of refractive index across the interface. This feature is attractive for applications that require a gradient of refractive index, such as GRIN lenses, and diffusive splitters. This graded profile is more difficult to generate, and almost impossible to modify in solid-state systems.

Diffusion, when sufficiently extended, flattens the contrast in chemical/physical composition (e.g., salt concentration, temperature) of the respective fluids, and therefore the contrast in the refractive index that defines the fluidic-optical interface. As described in Chap. 2, for solute ions flowing through a channel with width w = 100 μm at velocity $v = 100\ \mu\text{ms}^{-1}$, it would take only 5 s for the ions to diffuse across the width of the entire channel. That is, within 500 μm down the channel, the contrast in concentration and refractive index will be flattened. The use of a more viscous liquid, or a higher rate of flow of liquids, can mitigate this effect. Increasing the rate of flow reduces the residence time of the liquids inside the channel, and therefore reduces diffusive broadening for the same length of the channel. Figure 3-5 shows the simulations for the profile of refractive index at different rates of flow. In principle, the use of immiscible liquids can eliminate diffusion completely, but different wetting properties of the liquids on the PDMS wall and surface tension between the liquids (leading to droplet formation) can complicate the flow and make the manipulation of the L^2 interface more difficult.

3-5 Liquid-Liquid Optical Devices

3-5-1 L^2 Waveguides

Design and Construction

L^2 waveguides consist of two streams of liquids with lower refractive index (the cladding), sandwiching a stream of liquid with higher refractive index (the core) flowing in a microchannel [1] (Fig. 3-6). In principle, any liquid that does not swell PDMS [4] can be used in L^2 waveguides, as long as the contrast in index of refraction between the core and the cladding streams are large enough to sustain the propagation of light. In much of our work, we used a 5-M aqueous solution of calcium chloride ($n_d = 1.45$) as the core liquid, and water as the cladding ($n_d = 1.33$).

To introduce light into the device, an optical fiber is inserted into the PDMS device through a fiber port fabricated at the end of the channel. The guided light exits the L^2 waveguide when the core fluid is forced to turn by 90° with a radius of ~ 0.5 mm (much less than the critical radius) [10]. The output of the L^2 waveguide can then be

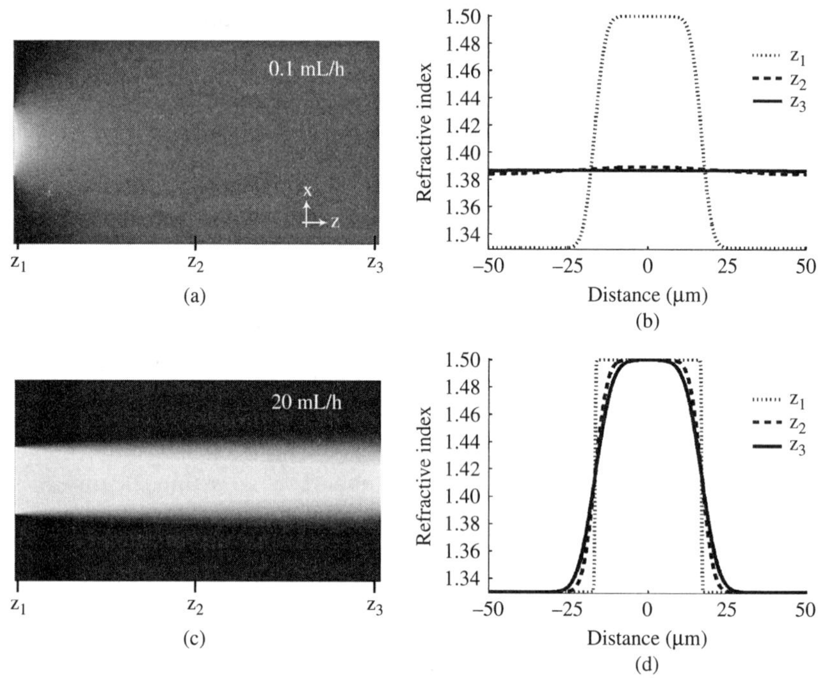

FIGURE 3-5 Simulated two-dimensional (*x-z*) distributions of refractive index in a 5-mm-long waveguide formed by water at total rates of flow of (a) 0.1 mL/h and (c) 20 mL/h. The refractive index of the injected core liquid n_{core} is 1.50, and is represented in white. The refractive index of the injected cladding liquid $n_{cladding}$ is 1.33, and is represented in black. Plot of the refractive index as a function of distance from the center of the waveguide in the transverse (*x*) direction for three longitudinal positions (z_1, z_2, and z_3) at total rates of flow of (b) 0.1 mL/h and (d) 20 mL/h. In this simulation, the width, height, and length of the channel are 100 μm, 100 μm, and 5 mm, respectively; the diffusivity is 10^{-9} m^2/s, and the viscosity is 8.90×10^{-4} Pa·s.

imaged and analyzed through an optically transparent window (Fig. 3-6) by using a microscope objective and a charge-coupled device, or through an additional inlet for an optical fiber at the end of the channel coupled to a photodetector.

Characterization

By controlling the relative rates of flow of the core and cladding liquids, it is possible to change the width of the core stream to achieve both single- and multimode guiding. Decreasing the ratio of flow rates of the core to the cladding streams decreases the core size from more than 100 μm to less than 10 μm, and thus switches the guiding from multi- to single-mode. At a rate of flow of 10 μL/min, the distance at which the L^2 waveguide can operate before complete diffusive mixing homogenizes the liquids is ~ 5 mm. This length scale is limited by diffusive broadening of the interface between streams, which decreases the contrast in refractive index between the core and the cladding.

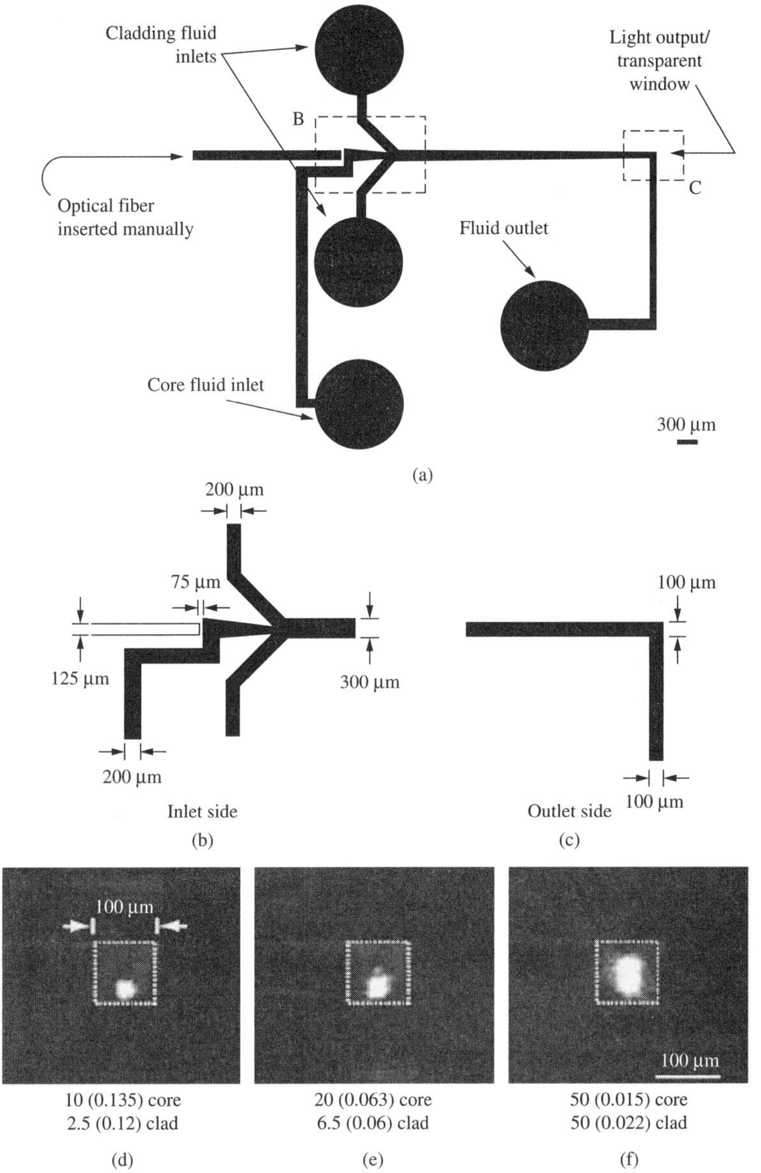

FIGURE 3-6 (a) Diagram of the design of a microfluidic channel used in these experiments. (b) and (c) Detailed diagrams of the regions of the microfluidic channel in (a) A highlighted by dashed box. (d), (e), and (f) Optical micrographs of the cross section of the outlet of the microfluidic channel viewed through the transparent window. The dashed box shows the location of the cross section of the microfluidic channel. The rates of flow (uL/min) (and the residence time in seconds) of the core and the cladding (clad) are listed. The guided light was from a fiber-coupled laser with $\lambda = 780$ nm. [(*D. J. Wolfe, R. S. Conroy, P. Garstecki, B. T. Mayers, M. A. Fischback, K. E. Paul, M. Prentiss, and G. M. Whitesides, "Dynamic control of liquid core/liquid-cladding optical waveguides," Proc. Natl. Acad. Sci, U.S.A, 101, (2004), 12434–12438.* (*Copyright 2004*) *National Academy of Sciences, U.S.A*).]

This unfavorable effect can be partially circumvented, however, by using a higher rate of flow as mentioned in the previous section.

The loss in the intensity of guided light in L^2 waveguides is around 0.1 dB/cm. The efficiency of coupling light from the L^2 waveguide into a multimode optical fiber (step-index fiber, numerical aperture = 0.22, core diameter = 105 μm, cladding diameter = 125 μm) is ~ 40%. Light exiting the L^2 waveguide remains polarized in the input direction to ~ 100:1; this ratio is indistinguishable from the light in the input fiber.

Complex Devices Derived from L^2 Waveguides

Based on the L^2 waveguide configuration, we have developed other functional optical devices in microfluidic systems (Fig. 3-7).

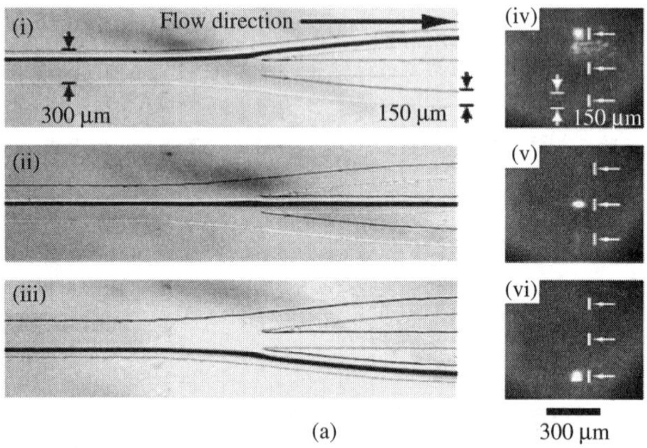

FIGURE 3-7 (a) *Optical switch.* (i), (ii), and (iii) Optical micrograph of the top view of the microfluidic channels. Dye in the core fluid makes it easily imaged; the dye is omitted in use. (iv), (v), and (vi) Optical micrograph of the cross section of the end of the channel showing light exiting the L^2 waveguides. The white arrows and lines represent the location of the ends of the branches of the microfluidic channel. (b) *Evanescent coupler.* Plot of the ratio of the intensity of the light emitted from the coupled guide (ICG) and the illuminated guide (IIG). (Insets) Shown are optical micrographs of the cross section of the output of the microfluidic channels viewed through the transparent window. (c) (i) Plot of the profile of the intensity of light output as a function of distance from the center of the channel. The light (λ = 780 nm) was coupled into the L^2 waveguide from a single-mode optical fiber. The rate of flow of the core fluids was 2.5 μL/min, of the central cladding fluids was 5 μL/min, and of the outer cladding fluids was 10 μL/min. (Inset) Optical micrograph of light exiting the microfluidic channel, viewed through the transparent window. The dashed box shows the walls of the channel. (ii) Contour plot of the refractive index as a function of the distance from the center of the width of the channel and of the distance along the length of the channel. The gradient of gray scale from black to white indicates values of the refractive index from 1.431 to 1.414. Only the main portion of the waveguide (1 cm × 0.005 cm, l × w) is simulated. [(a) and (b), *D. J. Wolfe, R. S. Conroy, P. Garstecki, B. T. Mayers, M. A. Fischback, K. E. Paul, M. Prentiss, and G. M. Whitesides, "Dynamic control of liquid core/liquid-cladding optical waveguides," Proc. Natl. Acad. Sci, U.S.A, 101, (2004), 12434–12438.* (*Copyright 2004*) *National Academy of Sciences, U.S.A*).]

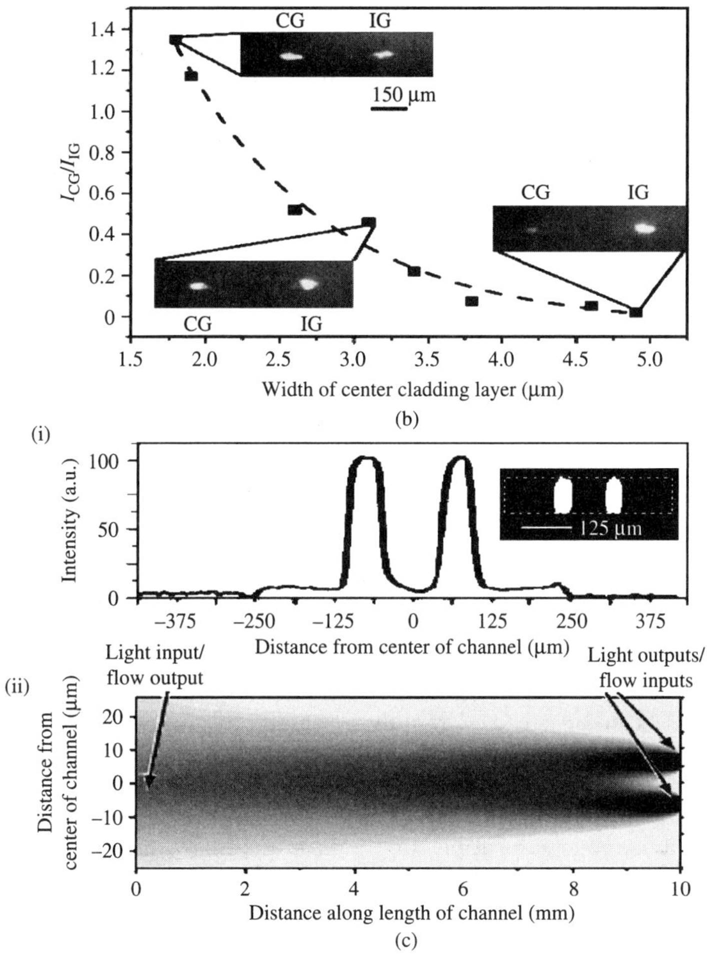

FIGURE 3-7 (*Continued*)

1. Optical switch [1]: An L^2 waveguide is branched into three separate outlet channels. The relative rates of flow of the cladding liquids determine the path of the core liquid, and therefore the path of the guided light.
2. Evanescent-wave coupler [1]: This device consists of two L^2 waveguides sharing an inner cladding stream with a width less than 5 μm. Light from an optical fiber is introduced into one of the L^2 waveguides. The rate of flow of the liquids adjusts the width of the inner cladding stream, and the efficiency of coupling of evanescent fields between the two cores of the L^2 waveguides. Efficient coupling is observed when the width of the inner cladding is below 2 μm.

Diffusion-Controlled Splitter

Diffusion-controlled splitter consists of two parallel L^2 waveguides [11]. The rate of flow is sufficiently low to allow complete diffusive mixing of the liquids as they reach the end of the channel. As a result, the two core streams merge smoothly into a single L^2 waveguide. Light propagates in a direction opposite to that of the flow of liquids, that is, in the direction of decreasing extent of diffusive mixing. This system has been demonstrated to split a single input beam into two output beams with equal intensities.

Advantages and Disadvantages of L^2 Waveguides

To conclude our discussion of these systems, L^2 waveguides have two main advantages:

1. They are dynamically reconfigurable. Their structure and function depend on a continuous, laminar flow of the core and cladding liquids, and can therefore be reconfigured and adapted continuously in ways that are not possible with solid-state waveguides.
2. They are simple to fabricate. The roughness of the wall of the channel does not affect the smoothness of the laminar interface between the core and the cladding streams, and does not lead to the scattering of light or degradation in the performance of waveguides. L^2 waveguides can therefore be fabricated easily and rapidly in organic polymers by using the convenient techniques of rapid prototyping [12].

The L^2 waveguides also have prominent disadvantages:

1. A constant supply of fluids is necessary to maintain the waveguiding streams (a supply of 144 mL is necessary to run one stream at 100 μL/min for 24 h).
2. L^2 systems using water and PDMS are unable to guide light in the infrared (λ = 1300–1600 nm) used in telecommunications applications because of large absorptive losses in both the fluids and in the PDMS.
3. The speed of optical switching is ~ 0.1 Hz. This value is much slower than switching in conventional planar waveguides (~ 1–100 GHz). Nevertheless, the system should meet the demands of applications that do not require fast switching, such as optical sensing and bioassays.

3-5-2 L^2 Lenses

Design

The design of the L^2 lens is similar to that of the L^2 waveguide: it is formed by laminar flow of three streams of fluids; the index of refraction of the central ("core") stream is higher than the index of the sandwiching ("cladding") streams [2]. The streams enter a microchannel containing an "expansion chamber"—a region in which the width of the channel expands laterally. Figure 3-8 shows a schematic diagram of this system. The expansion chamber is typically 10 times wider than its entrance and exit. For some rates of flow, the shape of the interface between the core and cladding streams in the expansion chamber is biconvex. This fluidic biconvex structure focuses light propagating in the plane of the expansion chamber, and perpendicular to the direction of flow of the liquids. By changing the relative rates of flow of the three streams, it is possible to change the curvature of the interface and thus the focal distance of the lens in real time.

To observe the focal point of the lens within the PDMS device (~ 2 cm × 2 cm), the contrast in refractive indices should be sufficiently large ($\Delta n_d >\sim 0.1$). Here, benzyl alcohol ($n_d = 1.54$) and benzothiazole ($n_d = 1.64$) have been used as the core liquid; and trifluoroethanol ($n_d = 1.29$) as the cladding. To facilitate beam tracing and determination of the focal point of the L^2 lens, an aperture can be included in front of the expansion chamber to block incident light from regions of the lens close to the inlet and outlet where the radius of curvature is highly nonuniform (Fig. 3-8). The aperture is formed

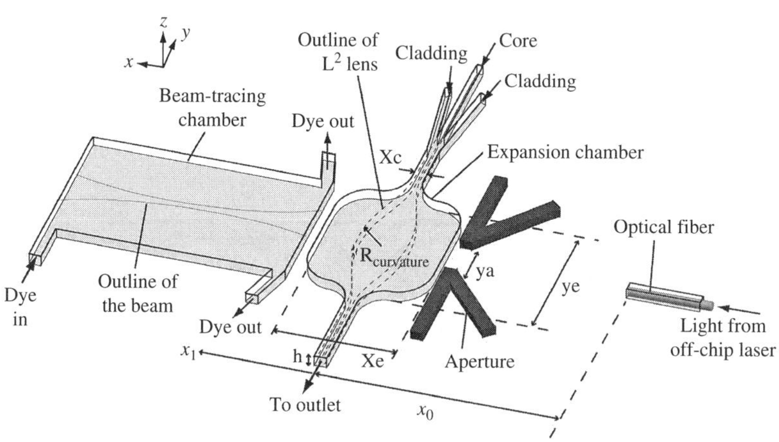

FIGURE 3-8 Schematic representation of the experimental setup for focusing light exiting an optical fiber through the liquid-core liquid-cladding (L^2) lens. The aperture is formed by two channels filled with black ink after fabrication. The channel for the formation of the L^2 lens contains a square expansion chamber. The solid lines show the walls of the channel, and the dashed lines show the interfaces between the core and the cladding streams. $R_{curvature}$ is the radius of curvature of this interface. The height (h) of the channel is about 100 mm. The beam-tracing chamber behind the L^2 lens is filled with solution of a fluorescent dye (2.5 μm Rhodamine 640 perchlorate in ethylene glycol) to make the optical path visible. (S. K. Y. Tang, C. A. Stan, and G. M. Whitesides, "Dynamically reconfigurable liquid-core liquid-cladding lens in a microfluidic channel," *Lab Chip*, 8, (2008), 395–401. *Reproduced by permission of the Royal Society of Chemistry.*)

by filling two separate channels with black ink. For applications that require higher intensity at the focus, the aperture can be removed. In order to visualize the optical path, a beam-tracing chamber filled fluorescent dyes (2.5 μM rhodamine 640 perchlorate in ethylene glycol) is incorporated behind the L^2 lens.

Characterization

Figure 3-9 shows the shapes of the L^2 lens under different flow conditions. Since the height of the expansion chamber was much smaller than its width and length, the flow was quasi-two-dimensional, and the L^2 lens is roughly cylindrical. When the rates of flow of the left and the right cladding streams were the same, the core stream, or the L^2 lens, was biconvex and symmetrical inside the expansion chamber. Varying the relative flow rates between the left and the right claddings varies the curvatures of the left and right interfaces separately. It is therefore possible to obtain an extensive range of lens shapes: meniscus, plano-convex, and biconvex.

The L^2 lens focused light; the FWHM (full width at half-maximum) of the beam at the focus achieved was ~ 16 μm, 20 times less than the initial beam width, using a 334-μm aperture. This beam size was limited by aberration due to the shape of the L^2 lens; the diffraction-limited width at the focal point is ~ 7 μm using this aperture. The enhancement factor (defined here as the ratio of the peak intensity of a focused beam to the intensity of an unfocused beam at the same point) achieved was 9 without any aperture (the enhancement factors were usually between 3 and 4 among previous works on microfabricated lenses).

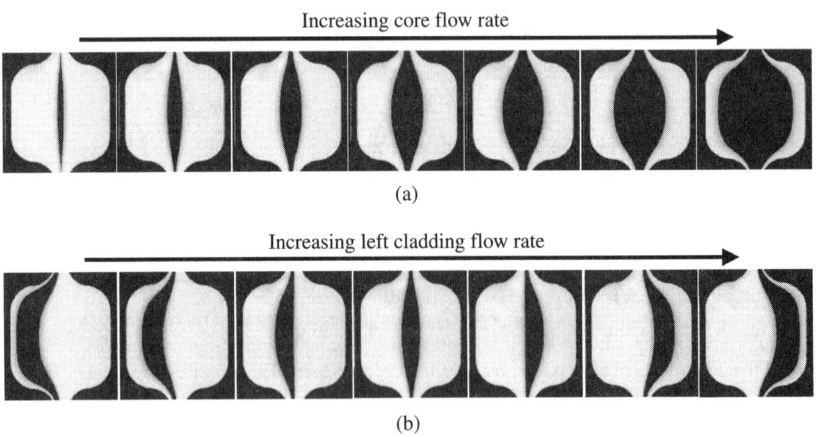

FIGURE 3-9 (a) Fluorescence images of the L^2 lens in the expansion chamber as the rate of flow of the core stream increases (from left to right). The cladding liquid was dyed to make it easily imaged; the dye was omitted in normal operation of the L^2 lens. (b) Fluorescence images of the L^2 lens as the rate of flow of the left cladding stream increases (from left to right).

Figure 3-10a shows the focal distance, measured from the center of the lens to the focal point, as a function of the core flow rate. The variation of the focal distance follows the variation of the curvature of the lens as expected from geometrical optics: a lens with a higher curvature focuses light at a shorter distance than one with a lower curvature. To achieve even shorter focal distances, one can use liquids with a larger contrast in refractive indices. Alternatively, one can use a smaller expansion chamber: at the same expansion ratio, the radius of curvature of the core-cladding interface is smaller in a smaller chamber; the focal distance achieved should also be shorter.

The beam-tracing chamber allows detailed analysis of the quality of the focused beam. Figure 3-10b and c compares the focused beam under the same flow conditions using a 500-μm aperture and a 334-μm aperture, respectively. The aberration of the L^2 lens was prominent in the former case: the areas of high light intensity were not limited to the paraxial focal point. This aberration is caused by the deviation of the shape of the core-cladding interface from the ideal lens shape. Making small adjustments to the shape of the expansion chamber and fine-tuning the shape of the lens should correct this aberration.

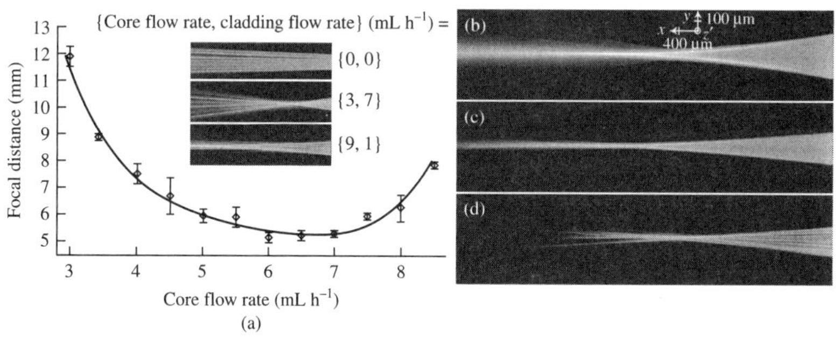

FIGURE 3-10 (a) Focal distance of the L^2 lens as a function of the rate of flow of the core stream. The core liquid was benzothiazole, and the cladding liquid was a mixture of ethylene glycol and ethanol with effective refractive index matched to that of PDMS. The total rate of flow of the core and cladding streams was fixed at 10 mL/h. The line is a guide to the eye only. The inset shows images of the focused beams in the beam-tracing chamber at the indicated flow rates. (*b*), (*c*), and (*d*) Optical micrographs of the focused beam using (b) a 500-μm aperture, and (c) a 334-μm aperture, respectively. The core liquid was benzothiazole ($n_d = 1.64$) and the cladding liquid was a mixture of ethylene glycol and ethanol with effective refractive index matched to that of PDMS ($n_d = 1.41$). The core flow rate was 6 mL/h, and the cladding flow rate was 4 mL/h. Aberration was more prominent in (b) using a 500-μm aperture. (d) Optical micrograph of the focused beam using trifluoroethanol ($n_d = 1.29$) as the cladding liquid. The core liquid was benzothiazole. The aperture size was 334 μm. The core flow rate was 3 mL/h, and the cladding flow rate was 7 mL/h. Compared with (c), beam quality decreased due to the scattering of light at the PDMS-cladding interface. (S. K. Y. Tang, C. A. Stan, and G. M. Whitesides, "Dynamically reconfigurable liquid-core liquid-cladding lens in a microfluidic channel," *Lab Chip*, 8, (2008), 395–401. *Reproduced by permission of the Royal Society of Chemistry.*)

Figure 3-10d shows the image of the focused beam using a L^2 lens with trifluoroethanol ($n_d = 1.29$) as the cladding liquid and benzothiazole as the core liquid. Due to the higher contrast in refractive index between the core and the cladding, the focal distance achieved was smaller. The quality of the beam was visibly worse than the case when the index of the cladding liquid was matched to that of PDMS (Fig. 3-10b and c). The streaks in the light beam were due to scattering of light from the rough channel wall.

3-5-3 L^2 Light Sources

We developed various on-chip fluidic light sources based on the L^2 waveguide systems for optical detection and spectroscopic analysis in integrated microanalytical systems (μTAS). In these systems, the liquid cores contain fluorescent dyes, excited by incident light from an external halogen bulb or a pump laser. Although external excitation sources are still necessary, integration of fluorescent light sources during device fabrication removes both the need for insertion and alignment of optical-fiber light sources and the constraints on channel size imposed by fiber optics.

Broadband Fluorescent Light Source

The construction of a microfluidic broadband light source is similar to that of a L^2 waveguide [13]. Solutions of multiple fluorescent dyes form the core streams, sandwiched by cladding streams with lower index of refraction. Excitation of these dyes by an external halogen bulb results in a broadband optical output with wavelength ranging from 450 to 750 nm.

Simultaneous use of multiple fluorophores in a common solution, in a single L^2 light source, is not possible, because of energy transfer from fluorophores emitting at shorter wavelength to fluorophores emitting at longer wavelength. Spatial separation of the fluorophores in different streams circumvents this problem. One design uses a cascade (series) of single-core, dye light sources of increasing absorption energy to generate a combined broadband output (Fig. 3-11a and b). The second approach uses a parallel array of single-core, dye light sources (Fig. 3-11c and d). The spectral content of the light output for both cascade and array light sources can be controlled through the choice of flow rates and dyes. Output intensity from these light sources is comparable to standard fiberoptic spectrophotometer light sources.

L^2 Microfluidic Dye Laser

Details about different microfluidic dye lasers can be found in Chap.10. Here we describe the use of L^2 waveguide for dye laser [14]. The construction of a microfluidic dye laser is similar to that of a L^2 waveguide. Solutions of fluorescent dye act as the gain media. They form the core streams, sandwiched by cladding streams with lower index of refraction, in a microchannel of length 5 to 20 mm where the

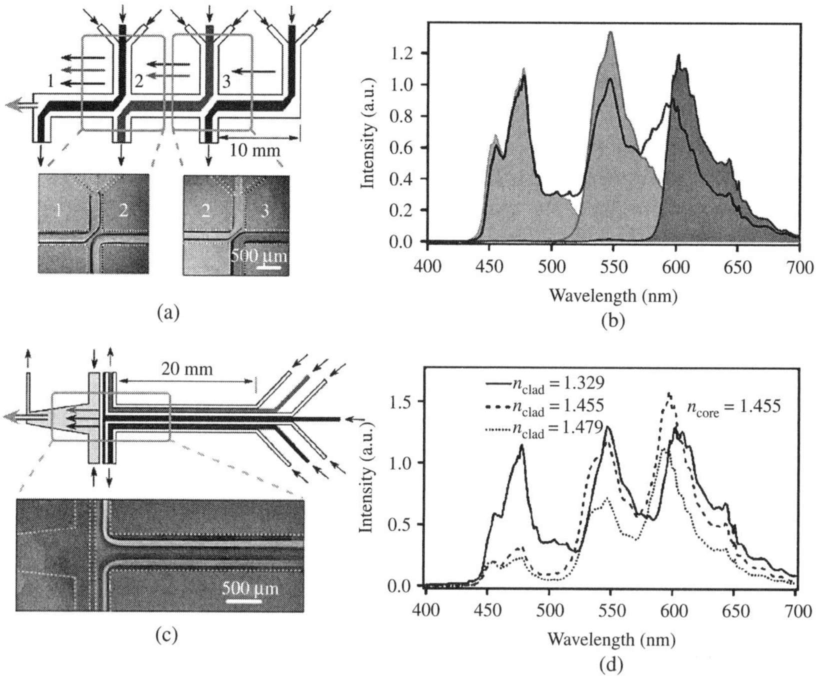

FIGURE 3-11 (a) Top-view scheme for a cascade of L^2 fluorescent light sources consisting of a series of microfluidic channels in PDMS. Multiple waveguides occupy the same central microfluidic channel. The flow of waveguide 2 displaces waveguide 3, and the flow of waveguide 1 displaces waveguide 2 at cross-junctions in the central channel. Light output is transferred between waveguides at these junctions where fluids take 90° turns. The dimensions of the central channel were 130 μm × 300 μm × 3 cm ($h \times w \times l$). Insets: Optical micrographs of the cross-junctions. The brightness and contrast have been adjusted for clarity. Dotted lines highlight the walls of the channels. (b) Spectral output (solid line) of a cascade of L^2 fluorescent light sources containing 0.5 mM solutions of perylene, fluorescein, and sulforhodamine B in DMSO/EG (1:1), when the entire central channel was irradiated with a single halogen source (uncollimated). Flow rates were 0.8, 2, and 5 mL/h for respective fluorescent cores (1, 2, and 3). Core/cladding rates were kept at a ratio of 2:1 for each waveguide. Selective illumination of discrete sections of the central microchannel with a collimated halogen source (each region of illumination was 4 mm in diameter) allowed selective excitation of individual fluorophores (shaded areas). (c) Top-view scheme for the array of L^2 fluorescent light sources, consisting of parallel L^2 waveguides in a single PDMS microchannel. An end-coupled, tapered, liquid-core waveguide filled with DMSO collected the total fluorescence output. Inset: Optical micrograph of the T-junction. Dotted lines outline the walls of the PDMS channels. (d) Spectral output (solid line) from an array of L^2 fluorescent light sources containing 0.5 mM solutions of perylene, fluorescein, and sulforhodamine B in DMSO/EG (1:1), with various cladding liquids: methanol ($n_{cladding} < n_{core}$); DMSO/EG (1:1, $n_{cladding} = n_{core}$); DMSO ($n_{cladding} > n_{core}$). Flow rates for all inputs were held constant at 4 mL/h each. (*Adapted with permission from* B. T. Mayers, D. V. Vezenov, V. I. Vullev, and G. M. Whitesides, "Arrays and cascades of fluorescent liquid-liquid waveguides: broadband light sources for spectroscopy in microchannels," *Anal. Chem.*, 77, (2005), 1310–1316. *Copyright 2005. American Chemical Society.*)

dye is pumped optically by an external laser source. The microchannel terminates at both ends with T-junctions coated with thin layers of gold or silver to act as mirrors for the optical cavity (Fig. 3-12).

Lasing Characteristics

Lasing was demonstrated with 2-mM rhodamine 640 perchlorate in methanol as the core stream, and pure methanol as the cladding stream. A 532-nm pump laser beam (frequency doubled Nd:YAG, 50-Hz repetition rate, 16-ns pulse) is elongated and focused onto the microchannel with a cylindrical lens. A high rate of flow (~ 30 mL/h) is applied to ensure a rapid replenishment of the dye. Coupled with the short (16 ns) pulse length, potential photobleaching of the dyes is avoided at the power levels used (up to 0.15 mJ/pulse).

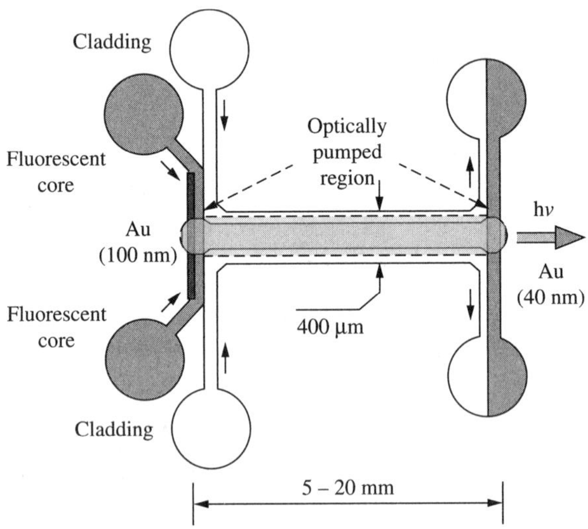

FIGURE 3-12 (a) Top-view scheme for a L^2 waveguide laser consisting of a liquid cladding and fluorescent liquid core flowing laminarly in a microfluidic channel in PDMS. The dimensions of the central channel were 100 mm × 400 mm × 10 mm (height × width × length). The 532-nm laser beam (frequency-doubled Nd:YAG, 50 Hz repetition rate, 16-ns pulse) was elongated with a cylindrical lens, and the optical pumping region covered the full length of the waveguiding region for the 10-mm-long channel.
(b) Dependence of the intensity of the optical output and line width (FWHM) on the energy of the pump pulse. The curves are drawn to guide the eye.
(c) Adjustment of the peak wavelength of the light output by changing the composition of the mixed methanol-DMSO liquid core with constant concentration of the dye (2 mM). (*Adapted with permission from* D. V. Vezenov, B. T. Mayers, R. S. Conroy, G. M. Whitesides, P. T. Snee, Y. Chan, D. G. Nocera, and M. G. Bawendi, "A low-threshold, high-efficiency microfluidic waveguide laser," *J. Am. Chem. Soc.*, 127, (2005), 8952–8953. *Copyright 2005 American Chemical Society.*)

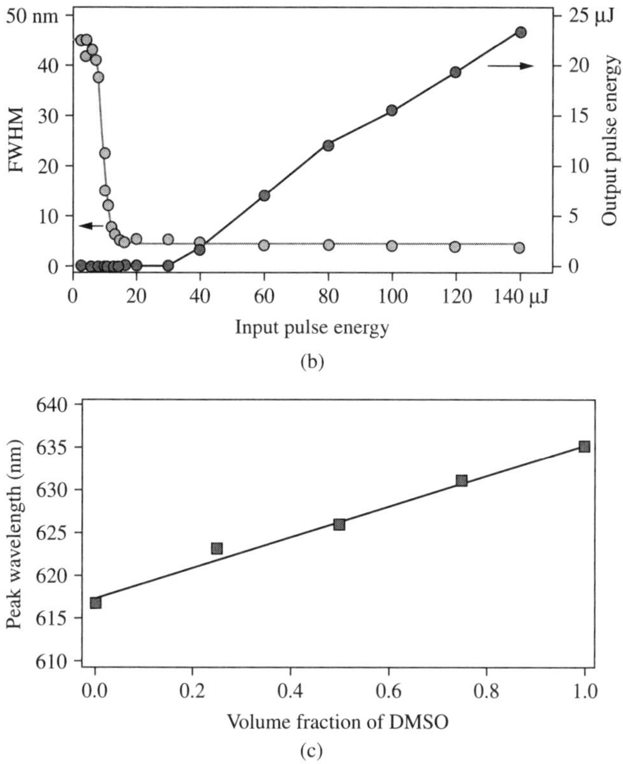

Figure 3-12 *(Continued)*

Figure 3-12b is a plot of output power and line width as a function of pump power. At low pump power (< 5 μJ/pulse) the line width (full width at half-maximum, FWHM) of the emission is ~ 45 nm, with λ_{max} centered at 625 nm. The line width drops to ~4 nm between pump powers of 7 and 16 μJ. The threshold for lasing occurs at 22 μJ pulse energy.

Wavelength Tunability

Like other dye lasers, the output wavelength can be changed by changing the choice and volume fraction of the solvent for the dye. Changing the solvent in the fluorescent core provides a simple way to adjust the wavelength of emission for a given dye without incorporating dispersive elements (prisms or gratings) into the optical cavity. For example, the wavelength of the light output for rhodamine 640 shifted by more than 20 nm by changing the composition of the core liquid; λ_{max} = 617 nm for methanol (MeOH), 631 nm for ethylene glycol, and 634 nm for dimethyl sulfoxide (DMSO). Adjusting the composition of a mixture of DMSO and MeOH in the core allows continuous tuning of λ_{max} (Fig. 3-12c).

3-5-4 Bubble Grating

So far we have only focused on laminar systems using miscible fluids. Another interesting system involves the use of immiscible fluids to form emulsions and foams in microfluidic channels. These dispersions form periodic, self-assembled geometries that can also be reconfigured in real time by varying fluid properties and flow conditions. Here we describe a tunable, fluidic, two-dimensional diffraction grating based on a microfluidic system comprising a flow-focusing device for generating bubbles, and a flowing, regular lattice of bubbles formed by dynamic self-assembly [15].

As described in Chap. 2, the flow-focusing device consists of two inlet channels for the liquid phase, and a single inlet channel for the gaseous phase (Fig. 3-13a). The gas and liquid phases meet at a junction upstream of a narrow orifice. The gaseous thread periodically

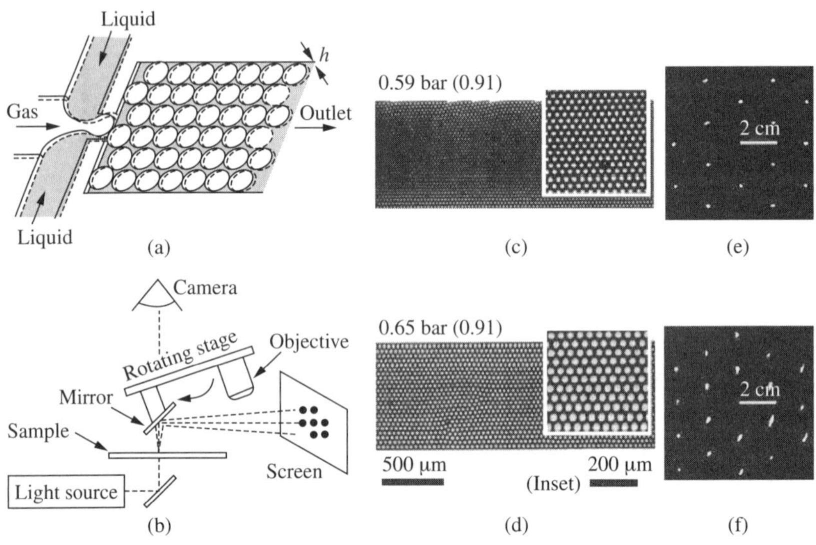

FIGURE 3-13 Schematic representations of the experimental setup. (a) The flow-focusing bubble generator and self-assembled lattice of bubbles. A tank of nitrogen gas connects to the gas inlet, and digitally controlled syringes connect to the liquid inlets. The system generates monodisperse bubbles that pack into a quasi-two-dimensional sheet. (b) Diffraction and display of incident laser beam. Diffracted laser light orthogonal to the plane of bubble lattice is displayed on a white screen. (c) and (d) Packing of bubbles in a straight outlet channel that was 1-mm wide and 16-μm high. The rate of flow of the continuous phase was 0.028 μL/s. The numbers denote the input gas pressure and the volume fraction of bubbles (φ_{vol}) in parentheses. (c) 0.59 bar, $\varphi_{vol} = 0.91$, and (d) 0.65 bar, $\varphi_{vol} = 0.91$; their corresponding diffraction patterns are shown in (e) and (f), respectively. (M. Hashimoto, B. Mayers, P. Garstecki, and G. M. Whitesides, "Flowing lattices of bubbles as tunable, self-assembled diffraction gratings," *Small*, 2, (2006), 1292–1298. *Copyright Wiley-VCH Verlag GmbH & Co. KGaA. Reproduced with permission.*)

enters the orifice, breaks, and releases a bubble into the outlet channel. As low gas pressures, the volume fraction (φ_{vol}) of the bubbles formed is low, and the bubbles flow in disordered packs. As φ_{vol} increases, the bubbles organize into hexagonally packed domains. As φ_{vol} approaches 0.91, the limit of packing of disks on the plane, the domains become a single lattice extending throughout the outlet channel. At $\varphi_{vol} \sim 0.91$, the bubbles fill the entire plane of the channel; the defects in the lattices are minimized.

Figure 3-13b shows the optical setup to characterize the diffraction patterns from the bubble lattices. A He/Ne laser ($\lambda = 632.8$ nm) illuminates the center of the bubble lattice. The direction of the beam is perpendicular to the plane of the device 1-cm downstream from the flow-focusing nozzle. Diffraction patterns are projected onto a white screen.

Figure 3-13c to f shows the bubble lattices and their corresponding diffraction patterns. These bubble lattice gratings can be modeled as both amplitude gratings and phase gratings. The menisci of the bubbles refract light radially, in a way that is similar to diffraction gratings formed from periodic arrays of dots or holes—that is, amplitude gratings. The bubbles and the carrying fluid also represent periodic arrays of alternating refractive indices—phase gratings. Changing the pressure of the gas and rate of flow of the liquid applied to the flow-focusing device changes the structure of the bubble lattices, and the diffraction patterns generated. The switching time is less than 10 s.

3-6 Conclusions

Dynamic optofluidic components based on liquid-liquid interfaces are simple to design, fabricate, and operate. They are adaptive and reconfigurable; the range of tuning is large, and only limited by the choice of liquids that can be injected into the microfluidic systems. Fluidic optical systems are also readily integrable with microanalytical and lab-on-a-chip systems for biochemical detection, where the analytes of interest are usually in the liquid phase.

The main disadvantage of these optofluidic components is the need for a constant supply of fluids. The range of refractive index available in fluids is also limited: the highest is around 1.75; this value is much lower than that in solids. They have limited transparency in the infrared, and are therefore mostly used in the visible region of the spectrum. In addition, the speed of optical switching is slow (on the order of seconds) compared to conventional optical devices. Nevertheless, these devices should still be useful for applications that do not require fast switching, such as optical sensing.

Optical systems based on liquid-liquid interfaces are still in their infancy of development. There are enough data to show that these systems "work" optically: one can make optofluidic analogs of various familiar devices, such as waveguides and lenses; one can manipulate light in ways that cannot be accomplished using conventional solid-state devices. The question now is "Who cares?"

It is unlikely that this class of optofluidic devices will compete with conventional, solid-state devices in optical communications, where durability and stability are of paramount importance. Optofluidic systems seem, however, to be well suited for bioanalysis and lab-on-a-chip systems, where the samples are usually present in aqueous solutions, and where it is possible to use the strategy of cofabrication to generate multiple useful functions, from analysis and generation of light to the manipulation of particles using magnetic fields, in devices made using a single step of fabrication [16,17]. A wide range of applications in biomedicine, food testing, environmental testing, biological research, drug testing, forensics, and homeland security all seem plausible.

Optics is an area that has followed a paradigm—solid-state fabrication focused on ultrahigh optical performance and durability, but with minimal adaptability. L^2 systems suggest another paradigm: systems that only function when they operate in dissipative mode—for example, with fluids flowing through them—and in which the systems are intrinsically unstable but highly adaptable. Time will tell the value of these characteristics.

References

1. D. B. Wolfe, R. S. Conroy, P. Garstecki, B. T. Mayers, M. A. Fischbach, K. E. Paul, M. Prentiss, and G. M. Whitesides, "Dynamic control of liquid-core/liquid-cladding optical waveguides," *Proc. Natl. Acad. Sci. U.S.A*, 101, (2004), 12434–12438.
2. S. K. Y. Tang, C. A. Stan, and G. M. Whitesides, "Dynamically reconfigurable liquid-core liquid-cladding lens in a microfluidic channel," *Lab Chip*, 8, (2008), 395–401.
3. H. G. Elias, in *Polymer Handbook*, eds., J. Brandrup, E. H. Immergut, E. A. Grulke, A. Abe, and D. R. Bloch, "Refractive Indices of Common Solvents," Wiley-Interscience, (1999), New York, p. III 55–58
4. J. N. Lee, C. Park, and G. M. Whitesides, "Solvent compatibility of poly(dimethylsiloxane)-based microfluidic devices," *Anal. Chem.*, 75, (2003), 6544–6554.
5. David R. Lide, (ed.), "Density, refractive index, freezing point depression, and viscosity of aqueous solutons," in *Handbook of Chemistry and Physics*, 77 ed, CRC, Boca Raton 8-56–8-78.
6. S. K. Y. Tang, B. T. Mayers, D. V. Vezenov, and G. M. Whitesides, "Optical waveguiding using thermal gradients across homogeneous liquids in microfluidic channels," *Appl. Phys. Lett.*, 88, (2006), 061112/061111–061112/061113.
7. R. S. Conroy, B. T. Mayers, D.V. Vezenov, D. B. Wolfe, M. G. Prentiss, and G. M. Whitesides, "Optical waveguiding in suspensions of dielectric particles," *Appl. Opt.*, 44, (2005), 7853–7857.
8. S. Y. Yang, J. J. Chieh, H. E. Horng, C.-Y. Hong, and H. C. Yang, "Origin and applications of magnetically tunable refractive index of magnetic fluid films," *Appl. Phys. Lett.*, 84, (2004), 5204–5206.

9. M. Brady and C. Pozrikidis, "Diffusive transport across irregular and fractal walls," *Proc. R. Soc. London, Ser. A*, 442, (1993), 571–583.
10. T. Tamir, *Guided-Wave Optoelectronics*, Springer, New York, (1998).
11. D. B. Wolfe, D. V. Vezenov, B. T. Mayers, G. M. Whitesides, R. S. Conroy, and M. G. Prentiss, "Diffusion-controlled optical elements for optofluidics," *Appl. Phys. Lett.*, 87, (2005), 181105/181101–181105/181103.
12. J. C. McDonald and G. M. Whitesides, "Poly(dimethylsiloxane) as a material for fabricating microfluidic devices," *Acc. Chem. Res.*, 35, (2002), 491–499.
13. B. T. Mayers, D. V. Vezenov, V. I. Vullev, and G. M. Whitesides, "Arrays and cascades of fluorescent liquid-liquid waveguides: broadband light sources for spectroscopy in microchannels," *Anal. Chem.*, 77, (2005), 1310–1316.
14. D. V. Vezenov, B. T. Mayers, R. S. Conroy, G. M. Whitesides, P. T. Snee, Y. Chan, D. G. Nocera, and M. G. Bawendi, "A low-threshold, high-efficiency microfluidic waveguide laser," *J. Am. Chem. Soc.*, 127, (2005), 8952–8953.
15. M. Hashimoto, B. Mayers, P. Garstecki, and G. M. Whitesides, "Flowing lattices of bubbles as tunable, self-assembled diffraction gratings," *Small*, 2, (2006), 1292–1298.
16. A. C. Siegel, S. S. Shevkoplyas, D. B. Weibel, D. A. Bruzewicz, A. W. Martinez, and G. M. Whitesides, "Cofabrication of electromagnets and microfluidic systems in poly(dimethylsiloxane)," *Angew. Chem., Int. Ed.*, 45, (2006), 6877–6882.
17. A. C. Siegel, D. A. Bruzewicz, D. B. Weibel, and G. M. Whitesides, "Microsolidics: fabrication of three-dimensional metallic microstructures in poly(dimethylsiloxane)," *Adv. Mater.*, 19, (2007), 727–733.

CHAPTER 4

Optofluidic Optical Components

Uriel Levy

Department of Applied Physics, The Benin School of Engineering and Computer Science, The Hebrew University of Jerusalem, Jerusalem, Israel

4-1 Introduction

The term optofluidic optical component (OOC) refers to a class of devices where micro-/nanofluidics is used to form an optical component by controlling its geometry, refractive index, and its optical functionalities, for example, transmission, reflection, absorption, or scattering. To date, the most widespread OOC is probably the liquid crystal display that is being incorporated in large variety of devices, including, for example, computers and TV screens, watches, and cell phones. In contrast to the liquid crystal display, which is available for many years, most of the OOCs are being investigated and developed only in recent years, and are expected to lie at the center of the emerging field of optofluidics, with the vision of integrating variety of OOCs to form miniaturized, on-chip optofluidic systems with potential applications in medicine, biology and biotechnology, chemical synthesis and controlled reactions, signal processing, communication, imaging, projection, storage, and military applications. Progress in optofluidics is now well documented by several recent review papers [1–3].

A key motivation for the implementation of OOCs is their ease of fabrication by rapid prototyping as well as the flexibility in forming variety of geometries and refractive index combinations, allowing the realization of almost any desired optical functionality. One of the fundamental terms in optics is the "optical path length." According to Fermat the path taken between two points by a ray of light is the path that can be traversed in the least time (the more accurate version of

Fermat's principle states that the optical path length must be extremal, that is, it can be either minimal, maximal, or a saddle point). The optical path length frequently determines the functionality of an optical component. It is calculated by integration of the incremental product of the physical path length and the refractive index of the medium along the path of the optical ray. Thus, the capability of forming large variety of geometries and refractive indices provides huge flexibility in the design and realization of OOCs with desired functionalities. In addition, the OOCs can be easily tuned by dynamically controlling their geometry and/or their refractive index. Most of the current OOCs are made of a soft elastomer, polydimethylsiloxane (PDMS). Besides the advantage of rapid prototyping, PDMS, being an elastic medium [typical Young's modulus < megapascals (MPa)] allows very large tunability by modifying the geometry of the optical device under the application of internal (usually in the form of gas pressure) or external forces. Flexible elastomer membranes are also key elements in pressure-actuated microvalves that can be integrated with optofluidic components. Geometrical tuning can also be achieved by the application of an electric field, resulting a change in the wetting angle of a liquid droplet via the electrowetting effect. The refractive index of OOCs is typically controlled simply by replacing the liquid forming the OOC with another liquid having different refractive index. This can be done either off-chip (e.g., by replacing the content of an external reservoir), or on-chip, by using a predesigned integrated mixer allowing the mixing of liquids having different refractive indices. Liquids are available in wide range of refractive indices spanning from ~1.33 to ~2.3, offering an incredibly large refractive index tuning of ~1. Even if the choice is limited to nontoxic liquids, refractive index tuning of ~0.3 is still achievable, and thus the tunability range of OOCs is orders of magnitudes larger than that achieved by solid optical components.

This chapter outlines and discusses some of the of the key OOCs required for the realization of integrated optofluidic systems, including waveguides that are being used for signal delivery, spectral filters, switches and splitters, and beam-steering devices.

4-2 Optofluidic Waveguides

A basic building block required for the realization of most on-chip integrated optofluidic systems is the optofluidic waveguide. In contrast to conventional waveguides, where the optical mode interacts with a solid core and with a solid/air clad, the optofluidic waveguide is based on the interaction (either partially of fully) of the optical mode with liquid (here we limit the discussion to interaction of light with liquid, although in broader perspective an optical-guided mode interacting with gas can also be considered as optofluidic waveguide).

In this section we describe in detail some of the major configurations used for the realization of optofluidic waveguides.

4-2-1 Solid-Core/Liquid Clad Waveguide

The most straightforward and simple optofluidic waveguide configuration is a structure made of solid core surrounded by a liquid clad (SCLC). A schematic sketch of such structure is shown in Fig. 4-1.

A solid core (typically made of glass, polymer, or semiconductor) is positioned on top of a lower solid clad and covered by a liquid forming the upper and side clad. Typical liquids being used for cladding material are distilled [DI] water (n~1.33 at visible wavelength), water-based solutions (n~1.33–1.45) and organic-based liquids (n~1.45–1.55). Liquids with higher refractive indices, up to ~2.3, are also available, although some of them are toxic. More details about commercially available liquid solutions and their properties can be found in Ref. 4. The solid lower clad is frequently made of silicon dioxide (SiO_2, n~1.45). In some specific cases the SiO_2 clad can be etched (e.g., by buffered HF), forming a bridged waveguide. Such a waveguide can be surrounded by liquid from all directions.

A silicon wafer is covered by a few microns–thick oxide layer. The solid core (SC) is positioned on top of the oxide. A microfluidic channel is defined in PDMS and filled with a liquid clad (LC), surrounding the SC. Liquids can be stationary or in constant flow. The guiding mechanism of most SCLC waveguides is based on total internal reflection (TIR). The optical mode in such waveguides extends beyond the core, and decays exponentially in the clad. This portion of the optical mode is known as the evanescent field. The interaction of light propagating through the SCLC waveguide with the liquid takes place via the spatial overlap between the liquid and the evanescent field in the clad. An important parameter for classifying and characterizing an optical waveguide is the confinement factor, defined as the portion of the optical mode that is confined to the core.

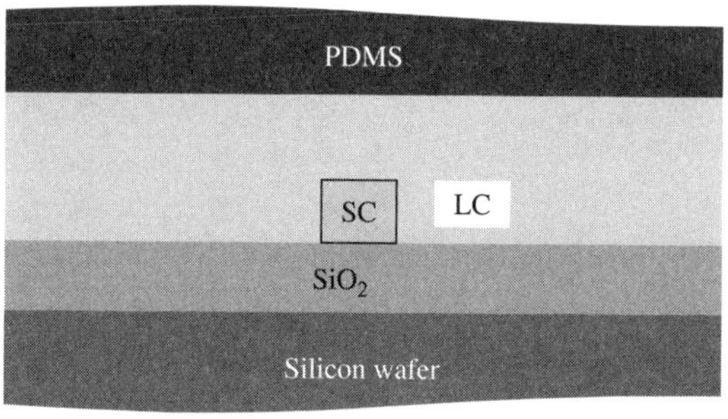

FIGURE 4-1 Schematic drawing showing a cross section of a typical SCLC waveguide.

Most standard waveguides are designed to maximize the confinement factor. However, if such designs are used for realizing SCLC waveguides, only small fraction of light interacts with the liquid cladding. The limited interaction of liquid with the optical mode is considered as one of the fundamental drawbacks of the SCLC configuration, limiting its usefulness for applications requiring large tuning range or high sensitivity sensors. This obstacle, however, can be overcome, at least partially, by proper design and optimization of the waveguide geometry and refractive index distribution. For example, one can increase the refractive index of the liquid, such that the optical mode expands much beyond the core. Alternatively, one can reduce the size of the core, resulting in a lower mode confinement and in turn larger interaction of the optical mode with the liquid clad. This, however, results in an increase in bending loss and sometimes (if the waveguide size or the refractive index difference goes down beyond a critical point) even an increase in mode size, posing a stringent limitation on the miniaturization of on-chip optofluidic integrated systems. Figure 4-2 shows the optical mode size and the mode confinement as a function of core size. A rectangular polymer bridge waveguide core ($n = 1.56$, corresponding to refractive index of commercially available SU8 polymer) surrounded by a liquid with refractive index of 1.45 is assumed. Wavelength is 1.55 μm.

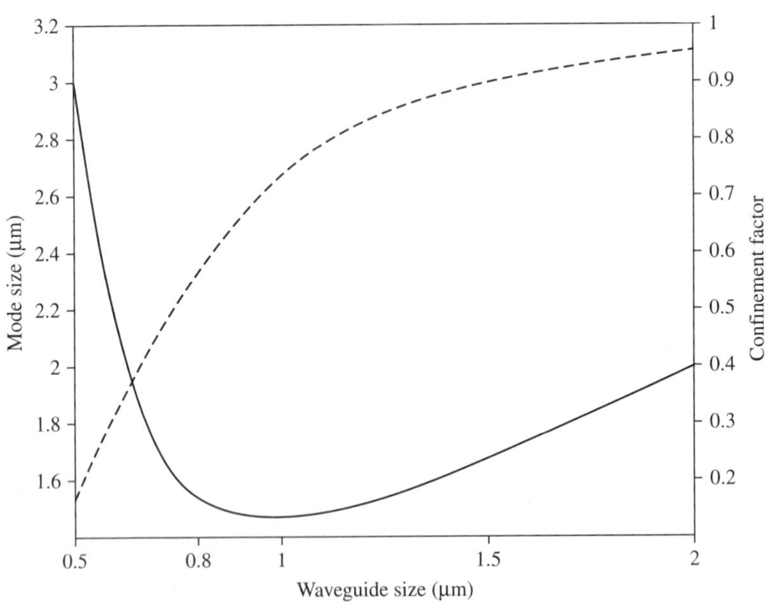

FIGURE 4-2 Mode size (solid line) and confinement factor (dashed line) vs. the size of the waveguide core. Refractive indices are 1.56 and 1.45 for the core and the clad, respectively. Wavelength is 1.55 μm.

As can be seen, mode confinement decreases from 95% for a 2-μm waveguide to 18% for a 0.5-μm waveguide, resulting in a significant increase in overlap between the optical mode and the liquid, from 5% to 82%. This, however, comes at the expense of an increase in mode size to more than 3 μm because the waveguide becomes weakly guided. Bending loss (not shown) is also increased drastically.

SCLC optofluidic waveguides can be integrated with other optofluidic components to support variety of applications. Among these applications, label-free biosensing is of increasing importance. A powerful method for optical biosensing is interferometry. A waveguide interferometric biosensing explores variations in the effective refractive index of a waveguide caused by biological analytes bound to the surface. Worth et al. [5] demonstrated a polarimetric waveguide interferometer based on silicon nitride on SiO_2 slab waveguide. With their approach, they could measure the differential effective index between the orthogonal waveguide modes, from which they could distinguish between specifically and nonspecifically bound particles.

The sensitivity and the tuning strength of an optofluidic device exploiting SCLC waveguides can be greatly enhanced by its coupling to an optical resonator. For example, Chao et al. [6] demonstrated homogenous and surface sensing by using a microring resonator (MRR) in SCLC waveguide configuration. The waveguide core was made of polystyrene on SiO_2, and was covered by the solution to be analyzed. With Q factor of ~20,000, their devices could detect effective index variations of ~10^{-7}. Binding of the specific biomolecules could be traced with a detection limit of 250 pg/mm^2 of mass coverage on the sensor surface. De Vos et al. [7] demonstrated the detection of protein concentrations down to 10 ng/mL using miniaturized (5-μm radius) silicon on insulator (SOI)–based MRR with liquid clad. This result demonstrates that the SCLC waveguide is promising for miniaturized optofluidic systems, as long as the limited interaction of the optical mode with the liquid can be tolerated.

4-2-2 Liquid-Core Waveguide

The disadvantage of insufficient interaction between the liquid clad and the optical mode propagating mostly in the core of the SCLC waveguide can be overcome by the use of liquid-core waveguides (LCW). The optical mode propagating in such waveguides is mostly confined to the liquid core; therefore the interaction of light with the liquid is enhanced tremendously.

Most of the early versions of LCWs were implemented by realizing a hollow-core structure surrounded by a solid clad. The hollow core can then be filled with liquid, forming a liquid-core waveguide. A major challenge in realizing such waveguides is the choice of cladding materials. Similarly to the SCLC waveguides, the guiding mechanism of the LCWs is based on TIR. As such, the refractive index of the clad needs to be lower than that of the core to avoid "leakage" of the optical mode. Moreover, in order to support the growing effort of miniaturization of optofluidic systems, large refractive index difference between the core and the clad is desired. For water-based LCWs, with core refractive index of ~1.33, achieving high refractive index contrast is very challenging. In fact, the refractive index of most solids rarely falls below 1.3. For example, glass, which is frequently used as cladding material, is not suitable as the cladding of water-based LCWs because its refractive index is higher than the liquid. An attractive cladding material for LCWs is Teflon AF, because of its low refractive index (n ~1.29). Various LCWs with Teflon AF as cladding material were demonstrated, with applications in Raman spectroscopy [8–9], fluorescence spectroscopy [10], and capillary electrophoresis [11]. Unfortunately, it is difficult to spin coat Teflon AF on substrates because it does not adhere well to most substrates. This technical obstacle may be overcome by surface treatment (e.g., by oxygen plasma). An alternative approach for realizing low-refractive-index cladding material is by using subwavelength nanoporous material. Because the dimensions of the pores are much smaller compared to the optical wavelength, scattering loss is minimized and the refractive index can be tuned by controlling the volume fraction of the pores. Based on this concept, a planar one dimensional waveguide having cladding material with effective refractive index ranging from 1.15 to 1.27 was demonstrated [12]. The nanoporous dielectrics were made by the sacrificial porogen approach, in which an organic macromolecular phase is selectively removed from a phase-separated organic/inorganic polymer hybrid, resulting in nanoscopic pores having a diameter in the range of 10 to 15 nm.

A different type of LCW is the antiresonant reflecting optical waveguide (ARROW). These waveguides were recently introduced as a promising approach for the realization of hollow-core integrated optics with very small core volumes. In contrast to the previous examples the guiding of light in these waveguides is not based on TIR. Instead, the ARROWs employ multiple dielectric cladding layers, and rely on the antiresonance of the transverse wave vector component for each layer, which yields quasi-guided modes [13]. Although these modes are leaky, a properly designed ARROW waveguide can guide light with loss as low as 1.1 dB/cm in the visible wavelength regime [14]. ARROW waveguides are typically fabricated by surrounding a sacrificial core with silicon dioxide and/or silicon nitride layers. The sacrificial layer is then removed by selective wet etching. The layers are grown to specific thicknesses such that ARROW-based optical confinement is obtained. Typical layer thickness is in the range of 100 to 200 nm. A variety of sacrificial materials can be used, including photosensitive polymers and metals. Different waveguide profiles, for example, rectangular, trapezoidal, and arch-shaped can be realized, depending on

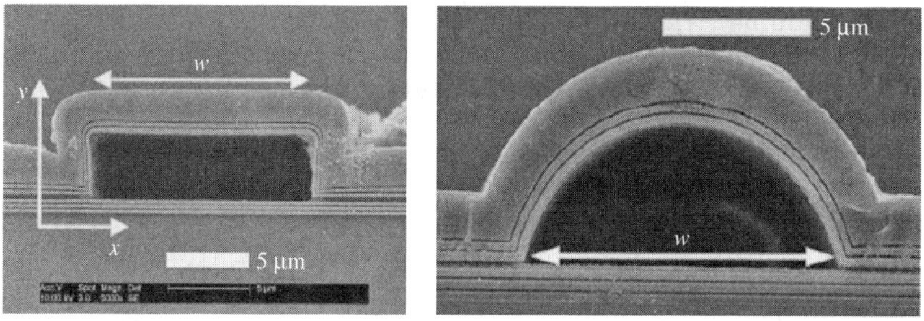

FIGURE 4-3 Scanning electron micrograph images of hollow-core ARROWs with rectangular (left) and arch-shaped (right) cross sections. (*D. Yin, J. P. Barber, E. J. Lunt, A. R. Hawkins, and H. Schmidt, "Optical characterization of arch-shaped ARROW waveguides with liquid cores,"* Opt. Exp., *13, (2005), 10564–10569.*)

the sacrificial layer and the fabrication process. ARROW waveguides having cross sections ranging from few microns to few 10s of microns were realized. Figure 4-3 shows an SEM picture of rectangular (left) and arch shaped (right) ARROW waveguides. Pictures were reprinted from Ref. 14. Such waveguides were recently demonstrated for applications such as fluorescence [15] and surface-enhanced Raman scattering (SERS) detection [16]. Two review papers describing the ARROWs were recently published [17,18].

Another type of LCW that is not based on guiding by TIR is the Bragg fiber, first demonstrated by Fink et al. [19]. The cladding of these fibers is made of dielectric mirrors surrounding the hollow core. The hollow core can be filled with liquids (although it was not demonstrated so far). Light cannot escape through the cladding because of the high reflectivity of the dielectric Bragg mirrors. The Bragg mirrors can be designed to be omnidirectional, that is, providing high reflection for all angles of incidence. A slightly different version of the Bragg fiber is the hollow-core photonic crystal fiber, described by Russell [20]. This fiber is made of a hollow core, typically in the range of few microns to 10s tens of microns. The hollow core is surrounded by a two-dimensional periodic structure made of air holes in silica, realizing a photonic band-gap and preventing the escape of light from the hollow core. With this configuration, liquids were injected into the hollow core to demonstrate light and particle guiding through the liquid-filled core [21], and detection of surface-enhanced Raman scattering from molecules in solution with silver nanoparticles [22].

Both the Bragg fiber and the photonic crystal fiber offer excellent control over photonic properties and low propagation loss, but cannot be monolithically integrated with on-chip optofluidic systems. An alternative type of LCWs, based on total internal reflection, is the liquid-liquid (L^2) waveguide demonstrated by the Whitesides group and others [23,24]. The L^2 approach allows the manipulation of light in waveguides that comprise a liquid core and a liquid cladding. The liquids are introduced into the channels of a microfluidic network designed to sandwich the flowing core liquid between flowing slabs of the cladding liquid. The core/clad boundary can be controlled by manipulating the rate of flow of the liquids, allowing the tunability of the optofluidic waveguides. More information about the L^2 waveguides is given in Chap. 3.

4-2-3 Hybrid-Core Waveguide

The optofluidic waveguides described up to now could be clearly defined either as solid-core waveguides or as liquid-core waveguides. In recent years, however, a new class of waveguides is emerging, where the waveguide core includes structures on the micro-nanoscale, with mixed regions of solid and air. The air regions can be filled with liquid, realizing special waveguides with a hybrid solid/liquid core. We thus use the term *hybrid core waveguides* (HCW) to describe them. Here we focus on a specific and attractive example of HCW, the slot waveguide.

The slot waveguide was first demonstrated by Xu et al. [25]. It was realized by etching a 100-nm vertical slot into a 540-nm wide, 250-nm thick silicon waveguide core, on top of SiO_2 lower cladding. The authors demonstrated a significant drop in effective index of the horizontal mode, leading to the conclusion that a significant portion of the mode was confined to the narrow slot. The operation concept of the slot waveguide can be explained as follows. If an optical mode with its electrical field (E) coincide with the horizontal axis is excited in this waveguide, a discontinuity in electric field is expected around the slot, whereas the electric displacement (D) across the slot boundary is continuous. Because the electric displacement is given by $D = \varepsilon E = n^2 E$, the discontinuity in the electric field is given by:

$$\frac{E_{\text{slot}}}{E_{\text{silicon}}} = \left(\frac{n_{\text{silicon}}}{n_{\text{slot}}}\right)^2$$

For air core waveguide, this ratio can go as high as 12. Even if the slot is to be filled with water, a high ratio of 7 is expected, making this waveguide very attractive for applications where small mode size and large overlap between the liquid and the optical mode is of interest. The slot waveguide was also realized with Si_3N_4 as a core material [26]. This material platform is less attractive in terms of field confinement because of the lower refractive index contrast, but on the other hand it can operate in the visible range, thus offering an important advantage for many biosensing applications. Si_3N_4 slot waveguides were realized with dimensions in the order of a single-micron width and 300-nm height. Typical slot widths are ~200 nm. Nitride-based slot waveguides were also used for the realization of MRRs for biosensing applications [27]. Resonance shift of 212 nm/RIU (refractive index unit) was reported. By using a similar platform, a label-free biosensing of bovine serum albumin (BSA) and anti-BSA was also demonstrated, with sensitivity limit in the range of 16 to 28 pg/mm^2.

4-3 Optofluidic Components for Manipulation of Optical Signals

In parallel to the rapid progress in optofluidic waveguides there is a growing effort to develop variety of other optofluidic components, with a prime goal of manipulating and processing optical signals. The integration of photonic components with liquids on the micro-/nanoscale paves the way to widen and enhance their optical functionalities, forming eventually a new class of optofluidic components for manipulating optical signals. Components such as tunable filters, switches, splitters and combiners, and beam deflectors were recently demonstrated. This section describes some of the recent work in the field, with a specific focus on tunable optofluidic filters. Other components, for example, switches, splitters, and beam-steering devices are covered in Chap. 8.

4-3-1 Optofluidic Filters

Optical filters are the subject of scientific and technological effort for many years, with applications in microscopy, avionics, spectroscopy, optical communication, sensing, astronomy, machine vision, laser range finders, and environmental monitoring, to name a few. Optofluidics is a promising approach for the realization of optical filters because (a) it offers a wide tunability range, much larger than can be achieved by most other physical effects, and (b) it allows the interaction of analytes carried by the liquid with the optical filter, thus enabling on-chip realization of optofluidic-filtering devices and systems. Two of the dominant mechanisms used for the realization of optical filters are absorption and interference. Optofluidic-absorption filters can be easily realized by introducing absorptive liquid into the filtering region. The spectral absorption properties of the liquid determine the spectral response of the filter. The function of tunability can be acquired by replacing the liquid with another liquid, having different spectral absorption properties. By mixing of liquids it is possible to achieve continuous tuning of such filters. Macroscopic liquid absorption filters were already demonstrated many years ago [28,29]. For example, Ref. 29 describes an optical cell with a variable path length designed for use in conjunction with liquid filters. A path length change from 1 mm to 14 mm changes the cutoff wavelength by typically 30 nm. The miniaturization and on-chip integration of absorption-based liquid filters holds great promise for the realization of flexible, high-performance, and integrated optofluidic systems.

In parallel to the absorption filters, various optofluidic interference filters were recently demonstrated; some of them are also tunable. For example, Mach et al. [30] demonstrated a tunable optofluidic microstructured fiber. This device combined long-period Bragg gratings and inner microchannels in the fiber. The tuning liquids consisted of adjacent segments of low index (n = 1.28) and high index (n = 1.73) immiscible microfluidic plugs. The liquids are pulled into the fiber one after another and positioned such that the interface between the liquids lies at the edge of the long-period Bragg grating. By using independent control mechanism based on microheaters it is possible to tune the transmission and the resonant wavelength independently. With this approach a tuning range of about 12 nm and attenuation of about 12 to 15 dB was demonstrated.

Another interference filtering scheme is based on the use of a diffraction grating [31]. With such an approach, Domachuk et al. [31] demonstrated an optofluidic on-chip spectrometer made by the integration of a diffraction grating with a microfluidic channel using soft lithography in PDMS. The device was calibrated by couple of spectral filters in different spectral regimes. Resolving power was estimated to be ~330. The functionality of the integrated device was demonstrated by performing a spectral analysis of chlorophyll probed using supercontinuum light source. The measured absorption data show reasonable agreement with previously reported absorption data.

Narrow-linewidth optical interference filters can be realized on a chip by the use of integrated resonators. Specifically, the microring resonator is of major importance for on-chip filtering applications. The MRR is very popular for on-chip realization of optical filters because of its robustness, flexibility, and the potential for dense integration of arrays of MRRs on chip. A modified version of the MRR is the microtoroid resonator, demonstrated by Armani et al. [32], with the advantage of ultrahigh Q factors. An MRR can operate in notch filtering mode or in add/drop filtering mode, depending on the number of bus waveguides coupled to the MRR. Recently, Levy et al. [33] demonstrated an on-chip tunable optofluidic notch filter by integrating a polymer MRR with a microfluidic channel network. The work was motivated by the need to achieve fine-tuning of an optical MRR. Tuning was obtained by dynamic variation of refractive index of the medium surrounding its waveguides. A magnified image showing a section of the fabricated device is shown in Fig. 4-4 (left).

The MRR was positioned at the bottom of a flow-through microchannel which is a part of a microfluidic chip. The liquid injected into the microchannel constitutes the upper cladding of the MRR waveguides. Variation of the refractive index of the liquid was achieved by on-chip mixing of two source liquids with different indices of refraction. The liquids injected into the inlets flow through a microchannel network of the type introduced by Whitesides [34]. The network generates repeated splitting and mixing, such that the concentration of the solute linearly varies across the stream emerging from the network (along the dashed line 1 in Fig 4-4 left). The stream further follows to a crossroad, where

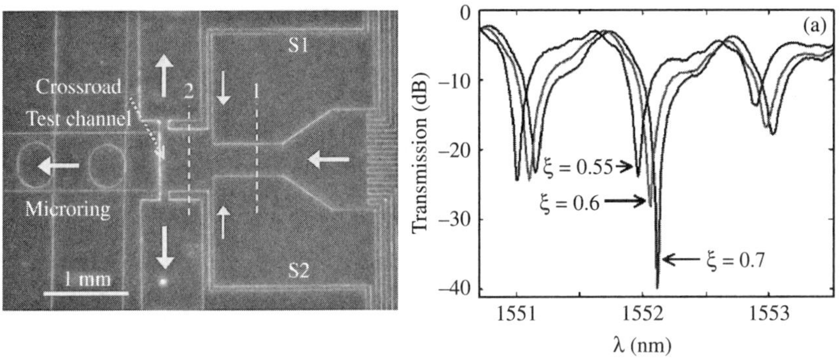

FIGURE 4-4 (Left)—A magnified microscope image showing a section of the fabricated device (left). Measured transmission versus wavelength for three different values of normalized pressure difference (right). (*Reprinted with permission from U. Levy, K. Campbell, A. Groisman, S. Mookherjea, and Y. Fainman, "On-chip microfluidic tuning of an optical microring resonator," Appl. Phys. Lett., 88, (2006), 111107. Copyright (2006), American Institute of Physics.*)

it is split between three outlets. By controlling the pressure difference between two outlets one can change the distribution of the flux directed to outlets and allow selecting the portion of the stream that flows in the test channel. This in turn, determines the concentration and the refractive index of the solute around the waveguides of the microring. The device was operated by injecting water and salt solution (having a refractive index higher by 0.07 compared with that of water) into the inlets. A normalized pressure difference term ξ was defined. Figure 4-4 (right) shows the measured transmission spectrum for three different values of ξ. Both the resonance wavelength and the extinction ratio are varied, because of the change in effective index, loss, and coupling strength. The authors reported a record extinction ratio of 37 dB, indicating operation conditions near critical coupling. This result demonstrates the advantage of the optofluidic platform in precisely controlling the refractive index over a wide tuning range.

On-chip optofluidic filter was also demonstrated in photonic crystal waveguide platform [35]. The filter was realized by the integration of a microfluidic channel made of PDMS with the photonic crystal waveguide made of silicon. The transmission spectrum of the photonic crystal waveguide was modified by allowing solutions of $CaCl_2$ ranging in composition from DI water (n~1.33) to 5 M (n~1.44), to flow over the photonic crystal waveguide. Tuning range of ~20 nm in wavelength and switching speed of few seconds was demonstrated.

The tunability of the devices described in Refs. 33 and 35 is based on the control of liquids either by pressure driven mixing or by the exchange of liquids off chip. The on-chip integration of such devices with microvalve networks [36] and micropumps [37] could be an important step toward enhancing and simplifying the control over on-chip tunable optofluidic filters. Alternatively, one can also consider the use of electrowetting for the realization of on-chip tunable optofluidic filter. Electrowetting is described in details in Chap. 19. Here, we focus on a recent demonstration of an on-chip tunable MRR that is actuated by electrowetting [38]. Tunability is achieved by controlling the wetting angle of a droplet that is partially covering an MRR made of polymer waveguide. By applying a voltage to the droplet, its wetting angle is modified, and the droplet covers larger area of the MRR. This results in an increase in the effective refractive index of the MRR waveguide, thus the resonant wavelength and the transmission through the device can be modified. In addition to the tuning of the resonant wavelength, the authors also demonstrated a significant tuning of the extinction ratio by positioning the droplet on top of the coupling region between the MRR and the bus waveguide, thus allowing controlling the coupling coefficient of the device. Figure 4-5 shows

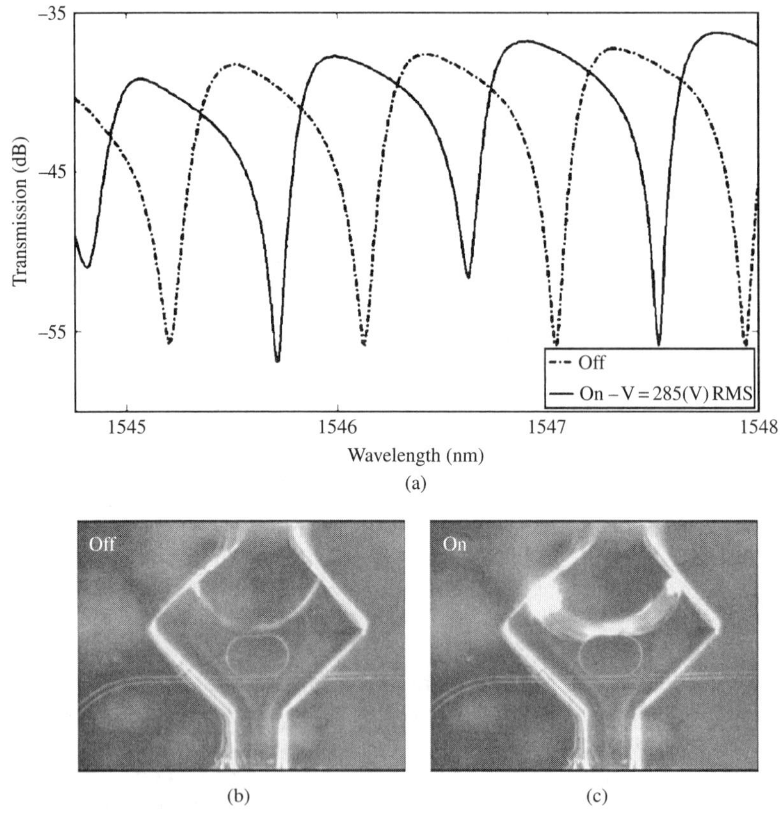

FIGURE 4-5 (a) Transmission spectrum of the device in the off (dashed curve) and the on (solid curve) states. (b) and (c) Microscope images show the MRR and the droplet in the off and the on states, respectively. (*R. Shamai and U. Levy, "On chip tunable micro ring resonator actuated by electrowetting," Opt. Exp. 17, (2009), 1116–1125.*)

the transmission spectrum of the device in the off (dashed curve) and the on (solid curve) states (a), together with microscope images (b) and (c) showing the MRR and the droplet in the off and the on states, respectively.

As can be seen, a significant shift in resonant wavelength is noticeable. However, variations in extinction ratio are relatively small. This is because the droplet is located far away from the coupling region. In contrast, Fig. 4-6 shows the transmission spectrum of the device for a case where the droplet covers the coupling region in the on state. As can be seen, extinction ratio varies drastically.

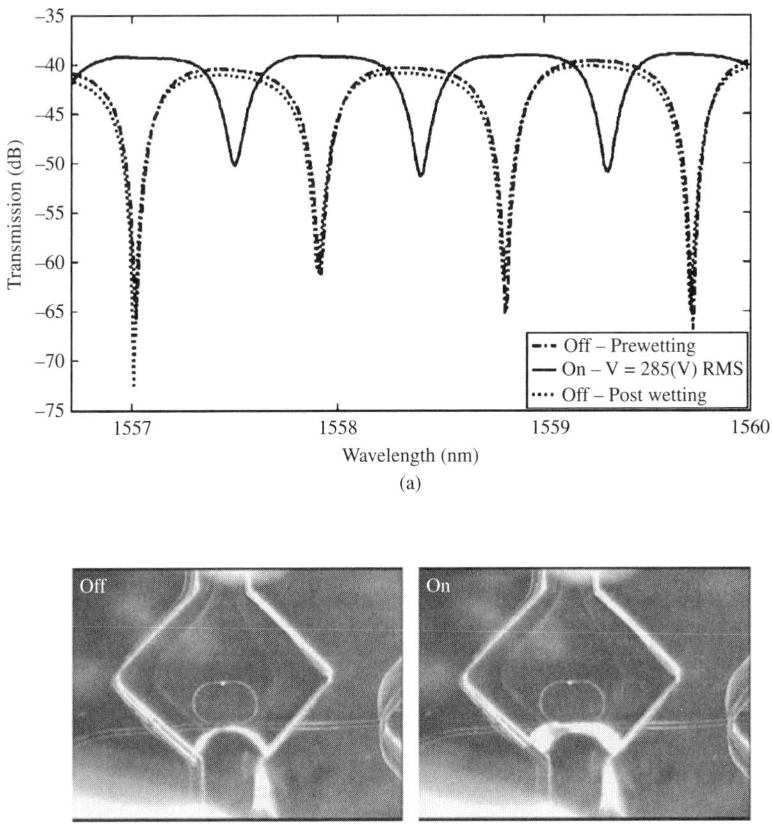

FIGURE 4-6 (a) Transmission spectrum of the device in the off (dashed curve and dotted curve) and the on (solid curve) states. (b) and (c) Microscope images show the MRR and the droplet in the off and the on states, respectively. (*R. Shamai and U. Levy, "On chip tunable micro ring resonator actuated by electrowetting," Opt. Exp. 17, (2009), 1116–1125.*)

4-4 Conclusions

Optofluidic optical components (OOCs) are promising candidates to serve as building blocks in future on chip integrated optofluidic systems. By considering their ease of design and fabrication, side by side with their great flexibility it is reasonable to predict that such components will play a major role in future optofluidic systems. In addition, the capability to deliver both analytes and optical signals in the same structure makes the OOCs promising for on chip biosensing applications. Finally, the OOCs offer very large tunabilty, both in geometry and in refractive index. The refractive index tuning range can be as high as ~1, orders of magnitude larger than the tunability that can be achieved by other approaches. Therefore, OOCs may become useful in application requiring tunablity and adaptation of optical components.

References

1. D. Psaltis D, S. R. Quake, and C. Yang, "Developing optofluidic technology through the fusion of microfluidics and optics," *Nature*, **442**, (2006), 381.
2. C. Monat, P. Domachuk, and B. J. Eggleton, "Integrated optofluidics: A new river of light," *Nat. Photon.*, **1**, (2007), 106.
3. G. M. Whitesides, "The origins and the future of microfluidics," *Nature*, **442**, (2006), 368.
4. http://www.2spi.com/catalog/ltmic/cargille-liquid.html.
5. C. Worth, B. B. Goldberg, M. Ruane, and M. S. Ünlü, "Surface desensitization of polarimetric waveguide interferometers," *IEEE J.* Sel. *Top in Quant. Electron.*, **7**, (2001), 874–877.
6. C. Y. Chao, W. Fung, and L. J. Guo, "Polymer microring resonators for biochemical sensing applications," *IEEE J. Sel.* Top. in *Quant. Electron.*, **12**, (2006), 134–142.
7. K. De Vos, I. Bartolozzi, E. Schacht, P. Bienstman, and R. Baets, "Silicon-on-insulator microring resonator for sensitive and label-free biosensing," *Opt. Exp.*, **15**, (2007), 7610–7615.
8. M. J. Pelletier and R. Altkorn, "Raman sensitivity enhancement for aqueous protein samples using a liquid-core optical-fiber cell," *Anal. Chem.*, **73**, (2001), 1393–1397.
9. M. Holtz, P. K. Dasgupta, and G. Zhang, "Small-volume Raman spectroscopy with a liquid core waveguide," *Anal. Chem.*, **71**, (1999), 2934–2938.
10. Q. Li, , K. J. Morris, P. K. Dasgupta, I. M. Raimundo, and H. Temkin, "Portable flow-injection analyzer with liquid-core waveguide based fluorescence, luminescence, and long path length absorbance detector," *Anal. Chem. Acta*, **479**, (2003), 151–165.
11. P. K. Dasgupta, Z. Genfa, J. Li, B. Boring, et al., "Luminescence detection with a liquid core waveguide," *Anal. Chem.*, **71**, (1999), 1400–1407.
12. W. Risk, H. Kim, R. Miller, H. Temkin, and S. Gangopadhyay, "Optical waveguides with an aqueous core and a low-index nanoporous cladding," *Opt. Exp.*, **12**, (2004), 6446–6455.
13. A. R. Hawkins, D. W. Deamer, and H. Schmidt, "Integrated optical waveguides with liquid cores," *Appl. Phys. Lett.*, **85**, (2004), 3477–3479.
14. D. Yin, J. P. Barber, E. J. Lunt, A. R. Hawkins, and H. Schmidt, "Optical characterization of arch-shaped ARROW waveguides with liquid cores," *Opt. Exp.*, **13**, (2005), 10564–10569.
15. D. Yin, J. P. Barber, A. R. Hawkins, D. W. Deamer, and H. Schmidt, "Integrated optical waveguides with liquid cores," *Appl. Phys. Lett.*, **85**, (2004), 3477–3479.

16. P. Measor, E. J. Lunt, L. Seballos, D. Yin, J. Z. Zhang, A. R. Hawkin, and H. Schmidt, "On-chip surface-enhanced Raman scattering (SERS) detection using integrated liquid-core waveguides," *Appl. Phys. Lett.*, **90**, (2007), 211107–211109.
17. H. Schmidt and A. R. Hawkins, "Optofluidic waveguides: I. Concepts and implementations," *Microfluid. Nanofluid.*, **4**, (2008), 3–16.
18. A. R. Hawkins and H. Schmidt, "Optofluidic waveguides: II. Fabrication and structures," *Microfluid. Nanofluid.*, **4**, (2008), 17–32.
19. Y. Fink, J. N. Winn, S. Fan, C. Chen, J. Michel, J. D. Joannopoulos, and E. L. Thomas, "A dielectric omnidirectional reflector," *Science*, **282**, (1998) 1679–1682.
20. P. Russell, "Photonic crystal fiber," *Science*, **299**, (2003), 358–362.
21. S. Mandal and D. Erickson, "Optofluidic transport in liquid core waveguiding structures," *Appl. Phys. Lett.*, **90**, (2007), 184103.
22. Y. Zhang, C. Shi, C. Gu, L. Seballos, and J. Z. Zhang, "Liquid core photonic crystal fiber sensor based on surface enhanced Raman scattering," *Appl. Phys. Lett.*, **90**, (2007), 193504.
23. D. B. Wolfe, R. S. Conroy, P. Garstecki, B. T. Mayers, M. A. Fischbach, K. E. Paul, M. Prentiss, and G. M. Whitesides, "Dynamic control of liquid-core/liquid-cladding optical waveguides," *PNAS*, **101**, (2004), 12434.
24. M. Brown, T. Vestad, J. Oakey, and D. W. M. Marr, "Optical waveguides via viscosity-mismatched microfluidic flows," *Appl. Phys. Lett.*, **88**, (2006), 134109.
25. Q. Xu, V. R. Almeida, R. R. Panepucci, and M. Lipson, "Experimental demonstration of guiding and confining light in nanometer-size low-refractive-index material," *Opt. Lett.*, **29**, (2004), 1626–1628.
26. C. A. Barrios, B. Sánchez, K. B. Gylfason, A. Griol, H. Sohlström, M. Holgado, and R. Casquel, "Demonstration of slot-waveguide structures on silicon nitride/silicon oxide platform," *Opt. Exp.*, **15**, (2007), 6846–6856.
27. C. A. Barrios, K. B. Gylfason, B. Sánchez, A. Griol, H. Sohlström, M. Holgado, and R. Casquel, "Slot-waveguide biochemical sensor," *Opt. Lett.*, **32**, (2007), 3080–3082.
28. K. A. Ingersoll, "Liquid filters for the ultraviolet, visible, and near infrared," *Appl. Opt.*, **11**, (1972), 2473–2476.
29. K. A. Ingersoll, "Tunable sharp cutoff liquid optical filter," *Appl. Opt.*, **12**, (1973), 1393–1394.
30. P. Mach, M. Dolinski, K. W. Baldwin, J. A. Rogers, C. Kerbage, R. S. Windeler, and B. J. Eggleton, "Tunable microfluidic optical fiber," *Appl. Phys. Lett.*, **80**, (2002), 4294.
31. P. Domachuk, H. Perry, M. Cronin-Golomb, and F. G. Omenetto, "Towards an integrated optofluidic diffractive spectrometer," *IEEE Phot. Tech. Lett.*, **19**, (2007), 1976–1978.
32. D. K. Armani, T. J. Kippenberg, S. M. Spillane, and Vahala, K. J., "Ultra-high-Q toroid microcavity on a chip," *Nature*, **421**, (2003), 925–928.
33. U. Levy, K. Campbell, A. Groisman, S. Mookherjea, and Y. Fainman, "On-chip microfluidic tuning of an optical microring resonator," *Appl. Phys. Lett.*, **88**, (2006), 111107.
34. N. L. Jeon, S. K. W. Dertinger, D. T. Chiu, I. S. Choi, A. D. Stroock, and G. M. Whitesides, "Generation of solution and surface gradients using microfluidic systems," *Langmuir*, **16**, (2000), 8311.
35. D. Erickson, T. Rockwood, T. Emery, A. Scherer, and D. Psaltis, "Nanofluidic tuning of photonic crystal circuits," *Opt. Lett.*, **31**, (2006), 59–61.
36. M. A. Unger, H. P. Chou, T. Thorsen, A. Scherer, and S. R. Quake, "Monolithic microfabricated valves and pumps by multilayer soft lithography," *Science*, **288**, (2000), 113.
37. D. J. Laser and J. G. Santiago, "A review of micropumps," *J. Micromech. Microeng.*, **14**, (2004), 35.
38. R. Shamai and U. Levy, "On chip tunable micro ring resonator actuated by electrowetting," *Opt. Exp.*, **17**, (2009), 1116–1125.

CHAPTER 5

Optofluidic Trapping and Transport Using Planar Photonic Devices

David Erickson

Sibley School of Mechanical and Aerospace Engineering, Cornell University, Ithaca, New York

Allen H. J. Yang

Chemical and Biomolecular Engineering, Cornell University, Ithaca, New York

Extended Abstract

In this chapter we introduce the concept of "optofluidic transport," which is shown conceptually in Fig. 5-1.

We review the use of near-field optical forces in the evanescent field of a waveguide to perform transport operations in lab-on-chip devices. Briefly, the near-field optical gradients (which serve to confine particles through a Lorenz force, F_{trap}) and concentrated optical energy (resulting in intense scattering and absorption forces for propulsion, F_{prop}) in these devices can be used to perform a series of particle-handling operations including transport and separation. The focus of this chapter is on describing the physics behind this form of transport and some of the potential advantages over the state of the art.

This represents a new method of performing optical transport in lab-on-chip devices, relying on the intense electromagnetic energy in

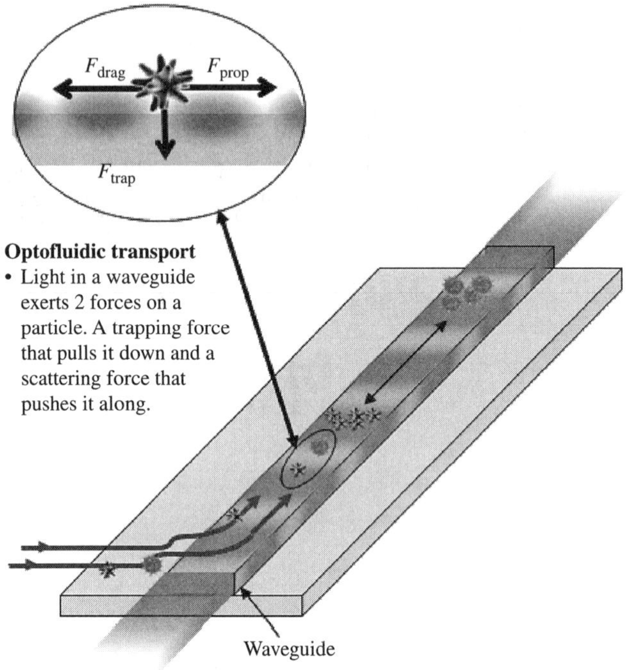

FIGURE 5-1 Schematic of the optofluidic transport of particles on a solid-core waveguide. The particles are trapped and then pushed along the waveguide surface via radiation pressure forces.

waveguiding devices rather than traditional free-space laser tweezing. Mechanistically, optofluidic transport is the combination of two unique phenomena: near-field optical trapping to attract a particle to the waveguide and radiation pressure to perform all forms of species handling including transport, concentration, and separation. The use of dielectric waveguides eliminates axial dispersion of the optical field, allowing us to apply the optical impulse over indefinitely long distances, as opposed to free-space systems, which are limited by the depth of focus of the objective lens. As we describe in detail in this chapter, optofluidic transport has a number of unique properties that give it several advantages over traditional microfluidic transport techniques, like pressure-driven flow and electrokinetics. The three most significant of these are

1. *Favorable transport scaling laws:* As the size of the device gets smaller, the propulsive velocity can increase.
2. *Extremely strong velocity dependence on particle size:* The propulsive velocity has as much as a fifth power dependence on particle size, which exceeds the state of the art in separation techniques by at least 600%.

3. *Insensitivity to surface/solution conditions:* Unlike electroosmosis, for example, this technique is largely independent of surface/solution conditions and can be used for a much broader class of bioanalytical operations.

In this chapter we begin by reviewing existing micro- and nanoscale transport mechanisms and discuss the advantages of optofluidic transport in the context of this state of the art. Following this we present a review of a number of recently published optofluidic transport architectures and introduce our own technique using SU-8 waveguides and polydimethylsiloxane (PDMS) microfluidics. A theoretical description of the transport is then provided and used to back up the advantages purported above. The final section discusses the application of this technique to a specific application area, namely, optofluidic chromatography.

5-1 Optically Driven Microfluidics

5-1-1 A Brief Review of Traditional Transport Mechanisms in Microfluidic Devices

On length scales relevant to transport in micro- and nanofluidic devices, fluid flow and species transport can be accomplished by a number of elegant techniques, a few of which include pressure-driven flow [1], electrokinetics [2–5], buoyancy [6], magnetohydrodynamics [7], capillarity, electrowetting [8], and thermocapillarity [9] (see Stone et al. [10] or textbooks by Nguyen and Wereley [11] or Li [5] for more details). Of these techniques the former two are the most commonly exploited, largely because of the relative ease with which they can be implemented. Pressure-driven flow is likely the simplest to implement, requiring only a pressure or vacuum source to generate flow, and is compatible with a broad range of fluid and surface conditions. On-chip valving techniques such as those used in multilayer soft lithography [1] enable precise and highly parallel flow control and sample routing down to the scale of approximaely 1 μm. Since the average velocity of a pressure-driven flow scales with the square of the critical channel dimension, controlled manipulation of length scales much smaller than this is exceptionally difficult. Another limitation of pressure-driven transport is that it exhibits a parabolic velocity profile meaning that the flow is faster in the center of the channel than at the edges near the walls. This causes an effect known as *dispersion* [12] (essentially the spreading out of a transported sample because parts of it are moving faster than others), which is undesirable in many separation and some transport applications. Electrokinetic transport, where flow is induced through the interaction of an applied electric field and the charge in the electrical double layer near a channel wall (electroosmosis) or a flowing particle (electrophoresis), exhibits a more favorable scaling ratio. Outside the limit where two electrical double layers overlap, the speed of electrokinetic transport is more or less independent of channel height. Within the double layer overlapped regime the velocity scales approximately with $1-1/\kappa d$ [2], where $1/\kappa$ is the double layer thickness (which varies in thickness between 10 nm and 1 μm depending on the ionic strength of the solution) and d is the channel half-height. As such when κd is on the order of 1 (as it is in the case of many nanofluidic systems) the flow can be nearly entirely impeded. In practice electrokinetic techniques are compatible only with a limited class of fluids (low-ionic strength aqueous solutions), exhibit extreme sensitivity to surface conditions, and generally cannot be used with semiconductor substrates such as silicon. Current flow through the channel results in significant Joule heating [13], which can lead to problems ranging from nonuniform viscosity fields to catastrophic boiling particularly in polymeric substrates.

5-1-2 Optical Manipulation in Microfluidic Devices

Free-space optical manipulation techniques in microfluidic systems have recently generated a significant amount of interest. Such techniques range from traditional optical tweezing (see a recent review by Grier [14], and some other papers of interest [14–18]) rotational manipulation of components based on form birefringence [19] to more recent electro-optic approach such as that by Chiou et al. [20]. As an example of a direct device integration, Wang et al. [17] developed an optical-force-based cell-sorting technique whereby radiation pressure was used to direct rare cells into a separate stream following a green florescent protein (GFP) detection event. Unlike the traditional transport techniques described above, the main advantage of these optical approaches lies in their ability to handle individual particles directly, as opposed to indirect manipulation of the surrounding flow field.

Broadly speaking, although very complex manipulations have been demonstrated, the majority of optical tweezer-based implementations tend to be "binary." This means that they rely on the ability either to trap or not to trap a particle based on whether the conditions for trapping stability are met [21–23]. Recently, however, a number of works have extended these ideas to exploit the dependence of this trapping potential on the particle properties, enabling much more advanced and subtle operations. As an example, MacDonald et al. [24] demonstrated an optical lattice technique where particles of different sizes were sorted into different streams depending on their strength of attraction or repulsion to regions of high optical intensity. In a series of papers, Imasaka and coworkers [25–28] provided the initial foundations for optically driven separation techniques, which they termed *optical chromatography*. In optical chromatography (see a recent review by Zhao et al. [29]) a loosely focused laser beam is incident on the particles of interest, resulting in the radiation pressure force that propels them forward. Because the net impulse imparted to a larger particle is greater than that imparted to a smaller particle, they will travel at different speeds and can thus be separated (this will be expanded on at several points in this chapter). Recent demonstrations of optical separations include those by Hart et al., who have demonstrated refractive index separation of colloids [30] and other bioparticles [31]. They have also recently integrated this into a microfluidic device format for pathogen detection [32], demonstrating very precise separation between very closely related bacteria *Bacillus anthracis* and *Bacillus thuringiensis* and millimeter scale separation [33].

5-1-3 Some Limitations of Traditional Optical Manipulation Systems

Despite these successes, the above optically based microfluidic transport systems are fundamentally limited in two ways. The first is by the diffraction limit. It is well known [34] that the diffraction limit places a lower bound on size to which light can be focused and is given by $d_{min} = 1.2\lambda/NA$, where NA is the numerical aperture and λ is the wavelength. In an aqueous environment and for an 850-nm wavelength (consistent with that used by others for optical chromatography [33]) and with a high numerical aperture, the minimum spot size is 550 nm. Since light intensity is given by the input power divided by the illuminated area, this places a fundamental limitation on the trapping and propulsive forces that can be applied to a particle. In practice this limits the size of targets we can trap to targets on the order of a few 100 nm in diameter and the speed with which we can transport them. The second (and ultimately more important here) is the light/species interaction length. From the diffraction limit equation given earlier, it is apparent that the simplest ways to decrease the area over which the optical energy is spread involve either reducing the wavelength of the laser (e.g., into the blue) or increasing the effective numerical aperture [(e.g., via solid immersion lenses (SIL)]. Decreasing the wavelength to 488 nm would reduce the spot size by slightly less than half. The SIL technique has been developed in a number of different flavors [35–37] with the general principle being that increasing the refractive index of the optical head gives one a nominal improvement in ultimate resolution ($1/n_i$). In either of these techniques the decrease in the spot size is necessarily offset by an equivalent decrease in the depth of focus. As such the light/species interaction length (i.e., the distance over which the optical impulse can be applied) becomes small, making it impossible to perform optical transport over long distances. The reason why the traditional channel-based transport techniques like pressure or electrokinetics are useful is not because they are particularly well suited to microfluidics (electrokinetics, arguably, is relatively weak, requiring often thousands of volts to impart a noticeable velocity) but because the impulse can be applied consistently over long distances (tens of centimeters).

5-1-4 Near-Field Optical Manipulation

One way to improve on the limitations imposed by the diffraction limit is through the use of near-field methods [38] such as those based on the use of surface plasmonic resonances [39,40] or other evanescent field techniques [41]. The advantage of these approaches is that the extremely high decay rate of the evanescent field leads to stronger trapping forces than can be achieved with free-space systems. Examples include the work of Cizmar et al. [42], who demonstrated short-range manipulation (on the order of 40 μm) of 350-nm polystyrene beads, and Grigorenko et al. [43], who used plasmonic resonance in surface bound metallic nanostructures to achieve high-quality trapping of dielectric particles as small as 200 nm in diameter. While in general these methods have in the past been successful at trapping and even assembling [44] small particles, similar to free-space trapping, they are limited by the distance through which they can transport objects, since the optical manipulation region is limited by the field of view of the focused laser, and the plasmon propagation distance is relatively short.

5-2 Optofluidic Transport

Though most readers of this book are likely to be at least somewhat familiar with the topic, *photonics* is defined as interaction of light with matter [45]. Photonic devices (e.g., waveguides, ring resonators, and photonic crystals, see Saleh and Teich [46] or Pollock and Lipson [47]) have found numerous applications in fields ranging from telecommunications and computing to biochemical sensing and detection.

5-2-1 Qualitative Description of Optofluidic Transport

For optofluidic transport, the photonic device we are primarily interested in is the dielectric waveguide. The reason for this is that they can confine light by total internal reflection over very long distances with very little lengthwise dilution of the optical energy. Though the light is confined to propagate in a single direction in a waveguide, a nonpropagating exponentially decaying component of this field (referred to as the evanescent field) extends outside the waveguide. The degree of this extension depends on the refractive index contrast between the waveguide and the surrounding media [46] but is typically on the order of a 100 nm. In Fig. 5-2 we compare the forces on a dielectric particle near an optically excited waveguide with those imparted by a traditional optical tweezer. As can be seen in Fig. 5-2b,

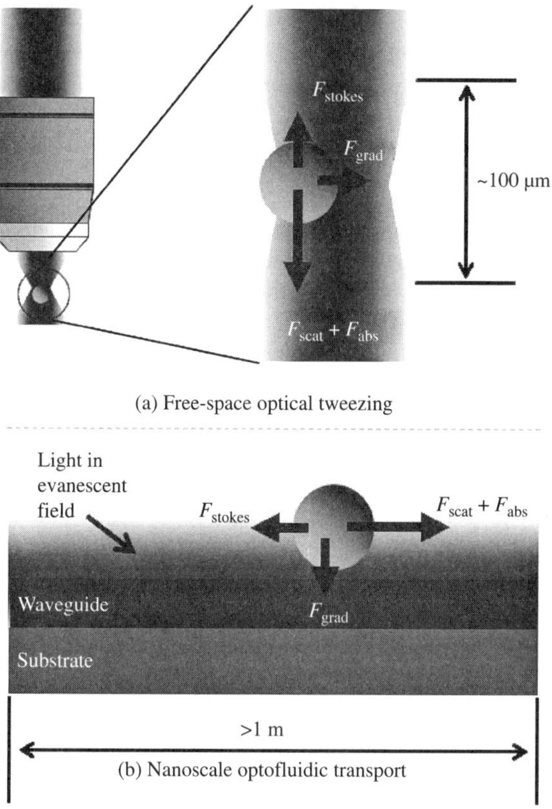

FIGURE 5-2 Comparison between (a) traditional optical tweezing and (b) optofluidic transport on a dielectric waveguide.

the evanescent mode extends outside the waveguide decaying exponentially into the surrounding medium with a portion of it interacting with the particle. This optical gradient partially polarizes the particle, resulting in a strong Lorenz force. This serves to attract the particle to the waveguide (F_{grad}). When this particle is trapped within the evanescent field, a certain percentage of the photons that flow through the waveguide are either scattered (radiated in a random direction) or absorbed when they contact the particle. Each of these photons has a momentum given by Planck's constant divided by the wavelength, h/λ. These scattering (F_{scatt}) and absorption (F_{abs}) events result in momentum transfer to the particle and a net forward velocity that is proportional to intensity and impeded by viscous drag (F_{stokes}). *In a sentence, what optofluidic transport allows us to do is simultaneously exploit the extremely high trapping strength available in the near field with the ability to apply a radiation pressure like transport force over indefinitely long distances.*

5-2-2 Why Is Optofluidic Transport Interesting?

We list here a number of the different fundamental and practical advantages of optofluidic transport over the traditional microfluidic techniques. However, before we go to this list, let's reexamine the limitations of optical transport described in Sec. 5-1-3 and how this method addresses them.

1. *Solution to diffraction limitation:* The high refractive index of the waveguide serves to confine the optical mode to a much smaller cross-sectional area than the free-space diffraction limit. As such the cross-sectional area is lower and the intensity of the light is greater for a given amount of optical power. As was demonstrated by Ng et al. [48] the waveguide can be designed such that the peak intensity occurs at the waveguide/liquid interface.
2. *Solution to light/species interaction length limitation:* Since the mode is confined by total internal reflection in the waveguide, the interaction length can be extended indefinitely. In telecommunications, for example, optical fibers carry signals over kilometer scale distances. As such it should be relatively easy to exploit this technology to create chip-based systems that enable optical transport over the distances required for microfluidic devices.

In addition to addressing these fundamental challenges with optical manipulation in microfluidic devices, we can also list a few additional advantages that optofluidic transport may have in comparison with some of the more traditional micro- and nanofluidic transport mechanisms introduced earlier. Some of these advantages are qualitative, whereas others are quantitative and rely on knowledge of some of the transport theory that is expanded on in Sec. 5-4. We summarize all these advantages here for continuity, but refer to the relevant sections in the rest of the text where they are expanded upon.

1. *Favorable transport scaling laws:* As the size of the photonic device gets smaller, the optical energy/intensity increases and with it the propulsive velocity. In Sec. 5-4-3, we will show that the transport velocity is directly proportional to intensity. As such as the cross-sectional area down to which the light is confined is decreased (thereby increasing the optical intensity) the transport velocity will increase. Pressure-driven flow and electroosmosis have the opposite scaling (smaller device sizes = slower transport).
2. *Strong dependence of velocity on particle size and optical properties:* As will be further explained in Sec. 5-5, we show that the optofluidic propulsive velocity has as much as a fifth power

dependence on particle radius, which is three powers greater than the best techniques currently used for microfluidic separations. We describe how this can be exploited to develop chromatography systems that are at least an order of magnitude more resolute than the state of the art.

3. *Extremely high optical trapping stability:* As alluded to in Sec. 5-1-4, the trapping force is proportional to the gradient in the intensity and the extremely high decay rate of the optical energy in the near field outside the waveguide can lead to a very high trapping force.
4. *Insensitivity to surface/solution conditions:* As mentioned in Sec. 5-1-1, electrokinetic techniques are compatible only with a limited class of fluids, exhibit extreme sensitivity to surface conditions, and are difficult to use with semiconductor substrates such as silicon (as it relies on an insulating substrate). Optofluidic transport is much less dependent on these conditions and can be used in a broader class of systems.
5. *Ability to exploit techniques and components already developed by the telecommunications industry:* Over the past 20 years, billions of dollars have been spent on research and development in the optical communications industry yielding very well-developed highly integrated device architectures and cheap low-power active components. Optofluidic transport allows us to exploit these already optimized techniques for microfluidics.

5-3 Demonstrations of Optofluidic Transport

Prior to expanding on the advantages in the next section (Sec. 5-4) we present a review of experimental literature on the subject in order to better familiarize the reader with the state of the art in the technology. Section 5-3-1 reviews the use of liquid-core and solid-core waveguides for optofluidic transport. In the final section we provide a more detailed review of our recently published [49] system with sufficient detail for the reader to develop their own implementations.

5-3-1 Optofluidic Transport within Solid- (and Liquid-) Core Waveguiding Device

Recently there have been a number of researchers who have published works on near-field optical manipulation methods (see Dholakia and Reece [38] for a recent review) such as those based on the use of surface plasmonic resonances [39,40] or other evanescent field techniques [41]. For example, Cizmar et al. [42] demonstrated the short-range manipulation (in the order of 40 μm) and sorting of polystyrene beads as small as 350 nm in diameter using interfering Gaussian beams reflected off a prism surface. Grigorenko et al. [43] also recently extended earlier approaches to surface plasmon resonance (SPR)-based optical manipulation by exploiting the localized plasmonic resonance in surface-bound metallic nanostructures. While in general these methods are successful at trapping and even assembling [44] small particles, they are limited by the distance through which they can transport objects, since the optical manipulation region is limited by the field of view of the focused laser, and the plasmon propagation distance is relatively short.

The first clear demonstrations of long-distance optical transport on waveguides focused on the use of solid-core, fluid-clad structures that relied on the evanescent field of the waveguide to both capture and transport suspended particles. These experiments featured the propulsion of a wide variety of materials, organic and inorganic, on waveguides. Kawata and Sugiura [50], for example, first demonstrated the use of an evanescent field-based optical trapping technique. This was further refined in 2000 by Tanaka and Yamamoto [51], who showed the propulsion of polystyrene spheres on a channel waveguide.

While these seminal papers demonstrated for the first time the potential for using evanescent field trapping as a potential mechanism for optofluidic transport, it was unknown if the method would have the same versatility demonstrated for optical tweezers. Gaugiran et al. [52] demonstrated the use of silicon nitride waveguides for trapping and propulsion of yeast and red blood cells, as shown in Fig. 5-3. The advantage in using silicon nitride waveguides is the ability to guide wavelengths of light at 1064 nm. Unlike silicon waveguides, which optimally guide light at telecom frequencies, at 1064 nm the light is not heavily absorbed by water, therefore reducing the impact on biological species. In addition, with a smaller

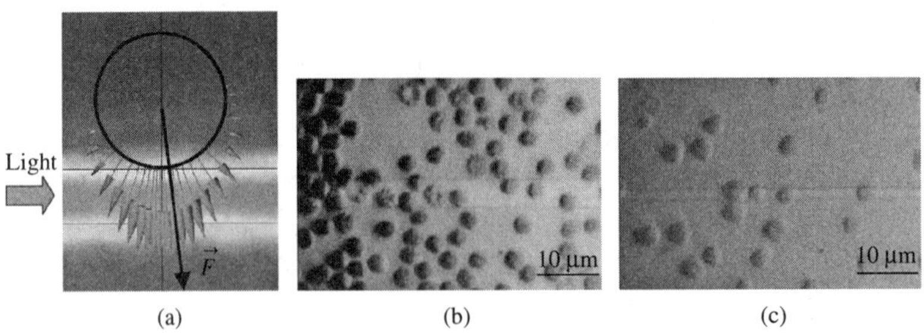

FIGURE 5-3 Optofluidic transport of biological species. (a) Finite element simulation of optical field in a channel waveguide and forces acting on a glass particle. (b) Image of radiation pressure transport of red blood cells. (c) Yeast cells. *(S. Gaugiran, S. Getin, J. M. Fedeli, G. Colas, A. Fuchs, F. Chatelain, and J. Derouard, "Optical manipulation of microparticles and cells on silicon nitride waveguides," Optics Express, 13(18), (2005), 6956–6963.)* (See also color insert.)

wavelength, the light in the waveguide is more strongly confined, leading to stronger trapping forces and higher propulsion velocities. Gaugiran et al. were also the first to propose an analysis of the optical propulsion and trapping forces for a particle near a waveguide using finite-element methods, in particular showing a deviation from the analytical Rayleigh particle assumption at larger particle sizes. Furthermore, the authors provided some of the first experimental quantification of numerical predictions of propulsion and trapping forces. Along similar lines, Ng et al. [53] demonstrated the propulsion of high-absorption gold nanoparticles, seeking to now exploit high-absorption materials to generate higher propulsion velocities and trapping forces. In combination, these two papers provided experimental evidence of the diverse materials that could be transported on waveguiding structures.

One of the advantages of using optically driven transport is that there are many types of devices that can be used to divert and alter the behavior of optical fields. Evanescent coupling can be used to cause light to effectively tunnel through a lower refractive index medium into an adjacent waveguide. Resonator devices allow for the attenuating properties of constructive and destructive interference to enable switching and/or creating highly focused hot spots in the guiding structure. Recently, there have been a few demonstrations of methods to create more complex optical fields for particle sorting/manipulation. Of particular note, Grujic et al. [54] was the first demonstration using Y-branch waveguides as a sorting mechanism for transported particles, shown in Fig. 5-4. The system consisted of CS^+ ion-exchange waveguides on class. Polystyrene microparticles were guided down the "upper" or "lower" waveguides at the Y-split by altering the physical position of the input fiber, creating preferential pathways for particles to follow. An improvement over this type of device would be one that accomplishes the sorting based on the intrinsic properties of the particle in question, as opposed to the arbitrary position of the input fiber.

Before moving on to a detailed example of optofluidic transport, it is important to at least briefly describe a slightly different architecture for optofluidic transport. The essential flaw with all the previously mentioned devices is that the majority of the guided optical energy is confined within the solid core of the waveguide and the particles only interact with the 10% to 20% of the energy that is accessible in the evanescent field. As such a number of recent works have investigated the possibility of using "liquid-core" waveguiding structures for optical transport. Since the overlap between the guided mode and the transported optical energy is stronger in these systems, the potential exists for greater transport speeds. As an example, Mandal and Erickson [55] recently demonstrated the use of a specially tailored hollow-core photonic crystal fiber (HCPCF) to propagate light within a liquid-core environment and levitate/transport dielectric particles.

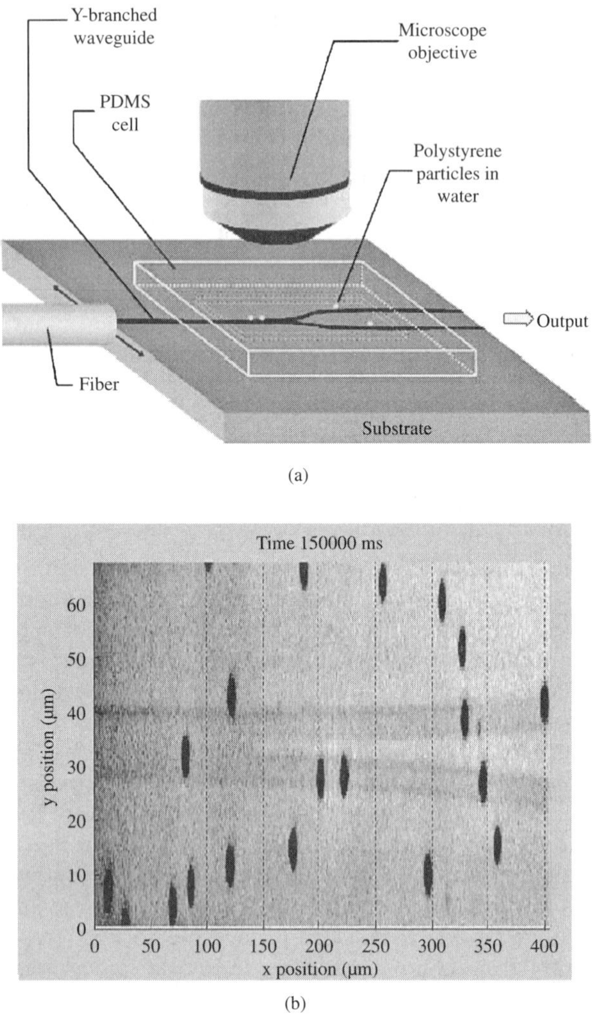

FIGURE 5-4 Particle sorting using Y-branch waveguide structure. (a) Schematic of experimental system. (b) Image of particle-sorting capture process for polystyrene microparticles. The eventual particle path is determined by the position of the input laser fiber at the point when the particle nears the Y-branch junction. (*K. Grujic, O.G. Helleso, J.P. Hole, and J.S. Wilkinson, "Sorting of polystyrene microspheres using a Y-branched optical waveguide," Optics Express, 13(1), (2005), 1–7.*)

In a more chip-friendly format, Measor et al. [56] demonstrated the use of particle transport within a planar liquid-core antiresonant reflective optical waveguides (ARROW) as a means of characterizing the optical performance of the waveguide.

5-3-2 A Detailed Example—Optofluidic Transport in PDMS Microfluidics Using SU-8 Waveguides

As mentioned earlier our goal in this section is not only to review the literature but also to provide the reader with sufficient information to exploit optofluidic transport within microfluidic devices of their own design. The technique we presented in this section is based on that presented by Schmidt et al. [49] and uses SU-8 waveguides with PDMS microfluidics. We choose to present this in detail here because of the relative ease with which both these types of devices can be manufactured.

The basic layout of our basic optofluidic transport architecture is shown in Fig. 5-5a to 5-5c. As mentioned earlier, the platform used here comprises SU-8 epoxy-based photonic structures, combined with PDMS microfluidics on a fused silica substrate. The fused silica substrate has a refractive index of 1.453, while the exposed SU-8 film has a measured refractive index of 1.554 at $\lambda = 975$ nm which, along with the water cladding with refractive index of 1.33, provides for significant refractive index contrast and strong evanescent field gradients. The waveguide dimensions were chosen to be a height of 560 nm and varied in width from 2.8 μm to as little as 500 nm. At the 975-nm excitation wavelength all these waveguide widths were found to be single mode in TM polarization.

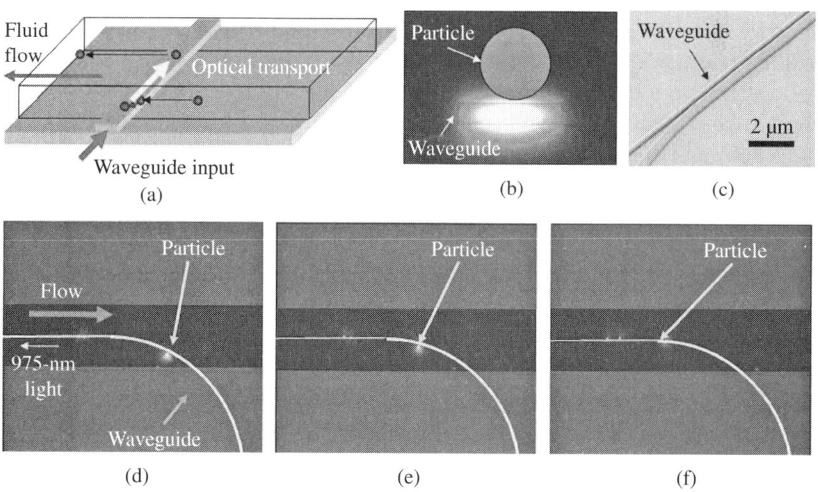

FIGURE 5-5 Optical trapping and transport in the evanescent field of an optical waveguide. (a,b) A particle flowing in a microchannel becomes captured in the evanescent field of the excited waveguide. (c) SEM of two waveguides. (d–f) Time step images showing transport of 3-μm polystyrene particles on a waveguide.

Both the waveguides and microfluidic devices were fabricated using common photolithographic and soft-lithography techniques. More details are available in the Schmidt et al. [49], however briefly; the waveguides were fabricated using SU-8 resist with MicroChem formulation 2000.5 by exposing the film with desired waveguide pattern using a standard photolithography arrangement, performing the recommended postexposure bake and developing procedure. Figure 5-5c shows an SEM image of two of these waveguides in close proximity. The input and output facets of the waveguides were diced from the backside with a dicing saw to a distance of 50 μm from the top surface and then cleaved by applying simple pressure to the substrate by hand. The microfluidics were made using a standard procedure for creating PDMS microfluidics by solution casting using a lithographically patterned mold [57,58]. The channels were designed to dimensions of 5 μm in height and 100 μm in width. We used relatively shallow channels to confine the flowing particles as close as possible to the waveguides in order to increase the capture rate.

To assemble the structure, the PDMS channels and the waveguide sample were both plasma-cleaned in air for several seconds and then bonded by placing them in conformal contact. As shown in Fig. 5-5a the arrangement was such that the channel ran perpendicular to the waveguides, though this is by no means a necessity. In the absence of a plasma cleaner (oxidizer), placing the two halves together will still form a temporary seal, sufficient to carry out most experiments that do not involve very high fluid pressures. The conformable nature of the PDMS greatly facilitated sealing of the microfluidics over the waveguides without greatly disturbing the optical mode. The use of other nonconforming materials complicates this process. In this case the channels were aligned perpendicularly to the waveguide inputs, leaving between 500 μm and 1 mm of space between the edge of the chip and the start of the PDMS. Leaving an air clad region at the edge of the chip facilitates coupling the light into the waveguides.

In the experiment shown in Fig. 5-5d through 5-5f, we flow fluorescently tagged dielectric particles in the main microfluidic channel toward the optically excited waveguide using pressure-driven flow. The particles used in our experiment were polystyrene spheres with refractive index n = 1.574 in a 100-mM phosphate buffer solution (PBS) with a regulated pH of 7.0. The light source used for testing was a fiber-coupled laser diode module with a wavelength of $\lambda = 975$ nm. To excite the waveguides we used a micrometer-controlled fiber positioning stage to position a lensed fiber near the end of the waveguide of interest. The light was considered optimally coupled into the waveguide when we received a maximum output power reading on a detector placed near the output end of the waveguide or by directly observing the scattered light (on a CCD camera sensitive to 1-μm light) from the waveguide using an upright microscope, which observed the chip from above.

When a flowing particle comes in contact with the optically excited waveguide, it may be captured in the evanescent field and begin moving in the direction of optical propagation. Figure 5-5d to 5-5f shows time step images of the particle becoming trapped on the waveguide and propelled in the direction opposite the initial flow. Using the system described here we observed particle trapping and optical transport velocities along the waveguide as high as 30 μm/s and capture particles flowing by as fast as 80 μm/s. We observed approximately linear behavior of the optical transport velocity and the guided optical power. This is roughly as expected from our qualitative description given earlier since the number of photon strikes should be proportional to the optical power in the waveguide. Movies showing the transport and many more results are available from Schmidt et al. [49].

Comments on Particle Capture Rate

As noted in the previous sections, the "capture rate" of particles flowing over the waveguide is relatively low in this arrangement at approximately 10% of all the particles, which overflow the waveguide (this is better illustrated in the movies) [49]. Experimentally, we observed that this capture rate increases as the flow rate decreases and the optical power increases. The reason is that in this experimental arrangement, a particle passing over a waveguide must be on a streamline that passes through a region of the evanescent field which exerts a force on the particle greater than the flow drag force in order to be captured (this is analogous to the condition that a flowing particle must be on a streamline that passes through the focal point of a free-space optical tweezer in order to be trapped). In a low-Reynolds number microfluidic flow, the only way in which a particle can hop streamlines is through diffusion or when acted upon by an external impulse. Since the average volume over which a particle will travel through diffusion increases with the amount of time it is observed, the probability that it will sample a streamline that passes through the evanescent field increases with the amount of time it takes for it to flow over it. As such the rate of capture can be increased by reducing the flow rate as observed. Increasing the optical power increases the strength of the trapping force at a given point in the evanescent field and, therefore, also increases the number of streamlines that pass through the "attraction basin." If greater trapping probabilities are desired, the simplest way of accomplishing this is to decrease the channel size (here we use a 5-μm-tall channel). This serves to physically confine the particles closer to the waveguide effectively reducing the number of streamlines that do not pass through the evanescent field.

5-4 Theory of Optofluidic Transport

In this section we present a theoretical description of optofluidic transport that will help to quantify some of the advantages described in Sec. 5-2 and the experimental observations made in Sec. 5-3. After a review of the relevant literature, we first present an overview of the relevant microscale fluid mechanics and the behavior of small particles suspended in a fluid environment. The second section covers the general electromagnetic and guided wave optics theory required to describe the relevant optical forces and how they are coupled with hydrodynamic theory. In the final sections we present a few analytical approximations for special cases and return to the aforementioned list in the context of the developed theory.

5-4-1 Overview and Recent Literature

The theory behind optofluidic transport has its basis in the fundamentals of electromagnetics and hydrodynamics. From this broad base, specific models have been developed to treat the specific geometries and cases that arise frequently. In the case of optofluidic transport, this often shows up in the form of analytical simplifications of more general phenomena. In the case of electromagnetics, the Rayleigh and Mie theories are often used to explain the propulsion and trapping forces exerted on particles in optofluidic systems by a present optical field. The influence of fluid forces on particle behavior is often summarized using the Stokes drag law or Faxen's law. Most of the studies up to date on optofluidic theory have focused on applying the mentioned theories to an optofluidic system. We summarize the results from these studies as follows.

The Mie and Rayleigh theories are specific toward evaluating the forces exerted on particles in the presence of an optical field. As might be expected, the major approximations of these theories assume a spherical scatterer and relatively noncomplex geometries. The main difference is that Rayleigh scattering theory [59] is designed to treat particles that are much smaller than the wavelength of light incident upon it, while Mie theory [60] treats larger particles, which exhibit different scattering behavior from Rayleigh particles. Both Almaas and Brevik [61] and Ng et al. [48] also deal specifically with the behavior of particles in evanescent fields. Figure 5-6 is adapted from the Ng et al. paper and illustrates the basic geometry used in their approach. A concise summary of both optical and hydrodynamic forces within the context of optical tweezing is provided by Svoboda and Block [62]. Readers interested in the behavior of metallic particles in optical fields are directed to a paper by Svoboda and Block [21] and another by Gaugiran et al. [63]. With the development of multiphysics-based simulation software packages, recent thrusts in understanding the behavior of particles have focused on using more general derivations

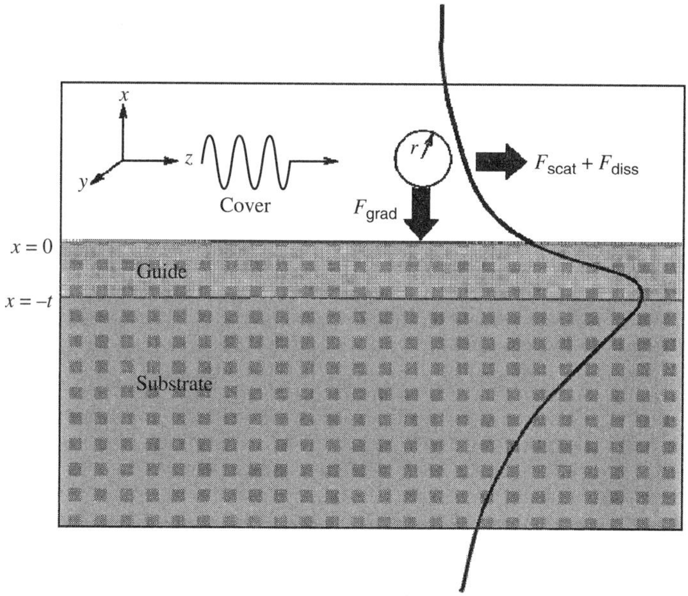

FIGURE 5-6 Schematic representation of an asymmetric planar waveguide. Radiation forces acting on a sphere of radius *r* are decomposed into gradient force in the transverse direction and a forward force in the direction of wave propagation. (*L.N. Ng, B.J. Luf, M.N. Zervas, and J.S. Wilkinson, Journal of Lightwave Technology, Copyright (2000) IEEE* [48])

of optical forces, such as the Maxwell stress tensor [64], and using simulation to evaluate optical and hydrodynamic forces in nontrivial geometries. In particular, Gaugiran et al. [52] first used finite element simulation to estimate the propulsion and trapping forces on rectangular waveguides.

5-4-2 Microscale Hydrodynamics and Particle Transport

The underlying principle behind continuum fluid dynamics is the conservation of two quantities: mass and momentum. In the most general sense these conditions are mathematically described by the conservation of mass and Navier-Stokes equations [65]. Solving this complete set of equations is very difficult, and analytical solutions are only available for a limited class of geometries and flow conditions. Fortunately, however, the nature of optofluidic transport allows us to make a few simple assumptions to reduce the complexity of the analysis without greatly sacrificing accuracy. The primary assumption we make is that the fluid is incompressible and of constant viscosity (i.e., Newtonian). This is generally valid for all liquids under the shear conditions likely to be encountered in the systems of interest here. The other assumption we make is that the transport occurs under conditions of low Reynolds number, $\mathrm{Re} = \rho U a/\mu$, where ρ is the fluid density, U is the characteristic transport speed, a is an appropriate size scale, and μ is the viscosity. For pure particle transport in a quiescent medium, U would be the particle speed and a would be its diameter. In water then a 1-μm particle transported at 100 μm/s would have a Reynolds number of approximately 10^{-4}. If one is considering an externally induced flow in a microchannel (say by the application of a pressure difference), U would be the average flow speed in the channel and a the channel height. In such a case Re can be as high as approximately 0.1 but is usually much less. In either case, physically this means that momentum transport occurs via diffusion rather than convection and that we can ignore the nonlinear terms in the Navier-Stokes equations. This also implies that the flow will reach its steady state velocity relatively quickly and that the transient period can be ignored. Under these assumptions the fluid dynamical equations reduce to conservation of volume [Eq. (5-1a)] and the Stokes equation [Eq. (5-1b)].

$$\nabla \cdot \mathbf{v} = 0 \tag{5-1a}$$

$$\mu \nabla^2 \mathbf{v} - \nabla P = 0 \tag{5-1b}$$

where $\mathbf{v}$ is the velocity field and P is the pressure.

Hydrodynamic Forces on a Particle in a Flow

Equations (5-1a) and (5-1b) are descriptive of the fluid velocity at every point in a flow. Generally speaking a particle in a flow will experience a net pressure force (caused by pressure drop across the particle) and a friction force (caused by the flow of a viscous liquid over the surface). In the most general case the net drag force can be written as

$$\mathbf{F}_D = \oint_S (\mathbf{T}_F \cdot \mathbf{n})\,dS \tag{5-2a}$$

where $\mathbf{F}_D$ = drag force
$\mathbf{T}_F$ = fluid stress tensor
$\mathbf{n}$ = normal vector to the surface of the particle.

For an incompressible Newtonian fluid, the stress tensor is written as

$$\mathbf{T}_F = -P\mathbf{I} + \mu\left(\nabla \mathbf{v} + \nabla \mathbf{v}^T\right) \tag{5-2b}$$

where $\mathbf{I}$ is the isotropic tensor and $\nabla\mathbf{v}$ is the gradient of the flow velocity.

The above forms of the hydrodynamic equations are appropriate for use in numerical simulations, but difficult to manipulate analytically. Simplified versions of these equations are, however, available for two important cases; the first being for spherical objects moving through a stagnant fluid in an infinite domain. In such a case Eq. (5-2a) reduces to the expression shown (often referred to as the Stokes drag equation):

$$\mathbf{F}_D = -6\pi\mu a U \tag{5-3}$$

where U is the velocity of the particle relative to the bulk flow and a is the particle radius.

The negative sign in the equation refers to the fact that the force acts opposite the direction of the particle velocity. This equation is only accurate when a particle is far from any no-slip boundaries (such as walls). It can be shown that a modification of the Stokes drag equation can be made to approximate the drag for a particle moving near an even solid surface. This equation (which is a form of Faxen's law [62,66]) is given as

$$\mathbf{F}_D = \frac{-6\pi\mu a U}{\left[1 - \frac{9}{16}\left(\frac{a}{h}\right) + \frac{1}{8}\left(\frac{a}{h}\right)^3 - \frac{45}{256}\left(\frac{a}{h}\right)^4 - \frac{1}{16}\left(\frac{a}{h}\right)^5\right]} \tag{5-4}$$

where h is the distance between the particle center and the wall surface.

5-4-3 Electromagnetic Forces on a Particle

As previously discussed, optical forces acting on particles can be separated into two main categories. The optical trapping force acts to pull a particle along the gradient of the electric field toward the region of highest optical intensity. The radiation pressure forces are due to the scattering and absorption of photons on the particle, which push particles in the direction of optical intensity. As described by Mishchenko et al. [67], this is an orthogonal decomposition of the total force that is more generally described by the surface integral of the time-averaged Maxwell stress tensor, $\langle \mathbf{T}_M \rangle$, as shown in Eq. (5-6a).

$$\langle \mathbf{T}_M \rangle = \mathbf{DE}^* + \mathbf{HB}^* - \frac{1}{2}(\mathbf{D} \cdot \mathbf{E}^* + \mathbf{H} \cdot \mathbf{B}^*)\mathbf{I} \tag{5-6a}$$

where $\mathbf{E}$ = electric field
$\mathbf{B}$ = magnetic flux field
$\mathbf{D}$ = electric displacement
$\mathbf{H}$ = magnetic field
$\mathbf{E}^*$ and $\mathbf{B}^*$ = complex conjugates
$\mathbf{I}$ = isotropic tensor.

We note that the use of the time-independent Maxwell stress tensor is justified here since the transport processes of interest occur on timescales much longer than the optical period (for more information interested readers are directed to a review article that discusses electromagnetic stress tensors [68]). When expanded out, Eq. (5-6a) becomes

$$\langle \mathbf{T}_M \rangle = \begin{pmatrix} D_x E_x^* + B_x H_x^* - \frac{1}{2}(\mathbf{D}\cdot\mathbf{E}^* + \mathbf{B}\cdot\mathbf{H}^*) & D_x E_y^* + B_x H_y^* & D_x E_z^* + B_x H_z^* \\ D_y E_x^* + B_y H_x^* & D_y E_y^* + B_y H_y^* - \frac{1}{2}(\mathbf{D}\cdot\mathbf{E}^* + \mathbf{B}\cdot\mathbf{H}^*) & D_y E_z^* + B_y H_z^* \\ D_z E_x^* + B_z H_x^* & D_z E_y^* + B_z H_y^* & D_z E_z^* + B_z H_z^* - \frac{1}{2}(\mathbf{D}\cdot\mathbf{E}^* + \mathbf{B}\cdot\mathbf{H}^*) \end{pmatrix} \tag{5-6b}$$

where the subscripts x, y, and z signify the coordinate directions. By integrating the time-independent Maxwell stress tensor on a surface enclosing the particle of interest, we can determine the total electromagnetic force acting on the system, F_{EM}, given by

$$\mathbf{F}_{\text{EM}} = \oint_S \left(\langle \mathbf{T}_M \rangle \cdot \mathbf{n}\right) dS \tag{5-7}$$

where $\mathbf{n}$ is the unit vector normal to the particle surface. As we [69] and others [52] have shown, the E and H fields can be computed either through a full solution to Maxwell's equations or by solving the time harmonic wave equation via the finite element method and the integration of Eq. (5-7) carried out numerically. For further information on how to carry out these computations, readers are referred to Refs. 52 and 69.

5-4-4 Solutions in Different Transport Regimes

The set of equations in the preceding section represent a relatively basic, but general, model for optofluidic transport, ignoring such effects as heating, surface friction, and electrical double layer repulsion. Despite this the basic model has proven to be relatively predictive of observed experimental behaviors [49]. In this section we discuss how to implement these models for two transport regimes of interest: (1) when the transported particle radius, a, is much smaller than the wavelength of light, λ, and (2) when the particle radius is approximately the same or much larger than λ.

Transport in the Development in the $a \ll \lambda$ Regime

This first regime is referred to as the Rayleigh regime and is defined by the assumption that the electromagnetic field is uniform as it impinges on the particle (hence the limitation that $a \ll \lambda$). For this case the scattering, adsorption, and trapping forces exerted on a particle (see Svoboda and Block [21], and others [22,48,70]) take the form

$$F_{\text{scat}} = \frac{8\pi^3 I_o \alpha^2 \varepsilon_m^{5/2}}{3c\lambda^4} \tag{5-8a}$$

$$F_{\text{abs}} = \frac{2\pi\varepsilon_m I_o}{c\lambda}\text{Im}(\alpha) \tag{5-8b}$$

$$F_{\text{trap}} = \frac{n_m \alpha}{2c}\nabla I_o \tag{5-8c}$$

where $\alpha = 3V(\varepsilon - \varepsilon_m)/(\varepsilon + 2\varepsilon_m)$
V = particle volume
c = speed of light
ε_p and ε_m = dielectric constants of the particle and material
I_o = optical intensity.

Equating F_{scat} and F_{abs} with Stokes drag Eq. (5-3) we obtain

$$U_o = \frac{n_m k_m I_o}{6\pi a c \eta}\left(\frac{k_m^3 \alpha^2}{6\pi} + \text{Im}(\alpha)\right) \tag{5-9}$$

where $k_m = 2\pi n_m/\lambda$ and is descriptive of the particle transport velocity in the Rayleigh regime.

For the case of a particle traveling very near the surface, we could improve the accuracy of Eq. (5-9) by equating the propulsion forces with Faxen's law, Eq. (5-4).

However, it is generally difficult to estimate the distance the particle is above the waveguide. A conservative estimate, however, could be obtained by assuming the particle was right near the surface, in which case $a = h$.

Transport in the $a \geq \lambda$ Regime

When the particle size is much larger than the wavelength of light, the assumption of a uniform electric field is no longer valid and we generally require a numerical method to obtain an estimate of the strength of the trapping forces and propulsion velocity. For details see Yang and Erickson [69]; however, generally speaking the E and H fields can be computed either through a full solution to Maxwell's equations or by solving the time harmonic wave equation via the finite element method. Once these solutions are obtained, Eq. (5-7) can be solved to obtain the net electromagnetic force on the particle, in all three coordinate directions, as a function of the optical power in the waveguide. The dynamic tracking of particle motion in a fluid is a relatively complex simulation, and thus to obtain the net drag on a particle moving along a waveguide, we shift reference frames by keeping the particle still and applying boundary conditions that simulate the flow moving by it. In the low-Reynolds number regime the drag force is linearly proportional to the flow velocity, so the simplest method is to apply a "slip" velocity boundary condition on all surfaces of some low flow speed (say 1 μm/s) opposite the direction one expects the particle to move. By integrating Eq. (5-2) over the particle surface a drag force is obtained, which can be scaled to the particle speed by simply dividing it by the imposed slip velocity. For cases where an imposed flow is incident on the particle, such as in [Ref. 69], it is appropriate to use standard microfluidic boundary conditions (see Erickson [71] for details).

Figure 5-7 shows the results of finite element-based computations performed on dielectric spheres of various sizes.

From Fig. 5-7b it is apparent that the propulsive force follows an approximate squared relation with particle size ($F_{scat} \propto a^2$) and from the experiments described earlier we know that F_{EM} is also proportional to the optical intensity. As such it is reasonable to qualitatively approximate $F_{EM, scat} = C_1 a^2 I_o$, where C_1 is a constant and an unknown function of the physical parameters of the system and I_o is the optical intensity. Using Faxen's law approximation $F_{flow} = 6\pi\eta a U_o/g(a/h)$, where $g(a/h)$ is the denominator in Eq. (5-4). If the distance between the bottom of the particle and the top of the waveguide is small compared to the particle diameter, $a/h \sim 1$ then $g(a/h)$ is a constant and we can derive the following approximate equation descriptive of the transport velocity:

$$U_o = \frac{C_1 a^2 I}{6\pi\eta a/g(a/h)} = C_2 \frac{a I_o}{\eta} \tag{5-10}$$

where C_2 is a different constant comprising of the same system physical parameters as C_1. We note from Eq. (5-10) that (based on these computations) the transport velocity in $a \geq \lambda$ appears to contain only a linear dependence on particle size (as opposed to the much stronger dependence in the Rayleigh regime).

5-4-5 Comments on the Influence of Brownian Motion and Trapping Stability

The transport path taken by a particle with a diameter greater than approximately 1 μm is largely deterministic and well described by Eq. (5-10). Below 1 μm, however, they begin to be significantly affected by Brownian motion and the hydrodynamic solution becomes predictive of the average motion of the transported particle (about which it will diffuse), rather than its actual pathline. The underlying cause of Brownian motion is the random fluctuating collisions of solvent molecules impacting larger microscopic particles. Brownian motion

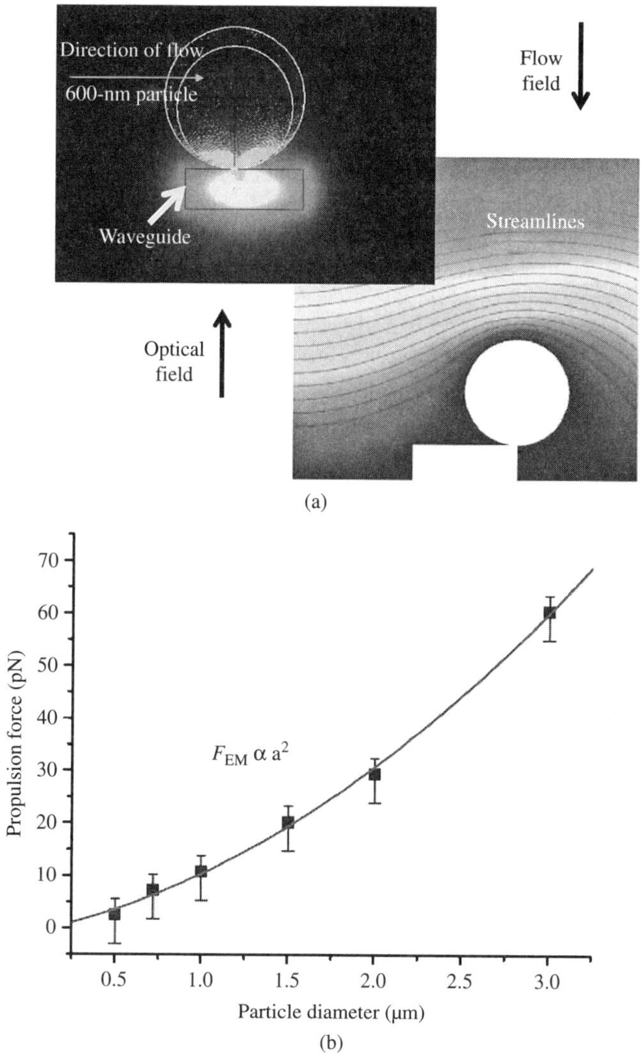

FIGURE 5-7 Forces on a particle trapped on a waveguide. (a) Cross section of guided mode in the waveguide and streamlines for a particle trapped on the waveguide and subject to a crossflow. (b) Propulsion force computed on particles trapped on the waveguide. (See also color insert.)

affects the transport of a particle along a waveguide in a number of ways. Most importantly it can serve to break the trap in that if the random thermal energy delivered to the particle exceeds the strength of the optical confinement, the trap is considered unstable and the particle will diffuse away. In this section we focus on describing how the conditions under which trap breaking will occur can be predicted, allowing us to define the conditions under which stable optofluidic transport can occur. The system we consider here is for a dielectric particle trapped in the evanescent field of a waveguide, subject to a microfluidic cross flow. Our approach, which is outlined in greater detail in a recent paper [69], is to compute the work required to remove a particle from a trapped state and determine a trapping stability parameter based on this. It can be easily shown that the downward trapping force far exceeds that in the direction of flow, so our stability condition is based on a particle being "swept" off the waveguide (since this is the most likely way it will be removed).

If we assume the spatial force variation in the flow direction is like that of the evanescent field, then a similar relationship would hold for a sufficiently wide waveguide:

$$\mathbf{F}_{\text{trap}}(x) = \mathbf{F}_{\text{T0}} \exp(-\gamma_f x) \tag{5-11}$$

where F_{T0} represents the transverse trapping force at a point $x = 0$, which we designate as the trailing edge of the waveguide, and γ_f is the trapping force decay rate. We will later show numerically that this relationship holds for a particle moving laterally from a rectangular waveguide. For a constant flow speed, we can assume that the drag forces on the particle are relatively constant compared to the steep decay of the transverse (x-direction) trapping force. Then, the net force acting on a particle is

$$\mathbf{F}_{\text{net}}(x) = \mathbf{F}_{\text{T0}} \exp(-\gamma_f x) + \mathbf{F}_D \tag{5-12}$$

where F_D is the drag force exerted on the particle. The nature of this equation is such that for any nonzero positive drag force, the net force acting on the particle will at some point become a net positive force. The crossing point can be analytically determined:

$$x_c = \gamma_f^{-1} \ln\left(\frac{\mathbf{F}_{\text{T0}}}{\mathbf{F}_D}\right) \tag{5-13}$$

where x_c is the crossing point where the net force acting on the particle becomes a positive value. The work necessary to release a particle from the trapping field can be shown to be

$$W_{\text{trap}} = \int_0^{x_c} F_{\text{trap}}(x)dx = F_{\text{T0}}\gamma_f^{-1}[1+\theta(\ln(\theta)-1)] \tag{5-14}$$

where θ is the ratio of the drag force to the transverse trapping force (F_D/F_{T0}). The equation is only valid for $0 \le \theta \le 1$, and it can be shown that the limit of the nonlinear term goes to unity as θ goes to zero. We can define the trapping stability of the particle by relating the work needed to release a particle compared to the random thermal motion of the particle, which is important on such size ranges [16]:

$$S = \frac{W_{\text{trap}}}{k_B T} = \frac{F_{T0}\gamma_f^{-1}}{k_B T}[1+\theta(\ln(\theta)-1)] \tag{5-15}$$

where S = stability number
k_B = Boltzmann number
T = temperature of the system.

S can take values greater than or equal to zero, with zero representing a critically unstable trap.

Figure 5-8 illustrates the stability diagram as presented for a waveguide with a refractive index of $n = 1.68$, dimensions of 800 × 400 nm and excitation wavelength of 1064 nm. From Eq. (5-15) the trapping stability number would need to be greater than one for successful trapping to occur. At stability numbers less than one, the trapping is considered to be weaker than the local Brownian motion of the particle; thus trapping would not occur. The regions highlighted

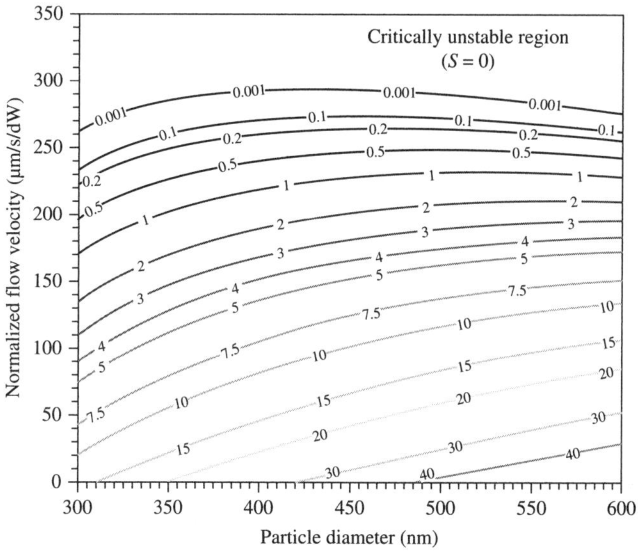

FIGURE 5-8 Stability diagrams as a function of experimental parameters. The system represented in the graph is a polymer waveguide ($n = 1.68$) on glass ($n = 1.45$) submersed in water ($n = 1.33$) with glass microparticles ($n = 1.45$). Stability numbers can be calculated for different power ratings (default is 100 mW) by adjusting the normalized flow rate and stability number by the power ratio [P_{actual} (in mW)/100 mW]. Higher stability numbers represent more stable trapping systems. *(A.J.H. Yang and D. Erickson, "Stability analysis of optofluidic transport on solid-core waveguiding structures," Nanotechnology, (2008).)*

as "critically unstable" represent regions where the drag force on the particle is stronger than the calculated trapping force. In such regions, particle trapping should not be possible, unless unrealistically powerful lasers (greater than 100 W) are employed.

As can be seen, increasing the fluid velocity proves to be detrimental to the trapping stability. It is interesting to note that successful particle trapping depends upon maintaining fluid flow below some critical limit, and passing the limit either renders the trapping ineffectual or enters the critically unstable region. This barrier is significantly lower for polymer waveguides, largely because the same amount of optical power translates into less trapping force than in a silicon waveguide. This limit can be adjusted, however, by increasing the available power to the waveguide, which would result in a larger region containing a stability number greater than one. Tuning the laser power coupled in the waveguide acts as one means of controllable release, where the trapping forces are reduced by a decrease in the available optical power. Our model system, as specified, provides another method of control, via tuning of fluidic flow rates. Tuning of flow speeds directly alters the drag force on the particle, allowing a nonoptical means of adjusting the ability of a particle to overcome the transverse trapping field.

5-5 Optofluidic Chromatography

As the final section in this chapter we discuss the application of optofluidic transport to particle chromatography. As alluded to in Sec. 5-1-2 one of the most interesting applications of optofluidic transport is in the development of a practical optical chromatography system. The basic idea behind such systems is that when an initially mixed group of particles is subjected to an intense optical field, they can be separated out (fractionated) into different homogeneous groups due to differences in the propulsion velocity based on size or dielectric constant. The major limitation of current free-space systems is related to light-particle interaction length problem discussed in Sec. 5-2-1. Optofluidic transport, as mentioned, avoids this and could allow us to apply these separation impulses over indefinitely long distances. The separation velocity relations shown in Table 5-1 allow us to directly compare the fractionalization resolution of this technique with the state of the art. As can be seen in the $a \ll \lambda$ regime the fifth power dependence on the particle size exceeds the nearest competitor (dielectrophoresis) by three powers suggesting an optofluidic system should be able to achieve higher resolution. We note, in the context of this table, that the reason why techniques such as electrophoresis are successful is not because they fundamentally are very size sensitive but rather that the impulse can be applied over very long distances (electrophoresis for example is carried out in capillaries that are tens of centimeters long).

The most common way to quantify separation resolutions is to compare the spatial separation between two species at some point downstream and divide the difference between the width of the peaks. If we assume that the peak widths will be roughly the same for all systems, it can be shown that the spatial separation divided by the distance traveled, describing the nondimensional separation resolution, R, is given by

$$R = |l_1(r = a) - l_2(r = a + \Delta a)| / l_1(r = a) \tag{5-16}$$

where l_1 and l_2 are the distances traveled by particles of radius a and $a + \Delta a$. Using the separation velocities from Table 5-1 the following relations can be derived for optical, R_{op}, and electrophoretic separations, R_{ep},

$$R_{op} = (1 + \Delta a/a)^5 - 1 \tag{5-17a}$$

$$R_{ep} = 1 - a/(a + \Delta a) \tag{5-17b}$$

which are valid for the $a << \lambda$ case. For $\Delta a/a = 0.01$ (1% size difference) we obtain $R_{op} = 0.051$ and $R_{ep} = 0.0099$, whereas for $\Delta a/a = 0.1$ (10% size difference) $R_{op} = 0.61$ and $R_{ep} = 0.09$. This represents a 500% improvement in fractionalization resolution over the state of the art in the small-size-difference regime and 680% improvement in the large-size-difference regime. In the $a > \lambda$ regime, it can be shown that R_{op} is approximately the same as for electrophoresis. As such in this regime there are likely to be only practical advantages as opposed to fundamental ones. Specifically optical forces in the 1-μm wavelength range are known to be biologically safe as opposed to the high electric fields required for electrophoresis.

As a result of this problem it is rare that electrophoresis is used to separate cellular systems, rather dielectrophoresis is preferred. Referring back to Table 5-1 and comparing dielectrophoresis and optical separation velocities (in the $a < \lambda$ regime) we can see that with regard to size V_{op} has a smaller dependence on size than V_{dep}. This, however, does not translate directly into separation resolution because the velocity is proportional to the gradient in the electric field and thus it is not appropriate for "long interaction length" separations (i.e., it has the same spatial limitation as the free-space optical separation systems described earlier). As such the optical method to be developed here is fundamentally more resolute than the state of the art in the $\lambda < a$ regime, also since it is the only technique that allows one to apply the separation impulse over an indefinitely long distance.

Technique	Operating principle	Separation velocity
Optical	Scattering force dependence on particle size and refractive index, $V_{op} \propto I$.	$V_{op} \propto a^5$, $V_{op} \propto (\varepsilon_p - \varepsilon_m)^2$ $(\lambda >> a)$ $V_{op} \propto a$ $(\lambda < a)$
Electrophoresis	Differences in species charge to drag ratio represented by electrophoretic mobility, $V_{ep} \propto E$.	$V_{ep} \propto \zeta \neq f(a)$* (thin EDL) $V_{ep} \propto 1/a$* (thick EDL)
Dielectrophoresis	Lorenz force dependence on particle size and polarizability, $V_{dep} \propto \nabla(E^2)$.	$V_{dep} \propto a^2$ $V_{dep} \propto (\varepsilon_p - \varepsilon_m)$
Centrifugation	Differences in density. Driving force is gravitational or centrifugal acceleration field, $V_{cent} \propto g$.	$V_{cent} \propto (\rho_p - \rho_m)a^2 \approx a$
Size exclusion chromatography	Size dependence on permeability through packed column. Typically, $V_{SEC} \propto \Delta P$	$V_{SEC} \propto -\log(a^3)$†

*For free solution electrophoresis. Gel-based electrophoresis results in a nonlinear drag that can increase the dependence of V_{ep} on a.
†Approximate, based on assumption of proportionality between hydrodynamic volume and molecular weight.

Table 5-1 Comparison of Bioanalytical Separation Techniques

5-6 Summary and Conclusions

In this chapter we have introduced the concept of "optofluidic transport" as a means for achieving long-interaction length optical transport of particles in a microfluidic format using planar dielectric waveguides. We have discussed the advantages of this approach in the context of other optical manipulation techniques and commonly used microfluidic transport methods such as pressure drive flow and electrokinetics. An experimental and theoretical overview of the technology has been developed with sufficient detail that most readers should be able to replicate the experiments (developing their own implementations) and perform equivalently detailed simulations. Although numerous technological advantages of optofluidic transport have been alluded to, we focused here on comparing the advantages of "optofluidic chromatography" with the state of the art.

References

1. M.A. Unger, H.P. Chou, T. Thorsen, A. Scherer, and S.R. Quake, "Monolithic microfabricated valves and pumps by multilayer soft lithography," *Science*, **288**(5463), (2000), 113–116.
2. R.J. Hunter, *Zeta Potential in Colloid Science: Principles and Applications*, Academic Press, (1981), London.
3. J. Lyklema, *Fundamentals of Interface and Colloid Science, Vol. 1*, Academic Press, (1991), London.
4. J. Lyklema, *Fundamentals of Interface and Colloid Science, Vol. 2, Solid Liquid Interfaces*, Academic Press (1995), London.
5. D. Li, *Electrokinetics in Microfluidics*, Elsevier Academic, (2004), Boston.
6. N. Krishnan, N. Agrawal, M.A. Burns, and V.M. Ugaz, "Reactions and fluidics in miniaturized natural convection systems," *Analytical Chemistry*, **76**(21), (2004), 6254–6265.
7. Y. Xiang and H.H. Bau, "Complex magnetohydrodynamic low-Reynolds-number flows," *Physical Review E*, **68**(1), (2003).
8. J. Zeng and T. Korsmeyer, "Principles of droplet electrohydrodynamics for lab-on-a-chip," *Lab on a Chip*, **4**(4), (2004), 265–277.
9. A.A. Darhuber, J.P. Valentino, S.M. Troian, and S. Wagner, "Thermocapillary actuation of droplets on chemically patterned surfaces by programmable microheater arrays," *Journal of Microelectromechanical Systems*, **12**(6), (2003), 873–879.
10. H.A. Stone, A.D. Stroock, and A. Ajdari, "Engineering flows in small devices: microfluidics toward a lab-on-a-chip," *Annual Review of Fluid Mechanics*, **36**, (2004), 381–411.
11. N. Nguyen and S. Wereley, *Fundamentals and Applications of Microfluidics*, Artech House, (2002), New York.
12. S. Ghosal, "Electrokinetic flow and dispersion in capillary electrophoresis," *Annual Review of Fluid Mechanics*, **38**, (2006), 309–338.
13. D. Erickson, D. Sinton, and D.Q. Li, "Joule heating and heat transfer in poly (dimethylsiloxane) microfluidic systems," *Lab on a Chip*, **3**(3), (2003), 141–149.
14. D. G. Grier, "A revolution in optical manipulation," *Nature*, **424**(6950), (2003), 810–816.
15. A. Ashkin, "Acceleration and trapping of particles by radiation pressure," *Physical Review Letters*, **24**(4), (1970), 156–159.

16. A. Ashkin, J. M. Dziedzic, J. E. Bjorkholm, and S. Chu, "Observation of a single-beam gradient force optical trap for dielectric particles," *Optics Letters*, **11**(5), (1986), 288–290.
17. M.M. Wang, E. Tu, D.E. Raymond, J.M. Yang, H.C. Zhang, N. Hagen, B. Dees, E.M. Mercer, A.H. Forster, I. Kariv, P.J. Marchand, and W.F. Butler, "Microfluidic sorting of mammalian cells by optical force switching," *Nature Biotechnology*, **23**(1), (2005), 83–87.
18. J.E. Curtis, B.A. Koss, and D.G. Grier, "Dynamic holographic optical tweezers," *Optics Communications*, **207**(1–6), (2002), 169–175.
19. S.L. Neale, M.P. MacDonald, K. Dholakia, and T.F. Krauss, "All-optical control of microfluidic components using form birefringence," *Nature Materials*, **4**(7), (2005), 530–533.
20. P.Y. Chiou, A.T. Ohta, and M.C. Wu, "Massively parallel manipulation of single cells and microparticles using optical images," *Nature*, **436**(7049), (2005), 370–372.
21. K. Svoboda and S.M. Block, "Optical trapping of metallic Rayleigh particles," *Optics Letters*, **19**(13), (1994), 930–932.
22. K.C. Neuman and S.M. Block, "Optical trapping," *Review of Scientific Instruments*, **75**(9), (2004), 2787–2809.
23. A. Ashkin and J.P. Gordon, "Stability of radiation-pressure particle traps—an optical earnshaw theorem," *Optics Letters*, **8**(10), (1983), 511–513.
24. M.P. MacDonald, G.C. Spalding, and K. Dholakia, "Microfluidic sorting in an optical lattice," *Nature*, **426**(6965), (2003), 421–424.
25. T. Hatano, T. Kaneta, and T. Imasaka, "Application of optical chromatography to immunoassay," *Analytical Chemistry*, **69**(14), (1997), 2711–2715.
26. T. Imasaka, "Optical chromatography. A new tool for separation of particles," *Analysis*, **26**(5), (1998), M53–M55.
27. J. Makihara, T. Kaneta, and T. Imasaka, "Optical chromatography: size determination by eluting particles," *Talanta*, **48**(3), (1999), 551–557.
28. T. Kaneta, Y. Ishidzu, N. Mishima, and T. Imasaka, "Theory of optical chromatography," *Analytical Chemistry*, **69**(14), (1997), 2701–2710.
29. B.S. Zhao, Y.M. Koo, and D.S. Chung, "Separations based on the mechanical forces of light," *Analytica Chimica Acta*, **556**(1), (2006), 97–103.
30. S.J. Hart and A.V. Terray, "Refractive index-driven separation of colloidal polymer particles using optical chromatography," *Applied Physics Letters*, **83**(25), (2003), 5316–5318.
31. S.J. Hart, A. Terray, K.L. Kuhn, J. Arnold, and T.A. Leski, "Optical chromatography of biological particles," *American Laboratory*, **36**(24), (2004), 13–17.
32. A. Terray, J. Arnold, and S.J. Hart, "Enhanced optical chromatography in a PDMS microfluidic system," *Optics Express*, **13**(25), (2005), 10406–10415.
33. S.J. Hart, A. Terray, T.A. Leski, J. Arnold, and R. Stroud, "Discovery of a significant optical chromatographic difference between spores of Bacillus anthracis and its close relative, Bacillus thuringiensis," *Analytical Chemistry*, **78**(9), (2006), 3221–3225.
34. M. Born and E. Wolf, *Principles of Optics*, 6th ed, Pergamon Press Ltd., (2003), London.
35. S.M. Mansfield, W.R. Studenmund, G.S. Kino, and K. Osato, "High-numerical-aperture lens system for optical storage," *Optics Letters*, **18**(4), (1993), 305–307.
36. J. Guerra, D. Vezenov, P. Sullivan, W. Haimberger, and L. Thulin, "Near-field optical recording without low-flying heads: integral near-field optical (INFO) media," *Japanese Journal of Applied Physics Part 1-Regular Papers Short Notes & Review Papers*, **41**(3B), (2002), 1866–1875.
37. I. Ichimura, S. Hayashi, and G.S. Kino, "High-density optical recording using a solid immersion lens," *Applied Optics*, **36**(19), (1997), 4339–4348.
38. K. Dholakia and P.J. Reece, "Near-field optical micromanipulation," in *Structured Light and Its Applications*, D.L. Andrews (ed.), Academic Press (2008), Amsterdam, 107–138.

39. M. Righini, A.S. Zelenina, C. Girard, and R. Quidant, "Parallel and selective trapping in a patterned plasmonic landscape," *Nature Physics*, **3**(7), (2007), 477–480.
40. R.F. Marchington, M. Mazilu, S. Kuriakose, V. Garces-Chavez, P.J. Reece, T.F. Krauss, M. Gu, and K. Dholakia, "Optical deflection and sorting of microparticles in a near-field optical geometry," *Optics Express*, **16**(6), (2008), 3712–3726.
41. M. Gu, J.B. Haumonte, Y. Micheau, J.W.M. Chon, and X.S. Gan, "Laser trapping and manipulation under focused evanescent wave illumination," *Applied Physics Letters*, **84**(21), (2004), 4236–4238.
42. T. Cizmar, M. Siler, M. Sery, P. Zemanek, V. Garces-Chavez, and K. Dholakia, "Optical sorting and detection of submicrometer objects in a motional standing wave," *Physical Review B*, **74**(3), (2006).
43. A.N. Grigorenko, N.W. Roberts, M.R. Dickinson, and ZhangY, "Nanometric optical tweezers based on nanostructured substrates," *Nature Photonics*, **2**(6), (2008), 365–370.
44. V. Garces-Chavez, R. Quidant, P.J. Reece, G. Badenes, L. Torner, and K. Dholakia, "Extended organization of colloidal microparticles by surface plasmon polariton excitation," *Physical Review B*, **73**(8), (2006).
45. P. Prasad, *Nanophotonics*, John Wiley and Sons, Inc., (2004), Hoboken.
46. B. Saleh and M. Teich, *Fundamentals of Photonics*, John Wiley and Sons, Inc., (1991), New York.
47. C. Pollock and M. Lipson, *Integrated Photonics*, Kluwer, (2003), Norwell.
48. L.N. Ng, B.J. Luf, M.N. Zervas, and J.S. Wilkinson, "Forces on a Rayleigh particle in the cover region of a planar waveguide," *Journal of Lightwave Technology*, **18**(3), (2000), 388–400.
49. B.S. Schmidt, A.H.J. Yang, D. Erickson, and M. Lipson, "Optofluidic trapping and transport on solid core waveguides within a microfluidic device," *Optics Express*, **15**(22), (2007), 14322–14334.
50. S. Kawata and T. Sugiura, "Movement of micrometer-sized particles in the evanescent field of a laser-beam," *Optics Letters*, **17**(11), (1992), 772–774.
51. T. Tanaka and S. Yamamoto, "Optically induced propulsion of small particles in an evenescent field of higher propagation mode in a multimode, channeled waveguide," *Applied Physics Letters*, **77**(20), (2000), 3131–3133.
52. S. Gaugiran, S. Getin, J. M. Fedeli, G. Colas, A. Fuchs, F. Chatelain, and J. Derouard, "Optical manipulation of microparticles and cells on silicon nitride waveguides," *Optics Express*, **13**(18), (2005), 6956–6963.
53. L.N. Ng, B.J. Luff, M.N. Zervas, and J.S. Wilkinson, "Propulsion of gold nanoparticles on optical waveguides," *Optics Communications*, **208**(1–3), (2002), 117–124.
54. K. Grujic, O.G. Helleso, J.P. Hole, and J.S. Wilkinson, "Sorting of polystyrene microspheres using a Y-branched optical waveguide," *Optics Express*, **13**(1), (2005), 1–7.
55. S. Mandal and D. Erickson, "Optofluidic transport in liquid core waveguiding structures," *Applied Physics Letters*, **90**, (2007), 184103.
56. P. Measor, S. Kuehn, E.J. Lunt, B.S. Phillips, A.R. Hawkins, and H. Schmidt, "Hollow-core waveguide characterization by optically induced particle transport," *Optics Letters*, **33**(7), (2008), 672–674.
57. D.C. Duffy, J.C. McDonald, O.J.A. Schueller, and G.M. Whitesides, "Rapid prototyping of microfluidic systems in poly(dimethylsiloxane)," *Analytical Chemistry*, **70**(23), (1998), 4974–4984.
58. J.C. McDonald, D.C. Duffy, J.R. Anderson, D.T. Chiu, H.K. Wu, O.J.A. Schueller, and G.M. Whitesides, "Fabrication of microfluidic systems in poly (dimethylsiloxane)," *Electrophoresis*, **21**(1), (2000), 27–40.
59. Y. Harada and T. Asakura, "Radiation forces on a dielectric sphere in the Rayleigh scattering regime," *Optics Communications*, **124**(5–6), (1996), 529–541.
60. H.Y. Jaising and O.G. Helleso, "Radiation forces on a Mie particle in the evanescent field of an optical waveguide," *Optics Communications*, **246**(4–6), (2005), 373–383.

61. E. Almaas and I. Brevik, "Radiation forces on a micrometer-sized sphere in an evanescent field," *Journal of the Optical Society of America B-Optical Physics*, **12**(12), (1995), 2429–2438.
62. K. Svoboda and S.M. Block, "Biological applications of optical forces," *Annual Review of Biophysics and Biomolecular Structure*, **23**, (1994), 247–285.
63. S. Gaugiran, S. Getin, J.M. Fedeli, and J. Derouard, "Polarization and particle size dependence of radiative forces on small metallic particles in evanescent optical fields. Evidences for either repulsive or attractive gradient forces," *Optics Express*, **15**(13), (2007), 8146–8156.
64. J.D. Jackson, *Classical Electrodynamics*, 2d ed, John Wiley and Sons, Inc., (1975), New York.
65. R.L. Panton, *Incompressible Flow*, 3d ed, John Wiley and Sons, Inc.,. (2005), Hoboken.
66. J. Happel and H. Brenner, *Low Reynolds Number Hydrodynamics, with Special Applications to Particulate Media*, 2d rev. ed, Noordhoff International Publishing, (1973), Leyden.
67. M. Mishchenko, L. Travis, and A. Lacis, *Scattering, Absorption, and Emission of Light by Small Particles*, Cambridge University Press (2002).
68. I. Brevik, T.A. Sivertsen, and E. Almaas, "Radiation forces on an absorbing micrometer-sized sphere in an evanescent field," *Journal of the Optical Society of America B-Optical Physics*, **20**(8), (2003), 1739–1749.
69. A.J.H. Yang and D. Erickson, "Stability analysis of optofluidic transport on solid-core waveguiding structures," *Nanotechnology*, **19**, (2008), 045704.
70. L.N. Ng, M.N. Zervas, J.S. Wilkinson, and B.J. Luff, "Manipulation of colloidal gold nanoparticles in the evanescent field of a channel waveguide," *Applied Physics Letters*, **76**(15), (2000), 1993–1995.
71. D. Erickson, "Towards numerical prototyping of labs-on-chip: modeling for integrated microfluidic devices," *Microfluidics and Nanofluidics*, **1**(4), (2005), 301–318.

CHAPTER 6

Optofluidic Colloidal Photonic Crystals

Seung-Man Yang, Shin-Hyun Kim, and Seung-Kon Lee

National Creative Research Initiative Center for Integrated Optofluidic Systems and Department of Chemical and Biomolecular Engineering, KAIST, Daejeon, Republic of Korea

Colloidal crystals have been extensively studied during the last two decades because their spatial regularity at the half-wavelength scale of interacting light induces a photonic bandgap. The bandgap properties appear as iridescent colors that are useful for many applications, such as optical waveguides, lasing resonators, and structural color pigments. However, tailoring colloidal crystals into structures of desired shape (with good physical and chemical resistance) is difficult, and controlling the bandgap position of the crystal in real time requires complicated fabrication processes. In spite of these limitations, there has been a recent breakthrough in colloidal-crystal research as such crystals can now be incorporated into optofluidic systems. Colloidal crystals embedded in microfluidic channels show tunable bandgaps that depend on the type of fluid flowing through the crystal interstices and may have important applications as sensing devices for biomolecules or chemicals. In addition, the optofluidic system allows the colloids to assemble into discrete and separated colloidal photonic crystals with desired shapes. In this chapter, we will discuss two main categories of colloidal-crystal-based optofluidic systems. The first is related to the integration of colloidal photonic crystals into microfluidic devices for optofluidic applications whereas the second involves the optofluidic synthesis of colloidal photonic crystals.

6-1 Introduction to Colloidal Crystals

6-1-1 Colloids and Colloidal Photonic Crystals

The term "colloid"—which means "glue" in Greek—was first used by Thomas Graham in 1861 to describe materials that could not penetrate through a membrane. Today, the word "colloid" is used to denote particulates, with sizes between 1 nm and 10 μm, dispersed in an immiscible continuous phase [1]. Therefore, an unlimited number of materials—ranging from sand and clay to micelles and carbon black—can be classified as colloids. The dispersion stabilities and rheological properties of such materials have been widely studied during the last two centuries. Recent advances in colloidal synthesis have accelerated the study of colloids—not only for their monodispersity, but also because many properties of the colloidal particles, including density, surface charge, and material affinity, can be controlled by varying the synthetic scheme. Even the design of particles with anisotropic shapes, internal structures, or chemical patterns can be achieved [2]. Based on colloidal particles with controlled properties, the crystallization into various lattices has been studied for two main applications, namely, the attainment of "visible" models for atomic or molecular assemblies and the development of photonic bandgap materials. Monodisperse colloidal particles with high surface charge density dispersed in a polar medium spontaneously form non-close-packed crystals, a process that is induced by the repulsive interparticle potential. Depending on the volume fraction of the colloids and the strength of the repulsion, the particles appear as either face-centered cubic (fcc) or body-centered cubic (bcc) structures in the thermodynamic equilibrium [3,4]. On the other hand, bidisperse colloidal systems with oppositely charged colloids enable the preparation of various crystal lattices, which have many similarities with atomic or molecular systems, although the valences of atoms are not consistent with those of colloidal systems [5,6]. In addition to these similarities in regard to the formation of crystals, bandgap properties are also observed in both atomic and colloidal crystals. At the atomic scale, because crystals exhibit a periodic modulation of the potential for the propagation of electrons, they may affect the conductivity of the electrons and sometimes even prevent their propagation at certain energy levels. It is well known that semiconductors have an electronic bandgap between the valence and conduction bands. Analogously, if the periodicity of a colloidal crystal lattice is comparable to the wavelength of light, the lattice will interact with the electromagnetic waves and induce a photonic bandgap. Photons with energy in this gap cannot propagate through the crystal. In this case, the crystal is a "photonic crystal" [7].

6-1-2 Photonic Characteristics of Colloidal Photonic Crystals

In fact, colloidal photonic crystals are a type of photonic crystal. Because colloidal crystallization is relatively simple and inexpensive—and due to the absence of an alternative effective technique to create subwavelength three-dimensional (3D) structures—the crystallization of colloids has become the most promising technique for preparing 3D photonic crystals. Unfortunately, typical colloidal photonic crystals, which are composed of monodisperse silica (with refractive index $n_{\text{silica}} \approx 1.45$) or polymeric particles ($n_{\text{polymer}} \approx 1.5$), do not show bandgaps in all directions—so called complete, full, or omnidirectional bandgaps. Instead, they exhibit "stop" or "pseudo" bands only in certain directions. In fact, sparkling opal gems composed of silica nanobeads packed into fcc lattices are natural photonic crystals. The vivid colors of opals are induced by the reflected lights corresponding to the bandgaps. As a typical example, the fcc lattice of close-packed polystyrene nanospheres ($n_{\text{PS}} = 1.591$) only exhibits a stop band, as is shown in its energy band structure (Fig. 6-1). The unit cell of the fcc structure is presented in Fig. 6-1a, where the gray triangle denotes the (111) plane, the densest hexagonal plane of fcc. In addition, scanning electron microscopy (SEM) images of the cross section and top surface of the crystal are shown in Fig. 6-1b and its inset, respectively, where the crystals are prepared on a glass substrate by evaporation-induced crystallization during vertical deposition. On the other hand, the energy band structure presented in Fig. 6-1c, which was constructed using *MIT photonic-bands* (MPB) for a water-filled fcc structure, shows a stop band (called L-gap) in the Γ L direction normal to the (111) plane [8]. The L-gap is the most useful stop band in colloidal photonic crystals, because the (111) planes of the fcc structures are formed along the wall of the confined geometry during crystallization, and the L-gap has a larger band width compared to other stop bands.

Tuning of the stop-band position can be achieved by infiltrating various fluids with different refractive indices into the interstices of the colloids. As the effective refractive index of the photonic crystal increases, so does the wavelength of the gap. In addition, a decrease in the index contrast leads to a reduction of the gap width. These tendencies are described in the calculation results of Fig. 6-1d, where a crystal with an fcc lattice composed of polystyrene nanospheres ($n_{\text{PS}} = 1.591$) with a diameter of 200 nm was infiltrated with various fluids, going from air ($n = 1$) to water ($n = 1.333$), ethanol ($n = 1.3614$), tetrahydrofuran ($n = 1.4072$), chloroform ($n = 1.4485$), and chlorobenzene ($n = 1.5248$). Since 74% of the space is occupied by the particles in an opaline fcc structure of close-packed spheres, the remaining 26% is index controllable. A rough estimation gives a wavelength shift of approximately 90-nm per unit change of the refractive index at

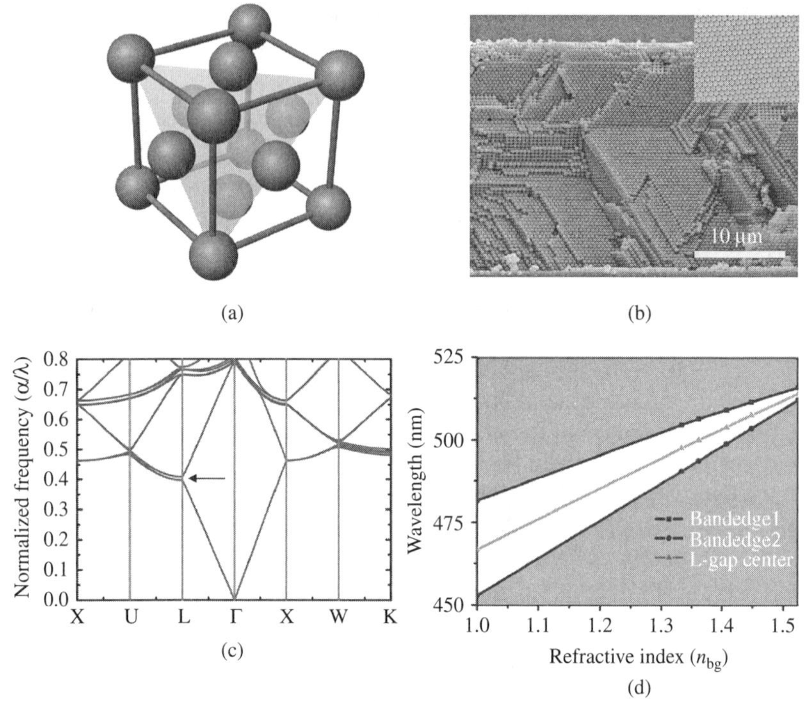

FIGURE 6-1 (a) Unit cell of face-centered cubic (fcc) lattice. The gray triangle denotes the (111) plane of fcc structure. (b) Scanning electron microscope (SEM) image of cross section of opaline colloidal crystal composed of 334-nm polystyrene particles. The inset shows surface of the crystal, (111) plane of fcc which shows hexagonal arrangement of the particles. (c) Energy band structure of water-filled fcc lattice composed of 200-nm polystyrene particles. The arrow denotes L-gap of a stop band. (d) Plot for position of L-gap center and two band edges which depend on the refractive index of fluid.

wavelengths close to 500 nm, although the refractive index of fluids does not vary linearly with the wavelength of the gap. The magnitude of the wavelength shift can be enhanced by increasing the volume fraction of free space through material inversion or etching [9–11]. This tunability of the bandgap—or the bandgap itself—makes the colloidal photonic crystals useful materials for a wide range of applications including displays, cosmetics, sensors, and lasers.

6-2 Integration of Colloidal Photonic Crystals into Microfluidic Systems

6-2-1 Crystallization of Colloids in the Microfluidic Systems

As discussed previously, crystals composed of monodisperse colloids exhibit photonic bandgap properties. It is well known that the characteristics of photonic crystals vary with the optical properties of the environment, for example, with the refractive index or the birefringence. Especially, if one can infiltrate a fluid into the photonic crystal, it will be easy to achieve good addressability and precise control of the photonic properties. Also, the introduction of fluids is important because it can transport readily an optical gain medium, such as a fluorescent dye or quantum dots in microfluidic channels and enable the fabrication of active optofluidic devices. This is the reason why the combination of colloidal crystals and fluids has become such an attractive and important subject in "optofluidics"—the rendezvous between photonics and fluidics.

Colloidal photonic crystals are organized by the self-assembly of building-block particles. However, the self-organization mechanism involves intrinsic problems related to undesirable grain boundaries and defects such as vacancies, faults, and cracks. In addition, controlling the crystal orientation and shaping the colloidal crystals in regular forms bring additional difficulties in practical applications. Therefore, in order to achieve optofluidic devices incorporating colloidal-crystal units, it is imperative to solve all the aforementioned problems.

At the microscale, fluids may exhibit quite different behavior compared to that observed at the macroscale. The Reynolds number ($\rho v D/\mu$) is very small in a microfluidic system, where ρ is the density of the fluid, v the mean fluid velocity, D the characteristic diameter of the capillary, and μ the fluid viscosity. This means that the flow will always show laminar motion. In multicomponent systems, mixing is restricted, because it is driven solely by molecular diffusion. However, if colloidal particles are packed in a microcapillary, the fluid motion changes dramatically. In this case, the fluid must pass through the interstices, whose sizes are much smaller than the particle diameter. Therefore, a large pressure drop—resulting from the capillary force and the large surface area which imposes no slip boundary condition—is required to drive fluid flow. The surface treatment of the colloidal particles represents an important point because it directly affects the surface energy, and finally, the capillary force. Technically, the aforementioned points are deeply related to the colloidal-crystal-based optofluidic systems. In addition, in order to fabricate integrated optofluidic devices with built-in colloidal crystals, we should be able to tailor the colloidal crystals into desired shapes and locate them at particular positions within the microfluidic system. Therefore, we must understand both the nature of colloidal crystallization and the behavior of fluids in small capillaries.

Evaporation-Induced Crystallization

The most basic and simple approach to the crystallization of colloidal particles inside a capillary is evaporation-induced crystallization. If we introduce a suspension of uniform colloidal particles at one end of a capillary, the solvent medium will evaporate at the other end. Therefore, the concentration of the colloidal suspension at the evaporation front will increase due to the convective transport of colloids from the bulk. Hard spheres, which do not interact with the neighboring particles until they touch each other, begin to crystallize from the suspension at a particle volume fraction of 0.494—this ordering is driven purely by entropy. Because the particles tend to occupy the largest free volume possible, they arrange into fcc or hexagonal close-packed (hcp) lattices, which have an identical maximum packing fraction of 0.7404, the highest among all the possible sphere arrangements. (In fact, the fcc structure is the equilibrium state, due to slight differences in energy, but both the fcc and hcp structures can be generated experimentally.) On the other hand, soft spheres show much more complex crystallization behavior, which is influenced by interparticle interactions, such as van der Waals forces, electrostatic potentials, and steric hindrance. Practically, most of the colloidal particles in polar solvents are soft spheres submitted to van der Waals attractions and electrostatic repulsions. Especially, the large electrostatic repulsion—overcoming the attractive forces—gives a high stability to the colloidal suspensions. Both inorganic and polymeric colloidal particles may have surface charges that originate from the functional groups on their surface, and thus can be dispersed in a polar medium without forming aggregates. Surface charges change depending on the pH conditions. The hydroxyl, sulfonic, and carboxylic groups are typical examples of functional groups that give negative charges under basic conditions. Oppositely, the amine group gives positive charges under acidic conditions. When decreasing or increasing the pH value of a solvent by titration, the surface charge will become zero at a certain point (called the isoelectric point). Uniform silica particles can be prepared by a sol–gel procedure, with the Stöber–Fink–Bohn method being the most famous technique [12]. During the sol–gel process, hydroxyl groups are intrinsically formed at the surface of the silica spheres. These hydroxyl groups can be substituted by different functional groups using silane coupling agents [13]. On the other hand, monodisperse polystyrene (PS) particles can be synthesized by emulsion polymerization—with or without surfactant—in polar media [14]. In this case, either a surfactant or linear copolymers with functional end groups form micelles and stabilize the colloids by anchoring on their surface. Functional groups are generated during the synthesis process by surfactants, comonomers, or initiators, which are mostly located at the surface.

Soft spheres with repulsive potentials can be crystallized at much lower concentrations, because their effective size is larger than that of hard spheres with the same particle size. Experimentally, the crystallization of soft spheres can proceed at extremely low concentrations as particle volume fraction of 1%, although the threshold concentration depends on the strength of the potential. Generally, colloids crystallize into the fcc structure at relatively high concentrations for a wide range of repulsive strengths. On the other hand, particles exhibiting very high repulsive strengths assemble into bcc structures at low concentrations (in equilibrium). According to recent reports, binary mixtures of oppositely charged small and large particles exhibit more complex crystal structures, which are generated by the complex action of both attractive and repulsive electrostatic interactions.

Usually, it is very difficult to fabricate uniform colloidal crystals on a large scale. Cracks, defects, and grain boundaries are formed during the crystallization process. The polydispersity of the particle size, the roughness of the surface, the low stability of suspension, and the presence of impurities may be partial reasons for this. Moreover, an inhomogeneous capillary force between the particles during the evaporation of the solvent is the main factor leading to the formation of poor colloidal crystals with cracks or uneven film thicknesses. However, crystallization in a capillary helps to overcome this problem. As mentioned earlier, evaporation occurs at one end of the capillary, and thus nucleation for crystallization also begins at that end due to the increased concentration. In addition, crystal growth is achieved by the supply of particles from the bulk through the convective flow induced by evaporation. In this way, high-quality colloidal crystals gradually grow inside the capillary. The prepared crystal has a low crack density. Moreover, the thickness of the crystal can be kept uniform because the solid wall of the capillary confines the crystallization.

Kim et al. fabricated colloidal crystals using a soft elastomer mold [15]. Soft lithography involves a family of techniques for replicating structures using elastomeric stamps and molds. By using the soft-lithography technique, the researchers could make various soft replica molds with grooves, patterns, and microchannels. After bonding a soft mold on a substrate, a colloidal suspension was injected and crystallized by the evaporation of the solvent at one end. In this case, soft molds with various patterns acted as replica matrices for the colloidal crystals. A scheme of the evaporation-induced crystallization presented in Fig. 6-2a and b shows the replicated colloidal crystal patterns fabricated using the soft molds. In addition, porous hierarchical structures could be prepared using the same soft-lithographic approach [16,17]. Yang et al. prepared colloidal crystals using PS particles, and then their interstices were filled with a titania precursor. Finally, a network of porous inverse opal structures was created after calcination and removal of the polymeric particles.

Also, Moon et al. and Kamp et al. prepared colloidal crystals in a circular glass capillary [18,19]. In these experiments, the silica suspension introduced in the capillary was evaporated at one of its ends, and cylindrical colloidal crystals could be obtained after several hours. In addition, inverse-opal-type porous crystals could be prepared by infiltration of the polymer through the interstices and subsequent particle removal. The colloidal crystals in the cylindrical capillary always have the (111) plane of the fcc structure at their surface. Figure 6-2c and d show SEM images of colloidal crystals

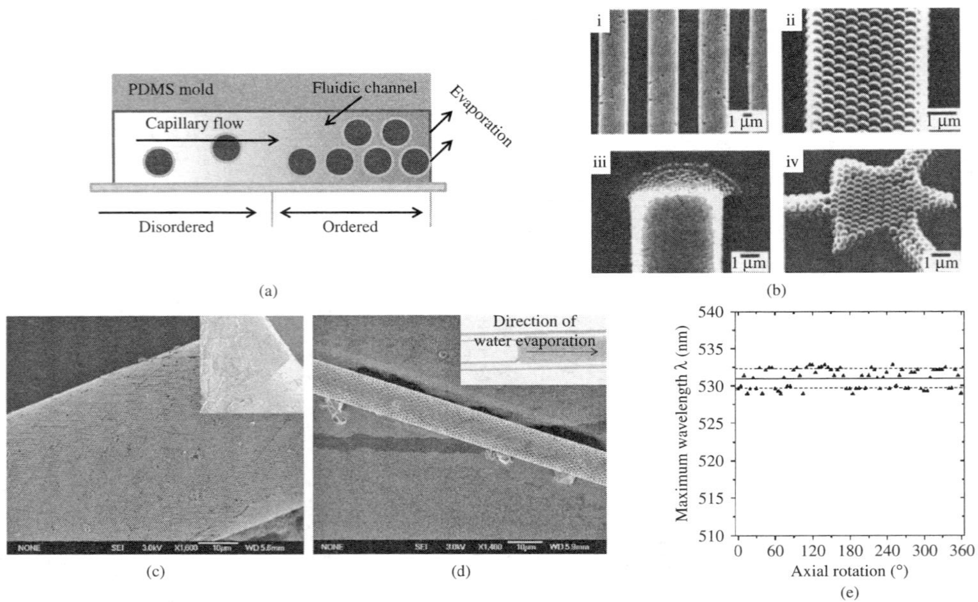

FIGURE 6-2 (a) Schematic illustration of evaporation-induced colloidal crystallization in microchannel with open ends. (b) SEM images of colloidal crystals with various shapes replicated by soft elastomer molds. (*E. Kim, Y. Xia, and G. M. Whitesides, "Two- and three-dimensional crystallization of polymeric microspheres by micromolding in capillaries," Advanced Materials, 8, (1996), 245–247. Copyright Wiley-VCH Verlag GmbH & Co. KGaA. Reproduced with permission*) (c and d) SEM images of cylindrical colloidal crystals with different cylinder diameters. Insets in (c) and (d) show the cross section of cylindrical colloidal crystal and the direction of crystallization in a glass capillary, respectively. (*Reprinted with permission from J. H. Moon, S. Kim, G.-R. Yi, Y.-H. Lee, and S.-M. Yang, "Fabrication of ordered macroporous cylinders by colloidal templating in microcapillaries," Langmuir, 20, (2004), 2033–2035. Copyright 2004 American Chemical Society.*) (e) Variation of reflectance peak positions depending on the rotation of the crystal along the axis of cylinders. (*U. Kamp, V. Kitaev, G. von Freymann, G. A. Ozin, and S. A. Mabury, "Colloidal Crystal Capillary Columns-Towards Optical Chromatography," Advanced Materials, 17, (2005), 438–443. Copyright Wiley-VCH Verlag GmbH & Co. KGaA. Reproduced with permission.)*

prepared inside capillary tubes of different diameters. Because the crystals exhibit a curved (111) plane of the fcc structure through the cylindrical surface, their photonic bandgap properties are independent of the crystal rotation along the axis of the cylinder. From Fig. 6-2e, we can find the invariability of the reflectance spectra taken from cylindrical colloidal crystals with rotation.

The aforementioned study on evaporation-induced crystallization in a capillary is very meaningful because today's microfluidic devices are based largely on soft lithography. In addition, this method offers several advantages, such as high-quality crystallinity and ease of shaping. However, evaporation-induced crystallization also has some intrinsic shortcomings. The principal problem is the long processing time. Since evaporation occurs only at the small exit of the capillary, several hours are required for crystallization to take place over the whole capillary length. Also, it is impossible to fabricate colloidal crystals only at a desired location. This is because the colloids begin to crystallize from the open end of the capillary. Moreover, highly concentrated colloidal suspensions are required because the volume shrinkage of the process is very large.

Centrifugal-Force-Induced Crystallization

Researchers have tried to solve the problems mentioned earlier related to crystallization in capillaries. Lee et al. first succeeded in accelerating the colloidal crystallization in microchannels using a centrifugal force [20]. To do this, they designed a centrifugal microfluidic device containing multiplex microfluidic chips in a disc. While the conventional evaporation process requires several hours for crystallization to occur, it is possible to crystallize the colloids within several minutes using the centrifugal system. The quality and properties of the centrifugally crystallized colloidal crystals are similar to those of the crystals obtained using the evaporation process. Figure 6-3a shows colloidal-crystal patterns incorporated into a microfluidic chip under a centrifugal force field.

When a centrifugal chip rotates, a force balance can be achieved between the centrifugal and capillary forces in the microfluidic channels. The centrifugal force pushes the suspension radially outward whereas the capillary force holds the solvent inside the channels. Colloidal particles dispersed in the stationary suspension precipitate in the radial direction. The capillary force depends on the surface tension, γ, and the hydraulic diameter, D_H, which is usually 4 times the cross-sectional area divided by the wetted perimeter of the channel. The centrifugal and capillary forces are in balance up to a certain radial frequency:

$$\rho \omega r \Delta r = a\left(\frac{4\gamma}{D_H}\right) + b \tag{6-1}$$

Here, ρ is the density of the fluid, r is the radius of rotation, and ω is the angular velocity. In this equation, a and b are constant for a given fluidic system. Above the threshold frequency, the centrifugal force exceeds the capillary force, causing the suspension to burst out from the channels. The threshold of the burst angular velocity ω_c decreases with increasing hydraulic diameter. In the case of the evaporation-induced crystallization, as the solvent medium evaporates, the colloidal particles crystallize, first from the noncontact mode, by electrostatic repulsion. Upon further evaporation of the solvent, the capillary force leads the particles to pack into the contact mode. At this moment, cracks are formed as a result of the volume shrinkage over the whole colloidal-crystal region. Instead of the capillary force, centrifugal sedimentation can stack the colloidal particles into a close-packed state. The centrifugal sedimentation helps to prevent crack formation, even over a scale of several hundred micrometers. During centrifugation, the centrifugal force packs and contacts the colloidal particles before the evaporation of the solvent medium. This medium is fully evaporated after crystallization of the colloids.

The Stokes equation enables us to approximate the sedimentation velocity of colloidal particles by balancing the centrifugal driving force $P_c = (\pi/6)d^3(\rho_p - \rho)C$ and the friction force $F = fv$. Here, d is the particle diameter, ρ_p is the particle density, $f = 3\pi\mu d$ is the friction coefficient for a spherical particle, and $C = (2\pi \text{ RPM})2r/60^2$ is the centrifugal acceleration (where r is the distance between the colloids and the axis of rotation):

$$\frac{v}{C} = \frac{d^2(\rho_p - \rho)}{18\mu} \tag{6-2}$$

Equation (6-2) includes several assumptions, namely, no interparticle interactions, a sufficiently large particle size relative to the solvent molecules, and no disturbances due to convection. From this equation, we notice that the sedimentation speed of the colloids can be enhanced by increasing the particle size, the rotation speed of the disc, the distance from the rotation axis, and the density contrast. Therefore, if the colloidal particles can be dispersed in a low-density medium, it is possible to reduce the processing time. This is particularly important for polymeric particles, which have the small density contrast from the aqueous medium. However, to disperse the particles into a low-density medium, such as ethanol, their surfaces must be modified to provide a high suspension stability; for example, we can disperse PS particles in ethanol through the physical adsorption of polyvinylpyrollidone (PVP) onto the PS beads and prepare crystals composed of the PS particles in a much shorter time than that required with an aqueous suspension. Since different kinds of particles are sequentially injected into the microchannel, on the other hand, it is possible to produce hybrid colloidal crystals composed of several blocks.

Figure 6-3b shows optical and electron microscopy images of sectioned colloidal crystals composed of silica particles of different sizes.

Electrically Addressable Crystallization

Electrowetting is the basic technique for actuating fluids in microfluidic chips. The wettability of a fluid on a substrate varies with the applied electric field. In the absence of an electric field, the fluid becomes stationary inside the hydrophobic microfluidic channels, whereas under a sufficiently large field, the surface tension changes and the fluid wets the surface. Eventually, the fluid is forced to flow inside the microfluidic channel by capillary forces.

Shiu et al. applied the electrowetting method to manipulate the flow of a colloidal suspension [21,22]. By using microfluidic chips with electrode patterns, they were able to induce colloidal crystallization in a desired region. In addition, the sequential crystallization of colloids of different sizes enabled the simple preparation of hybrid colloidal crystals inside microfluidic channels. Unlike the aforementioned centrifugal crystallization, in this case the length of the colloidal crystal blocks can be intentionally controlled by modulating the electrowetting. Figure 6-3c shows sectioned colloidal-crystal patterns composed of three different sizes of colloidal particles. In addition, combined with a multilayer pneumatic valve system, colloidal-crystal patterns with multiple reflection colors could be digitally fabricated in certain fluidic cells. Figure 6-3d shows the designs of the electrodes, the flow channels, and the control channels in a pixellated microfluidic chip. Using this complex microfluidic chip, it is possible to fabricate patterned colloidal-crystal arrays with freely controllable addressability. Figure 6-3e shows an optical image of the pixellated microfluidic chip with colloidal crystals. However, in this case, the complexity of the fabrication process still represents a significant problem. Therefore, the remaining issue should be "how to prepare colloidal crystals at a specific place within microfluidic chips in a simple way?" A possible means to achieve this is the use of photoresponsive colloidal particles [23]. Through proper surface modification, the colloidal particles may exhibit controllable surface properties, which can be modulated by photon irradiation. Therefore, it will be possible to crystallize colloids at desired areas by remotely changing the electrostatic interactions without the need for complex electrodes or valves integrated with the microfluidic chips.

6-2-2 Applications of Integrated Colloidal Photonic Crystals

Chemical and Biological Sensors

Microfluidic systems with built-in colloidal crystals can be applied to develop label-free chemical and biological sensors. As discussed earlier, colloidal crystals have peculiar reflectance properties that

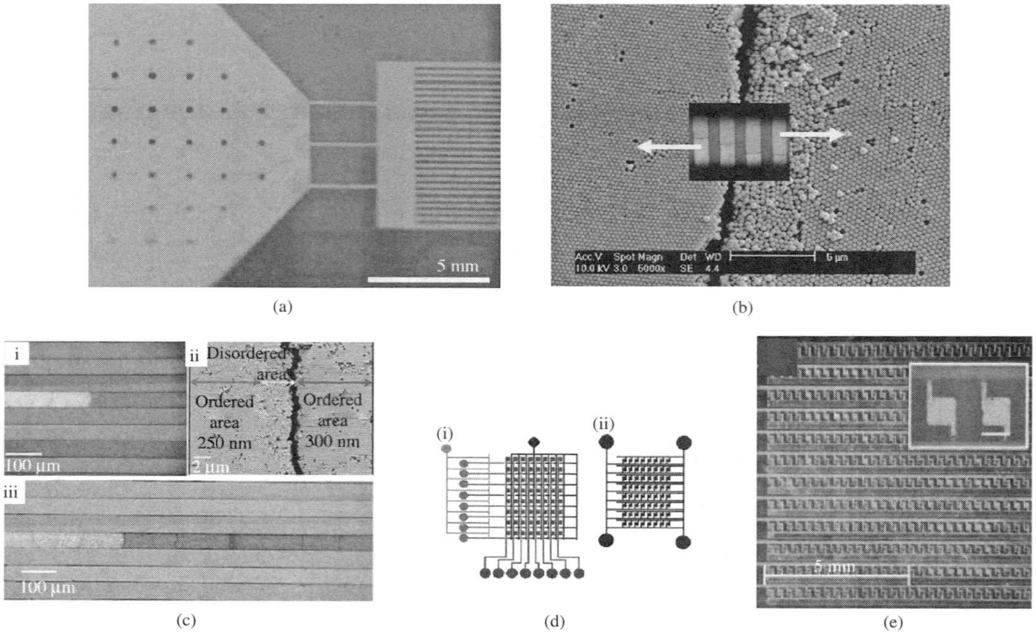

Figure 6-3 (a) Optical image of colloidal crystal pattern of silica particles prepared by the centrifugal microfluidic chips. (*S.-K. Lee, G-R. Yi, and S.-M. Yang, "High-speed fabrication of patterned colloidal photonic structures in centrifugal microfluidic chips," Lab on a Chip, 6, (2006), 1171–1177. Reproduced by permission of the Royal Society of Chemistry.*) (b) Optical and SEM images of sectioned colloidal crystal fabricated by sequential injection of colloidal suspensions with different sizes. (c) Optical and SEM images of sectioned colloidal crystals prepared by using electrowetting of suspensions. (*Reprinted with permission from J.-Y. Shiu, C.-W. Kuo, and P. Chen, "Actively controlled self-assembly of colloidal crystals in microfluidic networks by electrocapillary forces," Journal of the American Chemical Society, 126, (2004), 8096–8097. Copyright 2004 American Chemical Society.*) (d) Design of the electrodes, valves, and flow channels for the fabrication of pixellated microfluidic chip. (e) Pixellated microfluidic chip with colloidal crystals. (*J.-Y. Shiu and P. Chen, "Active patterning using an addressable microfluidic network," Advanced Materials, 17, (2005), 1866–1869. Copyright Wiley-VCH Verlag GmbH & Co. KGaA. Reproduced with permission.*)

originate from the photonic bandgaps, whose frequencies (or wavelengths) are determined by the effective refractive index or the lattice constant for a given crystal structure. In the case of close-packed crystals, which are prepared by evaporation or centrifugal-force-induced crystallization, the lattice constant cannot be changed because it is determined solely by the size of the colloidal building blocks. However, the refractive index can be easily controlled by introducing a fluid medium through the colloidal crystals. Kamp et al. reported optical chromatography results of alkanes that exhibited only small differences in their refractive indices [19]. Through a glass capillary packed with colloidal crystals—similar to a chemical chromatography column—they introduced liquid alkanes, such as octane, nonane, and decane. The position of the reflectance peak shifted to longer wavelengths upon reflecting the index changes of the interstices. Even though the refractive indices of the alkanes were distributed within a very small range (i.e., between 1.398 and 1.411), they could all be identified by the shift of the reflectance peak. Figure 6-4a shows the optical chromatography results of the alkanes.

Incorporation of colloidal crystals into a microfluidic chip leads to an enhanced performance compared to that of the crystals in a simple capillary. By using the microfluidics technology, it is possible to manipulate the flow characteristics of the fluid medium. In addition, a flow of single or multiple components can be introduced into the crystals. As shown previously, Lee et al. fabricated sectioned colloidal-crystal columns in a centrifugal microfluidic chip [20]. Using their technique, it is possible to create colloidal crystals composed of two different materials. Because the reflectance intensity of colloidal photonic crystals strongly depends on the refractive index contrast for crystals with a finite number of layers, it is difficult to obtain appreciable reflectance spectra for colloid/fluid combinations with very similar refractive indices. In Fig. 6-4b, the reflectance spectrum of a silica colloidal crystal is represented by solid lines. The spectra of the silica colloidal crystals containing various infiltrated fluids, such as water, hexadecane, decaline, ethanol, and isopropanol (IPA), are also shown. Among them, the silica colloidal crystal containing ethanol and IPA only show unclear reflectance signals due to their small refractive index contrast. Hybrid colloidal crystals composed of two materials with different refractive indices could solve this problem. Using PS colloids, which have a higher refractive index than silica, it is possible to compensate the blind region of the silica colloidal crystals. The reflectance spectra of PS colloidal crystals (dotted lines) are characterized by a significant signal, even for ethanol and IPA. Therefore, for a hybrid colloidal crystal composed of silica and PS crystal blocks, each complements the blind regions of the other.

Colloidal-Crystal Lasers

Many applications of colloidal crystals are based on the selectively reflected light (i.e., L-gap) induced by Bragg diffraction from the (111) plane of the fcc structure. On the other hand, if light emitters exist in the colloidal crystals, the situation becomes totally different because a periodic modulation of the refractive index in space can induce a different electromagnetic density of states within the crystals. According to Fermi's golden rule, the transition rate from one energy eigenstate to another is proportional to the density of the final states. Therefore, emitters, such as dye molecules or quantum dots embedded in the colloidal crystals, can show emission spectra that are quite different from those of the bulk. The density of states is low at the bandgap but high at the band edge or at defect modes. Therefore, spontaneous emission from the emitter in colloidal crystals is inhibited at the stop band but enhanced at the band edge. Especially, emission at the band edge can be stimulated, thus inducing lasing due to the high density of states and the low group velocity at the band edge. Shkunov et al. reported lasing phenomena at the band edge of a silica opal immersed in a dye solution (Fig. 6-4c and d) [24]. The silica opal was prepared slowly by sedimentation of silica particles. Due to the long attenuation length of the Bragg diffraction at small index contrasts (of the order of 10^{-2}), the thickness of the crystal should be on the millimeter scale. Also, Furumi et al. reported lasing at the defect mode of colloidal photonic crystals. Here, dye molecule-incorporated thin film is sandwiched with polydimethylsiloxane (PDMS)-infiltrated PS colloidal crystals [25].

These colloidal crystals for lasing applications can be incorporated into microfluidic devices. If the dye solution flows though the interstices of the colloidal crystal in the channel, the problem of dye bleaching, which is the major drawback of solid-state dye lasers, can be solved. In addition, tuning of the lasing wavelength can be achieved by changing the refractive index of the dye-laden fluid. More importantly, the emitted lasers can be directly used as light sources in optofluidics or micro-TAS (total analysis systems) applications by integration with other components.

6-3 Optofluidic Synthesis of Spherical Photonic Crystals

Colloidal crystals integrated in microfluidic chips exhibit an fcc structure with the (111) plane facing the wider walls of the microchannel. Thus, in a conventional rectangular channel, the (111) surfaces are aligned along the horizontal planes. According to Bragg's law, the angle of the incident light on the stacked planes determines the optical path length, and thus, the wavelength of constructive interference. Therefore, the colloidal crystals produced in the channel exhibit optical anisotropy. On the other hand, if spherical emulsion

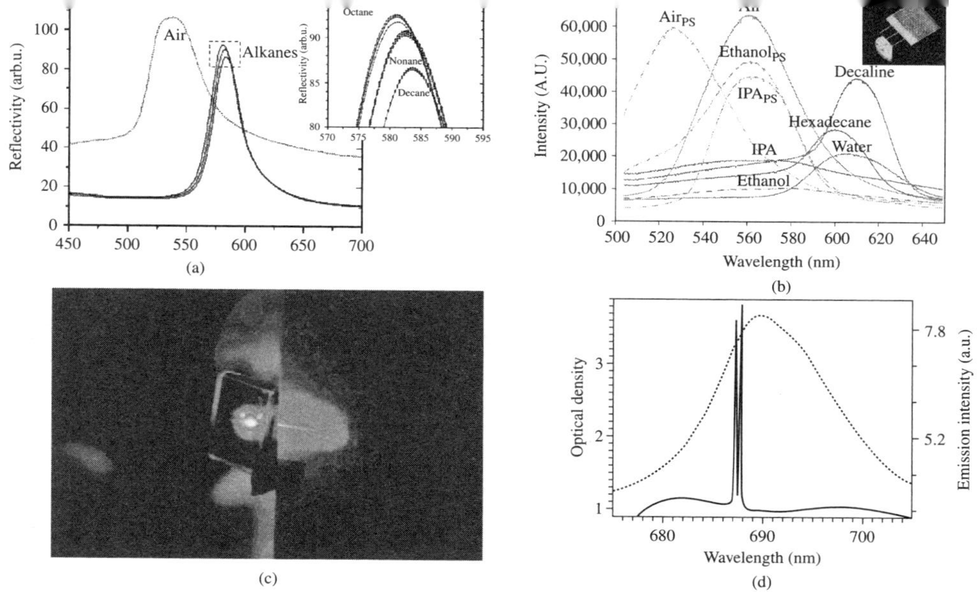

Figure 6-4 (a) The reflectance spectra showing optical chromatography results collected from the capillary with built-in colloidal crystal which is infiltrated with different types of alkanes (*U. Kamp, V. Kitaev, G. von Freymann, G. A. Ozin, and S. A. Mabury, "Colloidal Crystal Capillary Columns-Towards Optical Chromatography," Advanced Materials, 17, (2005), 438–443. Copyright Wiley-VCH Verlag GmbH & Co. KGaA. Reproduced with permission*). (b) The reflectance spectra from hybrid colloidal crystals composed of silica and PS particles. Here, the colloidal crystals were prepared in the centrifugal microfluidic chips. Solid and dotted lines represent the reflectance spectra of silica and PS colloidal crystals, respectively. (*S.-K. Lee, G-R. Yi, and S.-M. Yang, "High-speed fabrication of patterned colloidal photonic structures in centrifugal microfluidic chips," Lab on a Chip, 6, (2006), 1171–1177. Reproduced by permission of the Royal society of chemistry.*) (c) Optical image of laser emission at colloidal crystals immersed in dye solution. (d) Optical density (dotted line) of colloidal crystals and laser emission spectra at the band edge of lower wavelength. (*M. N. Shkunov, Z. V. Vardeny, M. C. DeLong, R. C. Polson, A. A. Zakhidov, and R. H. Baughman, "Tunable, gap-state lasing in switchable directions for opal photonic crystals," Advanced Functional Materials, 12, (2002), 21–26. Copyright Wiley-VCH Verlag GmbH & Co. KGaA. Reproduced with permission.*) (See also color insert.)

droplets are used as the confining geometry for crystallization, the (111) plane will be formed over the spherical interface [26–28]. This is because heterogeneous nucleation on the smooth wall of the confining geometry induces crystallization. If these crystals are generated, they will exhibit isotropic optical properties, that is, the rotation of the crystal will not influence the reflection wavelength. However, the (111) plane, which is characterized by a hexagonal arrangement of the colloids, cannot form a defect-free spherical surface because—as one may imagine—a perfect hexagonal arrangement of spheres can only fill the planes with zero Gaussian curvature. To inspect the generation of defects at the spherical interface, emulsion droplets containing anchored PS particles with a repulsive interparticle potential were used as a model system by Blausch et al. [29] As we can notice from the classical Euler formula ($V - E + F = 2$; where V, E, and F are the numbers of vertices, edges, and faces, respectively), closed systems will always have +12 defect charges if we count the charges as: $\cdots$, +2, +1, 0, −1, −2, $\cdots$ for $\cdots$, four, five, six, seven, eight, $\cdots$-fold particles, respectively. Emulsion model systems have shown that defect generation is related to the size of the system. If $R/l < 5$ (where R is the radius of the emulsion drop and l is the interparticle distance), only twelve fivefold particles will be observed. However, larger systems show excess dislocations, which appear as repetitive five- and sevenfold defects keeping the constraint of total charge of +12. The number of excess dislocations increases with the size of the system. On the other hand, particles in extremely large systems cannot feel the curvature of the emulsion droplet. Therefore, when particles are confined in a very larger droplet, they assemble spontaneously into fcc structures from the interface—similar to what happens in the case of crystallization in a rectangular confined geometry. Thus, a layered structure of concentric shells composed of hexagonal arrangements of colloids is formed.

In the following two sections, we will discuss the optofluidic synthesis of spherical colloidal crystals (called photonic balls) in the solid and liquid states using photocurable single- and double-emulsion droplets, respectively. The optofluidic scheme represents a simple and high-throughput technique for generating photonic balls.

6-3-1 Direct Synthesis of Photonic Balls in the Solid State

Emulsion droplets are useful templates for producing spherical materials. Especially, if they contain monodisperse colloidal particles, spherical colloidal crystals can be produced by diffusion-induced consolidation of the emulsion phase. However, the volume shrinkage of the emulsion takes a long time and requires complicated conditions. Kim et al. reported an in situ method for producing photonic balls, without the need of a diffusion process, using optofluidic devices composed of a monodisperse emulsion generator and a UV exposure unit [30,31].

If the colloidal particles confined in an emulsion droplet interact with each other (with a repulsive potential), they can form spherical non-close-packed crystals, thereby retaining the droplet volume. Especially, if the emulsion phase is photocurable, the spherical crystals can be captured by UV irradiation within a second. To achieve this, silica particles dispersed in a highly polar photocurable resin with a similar refractive index are used. Due to the diminishing van der Waals attractions by index matching, the repulsive potential dominates. This can be induced by either solvation films generated on the particle surface or electrostatics; thus, the particles adopt a polycrystalline form in suspension [32,33]. To generate monodisperse emulsion droplets, microfluidic devices composed of coaxial inner and outer glass capillaries are prepared. As inner and outer flows, the silica suspension and a surfactant-loaded aqueous solution are introduced using syringe pumps. If the outer flow is faster than the inner one, the monodisperse suspension droplets are generated in the dripping regime, without jetting. In this regime, the size and generation frequency of the droplets are determined by the outer and inner flow rates, respectively. Because the drag force by the outer flow and the capillary force by the inner tip are balanced at every moment during droplet generation, the size of the droplet can be estimated using the following equation:

$$3\pi\mu(d_{\text{drop}} - d_{\text{tip}})(v_{\text{outer}} - v_{\text{inner}}) \sim \pi d_{\text{tip}}\gamma \tag{6-3}$$

in which the drag force given by the Stokes equation is modified due to screening by the inner capillary. Here, μ, d_{drop}, d_{tip}, v_{outer}, v_{inner}, and γ are the viscosity of the continuous phase, the droplet diameter, the inner-tip diameter, the velocities of the outer and inner flows, and the interfacial tension, respectively [34].

As shown in Fig. 6-5a, the generated droplets are photopolymerized downstream by UV irradiation. The solidified emulsion droplets show Bragg diffraction colors which depend on the particle diameter and the volume fraction. Because the repulsive potential induces crystallization, the volume fraction of particles determines the lattice constant. The wavelength of the reflection color for normal-incident light on the (111) plane can be estimated by Bragg's law:

$$\lambda = \left(\frac{\pi}{3\sqrt{2}\phi}\right)^{1/3}\left(\frac{8}{3}\right)^{1/2} d\left(n_p{}^2\phi + n_m{}^2(1-\phi)\right)^{1/2} \tag{6-4}$$

This equation assumes a constant interparticle distance, with all the nearest neighbors within a given volume fraction and d, ϕ, n_m, and n_p being the diameter and volume fraction of the particles, and the refractive indices of the matrix and the particle, respectively. Figure 6-5b shows an arrangement of silica beads confined in droplets with layered structures. Silica particles with a diameter of 1 μm are used for visualization through an optical microscope. Also, optical microscopy images of blue, green, and red photonic balls and their reflection spectra are shown in Fig. 6-5c and d, respectively, where the positions of the reflection peaks are matched with the wavelength estimated from Eq. (6-4).

Moreover, using a modified device containing a pair of microcapillaries as the inner channel, it is possible to generate photonic Janus balls, as schematically shown in Fig. 6-5e. When differently colored suspensions are forced to flow through the paired inner capillaries, coalesced droplets with two hemispherical domains are generated at the end of the paired capillaries. Similar to the method described previously, the Janus droplets are solidified by UV irradiation and collected at the end of the outer capillary. The resulting balls show two distinctive reflection colors from their own hemispherical domains, as shown in Fig. 6-5f. Here, the two differently colored suspensions are composed of silica particles of different sizes at the same volume fraction. This enables a matching of the viscosities of both suspensions, thus providing stable Janus drop-forming conditions. Especially, the extremely high viscosity of the suspension compared to that of the continuous phase (i.e., the aqueous solution) prevents twin recirculatory flows in the droplets, which can induce severe mixing effects between the two suspensions, even for a slightly misaligned system. In addition, the diffusive mixing of colloids is also inhibited due to the high viscosity and the repulsive interparticle interactions. Without considering the interparticle potential at dilute suspension, the Stokes-Einstein law ($D_0 = kT/3\pi\mu d$) gives a very small diffusion coefficient of O (10^{-14} m^2/s) for 200-nm silica particles dispersed in a highly viscous photocurable resin at room temperature. This value indicates that the diffusion length of the particle is comparable to the particle size during 1 s, which is the time interval between coalesced droplet generation and UV irradiation. In the case of the repulsive interparticle potential in a concentrated suspension, the diffusion coefficient is close to zero because the structure is interlocked by the repulsive interparticle potential.

Photonic balls prepared by the optofluidic scheme are useful in many applications. The rotation-independent reflection colors and the wide reflection angles of the photonic balls enable their use as reflection color pigments for microdisplays in the reflection mode. Especially, Janus balls—which exhibit not only optical but also electrical anisotropies—can be used in full color E-paper with rotating balls (called Gyricon displays) [35]. In addition, photonic balls can be used as decorative color pigments in the cosmetic or jewelry industry.

6-3-2 Optofluidic Encapsulation of Crystalline Colloidal Arrays

Microfluidic devices composed of cylindrical glass capillaries have many advantages in comparison with devices based on rectangular

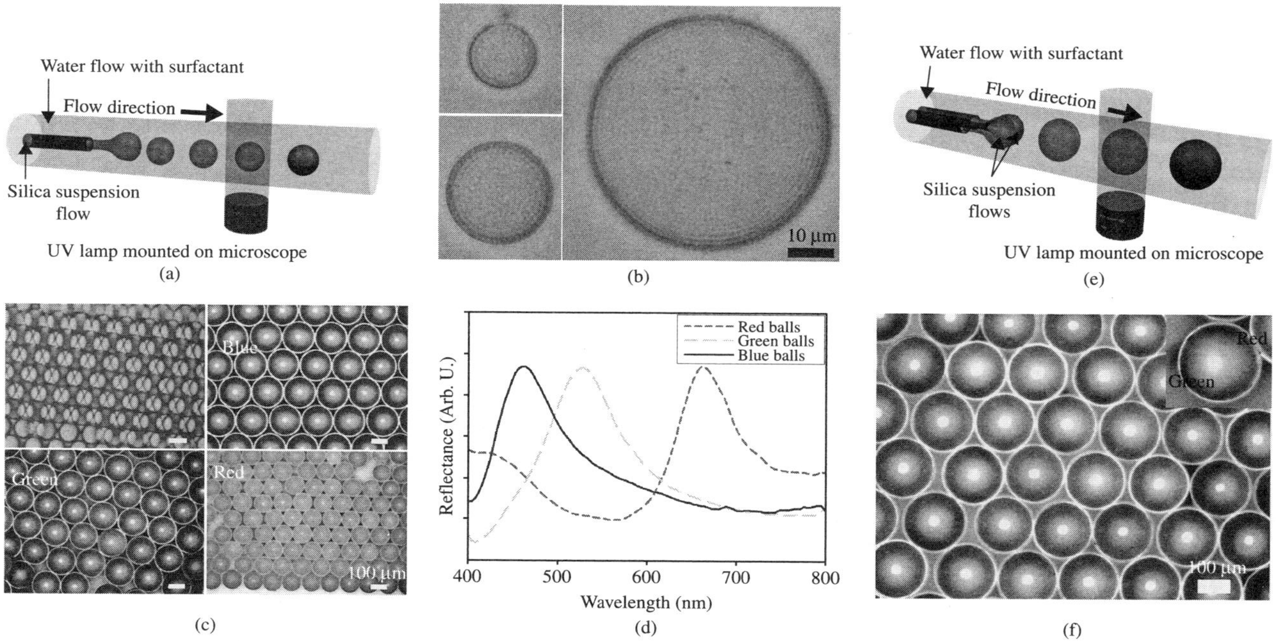

FIGURE 6-5 (a) Schematic illustration of optofluidic device for creating the photonic balls. (b) Optical microscope images of cross sections of emulsion droplets which contain 1-μm silica particles at 33% volume fraction. Layered structure from the outermost layers is observed. (c) Optical microscope image of monodisperse emulsion droplets and blue, green, and red photonic balls. (d) Normalized reflectance spectra from the blue, green, red photonic balls in (c). (e) Schematic of optofluidic device with paired inner capillaries for creating the photonic Janus balls. (f) Optical microscope image of photonic Janus balls which have green and red faces of a single photonic ball. (*S.-H. Kim, S.-J. Jeon, G.-R. Yi, C.-J. Heo, J. H. Choi, and S.-M. Yang, "Optofluidic assembly of colloidal photonic crystals with controlled sizes, shapes and structures," Advanced Materials, 20, (2008), 1649–1655. Copyright Wiley-VCH Verlag GmbH & Co. KGaA. Reproduced with permission.*) (See also color insert.)

PDMS channels. First, cylindrical channels have a relatively low hydrodynamic resistance (R_{hyd}), which is given by the ratio of volumetric flow rate to pressure difference. For a cylindrical channel with radius R_{cyl} and length L, the resistance can be expressed as $8\eta L/\pi R_{cyl}^4$, whereas a channel with a square cross section of equal sides h has a resistance of $28.4\ \eta L/h^4$. Considering the typical microfluidic channel dimensions and the similar length scales of both geometries, the cylindrical channel has a 10-times-smaller hydrodynamic resistance than the square channel [36]. In addition, the glass capillary can withstand various organic solvents whereas the PDMS channel can be deformed by some of them. Especially, the axial symmetry of the cylindrical channel is advantageous for generating emulsion droplets because the outer flow can induce a drag force on the inner flow, which is equal in all azimuthal directions. Therefore, the cylindrical capillary device can be used to generate complex emulsion system, such as multiphase emulsion droplets.

Double emulsion droplets, which are droplets in droplets, can be fabricated by one- or two-step drop breakup in a microcapillary device [37,38]. On account of the geometrical benefit, double emulsion droplets can be used as capsules for confining the materials dispersed in a core droplet. To ensure the long-term stability of the capsules, the core and shell phases should be stabilized by adequate surfactant molecules, and solidification of the shell phase is required to enforce the structure. One of the most novel and simple strategies is photopolymerization of the shell phase. Depending on the polymerized shell properties, the capsule can either completely prevent the penetration of small molecules through the membrane or permit the transfer in a controlled manner.

Optofluidic devices enable the fabrication of microcapsules with narrow size distributions by in situ photopolymerization of the shell phase in double emulsion droplets, as shown in Fig. 6-6a. Especially, if the core emulsion droplets contain PS particles with a high surface charge density, the particles can spontaneously assemble into the crystalline phase from the smooth inner wall of the shell [39]. In the case of rigid and compact shells, which do not permit the penetration of ionic species, the crystal phase in the capsule has long-term stability in spite of its fragility in an ionic environment. In Fig. 6-6b and c, still-shot images taken at the end of the middle capillary at the moment of double-emulsion-droplet generation and in the downstream are displayed, respectively. While the photocurable shell phase is transparent, the core droplet is opaque due to scattering by concentrated PS particles. If the polymerization occurs downstream, the solidified shell confines the PS particles without loss and the crystal structure of the particles shows Bragg diffraction colors (see Fig. 6-6d). On the other hand, an elaborate control of three flow rates enables the preparation of capsules containing a specific number of small core droplets, as shown in Fig. 6-6e and f. If the droplet-generation

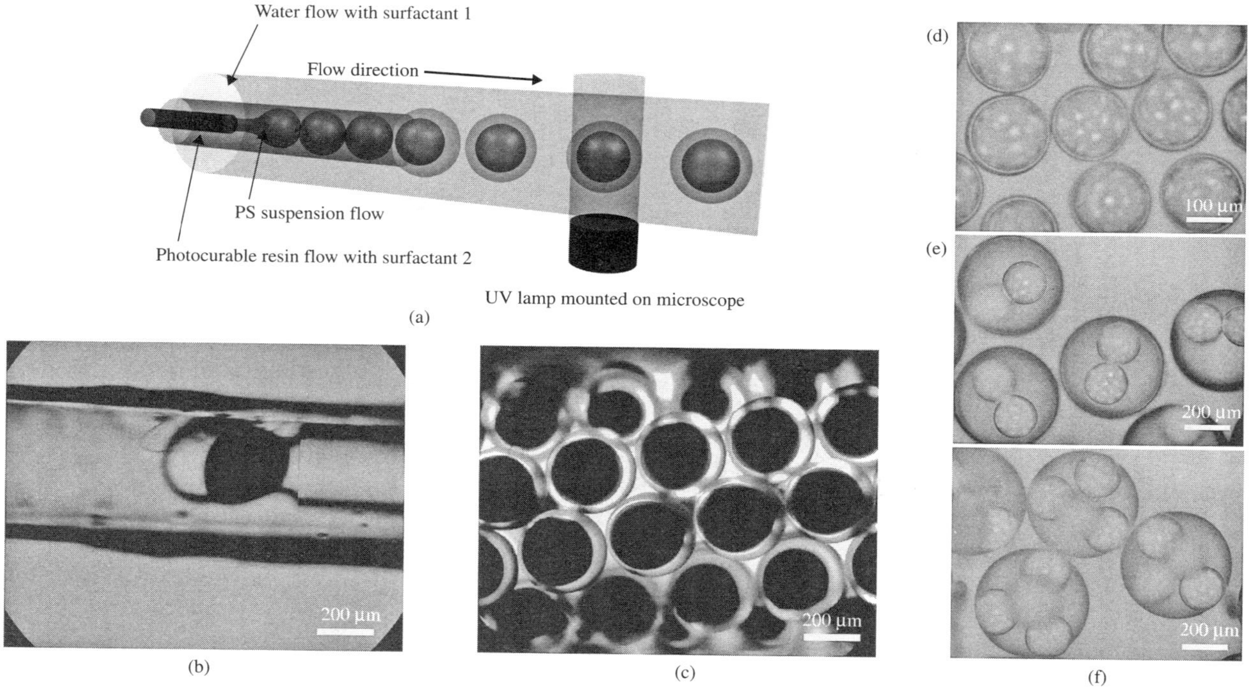

FIGURE 6-6 (a) Schematic illustration of optofluidic device for creating the microcapsules containing colloidal crystals in liquid state. (b and c) Still shot images taken at the end of the middle capillary at the moment of double-emulsion-droplet generation and in the downstream, respectively. (d, e, and f) Optical microscope images of microcapsules containing a large single core, small double cores, and small quadruple cores, respectively. (*S.-H. Kim, S.-J. Jeon, and S.-M. Yang, "Optofluidic encapsulation of crystalline colloidal arrays with spherical membranes," Journal of the American Chemical Society, 130, (2008), 6040–6046. Copyright 2008 American Chemical Society.*)

frequency at the end of the inner capillary is twice that at the middle capillary, two droplets will be confined into a single-shell drop. Likewise, four or more droplets can also be confined into the shell drop by controlling the relative generation frequency through the flow rates.

Microcapsules containing colloidal crystals in the liquid state can be used as tunable color pigments because the lattice constant of the crystal in the liquid state can be modulated by an external stimulus, such as a magnetic field, if the colloids contain magnetic nanoparticles [33,40]. In addition, high-throughput encapsulation without material loss will become a useful technique in many areas including drugs, cosmetics, and microreactions.

6-4 Conclusions and Outlook

Colloidal photonic crystals integrated in microfluidic devices are useful for many optofluidic applications, such as chemical and biological sensors or lasing cavities. This is because of their unique bandgap properties. However, there are many challenging issues that have to be solved to realize an optofluidic platform with colloidal crystals. First of all, the integration of colloidal crystals into optofluidic devices at a desired position is still difficult, although many integration techniques, such as evaporation-induced crystallization, centrifugal-force-induced crystallization, and electrically addressable crystallization, have been developed. In addition, the thickness of the crystal should be above the penetration depth of Bragg diffraction to achieve a high performance. This thickness should be on the order of 100 μm due to the small index contrast between the colloidal particles and the infiltrated fluid. Above all, the colloidal crystals should fill the channels completely, without leaving any gaps between the channel wall and the crystal. Because the hydrodynamic resistance in the gap is much smaller than that in the colloidal crystal, most of the fluid will pass through the gap without replacing the preoccupied fluid in the interstices if a gap exists. This means that effective tuning of the bandgap is impossible. Also, the crystal should have a large stiffness to prevent drift of the colloids by flow. Especially, a high flow rate of a fluid with a high affinity for colloids (for example, silica particles and ethanol) will easily wash away the particles. Therefore, an interconnection between the colloidal particles is required, which is achievable by annealing or neck formation through an etching process.

Once the colloidal crystals have been properly incorporated into the optofluidic devices, they can be used not only as bandgap materials but also as mixers or reactors for microfluidic units. A structured flow path for the colloidal crystals (at the submicron scale) can effectively induce mixing of the neighboring streams [41]. In addition, the high surface-area-to-volume ratios of colloidal crystals and their derivatives (such as inverse opal) lead to interesting applications in microfluidic continuous reactors by incorporating catalysts at the surfaces of the colloidal crystals [42]. Both mixers and reactors are important units in optofluidic systems to extend their applications to lab-on-a-chip or micro-TAS based on optics. Therefore, in the near future, we expect colloidal crystals to act as multifunctional units in integrated optofluidic systems.

6-5 Summary

Colloidal crystals have a photonic stop band that results from the periodic modulation of the refractive index at the half-wavelength scale of interacting light. This stop band can be controlled by infiltration of fluids into the crystal interstices. Therefore, the integration of colloidal crystals into microfluidic systems is important in order to exploit this property. Integration can be simply achieved by evaporating a colloidal suspension in microchannels or capillaries with one open end. Here, crystallization leads to a close-packed fcc structure, which has a volume fraction of colloids of 0.7404. However, evaporation-induced crystallization has many disadvantages. First of all, the crystallization process is too slow because evaporation occurs only at the small opening. In addition, soft spheres can induce cracks or gaps between the walls and the crystals because they can form non-close-packed crystals before the complete evaporation of the solvent. Moreover, the generation of open gaps between the colloidal crystals and the channel walls represents a severe problem in optofluidic applications. Because of the low hydrodynamic resistance at the gap, the majority of the fluid flows through the gap instead of through the crystal interstices. To solve these problems, centrifugal-force-induced crystallization was developed. Colloidal particles located in the rotating centrifugal chip move radially outwards. These particles are arranged into close-packed crystals much faster than with the evaporation-based method. Here, the crystallization time is determined by the particle size, the rotation speed, and the density contrast between the particles and the solvent. The colloidal crystals prepared in the centrifugal chips can be directly used as optofluidic devices, and hybrid colloidal crystals of different sizes and materials can also be prepared. However, in order to increase the flexibility of optofluidic systems containing colloidal crystals, crystallization should be located at a desired area. To achieve this, electrowetting is applied, whereby the microfluidic channel is combined with the electrode pattern. Electrowetting enables us to move the colloidal suspension into the desired position, and thus pixellate the colloidal crystals in electrically addressable microfluidic chips.

Colloidal crystals integrated in microfluidic devices can also be used as refractive index sensors because the reflection spectra of the crystals exactly respond to changes in the refractive indices. Especially, the detection of biomolecules could be possible using this scheme—without the need to label them. Although colloidal crystals do not show appreciable reflectance intensity for index-matching liquids, hybrid colloidal crystals composed of particles with different indices remove the blind index region. Unlike reflection-based applications, colloidal crystals can be used as lasing resonators. If the fluids contain active materials, such as dye molecules or quantum dots, the emission spectra change with respect to that corresponding to the bulk emission. This is a result of the modulated electromagnetic density of states in the photonic crystals. Because of the high density of states and the low group velocity at the band edge and at defect modes, it is possible to obtain an amplified stimulated emission. Especially, tuning the bandgap properties using fluids enables the control of the emission wavelength of the laser, which is important for a wide range of micro-TAS or lab-on-a-chip applications.

Unlike colloidal-crystal integrated systems, optofluidic devices can be used to continuously generate discrete colloidal crystals. Especially, if the emulsion droplets are used as templates for crystallization, photonic balls with optical isotropy can be generated. To achieve this, an optofluidic device composed of a microfluidic emulsion generator and a UV-exposure unit is prepared. In the emulsion-generator part of the device, photocurable emulsion droplets containing concentrated repulsive colloids are generated. The colloidal particles in the droplet spontaneously arrange into an onion-ring-like crystal structure, from the outermost layer, by minimizing the total repulsive energy. The droplets are then photopolymerized by passing them through the UV irradiated area. Finally, the photonic balls are generated. In addition, photonic Janus balls with two different colors on their own hemispherical domains can be generated with the paired inner capillaries.

Furthermore, double-emulsion droplets, which are droplets in droplets, could also be generated with the microfluidic device. Here, the middle phase of the emulsion was a photocurable resin and the inner droplets contained an aqueous suspension of PS particles with a high surface charge density. Using an optofluidic scheme similar to that described in the previous example, photonic balls in the liquid state were generated by encapsulating the suspension within a polymer shell. The prepared photonic balls exhibited long-term stability, even in the presence of high concentrations of ionic impurities, because the shell did not permit the penetration of impurities.

References

1. W. B. Russel, D. A. Saville, and W. R. Schowalter, *Colloidal Dispersion*, Cambridge University Press, (1999), New York.
2. S.-M. Yang, S.-H. Kim, J.-M.Lim, and G.-R. Yi, "Synthesis and assembly of structured colloidal particles," *Journal of Materials Chemistry*, **18**, (2008), 2177–2190.

3. Y. Monovooukas and A. P. Gast, "The experimental phase diagram of charged colloidal suspensions," *Journal of Colloid and Interface Science*, **128**, (1989), 533–548.
4. A. Yethiraj and A. van Blaaderen, "A colloidal model system with an interaction tunable from hard sphere to soft and dipolar," *Nature*, **421**, (2003), 513–517.
5. P. Bartlett and A. I. Campbell, "Three-dimensional binary superlattices of oppositely charged colloids," *Physical Review Letters*, **95**, (2005), 128302.
6. M. E. Leunissen, C. G. Christova, A.-P. Hynninen, C. P. Royall, A. I. Campbell, A. Imhof, M. Dijkstra, R. van Roij, and A.van Blaaderen, "Ionic colloidal crystals of oppositely charged particles," *Nature*, **437**, (2005), 235–240.
7. J. D. Joannopoulos, R. D. Meade, and Joshua N. Winn, "Photonic crystals: molding the flow of light," Princeton University Press, (1995), Princeton.
8. http://ab-initio.mit.edu/wiki/index.php/MIT_Photonic_Bands.
9. Y. A. Vlasov, X.-Z. Bo, J. C. Sturm, and D. J. Norris, "On-chip natural assembly of silicon photonic bandgap crystals," *Nature*, **414**, (2001), 289–293.
10. Y.-S. Cho, G.-R. Yi, J. H. Moon, D.-C. Kim, B.-J. Lee, and S.-M. Yang, "Connected open structures from close-packed colloidal crystals by hyperthermal neutral beam etching," *Langmuir*, **21**, (2005), 10770–10775.
11. G. von Freymann, S. John, V. Kitaev, and G. A. Ozin, "Enhanced coupling to slow photon modes in three-dimensional graded colloidal photonic crystals," *Advanced Materials*, **17**, (2005), 1273–1276.
12. W. Stober, A.Fink, and E. Bohn, "Controlled growth of monodisperse silica spheres in the micron size range," *Journal of Colloid and Interface Science*, **26**, (1968), 62–69.
13. A. van Blaaderen and A. Vrij, "Synthesis and characterization of colloidal dispersions of fluorescent, monodisperse silica spheres," *Langmuir*, **8**, (1992), 2921–2931.
14. P. H. Wang and C.-Y. Pan, "Preparation of styrene/acrylic acid copolymer microspheres: polymerization mechanism and carboxyl group distribution," *Colloid & Polymer Science*, **280**, (2002), 152–159.
15. E. Kim, Y. Xia, and G. M. Whitesides, "Two- and three-dimensional crystallization of polymeric microspheres by micromolding in capillaries," *Advanced Materials*, **8**, (1996), 245–247.
16. P. Yang, T. Deng, D. Zhao, P. Reng, D. Pine, B. F. Chmelka, G. M. Whitesides, and G. D. Stucky, "Hierarchically ordered oxides," *Science*, **282**, (1998), 2244–2246.
17. P. Yang, A. H. Rizvi, B. Messer, B. F. Chmelka, G. M. Whitesides, and G. D. Stucky, "Patterning porous oxides within microchannel networks," *Advanced Materials*, **13**, (2001), 427–431.
18. J. H. Moon, S. Kim, G.-R. Yi, Y.-H. Lee, and S.-M. Yang, "Fabrication of ordered macroporous cylinders by colloidal templating in microcapillaries," *Langmuir*, **20**, (2004), 2033–2035.
19. U. Kamp, V. Kitaev, G. von Freymann, G. A. Ozin, and S. A. Mabury, "Colloidal Crystal Capillary Columns-Towards Optical Chromatography," *Advanced Materials*, **17**, (2005), 438–443.
20. S.-K. Lee, G-R. Yi, and S.-M. Yang, "High-speed fabrication of patterned colloidal photonic structures in centrifugal microfluidic chips," *Lab on a Chip*, **6**, (2006), 1171–1177.
21. J.-Y. Shiu, C.-W. Kuo, and P. Chen, "Actively controlled self-assembly of colloidal crystals in microfluidic networks by electrocapillary forces," *Journal of the American Chemical Society*, **126**, (2004), 8096–8097.
22. J.-Y. Shiu and P. Chen, "Active patterning using an addressable microfluidic network," *Advanced Materials*, **17**, (2005), 1866–1869.
23. K. N. Plunkett, A. Mohraz, R. T. Haasch, J. A. Lewis, and J. S. Moore, "Light-regulated electrostatic interactions in colloidal suspensions," *Journal of the American Chemical Society*, **127**, (2005), 14574–14575.
24. M. N. Shkunov, Z. V. Vardeny, M. C. DeLong, R. C. Polson, A. A. Zakhidov, and R. H. Baughman, "Tunable, gap-state lasing in switchable directions for opal photonic crystals," *Advanced Functional Materials*, **12**, (2002), 21–26.

25. S. Furumi, H. Fudouzi, H. T. Miyazaki, and Y. Sakka, "Flexible polymer colloidal-crystal lasers with a light-emitting planar defect," *Advanced Materials*, **19**, (2007), 2067–2072.
26. O. D. Velev, A. M. Lenhoff, and E. W. Kaler, "A class of microstructured particles through colloidal crystallization," *Science*, **287**, (2000), 2240–2243.
27. J. H. Moon, G.-R. Yi, S.-M. Yang, D. J. Pine, and S. B. Park, "Electrospray-assisted fabrication of uniform photonic balls," *Advanced Materials*, **16**, (2004), 605–609.
28. S.-H. Kim, S. Y. Lee, G.-R. Yi, D. J. Pine, and S.-M. Yang, "Microwave-assisted self-organization of colloidal particles in confining aqueous droplets," *Journal of the American Chemical Society*, **128**, (2006), 10897–10904.
29. A. R. Bausch, M. J. Bowick, A. Cacciuto, A. D. Dinsmore, M. F. Hsu, D. R. Nelson, M. G. Nikolaides, A. Travesset, and D. A. Weitz, "Grain boundary scars and spherical crystallography," *Science*, **299**, (2003), 1716–1718.
30. S.-H. Kim, S.-J. Jeon, G.-R. Yi, C.-J. Heo, J. H. Choi, and S.-M. Yang, "Optofluidic assembly of colloidal photonic crystals with controlled sizes, shapes and structures," *Advanced Materials*, **20**, (2008), 1649–1655.
31. S.-H. Kim, S.-J. Jeon, W. C. Jeong, H. S. Park, and S.-M. Yang, "Optofluidic synthesis of electro-responsive photonic Janus balls with isotropic structural colors," *Advanced Materials*, (2008), DOI: 10.1002/adma.200801167.
32. S. R. Raghavan, "Rheology of silica dispersions in organic liquids: new evidence for solvation forces dictated by hydrogen bonding," *Langmuir*, **16**, (2000), 7920–7930.
33. J. Ge, Y. Yin, "Magnetically tunable colloidal photonic structures in alkanol solutions," *Advanced Materials*, (2008), DOI, 10.1002/adma.200800657.
34. P. B. Umbanhowar, V. Prasad, and D. A. Weitz, "Monodisperse emulsion generation via drop break-off in a coflowing stream," *Langmuir*, **16**, (2000), 347–351.
35. N. K. Sheridon, E. A. Richley, J. C. Mikkelsen, D. Tsuda, J. M. Crowley, K. A. Oraha, M. E. Howard, M. A. Rodkin, R. Swidler, and R. Sprague, "The gyricon rotating ball display," *Journal of the Society for Information Display*, **7**, (1999), 141–144.
36. H. Bruus, *Theoretical Microfluidics*, Oxford University Press, (2008), New York.
37. L.-Y. Chu, A. S. Utada, R. K. Shah, J.-W. Kim, and D. A. Weitz, "Controllable monodisperse multiple emulsions," *Angewandte Chemie International Edition*, **46**, (2007), 8970–8974.
38. A. S. Utada, E. Lorenceau, D. R. Link, P. D. Kaplan, H. A. Stone, and D. A. Weitz, "Monodisperse double emulsions generated from a microcapillary device," *Science*, **308**, (2005), 537–541.
39. S.-H. Kim, S.-J. Jeon, and S.-M. Yang, "Optofluidic encapsulation of crystalline colloidal arrays with spherical membranes," *Journal of the American Chemical Society*, **130**, (2008), 6040–6046.
40. J. Ge, Y. Hu, and Y. Yin, "Highly tunable superparamagnetic colloidal photonic crystals," *Angewandte Chemie International Edition*, **46**, (2007), 7428–7431.
41. G. H. Seong and R. M. Crooks, "Efficient mixing and reactions within microfluidic channels using microbead-supported catalysts," *Journal of the American Chemical Society*, **124**, (2002), 13360–13361.
42. P. Yang, T. Deng, D. Zhao, P. Feng, D. Pine, B. F. Chmelka, G. M. Whitesides, and G. D. Stucky, "Hierarchically ordered oxides," *Science*, **282**, (1998), 2244–2246.

CHAPTER 7

Optofluidic Photonic Crystal Fibers: Properties and Applications

P. Domachuk, P. Steinvurzel, B. Kuhlmey, and B. J. Eggleton

CUDOS, School of Physics, University of Sydney, Sydney, Australia

Optofluidics is an evolving design principle whereby aspects of microphotonics and microfluidics are combined to enable new device modalities [1,2]. This all-encompassing definition reflects the diversity of optofluidic devices in terms of photonic and microfluidic structures, functions, and applications. While often applied to planar photonics [3], optofluidic design can also be successfully applied to microstructured and photonic crystal optical fibers, or even simply using capillaries. Indeed, one of the first optical fiber waveguide designs entailed a hollow capillary filled with a high-refractive index fluid [4]. These fibers present a unique environment to perform optofluidics [2,5]. The fibers themselves provide very high quality optical waveguides culminating from decades of telecommunications research providing a photonic layer and associated light sources, detection equipment, and optical interfaces [6]. Synergistically, microstructured optical fibers (MOFs) composed of inclusions along the length of the waveguide provide a natural repository for microfluidic bodies [7,8]. These two factors enable optofluidics to provide natural extensions to MOF functionality. In this chapter we review the history, development, and directions of photonic crystal fiber (PCF) tuned using microfluidics. We review optofluidics in a variety of MOFs, including grapefruit fibers, transverse PCFs, fluid-filled PCFs transformed into bandgap-guiding fibers, and fiber-based ultracompact optofluidic interferometers.

7-1 Introduction

7-1-1 Optical Fibers

Since optical fibers are the defining characteristic of the optofluidic devices discussed here, a variety of relevant optical fiber designs warrant discussion. The ubiquitous silica telecommunications single-mode fiber (SMF) [6] is used for optical transport, while microstructured optical fibers (MOFs) [9–14], including photonic crystal fibers (PCFs) [15–19] are used for both optical and microfluidic transport and manipulation. In this section, both varieties will be discussed, with emphasis on the MOFs that form the optofluidic devices.

The most prevalent optical fiber design is the silica telecommunications SMF [6]. After many years of research and development, the SMF is deployed across the world, providing the physical backbone of modern communications. A consequence of this widespread use is that SMF and its associated light source and detection technologies provide high-quality, low-loss, modular means of generating, transporting, and analyzing light. This platform forms the basis for investigation into fiber-based optofluidic devices. SMFs are composed of highly pure fused silica whose circular core is a region, 8 μm in diameter silica doped using heavy ions during the fabrication process [20]. This doping provides a refractive index contrast against the surrounding silica cladding that is 125 μm in diameter and itself is protected by a variety of polymer coatings. Typically, these fibers support a single propagating mode at 1550 nm with a multimode cutoff wavelength of around 1100 nm. The single mode propagates with very low loss, approximately 0.2 dB/km [21].

Figure 7-1 shows several more recent developments in fiber waveguides: the MOF. MOFs are characterized by hollow inclusions that run the length of the fiber. These come in many forms, from fibers with a single hole forming a hollow core [22] to fibers with inclusions that surround an SMF-like core, known as a grapefruit fiber [23]. A special class of MOF is the PCF. These fibers consist of a periodic array of inclusions surrounding a core that may be either solid [16] or hollow [18,24]. These fibers display a wealth of optical phenomena [25–30]; however, it is the cladding/core inclusions comprising the fiber microstructure that are of interest in fiber-based optofluidics. These inclusions form a natural home for microfluids, and their overlap with the guided mode of the fiber allows them to influence the

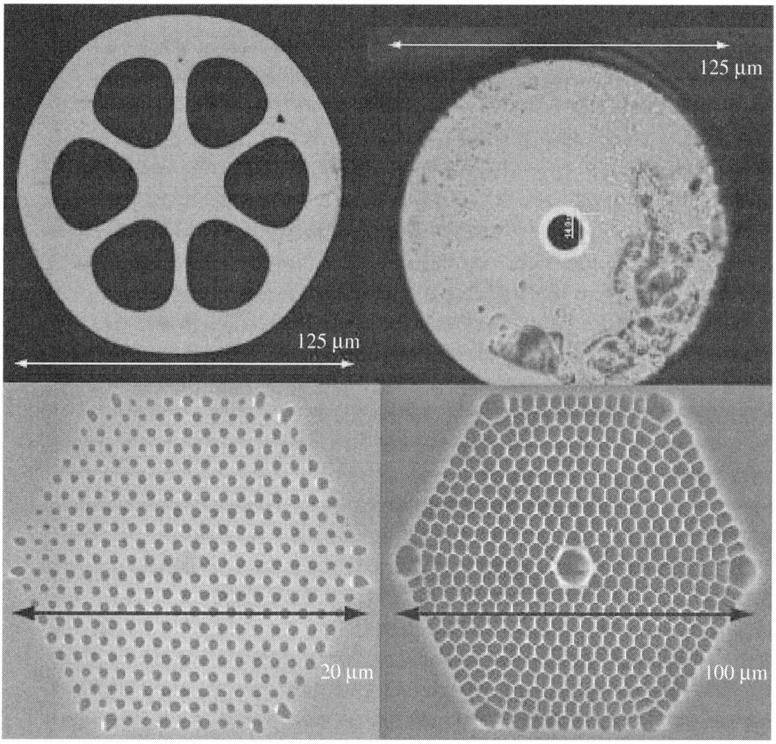

FIGURE 7-1 A selection of MOFs imaged in cross section. (from top left, clockwise) A grapefruit fiber whose inclusions shape the guided mode can be filled with a fluid; a simple hollow core fiber, with a ring of dopant surrounding the air core. The mode is ring shaped; a high air fill fraction, bandgap guidance hollow-core PCF; an effective index-guiding PCF. The microstructure lowers the refractive index of the medium, guiding a mode in the solid core.

fiber's optical properties. One of the earliest examples of an all-fiber optofluidic device that utilized the "grapefruit" fiber is mentioned in the preceding text [23].

7-1-2 Optical Fiber Postprocessing

A number of postprocessing techniques are used to improve optofluidic operation of existing fibers and MOFs. These include tapering to control the overlap between light and fluids, writing Bragg gratings and long-period gratings, for example, to couple two modes with higher overlap with the fluids, and surface treatments to improve microfluidic properties. Fiber tapering and grating writing play key roles in the development of fiber-based optofluidic structures, and are discussed here. Surface treatments will be discussed in Sec. 7-4.

Fiber Tapering

Fiber tapering is a mature technique of postprocessing the shape of optical fibers or, indeed, any capillary structure capable of being heated to a plastic state [31]. Very simply, the fiber to be tapered is heated along some length until it is softened, and then it is pulled in opposite direction, thus applying tension to the plastic glass. This results in the elongation of the fiber and the scaling of its internal structure to create waveguides with customized optical properties [32]. It is possible to fabricate tapers that continue to guide light despite having dimensions significantly below the wavelength of the said light [33,34]. Tapering also works on microstructured optical fibers and, with the right parameters, the fiber structure is maintained [32,35]. As such, fiber tapering is a simple and versatile process for customizing silica fiber waveguides.

Bragg Gratings

Fiber Bragg gratings (FBGs) are periodic refractive index modulations along the length of a fiber waveguide [36]. In silica telecommunications fibers, the germanium-doped cores are photosensitive to ultraviolet light: exposure causes a refractive index rise that is dependent on exposure time, light intensity, and wavelength. So in order to fabricate an FBG in SMF, all that needs to be done is expose the SMF core to a periodic spatially varying ultraviolet light field. The discovery of the FBG and the photosensitivity of germanium-doped silica arose from the propagation of short-wavelength visible light through a fiber that, upon reflection from the fiber ends, set up a standing wave, which provided the necessary periodic intensity modulation [37]. Typical modern FBGs are fabricated using a transverse beam of ultraviolet laser light passed through a phase mask that generates two beams of different diffraction orders. These beams then interfere creating the necessary periodic intensity modulation for writing FBGs.

The characteristic response of an FBG whose period is subwavelength is that it will resonantly reflect a given wavelength band while transmitting all others [38]. As such, the FBG can be thought of as an in-fiber wavelength-selective mirror. Two FBGs can be written spaced apart in SMF, and this structure acts essentially as a Fabry-Perot resonator [39]. Similarly, one continuous grating can have a defect, or phase shift, introduced somewhere along its length and also behave like a resonator [35]. The wavelength response of the FBG depends upon the period of the grating. If the grating is mechanically deformed or expanded through local heating, the resonant wavelength of the FBG will change. This effect is the basis for many designs of photonic environment sensors [36–38].

Long-Period Gratings

Another type of grating, the long-period grating (LPG), has a period of several hundred wavelengths and resonantly couples light forward into higher order core or cladding modes of the fiber [38]. In conventional fibers, the dispersion of cladding modes, and as a consequence the spectral response of LPGs, depends on the ambient refractive index. Fiber LPGs are thus another useful structure for fiber-based sensors [39]. In MOFs, the cladding modes LPGs couple can have a large overlap with an MOF's holes, so that the LPG spectral response can be very sensitive to the content of the holes. This enables all-in-fiber refractive index sensors, or microfluidic tunable filters, and is further discussed as follows.

When the fiber incorporates one or more germanium-doped cores, long-period gratings can be made using UV exposure in the same way as Bragg gratings. However, LPGs can also be imprinted using a large number of other techniques, discussed in Secs. 7-2 and 7-5.

7-1-3 Optofluidics: History and Development

Optofluidics is something of a derivative science, borrowing and combining techniques and technologies from the more established fields of microphotonics and microfluidics [40,41]. At first glance, these two fields would appear highly disparate: one pertaining to photon transport, usually associated with communications [6], and the other relating to highly confined fluid flows and small-volume chemical reactions [42]. However, closer comparison shows a wealth of similarities in scale, structure, and transport phenomenon, and a number of ways in which the nature of one may complement the other. Devices designed using optofluidic principles have a number of inherent advantages over photonic or microfluidic devices individually. These advantages arise from controlled fluids (microfluidic layer) that interact with controlled light (photonic layer). This interaction can then be used in a variety of ways: the optical properties depending on the exact fluids at play, integrated micro-optofluidic sensors are an obvious application, as are photonic devices (attenuators, polarization controllers, dispersion compensators, and delay lines) made tunable through the microfluidic layer.

Optofluidic design can be very generally applied to any photonic structure with a void near the optical field in the device; given suitable fluid properties and pressure gradients, fluids can be made to infiltrate practically any photonic structure that is open to the outside. This infused fluid interacts with the optical field in the device changing local optical properties. Fluids are also an inherently mobile phase. This allows a localized region of certain optical properties to be propagated to other parts of the device to allow modulation of the photonic structure. Importantly, this modulation can be achieved with little modification to the initial photonic structure upon which the optofluidic device is based. The use of discrete bodies of fluid in optofluidics also allows the air surrounding it to provide refractive index contrast greater than that available in the surrounding solid structure.

This enables compactness in devices that rely upon interference. The utilization of a fluid phase as an optical element enables further versatility; the optical properties of fluids, as a whole, are very broad and limited only by chemistry and composition. As such, optofluidic tuning enables a much broader range of optical properties in a given photonic structure than solid-state fabrication methods. Bringing fluids into contact with optical fields also enables a variety of sensing geometries. Optofluidically designed structures allow small, well, controlled volumes of reagents or analytes to interact efficiently with the optical field in the photonic layer effectively adding optical interrogation methods to the "lab-on-a-chip" analysis methods [42]. All these attributes of optofluidics combine to enable unique modulation geometries, device functionalities, and sensing platforms.

Potentially foreshadowing this modern development, the first observation of guided light was in a stream of water from a fountain by Jacques Babinet in 1840 [43]. From there, optofluidics again saw use in the development in some of the first purpose, built optical waveguides [4] whose cores were filled with fluid to provide the necessary refractive index contrast to guide light. From there, optofluidic development was relatively quiet until the emergence of microfluidics [40–42] and "lab-on-a-chip" technologies [43] that entail the control of fluids, typically chemical reagents or analytes, inside micron-scale flow channels. The length scales of microfluidic structures are similar to those of microphotonic structures, paving the way for the integration of both kinds of devices [44]. Optofluidic devices can be broadly classified by their underlying photonic technology that fall into either planar photonics or optical fibers. Both these varieties of optofluidics were developed essentially concurrently. Planar optofluidic devices use planar photonic structures such as integrated planar waveguides or photonic crystals [3,45–50] as their photonic layers. While highly compact and functional, these planar optofluidic devices require significant investment in design, experimentation, fabrication time, and cost.

7-1-4 Fiber-Based Optofluidics

In contrast to planar optofluidics, fiber-based optofluidics use optical fibers of various designs for photonic transport around the optofluidic environment, for microfluidic transport, or both. As such, there are two broad subcategories of fiber-based optofluidic devices. First is the "all-fiber" device, where both the microfluidic transport network and the photonic transport layers are provided exclusively by optical fibers. Second is the "semi-planar fiber" device, which uses optical fibers as the photonic transport layer, but relies on a more integrated, planar microfluidic environment. This device operates on a photonic and microfluidic level using optical fibers alone (typically MOFs). The reasons for doing this, aside from those outlined above, involve the quality of the microfluidic environment in silica microstructured optical fibers and the potential for integration into existing fiber-based specifications with minimal new development using commercially available components. Smoothness of the void surfaces in microstructured optical fibers is on the order of 100 pm [51]. As such, microstructured optical fibers provide an essentially atomically smooth environment for microfluidic flow and photon transport. This roughness is to be compared with that of a typical planar optofluidics material, silicon on insulator, undergoing an optimized fabrication process to remove roughness, for which the minimum roughness achieved was 1.4 nm [52]. This is almost an order of magnitude above that in a microstructured optical fiber. These smoothed surfaces enable very low loss optical waveguiding. As an additional benefit, microfluidic flow and interface behavior in these smooth structures are guaranteed to conform to classical theoretical behavior. Microstructured optical fibers are commercially available in a wide variety of designs, conforming to almost every conceivable photonic guidance requirement. The sheer variety of these fibers, coupled with standard fiber handling and probing techniques from communications technologies means that an almost limitless variety of device designs are possible using commercial components that are compatible with existing SMF photonic hardware. Further, creating all-fiber optofluidic devices typically involves laboratory postprocessing of commercial fibers using simple optofluidic and photonic techniques, circumventing the need for expensive fiber fabrication infrastructure.

Figure 7-2 shows an example of the all-fiber optofluidic fluid refractive index sensor [53], a measurement typical in optofluidic systems. This design runs a channel between two separated FBGs, acting as a Fabry-Perot resonator. Light propagation in the exposed core suffers a phase delay that is again dependent upon the index of the surrounding medium. Another design involves etching SMF [54] to expose the core to the ambient fluid. Yet another design involves exposing an FBG written into the core of an SMF, making the resonant wavelength dependent upon the index of the surrounding medium. These two designs provide a coarse and fine measurement of ambient refractive index, respectively. The components used for this device are standard optical communications equipment (SMF, holographically written UV FBGs, and hydrofluoric acid etching) probed using photonics standard tunable laser sources and spectral measuring equipment. Further, these devices are all in SMF, automatically guaranteeing compatibility with other SMF-based devices and networks. Similar work has been performed elsewhere [55] using different methods to expose the fiber-guided light to the fluid. The device shown in Fig. 7-2 is, in fact, a hybrid optofluidic device; a planar substrate is used for the microfluidic layer and optical fibers for optical control.

Figure 7-3 shows another all-fiber optofluidic technology—the selective filling of voids in a hollow-core PCF [56,57]. These devices use methods relying on differential capillary force across all holes in

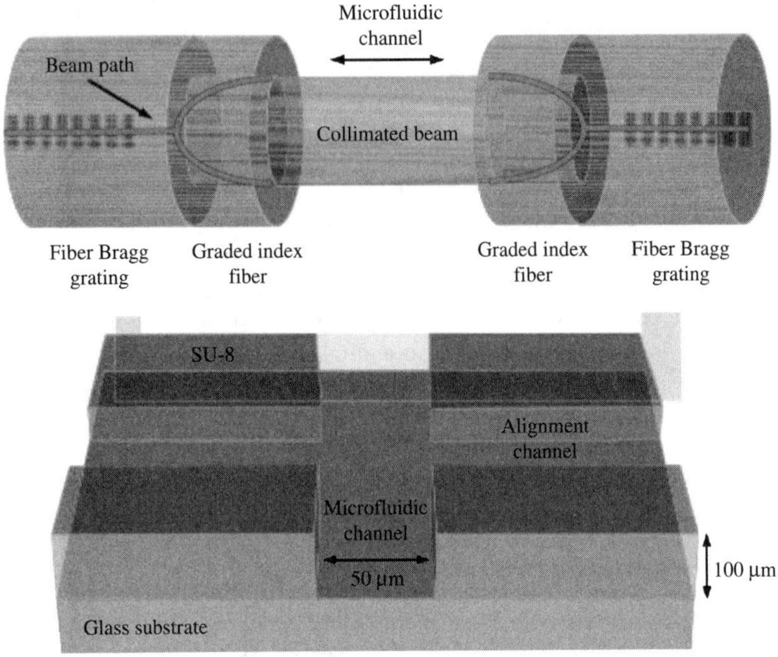

FIGURE 7-2 (top) A schematic of the fiber Fabry-Perot interferometer, the beam path of the device shown in red. (bottom) A photograph and schematic of the planar microfluidic substrate that the fiber Fabry-Perot refractometer inhabits. Clearly visible are the SMFs and microfluidic channel. (*Reprinted with permission from P. Domachuk, I. C. M. Littler, M. Cronin-Golomb, et al., "Compact resonant integrated microfluidic refractometer," Appl. Phys. Lett., 88, 093513 (2006). Copyright 2006, American Institute of Physics.*)

the PCF microstructure to infiltrate a UV curable adhesive (or other fluid) into the hollow core of the PCF. This kind of selective filling enables the PCF core to be composed of a wide variety of fluids even achieving low index guiding in the core material due to photonic bandgap confinement. The nature of the fluid may be chosen for enhanced optical nonlinearity, for instance, enabling nonlinear waveguides with much higher efficiencies than traditional silica waveguides. Also, the procedure is performed completely as a post-processing step in the lab, with no fiber-fabrication infrastructure being required. Again, splicing of PCFs to SMFs provides compatibility to existing SMF devices and infrastructure. Similar work is described using polymer microstructured optical fibers with a water core, using essentially the same filling technique [58].

If all the holes in a solid-core PCF are filled with a fluid whose refractive index is higher than the background silica, the core no longer supports modes guided by modified total internal reflection. However, it does support bandgap-guided modes [59]. As with the more familiar air-core photonic bandgap fibers (PBGFs) [18], fluidic PBGFs only

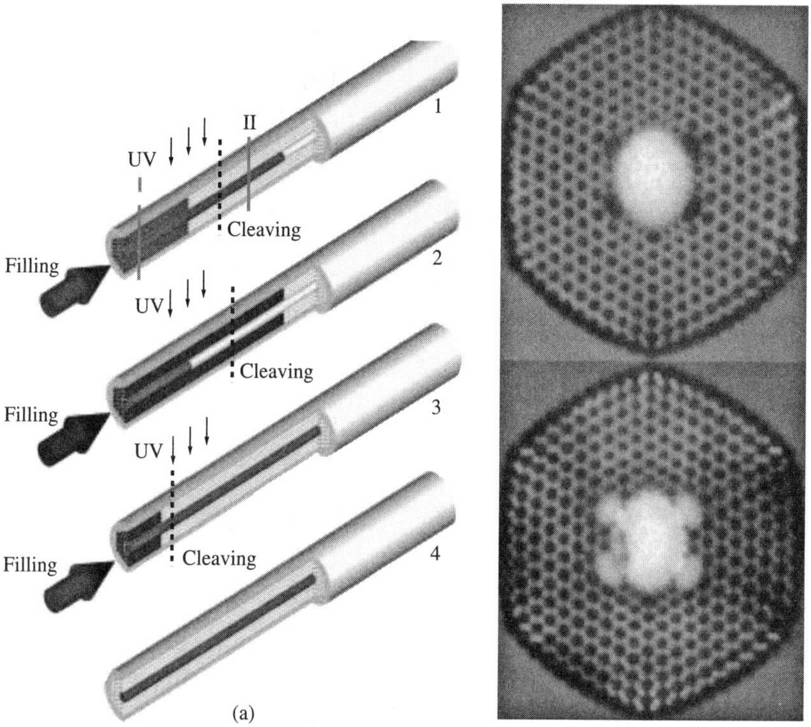

FIGURE 7-3 (left) Schematic method for selective filling PCFs. (*Reproduced with permission from Y. Y. Huang, Y. Xu, and A. Yariv, "Fabrication of functional microstructured optical fibers through a selective-filling technique," Appl. Phys. Lett., 85, 5182–5184 (2004). Copyright 2004, American Institute of Physics.*) (right) A PCF whose core and surrounding is selectively filled guiding light. (*K. Nielsen, D. Noordegraaf, T. Sørensen, et al., "Selective filling of photonic crystal fibers," J. Opt., A 7, L13–L20 (2005).*) This postprocessing method allows materials beyond the constituent silica to be used in the fiber core.

transmit discrete frequency bands and have strong wavelength-dependent modal dispersion. This latter property makes fluidic PBGFs an interesting platform for studying nonlinear pulse propagation [60] or LPG-mediated mode coupling [61]. In light of our earlier discussion, however, the use of fluidics also means that these fibers can be used as active devices or sensors, where the resonant nature of bandgap guiding design enhances the device tunability or sensitivity [61]. Fluid PBGFs are easy to fabricate by simply using a vacuum pump to fill a PCF with a desired liquid and, again, optical coupling can be done with standard SMF. This is yet another example of an optofluidic structure that may be created using simple laboratory postprocessing techniques and probed using standard photonics diagnostic tools. Fluidic PBGFs are discussed at length later in Sec. 7-5.

Figure 7-4 shows a semi-integrated optofluidic device that uses an LPG written inside SMF that is slung through the middle of a

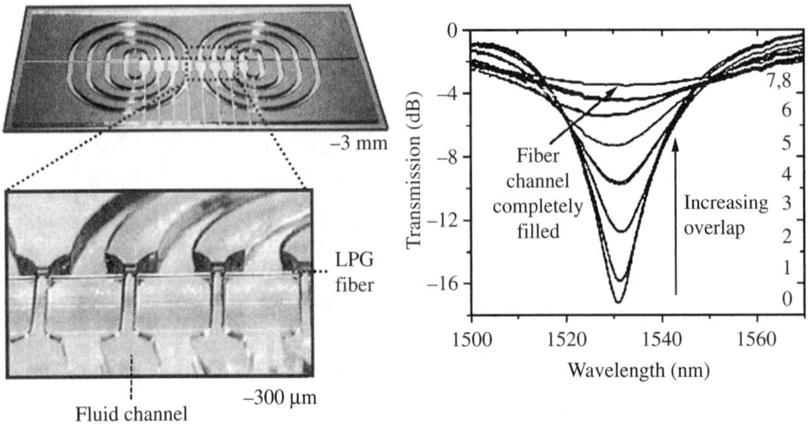

FIGURE 7-4 (left) An optofluidically tuned LPG fiber that utilizes external optofluidic tuning to alter the transmission of the LPG. (right) Spectrum of the tuned LPG demonstrating controllable extinction. (*F. Cattaneo, K. Baldwin, Shu Yang, T. Krupenkine, S. Ramachandran, J.A. Rogers, Digitally tunable microfluidic optical fiber devices," J. Microelectromech. Syst., Copyright 2003 IEEE.*)

planar microfluidic channel [62]. The channel contains integrated electrodes that are used to drive a fluid plug using electrowetting. Normally, the LPG couples to cladding modes exposed to air surrounding the fiber, but when the fluid plug is pumped onto the surface of fiber, the optical response of the LPG changes. Another, similar, device was demonstrated this time monitoring the back-reflected modes of an FBG [63]. The same structure has been serialized, providing a number of fluid intersections along the one LPG for a greater degree of control by allowing discrete overlaps of the fluid body with the fiber grating. These devices embody the advantages of the semiplanar device: a high-quality photonic structure embedded in a compact, planar microfluidic structure with advanced actuation technology leading to a high degree of compactness and functionality.

In the remainder of this chapter we discuss a variety of fiber-based optofluidic devices complementary to those described earlier. We begin with one of the earliest fiber-based optofluidic devices based upon fluid inclusion inside the microstructure of a grapefruit fiber. These inclusions serve to alter the local waveguiding properties of the fiber in a controllable fashion to provide significantly enhanced functionality over fiber devices alone, without compromising connectivity in SMF-based light paths. We go on to discuss optofluidic tuning of transverse PCFs. The transverse geometry renders the PCF microstructure into a planar two-dimensional (2-D) photonic crystal. The introduction of microfluidics into the microstructure allows for tuning the partial bandgap in a dynamic and reconfigurable fashion. This tuning is applied to make an optofluidic attenuator whose compactness and strong attenuation is enabled through the combination of optofluidics and photonic crystals. After this, we discuss another fiber-based optofluidic component—the ultracompact microfluidic interferometer. This device uses SMFs for optical transport and a tapered square fiber to hold a fluid meniscus in the beam path. This refractive index contrast between the fluid and the surrounding area provides a phase delay within the beam providing interferometric modulation. The use of optofluidics again enables the compactness and reconfiguration of the device. The last devices discussed are optofluidic photonic bandgap fibers based on PCFs that utilize optofluidic tuning to change their optical guidance mechanisms. We conclude with a discussion of potential future directions for fiber-based optofluidics.

7-2 Grapefruit-Fiber Optofluidic Devices

Figure 7-5 shows a cross section of a silica grapefruit fiber [9], so named for their microstructure shaped like the flesh of the aforementioned citrus fruit [64]. This fiber has a series of symmetric air inclusions surrounding a core of germanium-doped silica identical to a telecommunications SMF. This design makes the grapefruit fiber ideal for optofluidic application; the holes in the microstructure provide access to the guided light field for fluids infused therein while the doped core structure enables easy optical coupling to SMF and holographic writing of grating structures. Optofluidic tuning in these fibers enables a slew of in-fiber, reconfigurable components. Indeed, grapefruit fibers were among the first designs of optical fibers to be optofluidically tuned.

Figure 7-6 shows a schematic of the initial optofluidic grapefruit fiber used in combination with an LPG [9]. A liquid monomer was drawn into the fiber microstructure and cured in place using UV light. The resulting polymer has a tunable refractive index that decreases with increasing ambient temperature. When unmodified, the core of the grapefruit fiber is sufficiently far from the holes that the presence of polymer does not affect light propagation in the core. However, if the fiber is tapered, the core mode can be made to interact with the filled holes [31]. By tapering the grapefruit fiber and then filling its holes with polymer, an in-fiber variable attenuator can be made. As the polymer's refractive index is decreased, the modal field becomes less confined by the core in the tapered region, and increasingly leaks into the cladding. This controlled leakage allows for controlled attenuation inside the fiber.

More generally, shifting the polymer refractive index creates a tunable optical environment inside the fiber. This kind of local tunability can be used to influence other structures inside the fibers, such as holographically written long-period gratings (LPGs) in the fiber core. In that case, as the polymer in the fiber microstructure surrounding the LPG is heated, the refractive index of the polymer decreases

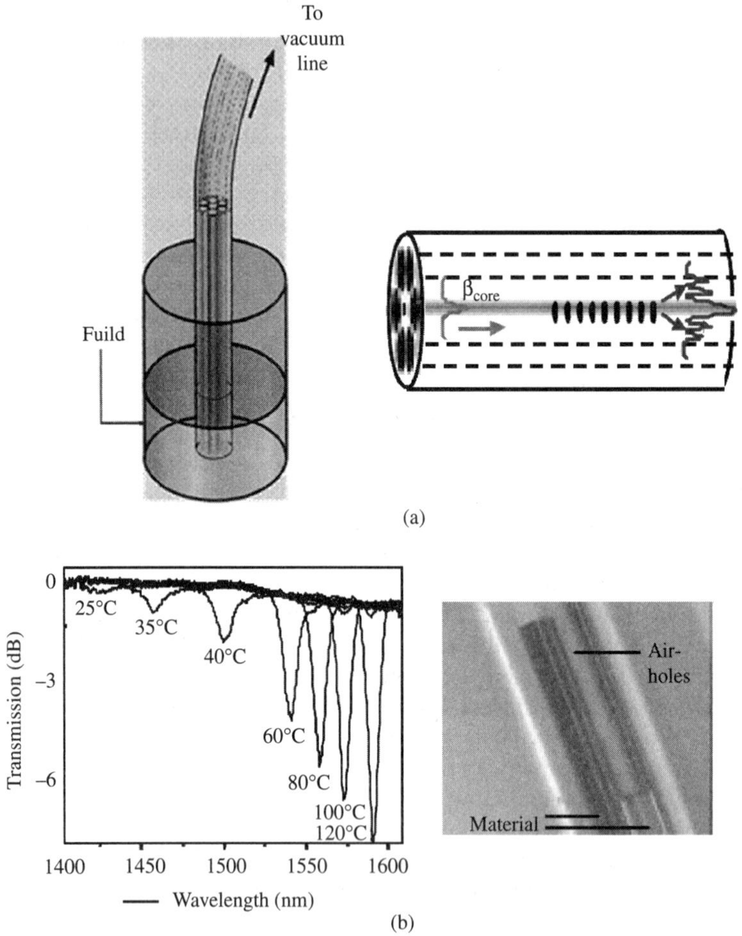

FIGURE 7-5 (top) An optofluidically tuned grapefruit fiber that utilizes external optofluidic tuning to alter the transmission of an LPG written in the grapefruit fiber core. (bottom) Spectrum of the tuned LPG demonstrating controllable extinction. (*B. J. Eggleton, C. Kerbage, P. S. Westbrook, et al., "Microstructured optical fiber devices," Opt. Express, 9, 698–713 (2001).*)

and the optical environment of the fiber cladding is changed, thereby tuning the response of the LPG in both wavelength and amplitude. Figure 7-5 shows the tunable response of this polymer surrounded LPG. Figure 7-6 shows a polymer-infiltrated fiber that only has some of the microstructure filled. When this selectively filled polymer is exposed to the core mode through tapering, the breaking of the symmetry of the microstructure refractive index profile introduces a birefringence into the fiber [65]. Again, the degree of birefringence can be tuned by changing the refractive index of the polymer through thermal control. The selective filling employed here is achieved by

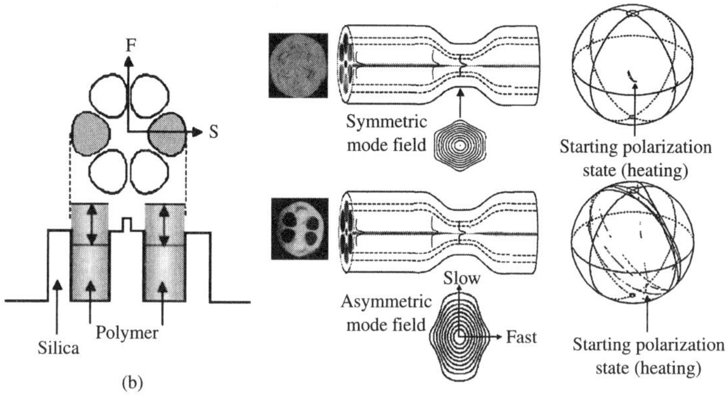

FIGURE 7-6 (left) A schematic of the selectively polymer filled grapefruit fiber. (right) The measured birefringence induced by the selectively polymer filled grapefruit fiber. (*C. Kerbage, P. Steinvurzel, P. Reyes, et al., "Highly tunable birefringent microstructured optical fiber," Opt. Lett., 27, 842–844 (2002).*)

manually blocking some of the microstructure apertures on the fiber face to allow filling of only the unblocked holes.

The previous examples of optofluidic grapefruit fibers have used optofluidic tuning as an aid to fabrication, yet once the fluid monomer phase is introduced and polymerized, the now-solid body is rendered immobile. While tunability is still conferred thermally in these cases, to fully take advantage of optofluidic tuning a mobile fluid phase is required. Figure 7-7 shows exactly how this can be achieved using a grapefruit fiber with a single inclusion filled with a mobile fluid whose refractive index of 1.42 is slightly below that of the surrounding silica [66]. Now mobile, this inclusion is moved controllably into and out of a tapered region of the fiber, thus controlling the exposure of the modal field to the fluid producing a variable amount of polarization phase change, again through broken symmetry in the microstructure [66]. Figure 7-8 also shows this tunable polarization. The fluid (introduced into the fiber by selective suction described above) is actuated by a heater, that lies away from the tapered region. The heater expands the air in the fiber microstructure that in turn pushes the fluid.

The use of fluids inside grapefruit fibers now allows fully tunable, reversible, and reconfigurable in-fiber devices to be realized. Figure 7-8 shows another LPG in a grapefruit fiber but now with separate low- and high-index fluids occupying contiguous spaces along the grapefruit fiber microstructure [67]. Two heaters are now used; one away from the LPG to move the microfluid bodies (the pump heater) and the other on top of the LPG to change its resonance wavelength. The pump heater affects the LPG spectral visibility by changing the overlap of the low-index fluid and the LPG, thereby altering the modal overlap with the fiber core. In combination, these heaters provide complete control over the LPG spectral response, all in fiber.

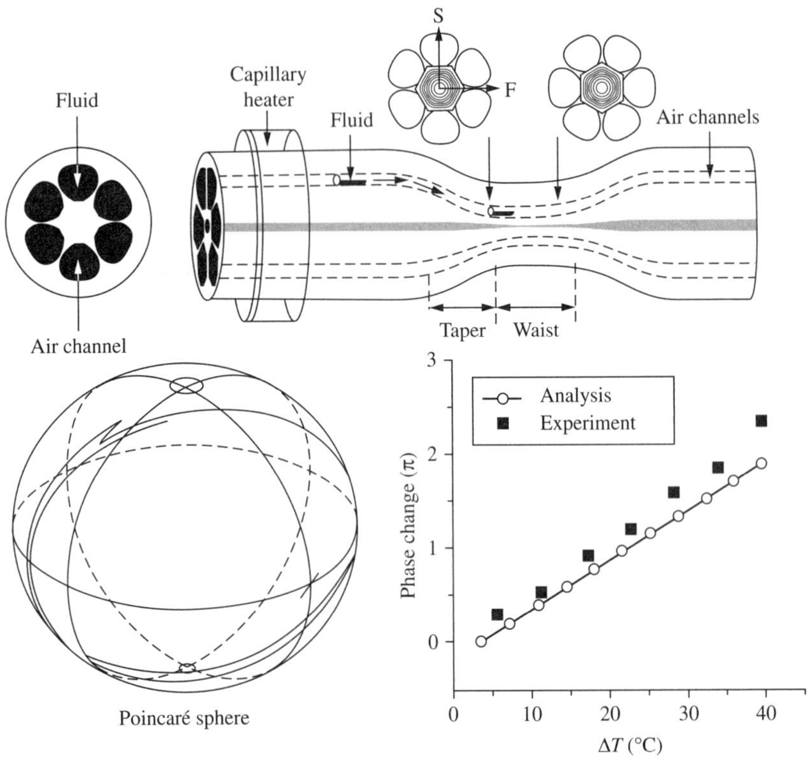

FIGURE 7-7 (top) A schematic of the selectively fluid-filled grapefruit fiber. (bottom) The measured birefringence induced by the selectively fluid-filled grapefruit fiber. (*C. Kerbage, M. Sumetsky, and B. J. Eggleton, "Polarisation tuning by micro-fluidic motion in air-silica microstructured optical fiber," Electron. Lett., Copyright 2002 IEEE.*)

The final evolution of optofluidic tuning inside grapefruit fibers is to consider the fluid body as an optical element in its own right, rather than as simply a way to modulate another structure in the fiber. To this end, an LPG was fabricated out of the microfluid itself, residing in the fiber microstructure [67]. Figure 7-9 shows a schematic of the fluid LPG. The grapefruit fiber is tapered to allow access to the light guided by the microstructure. Then fluid is suctioned into the microstructure as before; however, this time the fiber is rapidly, mechanically oscillated in and out of the reservoir from which the microfluid is drawn. This results in periodically spaced fluid and air bubbles that form the elements of refractive index contrast in the fluid LPG. Since the LPG is mobile, it can be moved away from the tapered region on demand, effectively turning off the LPG resonance. Figure 7-9 also shows that, upon applying an external heating source, the period of the LPG can be dynamically and reconfigurably tuned, allowing complete control over the wavelength of the LPG resonance.

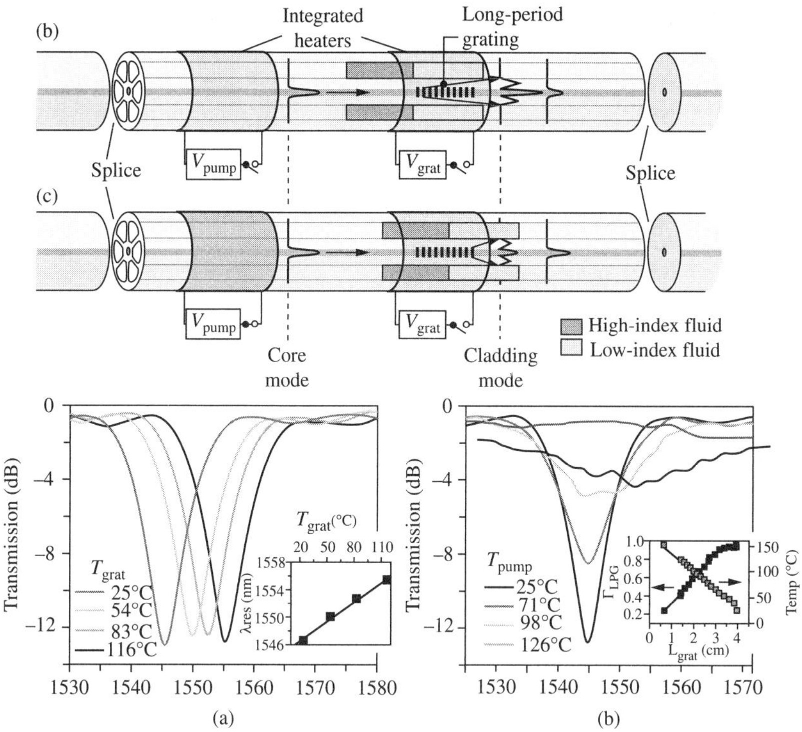

FIGURE 7-8 (top) A schematic of the fluid-filled grapefruit fiber with dual heaters to control the fluid position with respect to the LPG and a heater to tune the LPG directly. (bottom) Spectrum tunability as a function of both heaters' temperature. (*Reprinted with permission from P. Mach, M. Dolinski, K. W. Baldwin, et al., "Tunable microfluidic optical fiber," Appl. Phys. Lett., 80, 4294–4296 (2002). Copyright 2002, American Institute of Physics.*) (See also color insert.)

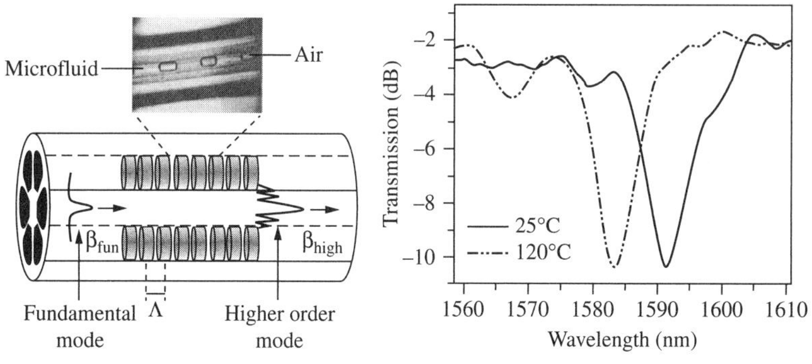

FIGURE 7-9 (left) A schematic of the optofluidic LPG inside the grapefruit fiber microstructure. (right) A graph of the wavelength tunability available through compression of the optofluidic LPG. (*Reprinted with permission from C. Kerbage and B. J. Eggleton, "Tunable microfluidic optical fiber gratings," Appl. Phys. Lett., 82, 1338–1340 (2003). Copyright 2003, American Institute of Physics.*)

Grapefruit fibers provide a wealth of potential optofluidic designs to be realized, thanks both to the easily accessible fiber microstructure and to the SMF compatibility. Combined with fiber post-processing techniques such as tapering and holographic grating writing, optofluidic tuning has not only allowed a broad range of optical components to be fabricated but also endowed those components with significantly more flexibility that is available through traditional fiber fabrication methods.

7-3 Optofluidic Transverse Fiber Quasi-2-D Photonic Crystals

PCFs are designed for use as waveguides. They guide light along their length through confinement provided by photonic bandgap [68] or effective refractive index [17] effects. However, if light is introduced into the side of the PCF, that is "probed transversely," the microstructure acts essentially as a planar photonic crystal [7], displaying a characteristic range of reflected wavelengths (the "band gap") [69]. Further, the hollow inclusions that define the PCF microstructure are a natural home for microfluids that can change the wavelength of the transverse bandgap or be moved to dynamically modulate the response of the transverse fiber [70]. In light of these considerations, transverse PCFs provide an excellent platform for optofluidic tuning.

7-3-1 Optofluidic Transverse PCF

Figure 7-10 presents a schematic of the transverse fiber probing experimental geometry [7]. An SMF is used to deliver probing light to

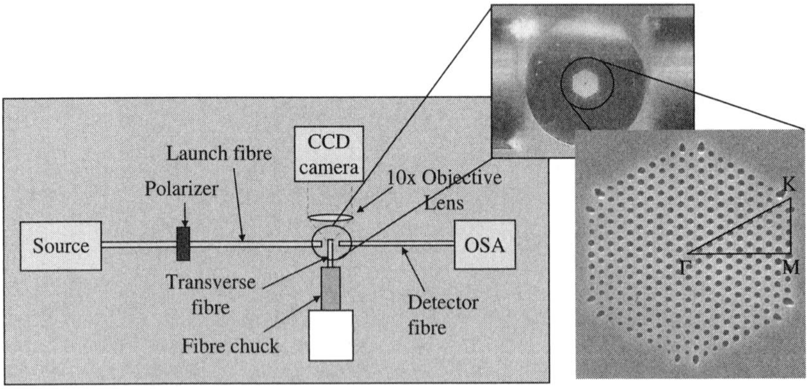

FIGURE 7-10 A schematic of the transverse PCF probing geometry. Shown also are visible light and electron micrographs of the PCF end face. Also shown on the electron micrograph are the crystalline high-symmetry points of the PCF microstructure. (*Reprinted with permission from P. Domachuk, H. C. Nguyen, B. J. Eggleton, et al., "Microfluidic tunable photonic band-gap device," Appl. Phys. Lett., 84, 1838–1840 (2004). Copyright 2004, American Institute of Physics.*)

the transverse fiber. Once the light has interacted with the transverse PCF, it is collected by another SMF and analyzed. These two SMFs are carefully aligned using 3-D positioning stages to ensure the most efficient possible coupling of light in the apparatus. The light source used is a broadband thermal halogen bulb. When probing PCFs the wavelength range of the source is chosen such that it intersects with the partial photonic bandgap of the transverse PCF. A polarizer is also inserted in-line and the SMF kept taut to control and maintain polarization. The collected output light is spectrally analyzed on an optical spectrum analyzer (OSA). Figure 7-10 also shows a close-up photograph of the apparatus being used to probe a transverse PCF. Shown are the two aligned SMFs used for probing and collection on either side of a PCF, whose hexagonal microstructure is plainly visible. The PCF has circular air inclusions of diameter 800 nm with a periodicity of 1.4 μm arranged in a hexagonal packing. The PCF sits between two SMFs. One SMF delivers light from the thermal halogen bulb with range 800 to 1700 nm. The polarizations are labeled TM for electric field parallel to the length of the transverse fiber and TE for the perpendicular orientation. The PCF is held in a rotational chuck to allow orienting various crystal axes to the optical axis between the SMFs and is visually aligned using a microscope to sit as central to the SMF optical axes as possible. Also shown in Fig. 7-10 are the high-symmetry points of the reciprocal lattice of the microstructure photonic crystal [71]. Figure 7-11 shows the method used to introduce fluid into the microstructure of a PCF. A drop of fluid is placed on the end of a separate SMF used as an applicator and is held there using surface tension. The SMF is moved closer and closer to the cleaved end of the PCF until the fluid is drawn into the PCF microstructure under the force of capillarity. Figure 7-11 shows a series of time-lapse photographs of this process. The point at which the PCF is probed is shown using a black arrow. The fluids used are a series of Cargille refractive index matching oils with refractive index between 1.45 and 1.75 in increments of 0.05. The fluid-infused PCF is probed over a wavelength range of 1.1 to 1.7 μm aligned in the Γ-M orientation.

To understand the principle of operation of the tranverse device, a numerical simulation of the effective 2-D photonic crystal is performed using the plane wave method [72] using a 2-D array of circular air inclusions ($n = 1.00$) with a diameter of 800 nm and a periodicity of 1.4 μm in a background of silica ($n = 1.45$). The plane wave method approximates the PCF microstrucure by assuming it is a photonic crystal of infinite extent and uniformity. While this is clearly not the case for the PCF, the results of the calculation provide approximate locations of the photonic crystal bandgaps. Figure 7-12 shows a comparison for the dispersion relations in the Γ-M direction in both TE and TM polarization for photonic crystal inclusions with refractive index of 1.00 (low index) and 1.75 (high index). The partial bandgaps are shown as solid color bars on the dispersion relation.

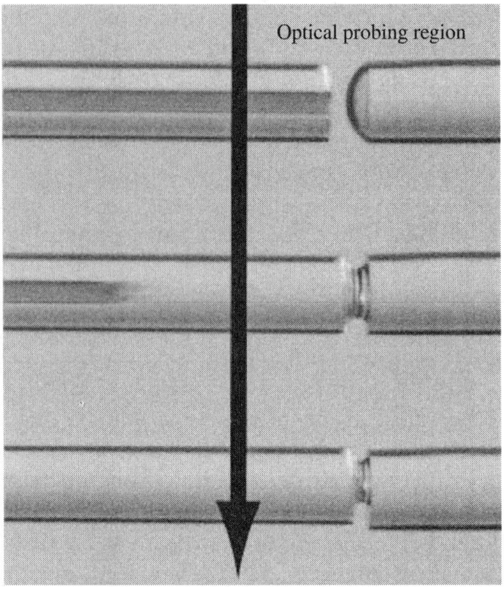

FIGURE 7-11 Feeding fluid into the microstructure of a PCF, utilizing the hydrophilic nature of silica. The optical probing position is shown using the black arrow. (*Reprinted with permission from P. Domachuk, H. C. Nguyen, B. J. Eggleton, et al., "Microfluidic tunable photonic band-gap device," Appl. Phys. Lett., 84, 1838–1840 (2004). Copyright 2004, American Institute of Physics.*)

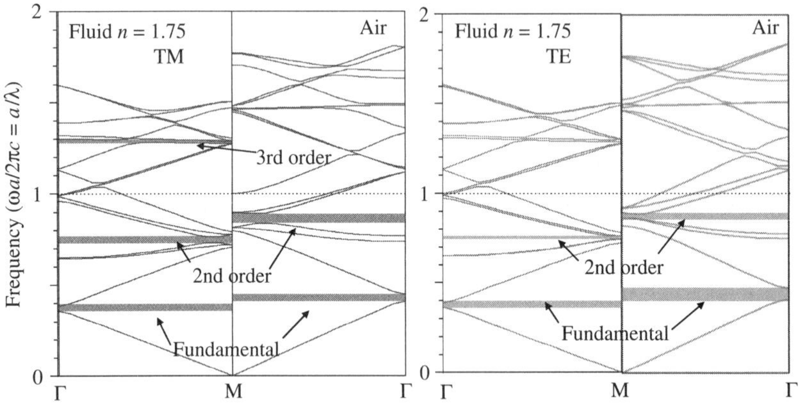

FIGURE 7-12 Dispersion relations for the transverse PCF in the Γ-M orientation, in low- and high-index inclusion regimes for both TE and TM polarization.

Several differences are noticeable between the high- and low-index regimes. First, the number of partial bandgaps visible in the TM increases to three once the PCF has high-index inclusions. Further, the bandgaps in both polarizations are shifted to longer wavelengths.

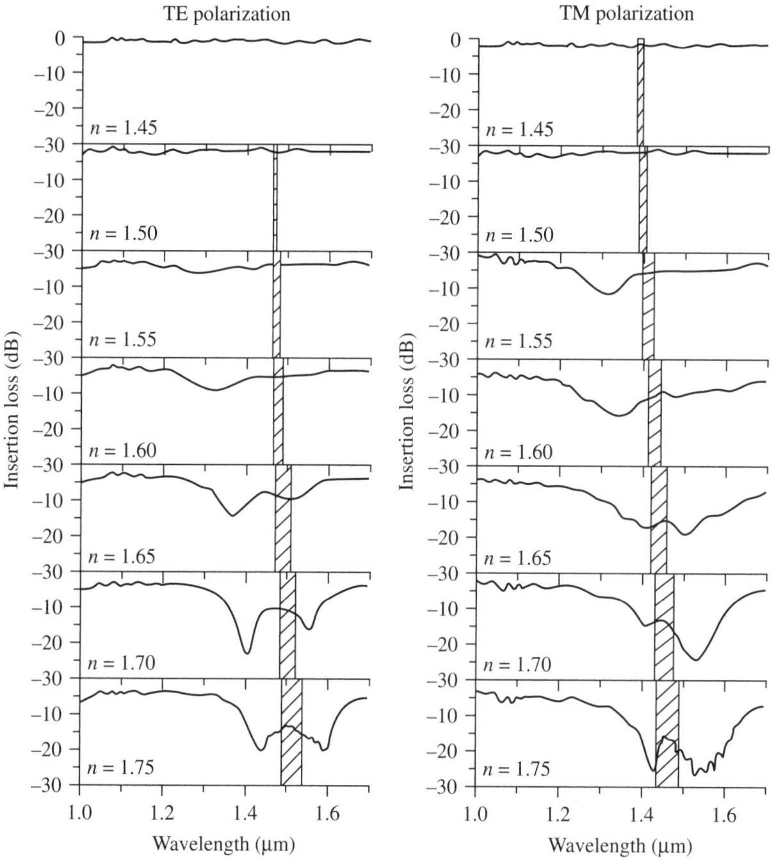

FIGURE 7-13 The emergence of the second order partial bandgap in the Γ-M orientation as the refractive indices of the inclusions are increased using fluids.

Figure 7-13 shows a series of spectra (in both TE and TM polarizations) of the transverse PCF in the Γ-M orientation as fluids of different refractive index are infiltrated into the microstructure. Also shown are the positions and widths of the partial gaps predicted by plane wave method for a given inclusion index. For a fluid refractive index of 1.45, the fluid renders the PCF practically transparent and the transmission spectrum is featureless. As the refractive index of the fluid and the inclusions increases, the partial bandgap begins to appear in the spectra of both polarizations. As the refractive index contrast of the PCF continues to increase, the bandgap becomes broader and moves to longer wavelengths, as predicted by the plane wave method.

7-3-2 Dynamic Optofluidic Attenuator

Since the fluid inside the PCF microstructure is mobile, this mobility may be used to create a reconfigurable photonic crystal switch [70]. Figure 7-14 shows the optical experimental setup: it is the

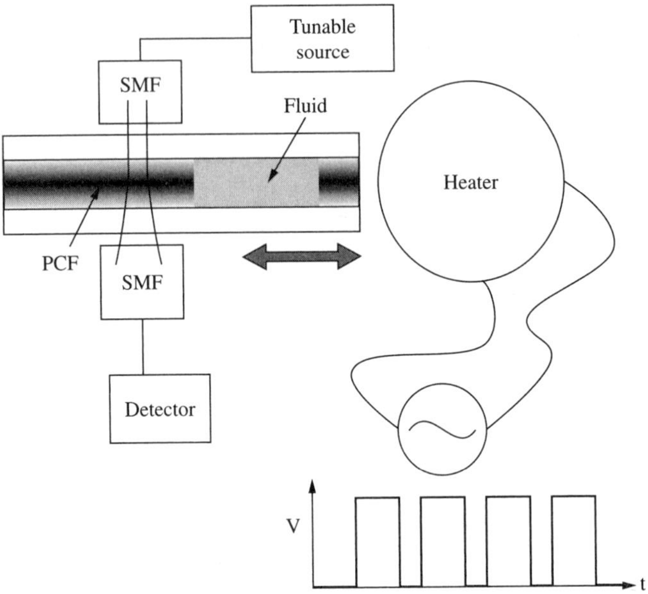

FIGURE 7-14 A schematic of the reconfigurable optofluidic switch, showing the PCF infused with fluid, the heat source, and the step driving voltage. (*P. Domachuk, H. C. Nguyen, and B. J. Eggleton, "Transverse probed microfluidic switchable photonic crystal fiber devices," IEEE Photon. Technol. Lett., Copyright 2004 IEEE.*)

same as that used in the experiment described earlier. The same PCF is probed along the Γ-M orientation mounted between two SMFs. However, a single fluid with refractive index 1.50 is used. The PCF is probed at a wavelength within the second order bandgap of the empty PCF $\lambda = 1.5\ \mu m$, which is strongly reflected, and the output is monitored on a detector. The fluid body is moved along the PCF using suction from a syringe. Once there is a body of air between the fluid and the end of the PCF, it is sealed with two-part epoxy. This pocket of air is heated to provide motivation to the fluid body inside the PCF. The fluid is moved in and out of the optical probing region, rendering the PCF alternatively transparent and highly reflective. The heater is manufactured by sputtering a thin film of gold onto a silica capillary. A voltage passed through the film causes resistive heating. In this experiment, voltage across the heater was alternated between 7 and 0 V using a square wave modulation with a period of 4.5 s, providing a temperature fluctuation of 50°C. This causes the fluid body to be displaced 300 μm over the course of the second heating cycle.

Figure 7-15 shows, at top, the temporal response of the reconfigurable optofluidic switch and, at bottom, a detail of one period of the response. The device is driven using the square wave with

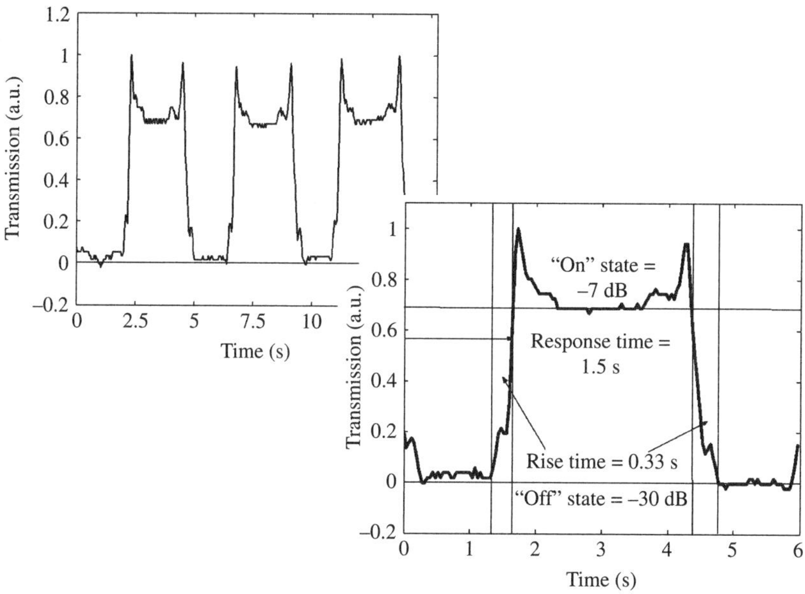

FIGURE 7-15 The temporal response of the device (top) for a number of periods and (bottom) detail of one period. (*P. Domachuk, H. C. Nguyen, and B. J. Eggleton, "Transverse probed microfluidic switchable photonic crystal fiber devices," IEEE Photon. Technol. Lett., Copyright 2004 IEEE.*)

period described above. After the initial rise of voltage, the response of the device is approximately 2 s with a rise time of 0.5 s. The discrimination between the transmission and reflection of the device is 20 dB, arising from the high reflectivity of the photonic crystal microstructure. The large response time of the structure is due to the high thermal mass of the silica PCF [73].

7-4 Ultracompact Microfluidic Interferometer

Future photonic devices need broad tunability and reconfigurability. Such functionality enables higher performance out of a given device with clear integration and time/material investment benefits. Optofluidic tuning enables these device attributes. The high refractive index contrast between fluids or between a fluid and the surrounding air enables very short device lengths. The inherent mobility of the fluid phase imparts tunability and reconfigurability to such devices as well. In this section, a reconfigurable, optofluidic, and compact interferometer is discussed [74,75]. The high refractive index contrast of a fluid-air meniscus is used to enable a very compact interferometric device with a path length under 8 wavelengths. The air-fluid interface (meniscus) of a fluid plug inside a square silica capillary is placed into the beam between two SMFs, generating an interferometric response that is tunable in amplitude due to the inherent mobility of the fluid plug. The capillary is treated using silanization chemistry to manipulate the meniscus shape for optimal optical performance.

Figure 7-16 (top) shows a generic Mach-Zehnder interferometer. The signal entering the device is split into two arms, one of which induces a comparative delay resulting in a phase difference that underlies the characteristic attenuation of the interferometers spectral structure. Figure 7-16 (bottom) shows a schematic of the single-beam interferometer in which the incident beam is split in two by an interface, on either side of which is a material of differing refractive index, effectively resulting in a Mach-Zehnder interferometer. As opposed to what is usually found in a Mach-Zehnder interferometer, the

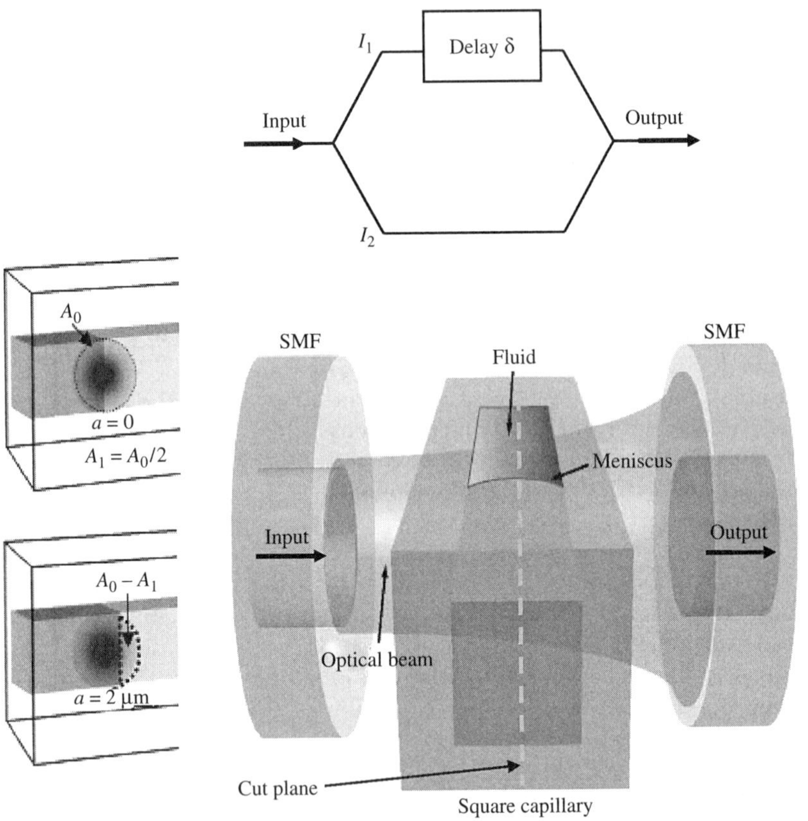

FIGURE 7-16 (top) A schematic of a Mach-Zhender interferometer showing the relative light paths and delay in one arm. (bottom) A schematic of the compact single beam optofluidic interferometer. (*C. Grillet, P. Domachuk, V. Ta'eed, et al., "Compact tunable microfluidic interferometer," Opt. Express, 12, 5440–5447 (2004).*)

intensity of the two beams in the device is not necessarily equal. Since a fluid-air interface is used to provide optofluidic tuning of the device, the mobility of the fluid allows the visibility of the interferometer modulation to be tuned simply by moving the meniscus with respect to the center of the beam. To change the wavelength of the Mach-Zehnder resonances, the diameter of the square capillary or the refractive index of the fluid is changed.

Figure 7-17 shows photographs of the tunable microfluidic interferometer. The meniscus used to introduce the optical path difference is formed between a short length of deionized water of index 1.33 and air inside a section of square silica capillary. Light is coupled in the transverse direction in the manner described in Sec. 7-3. The SMFs shown in Fig. 7-17 are used to transversely probe the meniscus with an 80-μm spacing between them. The square capillary is surrounded by index matching fluid to minimize reflections at the component interfaces. The square capillary was tapered using the flame brush method [31]. Its dimensions were reduced to an inner width of 10 μm and an outer width of 80 μm. The tapering was performed to minimize beam divergence through the device and, therefore, enhance the coupling between the excitation and collection fibers as well as to ensure that the fiber mode did not interact excessively with the interior walls of the capillary core. The square capillary was seen, upon examination, to retain its square profile after tapering.

The meniscus in the square capillary is inherently curved, due to the balance of surface tensions between the silica, water and surrounding air. This curved surface forms a contact angle of 40 degrees where it intersects with the capillary surface. Figure 7-18 shows this behavior in the square capillaries used in the device. This curvature of the meniscus presents an issue optically due to incident light being

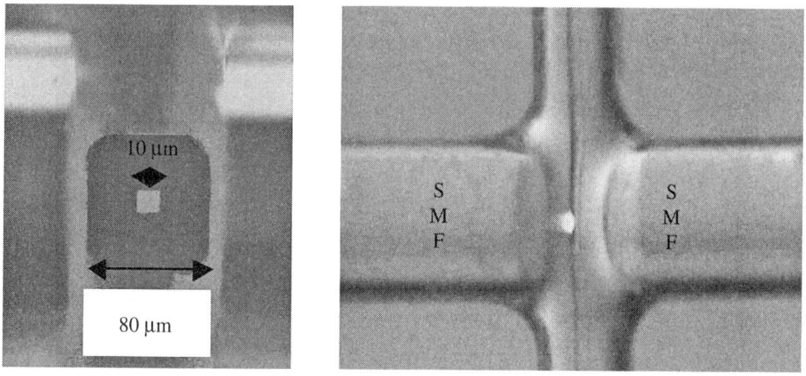

FIGURE 7-17 Photographs of the compact optofluidic single-beam interferometer with views from the front (left) and top (right), showing the refractive-index-matching oil surrounding the square capillary. (*C. Grillet, P. Domachuk, V. Ta'eed, et al., "Compact tunable microfluidic interferometer," Opt. Express, 12, 5440–5447 (2004).*)

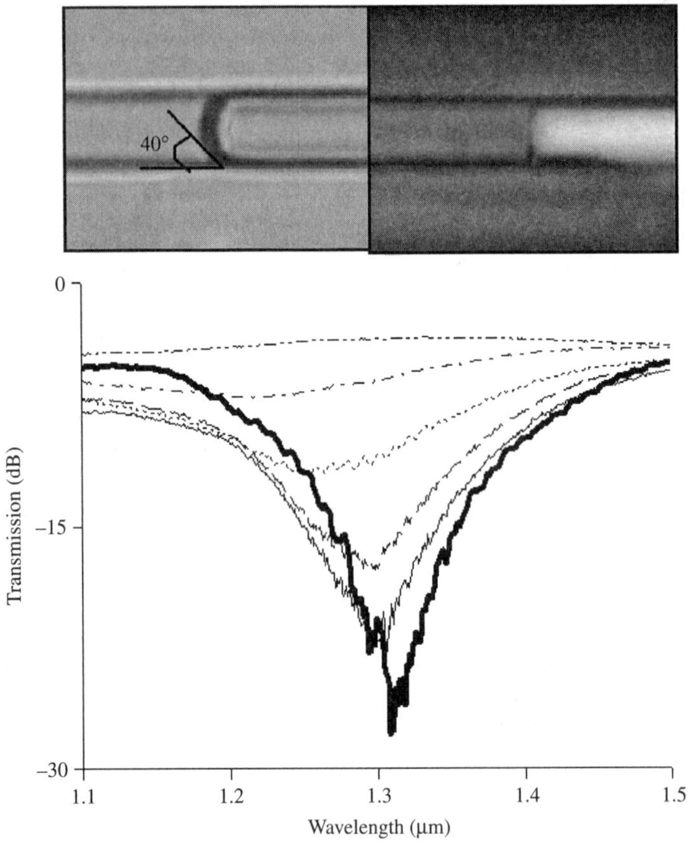

FIGURE 7-18 (top) The behavior of the fluid meniscus in an untreated square capillary (left) and a flat meniscus in an organosilane treated capillary (right). (bottom) The transmission of the device as a function of meniscus position from completely outside the beam to optimally inserted. (*C. Grillet, P. Domachuk, V. Ta'eed, et al., "Compact tunable microfluidic interferometer," Opt. Express, 12, 5440–5447 (2004).*)

scattered off the curved surface. Thus a flat meniscus, one with a contact angle of 90 degrees, is desirable. A flat water meniscus is attainable on silica using a silanization monolayer surface treatment [76]. This process involves the chemical addition of organosilanes to the silica surface to form an atomic monolayer. If this chemistry is applied all over the surface, the total surface energy of the silica is reduced by this molecular monolayer. Such a monolayer is appropriate for surface coatings in an optical environment since its thickness is considerably subwavelength, as opposed to polymer surface treatments that are significantly thicker. In this case, the organosilane used is dodecyltrichlorosilane ($C_{12}H_{25}SiCl_3$)—an organic molecule with a 12-carbon chain attached to a chlorosilane group. Experimentally, this is found to adjust the surface energy of silica so that the water meniscus contact angle is close to the desired 90 degrees. Figure 7-18 shows the flattening effect of the organosilane layer on the water meniscus in the square capillary.

The now flat meniscus is moved using pressure applied from a syringe coupled to the square capillary using silicone rubber tubing sealed with epoxy. A video microscope is used to track the position of the meniscus. The SMF probe is connected to a thermal halogen broadband white light source with a wavelength range of 1.0 to 1.8 μm while the collection SMF is connected to an OSA. Figure 7-18 shows the device transmission as the meniscus is pushed across the beam. The transmission starts out spectrally flat, and then develops a pronounced resonance as the meniscus traverses the beam. Figure 7-19 shows (solid line) the transmission spectrum when the meniscus crosses the center of the beam. A strong resonance is observed, centered at 1.31 μm. The resonance depth reaches almost −30 dB, whereas the out-of-resonance loss is maintained at −4 dB. This out-of-resonance loss is related to beam divergence across the total extent of the square capillary and could be further reduced. Figure 7-19 also shows (dots) the experimental transmission of the device without organosilane coatings in the silica capillary. There is a marked difference (some 10 dB) in the out-of-resonance insertion loss between the silanized and non-silanized capillaries, due to the curvature of the meniscus deflecting the beam away from the output.

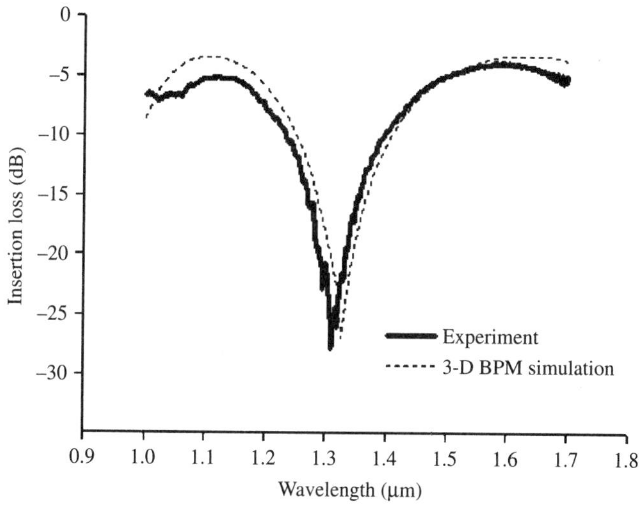

Figure 7-19 Experimental (solid line) spectral response of the device as compared to 3-D BPM numerical simulation (dashed line) when the meniscus is well centered. (*C. Grillet, P. Domachuk, V. Ta'eed, et al., "Compact tunable microfluidic interferometer," Opt. Express, 12, 5440–5447 (2004).*)

The compact, single-beam interferometer provides natural exposition of the advantages of optofluidic tuning. The very high refractive index contrast between the water and its meniscus provides a 2π phase shift in a path length of only 10 μm or 7.7 wavelengths at the resonance considered above. Creating a vertical feature on the micron scale is rendered a routine task using optofluidics. Even though the meniscus is curved naturally in a silica capillary, established surface chemistry can provide almost arbitrary control over the meniscus shape. The mobility of the fluid automatically supplies the optofluidic interferometer with additional functionality. This intensity modulation of the device resonance is exactly the kind of enhancement enabled by optofluidics that is impossible using solid-state materials.

7-5 Fluidic Photonic Bandgap Fiber

In Sec. 7-3, we showed how photonic bandgap structures may be used as wavelength-selective reflectors. Such structures may then also be used for light confinement (indeed, photonic crystals were first proposed for this purpose), either in a cavity or in a waveguiding geometry. In photonic bandgap fibers (PBGFs), light propagating perpendicular to the plane of periodicity is confined in a 2-D defect core [77]. Unlike the in-plane propagation experiment we described earlier, however, the wavelengths transmitted through PGBFs lie in the bandgap, and the transmission spectrum consists of a passband (or a series of passbands) rather than a notch. Also, we generally observe bandgap guidance in fibers only when the core index is lower than the effective cladding index (bandgap-guided modes can still exist if this is not the case, but index-guided modes dominate [78, 79]); unlike the case of transverse propagation, whether we observe bandgap effects depends on whether the refractive index of the fluid is greater or less than that of the fiber background material.

Here we focus on PBGFs consisting of silica-core PCFs filled with a high-index fluid, shown for low-index fluid in Fig. 7-11. We note that fluidic PBGFs may also be made by filling hollow-core PCFs with low-index fluids [80–82]. These are technologically significant because low-index fluids include water and most organic solvents, and since the mode is confined in the fluid, rather than just its evanescent tail, one can make very efficient chemicals or biosensors (e.g., using Raman, fluorescence, or absorption spectroscopy) with long interaction lengths. Of course, one may use selective filling techniques [57,65,83–85] to achieve index guidance in a low-index fluid core [83–90], and the increased transmission bandwidth of such a design is often desirable when the primary application is efficient liquid/light interaction. However, for many of the device designs we will describe, the inherently resonant nature of bandgap guidance combined with the enhanced tunability of liquid phase materials provides an advantage over index-guided fluidic fibers. Furthermore, our specific choice of a PBGF consisting of an array of high index rods in a low index background is amenable to a quasi-analytic model based on the resonant scattering spectrum of a single high index rod [59,91–94] (an analogous, though nonanalytic, model exists for hollow PBGFs with a honeycomb PC cladding [95]). This is closely linked to the analytic waveguide theory of dielectric cylinders [59,92–94,96,97] and makes it very easy to design the transmission properties of the fiber, and by extension, the tuning properties of devices formed from such fibers [98,99].

Representative transmission spectra of fluidic PBGFs are shown in Fig. 7-20a, where the high transmission regions correspond to photonic bandgaps of the cladding. As we mentioned, this fiber is inherently a bandpass filter, and since the center frequency of the bandgaps depends on the index contrast (see Sec. 7-3), we can tune the passbands by appropriate choice of filling fluid using the same PCF template. The thermo-optic coefficients of most fluids are very large relative to that of silica (~10–100× greater), so one can make the PBGF into a tunable filter by uniformly heating it [100]. In a variant on this design, one applies a thermal gradient to the fiber, so that the center frequencies of the bandgaps vary as a function of propagation distance along the fiber [101]. Only those wavelengths that lie in the bandgap of both the hottest and coldest points of the fiber are transmitted, so by increasing the thermal gradient, one can reversibly narrow the passbands or completely close them off (Fig. 7-20b). From a resonant scattering point of view, axial variations in the diameter of the high-index inclusions are equivalent to variations in the index contrast, so one may also shape the passbands by weakly tapering the fiber [102]. Highly tunable filtering based on dual-core fluidic PBGF directional couplers has also been demonstrated [103]. One may further enhance the functionality by employing fluidic PBGF filters based on liquid crystals (LCs), which have far stronger thermal tunability than isotropic fluids [102,104–108] and can be tuned electrically [108–110] or photochemically [110].

Though we have focused thus far on bandpass filtering, we can make a fluidic PBGF band rejection filter by using an LPG, as shown in Fig. 7-21. Silica/fluid fibers are not photosensitive like germanosilicate in SMF or grapefruit fiber, but LPGs can be formed in fluidic PBGFs by applying spatially periodic mechanical stress [111,112], electric arcs [113], or electric fields (in the case of LC-PBGFs) [112], or by launching an acoustic wave along the fiber [114]. LPGs are of interest for their own sake in that they allow one to experimentally probe the cladding mode and higher order mode properties of PBGFs [111,115] that are very different from those of index-guided fibers. From a device perspective, fluidic PBGF LPGs have very large thermal or index sensitivity [116] (~1.5 nm/°C, or ~3500 nm/R.I.U., has been demonstrated for isotropic fluids, even larger for LC-PBGFs [112]) owing not so

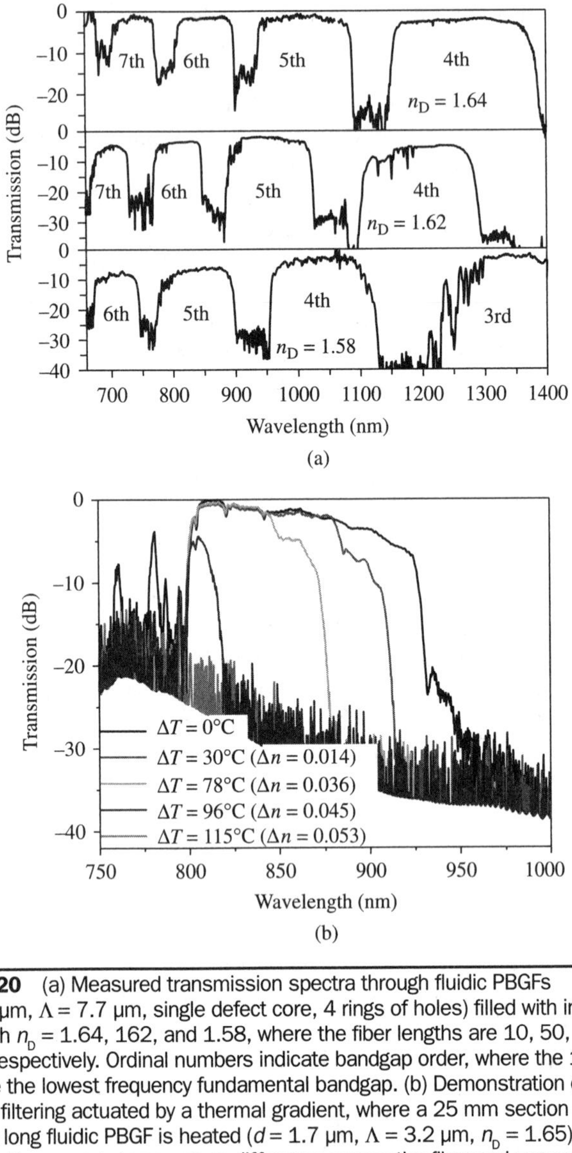

FIGURE 7-20 (a) Measured transmission spectra through fluidic PBGFs (d = 3.5 μm, Λ = 7.7 μm, single defect core, 4 rings of holes) filled with index fluids with n_D = 1.64, 162, and 1.58, where the fiber lengths are 10, 50, and 38 cm, respectively. Ordinal numbers indicate bandgap order, where the 1st order would be the lowest frequency fundamental bandgap. (b) Demonstration of dynamic filtering actuated by a thermal gradient, where a 25 mm section of a 12.4 cm long fluidic PBGF is heated (d = 1.7 μm, Λ = 3.2 μm, n_D = 1.65); legend indicates the average temperature difference across the fiber and corresponding index gradient in the fluid. (*P. Steinvurzel, B. J. Eggleton, C. M. de Sterke, et al., "Continuously tunable bandpass filtering using high-index inclusion microstructured optical fiber," Electron. Lett., Copyright 2005 IEEE.*) (See also color insert.)

much to the properties of the fluid but to the resonant shift of the transmission bands we described earlier. This is markedly different from the case of conventional fibers, where one must carefully design the fiber waveguide dispersion and choose the appropriate mode pair to achieve comparable tunability [117]. The index sensitivity of the

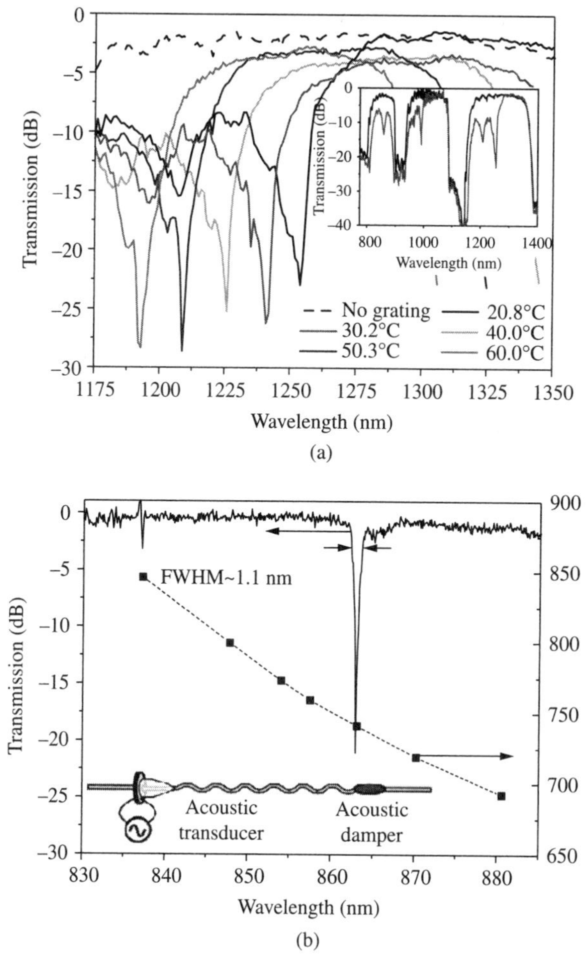

FIGURE 7-21 (a) Inset: transmission spectrum through a fluidic PBGF with (red) and without (black) a mechanical stress-induced LPG with a period of 660 μm. (*P. Steinvurzel, E. D. Moore, E. C. Mägi, et al., "Long period grating resonances in photonic bandgap fiber," Opt. Express, 14, 3007–3014 (2006).*) Main figure shows temperature tuning of the LPG resonance at 1250 nm [116]. (b) Spectrum of an acousto-optic LPG in a similar fiber for an acoustic wavelength of 740 μm (left axis). Squares indicate the optical beat length as a function of wavelength (right axis), where the large slope of this curve (proportional to the group index difference between the coupled modes) is what makes the resonance narrow. (*D.-I. Yeom, P. Steinvurzel, B. J. Eggleton, et al., "Tunable acoustic gratings in solid-core photonic bandgap fiber," Opt. Express, 15, 3513–3518 (2007).*) (See also color insert.)

resonant loss in PBGFs depends on the geometry and refractive index contrast between the fluid and the background material [116], and PBGFs optimized for sensing could in principle have at least one, possibly two orders of magnitude higher sensitivity than that demonstrated in Ref. [116], making them competitive with surface plasmon resonance sensors [134]. The dispersive properties of the fiber modes also determine the width of the rejection band (bandwidth inversely proportional to the group index difference of the coupled modes [118]), and in order to achieve a narrow rejection band, one must design the fiber around that goal [119]. PBGFs inherently have the necessary dispersive properties for narrow-band coupling, and a rejection bandwidth of 1.1 nm is shown in Fig. 7-21b [114].

The resonant dispersive properties of PBGFs are also useful for shaping optical pulse propagation. The waveguide dispersion of a PBGF-core mode can be understood in terms of a simple Kramers-Kronig model [120]; the frequency bands where the PBGF does not transmit may be modeled as a resonant loss, so the mode index is rapidly increasing as one approaches the loss band from low frequency and rapidly decreasing on the approach from high frequency (see Fig. 7-22a and 7-22b). Within each transmission band, then, the mode index traces out a sideways-s-shaped curve, and more importantly, the group velocity dispersion (GVD) goes from normal to zero to anomalous within each band [121]. Therefore, tuning the bands (e.g., by applying heat) not only changes the power transmission spectrum but

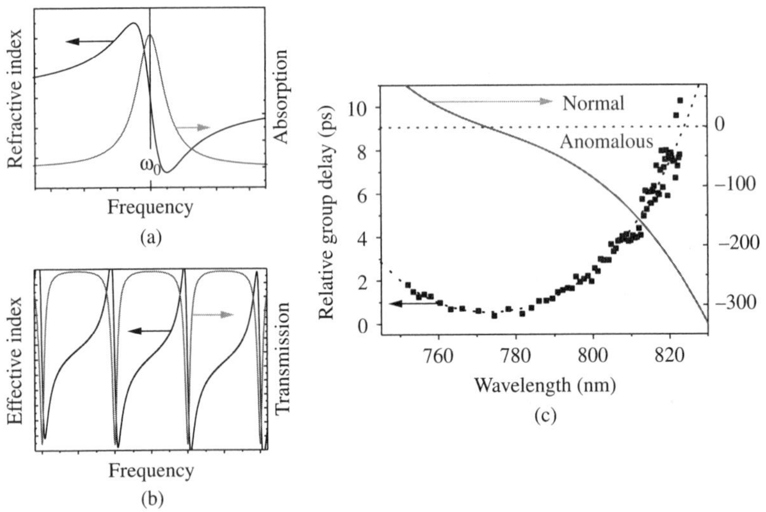

FIGURE 7-22 Heuristic model for waveguide dispersion in PBGFs. (a) Lorentzian absorption peak at ω_0 gives rise to dispersion on either side of the peak. (b) Low-transmission bands of PBGFs (out of bandgaps) may be modeled as resonant loss with a similar effect on dispersion away from the resonance (in the bandgaps), resulting in strong waveguide dispersion and GVD. (c) Interferometrically measured group delay (squares) and polynomial fit (black line) through a 35-cm-long fluidic PBGF (left axis), and corresponding GVD (red line, right axis) showing regions of normal, zero, and anomalous dispersion in a single transmission band (see Ref. 127 for details).

also changes the dispersion profile, and the bandpass filters we described earlier also function as tunable delay lines [122]. The resonant nature of the PBGF modal dispersion enables one to achieve zero or anomalous dispersion at short wavelengths (Fig. 7-22c) [123,124] without the need for a small-core and high-index contrast as with index-guided fibers. Fluidic PBGFs then offer an attractive platform for investigating nonlinear pulse propagation at wavelengths below 1 μm [60,125]. In Fig. 7-23 we show that for femtosecond pulse

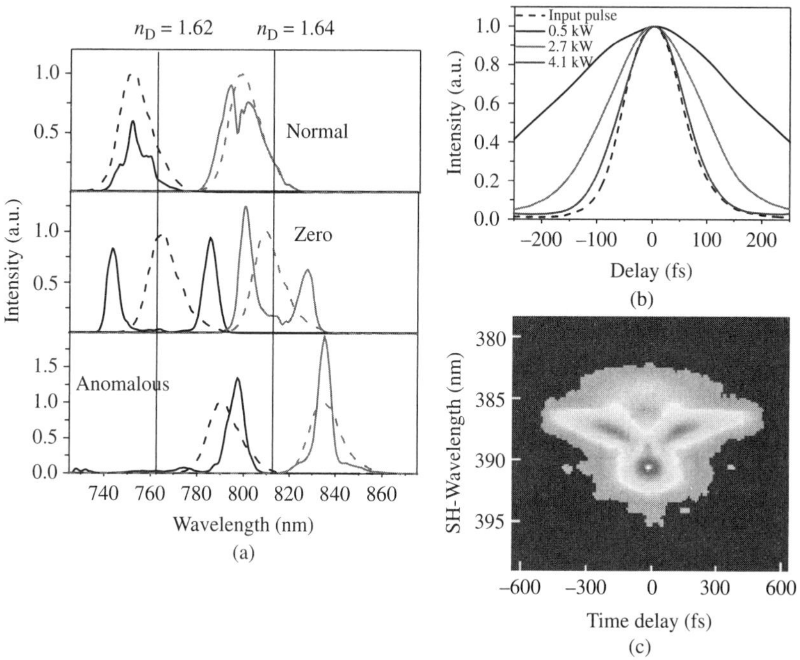

FIGURE 7-23 (a) Measured spectra of 70–100 fs pulses showing the effect of index scaling on waveguide dispersion, where ~0.02 index change shifts the dispersion profile by ~50 nm. Dashed lines correspond to input pulse spectrum; solid lines to spectrum after 40–60 cm length of PBGF with nonlinear parameter γ~16.2 (kW·m)$^{-1}$; black = n_D 1.62, 4.1 kW peak power; red = n_D 1.64, 3.7 kW peak power; center pulse wavelengths are as indicated and vertical lines indicate dispersion zero. Top spectra show SPM induced broadening, middle show dispersive wave radiation and soliton recoil, and bottom show soliton propagation with Raman self-frequency shift (see Ref. 126). (b) Autocorrelation time traces showing soliton formation with increasing peak power for n_D = 1.62 fiber at 780 nm. (*A. Fuerbach, P. Steinvurzel, J. A. Bolger, et al., "Nonlinear propagation effects in anti-resonant high-index inclusion photonic crystal fibers," Opt. Lett., 30, 830–832 (2005).*) (c) Measured-time and frequency-resolved spectrograph of pulse propagation at the dispersion zero, where the short wavelength band elongated along the time axis corresponds to the dispersive waves and the long wavelength band compressed along the time axis is the soliton. (*A. Fuerbach, P. Steinvurzel, J. A. Bolger, et al., "Nonlinear pulse propagation at zero dispersion wavelength in anti-resonant photonic crystal fibers," Opt. Express, 13, 2977–2987 (2005).*) (See also color insert.)

propagation from a Ti:sapphire laser dominated by self-phase modulation (SPM) [126], we can observe nonlinear broadening [126], soliton propagation [126], or dispersive wave radiation [127,128] depending on whether the dispersion is normal, anomalous, or zero at the launch wavelength. The modal dispersion of fluidic PBGFs is also of interest for phase-matched nonlinear processes; the fact that the modal dispersion profile is similar in each transmission band can enable wideband parametric amplification across different bandgaps [129], where temperature tuning may enhance the wavelength range or conversion efficiency. Finally, whereas the experiments described here rely on the optical nonlinearity of the silica core, the nonlinear properties of the fluid itself offer another rich area of investigation [90,131], especially with regard to the PBGF cladding modes [132–133].

7-6 Future Directions

7-6-1 Photonic Devices

Many of the device geometries discussed in this chapter are academic proofs of concepts, somewhat remote from real-world applications. It is hard to imagine that a thermally driven microfluidic optical switch with a 2-s response time can compete against the many other, much faster optical switching technologies that are already commercially available. Speed, however, need not be an issue: in principle, a fluid can be driven up to the speed of sound. As seen in some of the examples given earlier, the distances over which a fluid needs to be displaced to achieve an optical response can be as short as a few microns. This limits the reconfiguration time to a few nanoseconds, which in many instances is more than enough. We nevertheless believe the greatest potential for microfluidic PCF devices lies not in devices needing ultrafast reconfiguration times, but rather in the possibility to tune optical properties such as dispersion or filter characteristics over wide ranges within milliseconds.

In particular, one of the most promising aspects of microfluidic optical fibers is the demonstrated ability to tune their dispersion properties over wide ranges—be it through the thermal or electrical tuning of bandgaps in fluid or liquid crystal filled photonic bandgap fibers, or through adjusting the refractive index in the holes of an index-guiding fiber. While this tunable dispersion has already been used to demonstrate some nonlinear phenomena at unusual wavelengths, as discussed in Sec. 7-5, many more applications can be foreseen. It has been numerically predicted that some PCF geometries should allow second and third harmonic generation within the fundamental mode (as opposed to harmonic generation into a higher order mode), allowing very large conversion efficiencies [135]. All-in-fiber degenerate four-wave mixing for efficient wavelength conversion from 532 nm to shorter wavelengths has also been suggested using fluid-filled bandgap fibers [129], as has multiorder dispersion engineering assisted by microfluidics for optimal four-wave mixing [136]. In both cases, the control over the dispersion required to achieve phase matching is rather stringent, but realistic using tunable fluid-filled PCFs. Recent development in the understanding of supercontinuum generation also shows that a close control over dispersion properties can increase dispersive wave generation and hence improve the power density in the blue part of the spectrum [137,138]. Fluid-filled PCFs further offer the possibility to locally tune the dispersion along the length of the fiber. This could be used for soliton compression, to gain further control over supercontinuum generation [139], or even for harnessing optical rogue waves [140,141]. Using liquid cores to generate supercontinuum over longer wavelength ranges has also been suggested [90,130]. Finally, PCFs in general have been proposed for creating ultraflat dispersion fibers, or for dispersion compensation in telecommunications links [142–146]. However, many of the PCF designs suggested for these purposes require unrealistically stringent fabrication tolerances. By adding the tunability that fluids provide in PCFs, fabricating these devices could become more realistic [147].

Fluid-filled PCFs used as tunable spectral filters also offer unique characteristics that are bound to be used in future work. We have already discussed the tunable band and notch filters that can be achieved combining PCFs, their bandgaps, LPGs, and microfluidics. Tunable short-pass filters can also be made, using the refractive index dependence of the fundamental core mode cutoff within a PCF taper [32,148]. However, it is perhaps the fact that these filters can be distributed along an appropriately designed fiber that will attract the most interesting applications and further work. Indeed, gain-doped solid-core PBGFs inherently suppress unwanted optical emission in fiber amplifiers and lasers [149]. Making this effect tunable through microfluidics should allow unprecedented control over amplifier noise and enable higher power all-fiber tunable lasers.

We briefly mentioned in the preceding section that the nonlinearity of the fluids infiltrated in PCFs offers new possibilities that have only just started to be explored [90,131–133]. PCFs, with their 2-D periodic arrangement of very long holes, indeed offer a unique platform for the experimental analysis of 2-D discrete nonlinear dynamics: when filled with high-index fluids, each hole of the PCF becomes an individual waveguide, so that the fiber becomes an almost ideal array of coupled waveguides. Other work [132] has shown that the thermo-optical nonlinearity of the fluids creates nonlocal, nonlinear coupling between waveguides, and has demonstrated nonlocal gap solitons in such a geometry [132]. It has also been shown that nonlinear localization in space and time (space-time solitons, or light "bullets") can occur in the fluid-filled PCF geometry [133], a phenomenon that cannot exist in continuous nonlinear media. Fluid-filled PCFs thus enable entirely new (and relatively simple) ways of experimentally studying 2-D discrete nonlinear dynamics, which offer a wealth of unusual physical phenomena, in large part identical to those of cold atoms in optical lattices [150], which are experimentally much more challenging.

While we have only discussed properties emerging from microfluidics using liquids, the infusion and flow of gases within fibers is emerging as a field in its own. Hollow-core PCFs filled with gases have been used for sensing [150,151] and as gas cells for frequency references [152], but also as a platform to exploit optical nonlinearity of gases far more efficiently than can be done in gas-filled capillaries or free-space tight focusing geometries. This offers particularly intriguing prospects, such as the generation of frequency combs using cascaded Raman effects in a hydrogen-filled PCF [153], or in-fiber electromagnetically induced transparency [154]. One natural extension of this work is to use optical forces to transport, trap, or accelerate atomic gases, single atoms, particles [155,156], or even Bose-Einstein condensates within the fiber. In a related context, optical forces have been used to transport microbeads along hollow-core PCFs [152], and such optically displaced microspheres have in turn been used to write reconfigurable LPGs in fluid-filled PCFs [157].

7-6-2 Sensing

One of the most obvious—and in many aspects most promising—applications of microfluidics in PCFs, and one that has been discussed throughout this chapter, is sensing, and in particular biochemical sensing. Given their tremendous potential impact on medical diagnostics, environmental monitoring, and threat detection, biosensors are a very active field of research. Detection by optical techniques is by no means the only approach in the field, but it may be the one likely to achieve the highest sensitivities [134], as demonstrated by surface plasmon resonance (SPR) sensors [158], surface-enhanced Raman scattering (SERS) [159] or microcavity resonance sensors, which can detect single molecules [160]. It is in the context of this strong competition in terms of sensing techniques that the potential of microfluidic PCF-based sensors should be discussed.

While no one has yet demonstrated PCF sensors with sensitivities comparable to those of SPR, SERS, or microcavity resonantors, PCFs have a number of benefits that make them a platform worthwhile of further exploration. The main advantage is certainly that PCFs can be mass produced much more readily than microresonators or SPR sensors. In the simple geometry of "dip-sensors," a short piece of PCF with its own microfluidic channels is simply dipped in the solution to be analyzed, with light being injected and analyzed through the opposite end. Information on the content of the analytes is then gathered for example from absorption lines, florescence spectra [161], optical activity [58], or refractive index measurements using the shift of bandgaps [99] and/or of LPG resonances [61,162,163]. Such dip-sensors could in principle be manufactured by the millions in one single drawing process, and thus would be cheap, disposable (avoiding the issues of sample cross contamination in the microfluidic channels), safe, and biocompatible. Polymer MOFs, which are more readily coated with biosensitive organic molecules and avoid the risk of leaving glass shards in the sample (possibly a living organism) [58], are especially well suited for this application.

The shift of LPG resonances in fluidic photonic bandgap fibers is one of the most sensitive in-fiber refractive-index-sensing schemes [61]. However, this mechanism cannot be used for fluids with refractive indices below that of the fiber material (as the fibers then become index guiding), de facto excluding most biomedically relevant water-based analytes. One solution to circumvent this problem is to add a high-refractive-index coating to the holes of the PCFs, which can restore bandgap guidance [164].

Some of the more noticeable recent efforts to increase the sensitivity of PCF sensors include the combination of PCFs with existing highly sensitive techniques, such as SPR and SERS. Indeed, compelling arguments can be made in favor of including SPR techniques into PCFs—for the small sample volumes, propagation constant engineering, and ready access to waveguide fields [164–167]—however, achieving metallic coatings of sufficient quality (~50-nm thickness and negligible surface roughness) to allow unhindered propagation of surface plasmons at the metal/analyte interface remains a challenge. So far the only surface plasmon resonances that have been clearly demonstrated in PCFs have used bulk metallic inclusions (fully filled holes), which can be very smooth at the metal/silica interface, but can hardly be used for sensing [168, 169]. A number of metallic coating techniques have been applied to PCFs, mostly in a non-SPR context: liquid phase deposition is reasonably easy to implement and can coat several meters of PCF holes at a time, but leads to surfaces too rough for SPR [170, 171]. Sazio et al. demonstrated high-pressure chemical vapor deposition of thin gold coatings in PCF holes with what appear to be extremely smooth interfaces [170]. This seems to be the most promising technique to achieve SPR capable PCFs, but is not easily implemented, and to the best of our knowledge no one has yet tried to use such fibers in the context of SPR sensing.

While the silver surfaces obtained with liquid phase depositions are not suitable for SPR, their roughness makes them a very good candidate for SERS. "Conventional" SERS techniques exploit hot spots (spots of extreme plasmonic field enhancement) to locally increase Raman scattering cross sections by up to 14 orders of magnitude. These hot spots are obtained using either a suspension of metal nanoparticles or a rough gold or silver surface. The difficulty is that there are only few good hot spots per unit volume or surface, and their position is unpredictable. In conventional beam-optics-based SERS sensors only a few of those hot spots can be exploited, the random fluctuation of which makes it difficult to obtain reproducible quantitative SERS measurements. In a PCF geometry, light can interact over long lengths with the metallic rough surface [170] or with nanoparticle suspensions in the PCF holes [88,172], maximizing the likelihood of encountering hot spots and making SERS measurements more reliable.

From the various methods described here, it is clear that the spectrum of applications achievable with microfluidic PCFs can be greatly extended by coating their holes, be it with metals, dielectrics or even metamaterials [173] to alter the optical properties, fluorphores and chemically active materials to add specific chemical sensitivity [89,163], and surface treatments to improve wetting properties [74,161]. Coating techniques, be they at the drawing or at postprocessing stages, are thus bound to be explored much further in the future—experimentally as well as theoretically—adding an entirely new dimension to microfluidic PCF research.

Another postprocessing technique that has emerged over the last two years and will no doubt be explored further in the future is that of enabling side access to PCFs, to optimize filling times. Indeed, to fully exploit PCF sensitivities, long interaction lengths are desirable; however, filling more than a few centimeters of PCF with fluids can be a very lengthy process when only capillary forces or small pressure differences are used. By adding regularly spaced small holes between the PCF holes and the outside, filling times can be reduced dramatically. Techniques explored for realizing side access to PCFs include focused ion beam etching [174], inflating techniques [84], and also manufacturing fibers with continuous lateral access—in other words, replacing PCF holes by trenches, at the preform stage, be it for hollow- or solid-core PCFs [175,176].

7-7 Summary

Optofluidics has emerged as a versatile design principle for enabling highly functional, compact, and micron-scale devices that bring together aspects of photonics, microfluidics, and various other disciplines. While typically applied to planar photonic structures, microstructured fibers provide an ideal optofluidic platform. A large body of research already exists in the design and fabrication of high-quality, low-loss, microstructured optical fibers. These fibers provide a high-quality optical environment for waveguiding, and the fiber microstructure provides a natural location for the microfluidic bodies used in optofluidic tuning. Optofluidic microstructured optical fibers thus provide a wealth of functional and compact device platforms for a multitude of applications while at the same time inherently interfacing using high-quality fiber optics and associated support apparatus.

References

1. D. Psaltis, S. R. Quake, and C. H. Yang, "Developing optofluidic technology through the fusion of microfluidics and optics," *Nature*, 442, 381–86 (2006).
2. C. Monat, P. Domachuk, and B. J. Eggleton, "Integrated optofluidics: a new river of light," *Nature Photon.*, 1, 106–114 (2007).
3. D. Erickson, T. Rockwood, T. Emery, et al., "Nanofluidic tuning of photonic crystal circuits," *Opt. Lett.*, 31, 59–61 (2006).
4. J. Stone, "Optical transmission in liquid-core quartz fibers," *Appl. Phys. Lett.*, 20, 239–241 (1972).
5. C. Monat, C. Grillet, R. Domachuk, et al., "Frontiers in microphotonics: tunability and all-optical control," *Laser Phys. Lett.*, 4, 177–186 (2007).
6. R. D. Maurer, "Glass fibers for optical communications," *Proc. IEEE*, 61, 452–462 (1973).
7. P. Domachuk, H. C. Nguyen, B. J. Eggleton, et al., "Microfluidic tunable photonic band-gap device," *Appl. Phys. Lett.*, 84, 1838–1840 (2004).
8. C. Kerbage, M. Sumetsky, and B. J. Eggleton, "Polarisation tuning by microfluidic motion in air-silica microstructured optical fiber," *Electron. Lett.*, 38, 1015–1017 (2002).
9. B. J. Eggleton, C. Kerbage, P. S. Westbrook, et al., "Microstructured optical fiber devices," *Opt. Express*, 9, 698–713 (2001).
10. T. M. Monro, W. Belardi, K. Furusawa, et al., "Sensing with microstructured optical fibers," *Meas. Sci. Technol.*, 12, 854–858 (2001).
11. B. J. Eggleton, P. S. Westbrook, R. S. Windeler, et al., "Grating resonances in air-silica microstructured optical fibers," *Opt. Lett.*, 24, 1460–1462 (1999).
12. L. Provino, J. M. Dudley, H. Maillotte, et al., "Compact broadband continuum source based on microchip laser pumped microstructured fiber," *Electron. Lett.*, 37, 558–560 (2001).
13. J. C. Knight, T. A. Birks, B. J. Mangan, et al., "Microstructured silica as an optical-fiber material," *MRS Bull.*, 26, 614–617 (2001).
14. J. K. Ranka, R. S. Windeler, and A. J. Stentz, "Optical properties of high-delta air-silica microstructure optical fibers," *Opt. Lett.*, 25, 796–798 (2000).
15. J. C. Knight, "Photonic crystal fibers," *Nature*, 424, 847–851 (2003).
16. J. C. Knight, T. A. Birks, P. St. J. Russell, et al., "All-silica single-mode optical fiber with photonic crystal cladding," *Opt. Lett.*, 21, 1547–1549 (1996).
17. T. A. Birks, J. C. Knight, and P. S. Russell, "Endlessly single-mode photonic crystal fiber," *Opt. Lett.*, 22, 961–963 (1997).
18. R. F. Cregan, B. J. Mangan, J. C. Knight, et al., "Single-mode photonic band gap guidance of light in air," *Science*, 285, 1537–1539 (1999).
19. F. Benabid, J. C. Knight, G. Antonopoulos, et al., "Stimulated Raman scattering in hydrogen-filled hollow-core photonic crystal fiber," *Science*, 298, 399–402 (2002).
20. W. G. French, J. B. MacChesney, and A. D. Pearson, "Glass fibers for optical communications," *Ann. Rev. Mat. Sci.*, 5, 373–394 (1975).
21. T. Miya, Y. Terunuma, T. Hosaka, et al., "Ultimate low-loss single-mode fiber at 1.55 μm," *Electron. Lett.*, 15, 106–108 (1979).
22. R. L. Kozodoy, A. T. Pagkalinawan, and J. A. Harrington, "Small-bore hollow waveguides for delivery of 3 μm laser radiation," *Appl. Opt.*, 35, 1077–1082 (1996).
23. B. J. Eggleton, P. S. Westbrook, C. A. White, et al., "Cladding-mode-resonances in air-silica microstructure optical fibers," *J. Lightwave Technol.*, 18, 1084–1100 (2000).

24. D. G. Ouzounov, F. R. Ahmad, D. Muller, et al., "Generation of megawatt optical solitons in hollow-core photonic band-gap fibers," *Science*, 301, 1702–1704 (2003).
25. J. M. Dudley, G. Genty, and S. Coen, "Supercontinuum generation in photonic crystal fiber," *Rev. Mod. Phys.*, 78, 1135–1184 (2006).
26. P. St. J Russell, "Photonic-crystal fibers," *J. Lightwave Technol.*, 24, 12 (2006).
27. W. H. Reeves, D. V. Skryabin, F. Biancalana, et al., "Transformation and control of ultrashort pulses in dispersion-engineered photonic crystal fibers," *Nature*, 424, 511–515 (2003).
28. J. G. Rarity, J. Fulconis, J. Duligall, et al., "Photonic crystal fiber source of correlated photon pairs," *Opt. Express*, 13, 534–44 (2005).
29. J. C. Knight, J. Arriaga, T. A. Birks, et al., "Anomalous dispersion in photonic crystal fiber," *IEEE Photon. Technol. Lett.*, 12, 807–809 (2000).
30. J. Broeng, D. Mogilevstev, S. E. Barkou, et al., "Photonic crystal fibers: A new class of optical waveguides," *Opt. Fiber Technol.*, 5, 305–330 (1999).
31. T. A. Birks and Y. W. Li, "The shape of fiber tapers," *J. Lightwave Technol.*, 10, 432–438 (1992).
32. H. C. Nguyen, B. T. Kuhlmey, E. C. Mägi, et al., "Tapered photonic crystal fibers: properties, characterisation and applications," *Appl. Phys.*, B 81, 377–387 (2005).
33. M. Sumetsky, Y. Dulashko, P. Domachuk, et al., "Thinnest optical waveguide: experimental test," *Opt. Lett.*, 32, 754–756 (2007).
34. L. M. Tong, L. L. Hu, J. J. Zhang, et al., "Photonic nanowires directly drawn from bulk glasses," *Opt. Express*, 14, 82–87 (2006).
35. P. Domachuk, A. Chapman, E. Mägi, et al., "Transverse characterization of high air-fill fraction tapered photonic crystal fiber," *Appl. Opt.*, 44, 3885–3892 (2005).
36. A. Othonos, "Fiber Bragg gratings," *Rev. Sci. Instrum.*, 68, 4309–4341 (1997).
37. K. O. Hill, Y. Fujii, D.C. Johnson, et al., "Photosensitivity in optical fiber waveguides: Applications to reflection filter fabrication," *Appl. Phys. Lett.*, 32, 647–649 (1978).
38. T. Erdogan., "Fiber grating spectra," *J. Lightwave Technol.*, 15, 1277–1294 (1997).
39. Y. J. Rao, "In-fiber Bragg grating sensors," *Meas. Sci. Technol.*, 8, 355–375 (1997).
40. T. M. Squires and S. R. Quake, "Microfluidics: Fluid physics at the nanoliter scale," *Rev. Mod. Phys.*, 77, 977–1026 (2005).
41. G. M. Whitesides, "The origins and the future of microfluidics," *Nature*, 442, 368–373 (2006).
42. H. A. Stone, A. D. Stroock, and A. Ajdari, "Engineering flows in small devices: Microfluidics toward a lab-on-a-chip," *Ann. Rev. Fluid Mech.*, 36, 381–411 (2004).
43. J. Hecht, *City of Light: The Story of Fiber Optics.*, Oxford University Press, New York (1999).
44. C. Monat, P. Domachuk, C. Grillet, et al., "Optofluidics: a novel generation of reconfigurable and adaptive compact architectures," *Microfluidics and Nanofluidics*, 4, 81–95 (2008).
45. S. H. Kim, J. H. Choi, S. K. Lee, et al., "Optofluidic integration of a photonic crystal nanolaser," *Opt. Express*, 16, 6515–6527 (2008).
46. H. C. Hunt and J. S. Wilkinson, "Optofluidic integration for microanalysis," *Microfluid. Nanofluidics*, 4, 53–79 (2008).
47. X. Heng, D. Erickson, L. R. Baugh, et al., "Optofluidic microscopy—a method for implementing a high resolution optical microscope on a chip," *Lab. Chip.*, 6, 1274–1276 (2006).
48. Z. Y. Li, Z. Y. Zhang, T. Emery, et al., "Single mode optofluidic distributed feedback dye laser," *Opt. Express*, 14, 696–701 (2006).
49. C. Peroz, J. C. Galas, L. Le Gratiet, et al., "Compact dye laser on a chip fabricated by ultraviolet nanoimprint lithography," *Appl. Phys. Lett.*, 89, 243109 (2006).

50. F. B. Arango, M. B. Christiansen, M. Gersborg-Hansen, et al., "Optofluidic tuning of photonic crystal band edge lasers," *Appl. Phys. Lett.*, 91, 223503 (2007).
51. P. J. Roberts, F. Couny, H. Sabert, et al., "Ultimate low loss of hollow-core photonic crystal fibers," *Opt. Express*, 13, 236–244 (2005).
52. D. K. Sparacin, S. J. Spector., and L. C. Kimerling, "Silicon waveguide sidewall smoothing by wet chemical oxidation," *J. Lightwave Technol.*, 23, 2455–2461 (2005).
53. P. Domachuk, I. C. M. Littler, M. Cronin-Golomb, et al., "Compact resonant integrated microfluidic refractometer," *Appl. Phys. Lett.*, 88, 093513 (2006).
54. W. Liang, Y. Huang, Y. Xu, et al., "Highly sensitive fiber Bragg grating refractive index sensors," *Appl. Phys. Lett.*, 86, 151122 (2005).
55. A. Iadicicco, S. Campopiano, A. Cutolo, et al., "Refractive index sensor based on microstructured fiber Bragg grating," *IEEE Photon. Technol. Lett.*, 17, 1250–1252 (2005).
56. Y. Y. Huang, Y. Xu, and A. Yariv, "Fabrication of functional microstructured optical fibers through a selective-filling technique," *Appl. Phys. Lett.*, 85, 5182–5184 (2004).
57. K. Nielsen, D. Noordegraaf, T. Sørensen, et al., "Selective filling of photonic crystal fibers," *J. Opt.*, A 7, L13–L20 (2005).
58. F. M. Cox, A. Argyros, and M. C. J. Large, "Liquid-filled hollow core microstructured polymer optical fiber," *Opt. Express*, 14, 4135–4140 (2006).
59. N. M. Litchinitser, S. C. Dunn, B. Usner, et al., "Resonances in microstructured optical waveguides," *Opt. Express*, 11, 1243–1251 (2003).
60. A. Fuerbach, P. Steinvurzel, J. A. Bolger, et al., "Nonlinear pulse propagation at zero dispersion wavelength in anti-resonant photonic crystal fibers," *Opt. Express*, 13, 2977–2987 (2005).
61. P. Steinvurzel, E. D. Moore, E. C. Mägi,et al., "Tuning properties of long period gratings in photonic bandgap fibers," *Opt. Lett.*, 31, 2103–2105 (2006).
62. B. R. Acharya, T. Krupenkin, S. Ramachandran, et al., "Tunable optical fiber devices based on broadband long-period gratings and pumped microfluidics." *Appl. Phys. Lett.*, 83, 4912–4914 (2003).
63. F. Cattaneo, K. Baldwin, S. Yang, et al., "Digitally tunable microfluidic optical fiber devices," *J. Microelectromech. Syst.*, 12, 907–912 (2003).
64. D. G. Bailey, J. Malcolm, O. Arnold, et al., "Grapefruit juice-drug interactions," *Br. J. Clin. Pharmacol.*, 46, 101–110 (1998).
65. C. Kerbage, P. Steinvurzel, P. Reyes, et al., "Highly tunable birefringent microstructured optical fiber," *Opt. Lett.*, 27, 842–844 (2002).
66. C. Kerbage, M. Sumetsky, and B. J. Eggleton, "Polarisation tuning by microfluidic motion in air-silica microstructured optical fiber," *Electron. Lett.*, 38, 1015–1017 (2002).
67. P. Mach, M. Dolinski, K. W. Baldwin, et al., "Tunable microfluidic optical fiber," *Appl. Phys. Lett.*, 80, 4294–4296 (2002).
68. C. Kerbage and B. J. Eggleton, "Tunable microfluidic optical fiber gratings," *Appl. Phys. Lett.*, 82, 1338–1340 (2003).
69. H. C. Nguyen, P. Domachuk, B. J. Eggleton, et al., "New slant on photonic crystal fibers," *Opt. Express*, 12, 1528–1539 (2004).
70. P. Domachuk, H. C. Nguyen, and B. J. Eggleton, "Transverse probed microfluidic switchable photonic crystal fiber devices," *IEEE Photon. Technol. Lett.*, 16, 1900–1902 (2004).
71. J. D Joannopoulos, R. D. Meade, and J. N. Winn, *Photonic Crystals: Molding the Flow of Light*, Princeton University Press, Princeton, NJ (1995).
72. M. J. Steel, *BandSOLVE User Guide* 2d ed., RSoft Design, Inc., Ossining (2005).
73. T. R. Salamon, J. A. Rogers, and B. J. Eggleton, "Analysis of heat flow in optical fiber devices that use microfabricated thin film heaters," *Sens. Actuators*, A 95, 8–16 (2001).
74. P. Domachuk, C. Grillet, V. Ta'eed, et al., "Microfluidic interferometer," *Appl. Phys. Lett.*, 86, 024103 (2005).
75. C. Grillet, P. Domachuk, V. Ta'eed, et al., "Compact tunable microfluidic interferometer," *Opt. Express*, 12, 5440–5447 (2004).

76. J. Cooper-White and L. E Rodd, Silanization Methods for Glass, Univeristy of Queensland, Brisbane (2005).
77. P. St. J. Russell, "Photonic crystal fibers," *Science*, 299, 358–362 (2003).
78. M. Yan and P. Shum, "Guidance varieties in photonic crystal fibers," *J. Opt. Soc. Am.*, B 23, 1684–1691 (2006).
79. J. C. Flanagan, R. Amezcua-Correa, F. Poletti, et al., "Parasitic modes in large mode area microstructured fibers," *Opt. Fiber Comm. Conf. (OFC)*, paper OML4, Anaheim, (2007).
80. G. Antonopoulos, F. Benabid, T. A. Birks, et al., "Experimental demonstration of the frequency shift of bandgaps in photonic crystal fibers due to refractive index scaling," *Opt. Express*, 14, 3000–3006 (2006).
81. S. Lebrun, P. Delaye, R. Frey, et al., "High-efficiency single-mode Raman generation in a liquid-filled photonic bandgap fiber," *Opt. Lett.*, 32, 337–339 (2007).
82. J. Sun, C.-C. Chan, X.-Y. Dong, et al., "High-resolution photonic bandgap fiber-based biochemical sensor," *J. Biomed. Opt.*, 12, 044022 (2007).
83. L. Xiao, W. Jin, M. Demokan, et al., "Fabrication of selective injection microstructured optical fibers with a conventional fusion splicer," *Opt. Express*, 13, 9014–9022 (2005).
84. C. M. B. Cordeiro, E. M. dos Santos, C. H. B. Cruz, et al., "Lateral access to the holes of photonic crystal fibers—selective filling and sensing applications," *Opt. Express*, 14, 8403–8412 (2006).
85. C. Martelli, J. Canning, K. Lyytikainen, et al., "Water-core Fresnel fiber," *Opt. Express*, 13, 3890–3895 (2005).
86. S. Yiou, P. Delaye, A. Rouvie, et al., "Stimulated Raman scattering in an ethanol core microstructured optical fiber," *Opt. Express*, 13, 4786–4791 (2005).
87. C. J. De Matos, C. M. B. Cordeiro, E. M. dos Santos, et al., "Liquid-core, liquid-cladding photonic crystal fibers," *Opt. Express*, 15, 11207–11212 (2007).
88. Y. Zhang, C. Shi, C. Gu, et al., "Liquid core photonic crystal fiber sensor based on surface enhanced Raman scattering," *Appl. Phys. Lett.*, 90, 193504 (2007).
89. S. Smolka, M. Barth, and O. Benson, "Highly efficient fluorescence sensing with hollow core photonic crystal fibers," *Opt. Express*, 15, 12783–12791 (2008).
90. A. Bozolan, C. J. de Matos, C. M. B. Cordeiro, et al., "Supercontinuum generation in a water-core photonic crystal fiber," *Opt. Express*, 16, 9671–9676 (2008).
91. N. M. Litchinitser, A. K. Abeeluck, C. Headley, et al., "Antiresonant reflecting photonic crystal optical waveguides," *Opt. Lett.*, 27, 1592–1594 (2002).
92. T. P. White, R. C. McPhedran, C. M. de Sterke, et al., "Resonance and scattering in microstructured optical fibers," *Opt. Lett.*, 27, 1977–1979 (2002).
93. P. Steinvurzel, B. T. Kuhlmey, T. P. White, et al., "Long wavelength antiresonant guidance in high index inclusion microstructured fibers," *Opt. Express*, 12, 5424–5433 (2004).
94. P. Steinvurzel, C. M. de Sterke, M. J. Steel, et al., "Single scatterer Fano resonances in solid core photonic bandgap fibers," *Opt. Express*, 14, 8797–8811 (2006).
95. F. Couny, F. Benabid, P. J. Roberts, et al., "Identification of Bloch-modes in hollow-core photonic crystal fiber cladding,"*Opt. Express*, 15, 325–338 (2007).
96. J. Lægsgaard, "Gap formation and guided modes in photonic bandgap fibers with high-index rods," *J. Opt.*, A 6, 798–804 (2004).
97. T. A. Birks, G. J. Pearce, and D. M. Bird, "Approximate band structure calculation for photonic bandgap fibers," *Opt. Express*, 14, 9483–9490 (2006).
98. N. M. Litchinitser, S. C. Dunn, P. Steinvurzel, et al., "Application of an ARROW model for designing tunable photonic devices," *Opt. Express*, 12, 1540–1550 (2004).
99. N. M. Litchinitser and E. Poliakov, "Antiresonant guiding microstructured optical fibers for sensing applications," *Appl. Phys.*, B 81, 347–351 (2005).
100. R. T. Bise, R. S. Windeler, K. S. Kranz, et al., "Tunable photonic band gap fiber," *Opt. Fiber Comm. Conf. (OFC)*, paper ThK3, Anaheim, (2002).

101. P. Steinvurzel, B. J. Eggleton, C. M. de Sterke, et al., "Continuously tunable bandpass filtering using high-index inclusion microstructured optical fiber," *Electron. Lett.*, 41, 463–464 (2005).
102. L. Scolari, T. T. Alkeskjold, and A. Bjarklev, "Tunable Gaussian fillter based on tapered liquid crystal photonic bandgap fiber," *Electron. Lett*,. 42, 1270–1271 (2006).
103. J. -B. Du, Y. -G. Liu, Z. Wang, et al., "Thermally tunable dual-core photonic bandgap fiber based on the infusion of a temperature-responsive liquid," *Opt. Express*, 16, 4263–4269 (2008).
104. T. T. Larsen, A. Bjarklev, D. S. Hermann, et al., "Optical devices based on liquid crystal photonic bandgap fibers," *Opt. Express*, 11, 2589–2596 (2003).
105. T. T. Larsen, J. Broeng, D. S. Hermann, et al., "Thermo-optic switching in liquid crystal infiltrated photonic bandgap fibers," *Electron. Lett.*, 39, 1719–1720 (2003).
106. T. T. Alkeskjold, J. Lægsgaard, A. Bjarklev, et al., "All-optical modulation in dye-doped nematic liquid crystal photonic bandgap fibers," *Opt. Express*, 12, 5857–5871 (2004).
107. J. Li, S. Gauza, S. T. Wu, et al., "High dn_o/dT liquid crystals and their applications in a thermally tunable liquid crystal photonic crystal fiber," *Molecular Cryst. Liq. Cryst.*, 453, 355–370 (2006).
108. T. T. Alkeskjold, J. Lægsgaard, A. Bjarklev, et al., "Highly tunable large-core single-mode liquid-crystal photonic bandgap fiber," *Appl. Opt.*, 45, 2261–2264 (2006).
109. T. R. Woli ski, K. Szaniawska, S. Ertman, et al., "Influence of temperature and electrical fields on propagation properties of photonic liquid-crystal fibers," *Meas. Sci. Technol.*, 17, 985–991 (2006).
110. M. W. Haakestad, T. T. Alkeskjold, M. D. Nielsen, et al., "Electrically tunable photonic bandgap guidance in a liquid-crystal-filled photonic crystal fiber," *IEEE Photon. Technol. Lett.*, 17, 819–821 (2005).
111. L. Scolari, T. T. Alkeskjold, J. Riishede, et al., "Continuously tunable devices based on electrical control of dual-frequency liquid crystal filled photonic bandgap fibers," *Opt. Express*, 13, 7483–7496 (2005).
112. V. K. Hsiao and C. -Y. Ko, "Light-controllable photoresponsive liquid-crystal photonic crystal fiber," *Opt. Express*, 16, 12670–12676 (2008).
113. P. Steinvurzel, E. D. Moore, E. C. Mägi, et al., "Long period grating resonances in photonic bandgap fiber," *Opt. Express*, 14, 3007–3014 (2006).
114. D. Noordegraaf, L. Scolari, J. Lægsgaard, et al., "Electrically and mechanically induced long period gratings in liquid crystal photonic bandgap fibers," *Opt. Express*, 15, 7901–7912 (2007).
115. T. B. Iredale, P. Steinvurzel, and B. J. Eggleton, "Electric arc-induced long period gratings in fluid-filled photonic bandgap fiber," *Electron. Lett.*, 42, 739–740 (2006).
116. D. -I. Yeom, P. Steinvurzel, B. J. Eggleton, et al., "Tunable acoustic gratings in solid-core photonic bandgap fiber," *Opt. Express*, 15, 3513–3518 (2007).
117. B. T. Kuhlmey, F. Luan, L. B. Fu, et al., "Experimental reconstruction of bands in solid core photonic bandgap fibers using acoustic gratings," *Opt. Express*, 16, 13845–13856 (2008).
118. P. Steinvurzel, E. D. Moore, E. C. Mägi, et al., "Tuning properties of long period gratings in photonic bandgap fibers," *Opt. Lett.*, 31, 2103–2105 (2006).
119. X. Shu, L. Zhang, and I. Bennion, "Sensitivity characteristics of long-period fiber gratings," *J. Lightwave Technol.*, 20, 255–266 (2002).
120. D. Östling and H. E. Engan, "Narrow-band acousto-optic tunable filtering in a two-mode fiber," *Opt. Lett.*, 20, 1247–1249 (1995).
121. D. I. Yeom, H. S. Kim, M. S. Kang, et al., "Narrow-bandwidth all-fiber acoustooptic tunable filter with low polarization-sensitivity," *IEEE Photon. Technol. Lett.*, 17, 2646–2648 (2005).
122. M. W. Haakestad and J. Skaar, "Causality and Kramers-Kronig relations for waveguides," *Opt. Express*, 13, 9922–9934 (2005).

123. J. Jasapara, T. H. Her, R. T. Bise, et al., "Group-velocity dispersion measurements in a photonic bandgap fiber," *J. Opt. Soc. Am.*, B 20, 1611–1615 (2003).
124. R. T. Bise, J. Jasapara, P. Steinvurzel, et al., "Temperature tuning of dispersion in a photonic band gap fiber," *Conf. Lasers Electro-Opt./Quantum Electron. Laser Sci. (CLEO/QELS)*, paper JThA1, Long Beach (2002).
125. J. Lægsgaard and A. Bjarklev, "Doped photonic bandgap fibers for short-wavelength nonlinear devices," *Opt. Lett.*, 28, 783–785 (2003).
126. J. Riishede, J. Lægsgaard, J. Broeng, et al., "All-silica photonic bandgap fiber with zero dispersion and a large mode area at 730 nm," *J. Opt.*, A 6, 667–670 (2004).
127. A. Fuerbach, P. Steinvurzel, J. A. Bolger, et al., "Nonlinear propagation effects in anti-resonant high-index inclusion photonic crystal fibers," *Opt. Lett.*, 30, 830–832 (2005).
128. G. P. Agrawal, *Nonlinear Fiber Optics*, 2d ed., Academic Press, San Diego (1995).
129. P. K. A. Wai, C. R. Menyuk, H. H. Chen, et al., "Soliton at the zero-group-dispersion wavelength of a single model fiber," *Opt. Lett.*, 12, 628–630 (1987).
130. N. Akhmediev and M. Karlsson, "Cherenkov radiation emitted by solitons in optical fibers," *Phys. Rev.*, A 51, 2602–2607 (1995).
131. P. D. Rasmussen, J. Lægsgaard, and O. Bang, "Degenerate four wave mixing in solid core photonic bandgap fibers," *Opt. Express,* 16, 4059–4068 (2008).
132. R. Zhang, J. Teipel, and H. Giessen, "Theoretical design of a liquid-core photonic crystal fiber for supercontinuum generation," *Opt. Express,* 14, 6800–6812 (2006).
133. K. A. Brzdąkiewicz, U. A. Laudyn, M. A. Karpierz, et al., "Linear and nonlinear properties of photonic crystal fibers filled with nematic liquid crystals," *Opto-Electron. Rev.*, 14, 287–292 (2006).
134. C. R. Rosberg, F. H. Bennet, D. N. Neshev, et al., "Tunable diffraction and self-defocusing in liquid-filled photonic crystal fibers," *Opt. Express,* 15, 12145–12150 (2007).
135. P. D. Rasmussen, A. A. Sukhorukov, D. N. Neshev, et al., "Spatiotemporal control of light by Bloch-mode dispersion in multicore fibers," *Opt. Express,* 16, 5878–5891 (2008).
136. X. Fan, I. M. White, S. I. Shopova, et al., "Sensitive optical biosensors for unlabeled targets: A review" *Analytica Chimica Acta.*, 620, 8–26 (2008).
137. A. Bétourné, Y. Quiquempois, G. Bouwmans, et al., "Design of a photonic crystal fiber for phase-matched frequency doubling or tripling," *Opt. Express,* 16, 14255–14262 (2008).
138. M. R. Lamont, B. T. Kuhlmey, and C. M. de Sterke, "Multiorder dispersion engineering for optimal four-wave mixing," *Opt. Express,* 16, 7551–7563 (2008).
139. D. R. Austin, C. M. de Sterke, B. J. Eggleton, et al., "Dispersive wave blue-shift in supercontinuum generation," *Opt. Express,* 14, 11997–12007 (2006).
140. J. M. Stone and J. C. Knight, "Visibly white light generation in uniform photonic crystal fiber using a microchip laser," *Opt. Express,* 16, 2670–2675 (2008).
141. J. C. Travers, S. V. Popov, and J. R. Taylor, "Extended blue supercontinuum generation in cascaded holey fibers," *Opt. Lett.*, 30, 3132–3134 (2005).
142. D. R. Solli, C. Ropers, P. Koonath, et al., "Optical rogue waves," *Nature,* 450, 1054–1057 (2007).
143. J. M. Dudley, G. Genty, and B. J. Eggleton, "Harnessing and control of optical rogue waves in supercontinuum generation," *Opt. Express,* 16, 3644–3651 (2008).
144. A. Ferrando, E. Silvestre, J. J. Miret, et al., "Nearly zero ultraflattened dispersion in photonic crystal fibers," *Opt. Lett.*, 25, 790–792 (2000).
145. G. Renversez, B. Kuhlmey, and R. McPhedran, "Dispersion management with microstructured optical fibers: ultraflattened chromatic dispersion with low losses," *Opt. Lett.*, 28(12) 989–991 (2003).

146. W. H. Reeves, J. C. Knight, P. St. J. Russell, et al., "Demonstration of ultra-flattened dispersion in photonic crystal fibers," *Opt. Express*, 10, 609–613 (2002).
147. P. J. Roberts, B. J. Mangan, H. Sabert, et al., "Control of dispersion in photonic crystal fibers" *J. Opt. Fiber Commun. Rep.*, 2, 435–461 (2005).
148. F. Gérôme, J. -L. Auguste, and J. -M. Blondy, "Design of dispersion-compensating fibers based on a dual-concentric-core photonic crystal fiber," *Opt. Lett.*, 29, 2725–2727 (2004).
149. K. M. Gundu, M. Kolesik, J. V. Moloney, et al., "Ultraflattened-dispersion selectively liquid-filled photonic crystal fibers," *Opt. Express*, 14, 6870–6878 (2006).
150. T. Kuhlmey, H. C. Nguyen, M. J. Steel, et al., "Confinement loss in adiabatic photonic crystal fiber tapers," *J. Opt. Soc. Am.*, B 23, 1965–1974 (2006).
151. R. Goto, K. Takenaga, K. Okada, et al., "Cladding-pumped Yb-doped solid photonic bandgap fiber for ase suppression in shorter wavelength region," *Opt. Fiber Comm. Conf. (OFC)*, paper OTuJ5, San Diego (2008).
152. V. S. Shchesnovich, A. S. Desyatnikov, and Yu. S. Kivshar, "Interband resonant transitions in two-dimensional hexagonal lattices: Rabi oscillations, Zener tunnelling, and tunnelling of phase dislocations," *Opt. Express*, 16, 14076–14094 (2008).
153. T. Ritari, J. Tuominen, H. Ludvigsen, et al., "Gas sensing using air-guiding photonic bandgap Fibers," *Opt. Express*, 12, 4080–4087 (2004).
154. V. P. Minkovich, D. Monzón-Hernández, Joel Villatoro, et al., "Microstructured optical fiber coated with thin films for gas and chemical sensing," *Opt. Express*, 14, 8413–8418 (2006).
155. F. Benabid, F. Couny, J. C. Knight, et al., "Compact, stable and efficient all-fiber gas cells using hollow-core photonic crystal fibers," *Nature*, 434, 488–491 (2005).
156. F. Couny, F. Benabid, P. J. Roberts, et al., "Generation and photonic guidance of multioctave optical-frequency combs," *Science*, 318, 1118–1121 (2007).
157. F. Benabid, P. S. Light, F. Couny, et al., "Electromagnetically-induced transparency grid in acetylene-filled hollow-core PCF," *Opt. Express*, 13, 5694–5703 (2005).
158. X. E. Lin, "Photonic band gap fiber accelerator," *Phys. Rev. Special Topics*, 4, 051301 (2001).
159. T. Takekoshi and R. J. Knize, "Optical guiding of atoms through a hollow-core photonic band-gap fiber," *Phys. Rev. Lett.*, 98, 210404 (2007).
160. A. Ozcan and U. Demirci, "Rewritable self-assembled long-period gratings in photonic bandgap fibers using microparticles," *Opt. Comm.*, 270, 225–228 (2007).
161. J. Homola, S. S. Yee, and G. Gauglitz, "Surface plasmon resonance sensors: review," *Sens. Actuators*, B 54, 3–15 (1999).
162. C. L. Haynes, C. R. Yonzon, X. Zhang, et al., "Surface-enhanced Raman sensors: early history and the development of sensors for quantitative biowarfare agent and glucose detection," *J. Raman Spectrosc.*, 36, 471–484 (2005).
163. A. M. Armani, R. P. Kulkarni, S. E. Fraser, et al., "Label-free, single-molecule detection with optical microcavities," *Science*, 317, 783–787 (2007).
164. E. P. Schartner, Y. Ruan, P. Hoffmann, et al., "An optical fiber protein sensor," *Australian Conf. Opt. Fiber Technol. (ACOFT)*, paper WeB1–3, Melbourne (2007).
165. L. Rindorf, J. B. Jensen, M. Dufva, et al., "Photonic crystal fiber long-period gratings for biochemical sensing," *Opt. Express*, 14, 8224–8231 (2006).
166. L. Rindorf and O. Bang, "Sensitivity of photonic crystal fiber grating sensors: biosensing, refractive index, strain, and temperature sensing," *J. Opt. Soc. Am.*, B 25, 310–324 (2008).
167. B. T. Kuhlmey, K. Pathmanandavel, and R. C. McPhedran, "Multipole analysis of photonic crystal fibers with coated inclusions," *Opt. Express*, 14, 10851–10864. (2006).

168. A. Hassani and M. Skorobogatiy, "Design of the microstructured optical fiber-based surface plasmon resonance sensors with enhanced microfluidics" *Opt. Express*, 14, 11616–11621 (2006).
169. A. Hassani and M. Skorobogatiy, "Design criteria for the microstructured optical fiber-based surface plasmon resonance sensors," *J. Opt. Soc. Am.*, B 24, 1423–1429 (2007).
170. M. Hautakorpi, M. Mattinen, and H. Ludvigsen, "Surface-plasmon-resonance sensor based on three-hole microstructured optical fiber," *Opt. Express*, 16, 8427–8432 (2008).
171. J. Hou, D. Bird, A. George, et al., "Metallic mode confinement in microstructured fibers," *Opt. Express*, 16, 5983–5990 (2008).
172. M. A. Schmidt, L. N. Prill Sempere, H. K. Tyagi, et al., "Waveguiding and plasmon resonances in two-dimensional photonic lattices of gold and silver nanowires," *Phys. Rev.*, B 77, 033417 (2008).
173. A. Amezcua-Correa, J. Yang, C. E. Finlayson, et al., "Surface-enhanced Raman scattering using microstructured optical fiber substrates," *Adv. Funct. Mater.*, 17, 2024–2030 (2007).
174. X. Zhang, R. Wang, F. Cox, et al., "Selective coating of holes in microstructured optical fiber and its application to in-fiber absorptive polarizers," *Opt. Express*, 15, 16270–16278 (2007).
175. F. M. Cox, A. Argyros, M. C. J. Large, et al., "Surface enhanced Raman scattering in a hollow core microstructured optical fiber," *Opt. Express* 15, 13675–13681 (2007).
176. B. T. Kuhlmey and R. C. McPhedran, "Photonic crystal fibers with resonant coatings," *Physica* B 394, 155–158 (2007).
177. C. M. B. Cordeiro, C. J. S. de Matos, E. M. dos Santos et al., "Towards practical liquid and gas sensing with photonic crystal fibers: side access to the fiber microstructure and single-mode liquid-core fiber," *Meas. Sci. Technol.*, 18, 3075–3081 (2007).
178. F. M. Cox, R. Lwin, M. C. J. Large, et al., "Opening up optical fibers," *Opt. Express*, 15, 11843–11848 (2007).
179. J. A. West, E. M. Kosik Williams, and K. W. Koch, "Microstructured hollow-core rib waveguides," *Conf. Lasers Electro-Opt./Quantum Electron. Laser Sci. (CLEO/QELS)*, paper JFA6, San Jose (2008).

CHAPTER 8
Adaptive Optofluidic Devices

Steve Zamek and Yeshaiahu Fainman

Department of Electrical Engineering, University of California, San Diego, California

The primary advantages of fluids are their ability to easily change their shape, to mix and dissolve, and to form very smooth solid-fluid and fluid-fluid interfaces. Moreover, various fluids provide very wide range of refractive indices, and by mixing fluids with large refractive index difference, we can create tunable index fluids with wide dynamic range, high resolution, and ease in control. Superiority of fluids over solids in this regard is obvious. First, geometry variations are very limited with solids, for which the stresses inherent in deformations result in undesired birefringence and aging. Second, controllable real-time mixing of liquids allows tuning of the refractive index of the mixture by ±0.1. This tuning range is several orders of magnitude wider than the one obtained in solids with electro-optic, magneto-optic, thermo-optic, photorefractive, and other effects. In fact, even wider tuning range is obtained by introduction and displacement of fluids with very different optical properties into and out of the region of interest. These unique features enabled by fluids gave rise to two primary approaches in optofluidics: varying the geometry and tuning the refractive index of the optical medium.

In the beginning of the 1980s, an intersection of physics, chemistry, and nanotechnologies laid the foundations for microfluidics. Microfluidics allowed manipulation of very small volumes of fluid in a fast controllable fashion, and these capabilities opened new avenues in optics. Integration of fluidics with optoelectronic components became known as *optofluidics* [1,2]. This integration throve twofold. First, it allowed integration of optical components into lab-on-a-chip devices, giving a clear path for miniaturization of biomedical devices, known also as *micro total analysis systems* (μTAS). Second, it inherited methods used in microfluidics to build new optical elements and attain new functionalities. In this chapter we focus on those optical elements and devices that are based on integrated optofluidic components. Throughout this chapter we use the term *fluid* in its broad sense, meaning liquid or gaseous phases of the substances and pure or mixed liquids including solutions and colloids.

The diverse field of optofluidics has been steadily penetrating application areas of optical communications, data storage, display technologies, bioengineering, medical devices, imaging, metrology, computing, and many others. Ever-growing field of microfluidics enabled fast and easy fabrication, versatile and modular design, simulation tools, and robust integration of fluids into optoelectronic components. In the following sections we discuss areas in optofluidics, which have been under our thorough investigation. Specifically, we cover fluidic lenses, optofluidic switches, and integrated tunable devices.

8-1 Switching and Beam Deflection

Optical switching technologies were advanced by the fast-developing field of telecommunication. Various physical phenomena were employed for optical switching applications including electro-optic [3–6], acousto-optic [7,8], magneto-optic [9], and thermo-optic [10,11] effects and micromechanical components [12,13].

One of the first fluid-based switches was magneto-optic fluidic switch. When a magnetic fluid thin film is subjected to an external magnetic field parallel to the plane of the film, the particles in the film agglomerate and form chains. As the strength of the field increases, the chains evolve from a disordered phase to structured patterns, exhibiting optical anisotropy. These magneto-optic fluids were extensively exploited in magneto-optic fluidic switches [14–19].

All-optical switching based on changing the physical properties of black oils was suggested in 1986 [20]. The surface of a liquid film is deformed using an optical beam. These modifications alter the phase and intensity distribution of the reflected and transmitted laser beams. Surface deformation of a laser-heated liquid film and time evolution of the geometry of the surface were theoretically studied [21].

Despite numerous works on optofluidic switches, these devices are still in their embryonic stage. Ever-growing field of communications requires fast multiport switching with short delays, wide bandwidth, and low insertion losses. Very compelling optofluidic technology set a few records trying to address these requirements during the last decade. Broad scope of effects was employed to perform optical switching using fluids. These include total internal reflection on solid-fluid interfaces, diffraction from tunable gratings, and reconfigurable liquid-core waveguides.

Since the timescales on which fluids can be displaced (replaced) are commonly on the order of milliseconds, these components promise to benefit optical protection switches. Such switches reconfigure the interconnection of $N \times N$ input/output ports in case of an interruption of service, while one or more faulty optical transmission lines are repaired or replaced. The major advantage of optofluidic switches in such applications is a transparency over a wide bandwidth and insensitivity to polarization.

8-1-1 Switches Based on Total Internal Reflection

Total internal reflection (TIR) off an otherwise transparent surface was achieved by replacement of one fluid by another [22], bubble generation [23] and manipulation [24–27], fluid-fluid interface control using electrowetting [28–31], thermocapillary effect [32], and hydrodynamic spreading [33].

A concept of TIR by bubbles was commercially developed by Agilent Technologies (Palo Alto, CA) [34,35]. Multiple waveguides are created in the planar-lightwave circuits, intersecting at several cross points (see Fig. 8-1a). At those cross points, the light travels through a fluid whose refractive index is matched to the waveguide. As a result, the optical mode travels unimpeded through the cross point. When a bubble is inserted into the cross point, the light is reflected into another waveguide. These bubbles can be formed and removed hundreds of times per second, providing a fast and reliable switching function. The technology is similar to that used in ink-jet printers, indicating that such bubble switches should be mass-producible. Traditional bubble generation by resistive heaters was recently replaced by more efficient laser-activated heaters [36]. Heat and fluid flow models provided insights into the behavior of the bubble [27].

Another approach was based on oil latching interfacial tension variation effect (OLIVE) [32]. The switch is based on thermal-capillary effect to move trapped bubbles. The light path is switched when the refractive-index-matching oil moves in the slit due to surface-tension variation caused by heating (thermocapillarity). High extinction ratio (>50 dB), low crosstalk (<–50 dB), and a response time below 10 ms were achieved in 16×16 switch [37,38]. Surface wettability was shown to be crucial for fast bubble manipulation [26].

A 2×2 TIR optical switch was demonstrated by Campbell et al. and operated in free-space configuration [22]. The switch had an insertion loss smaller than 1 dB and extinction ratio on the order of 20 dB. The device could switch between transmission (bypass) and reflection (exchange) modes within less than 20 ms. The device, shown schematically in Fig. 8-1c and 8-1d, has two distinct layers of microchannels made in polydimethylsiloxane (PDMS). Channels of one layer (the flow layer) are used to deliver the liquids into the mirror channel. Channels of the second layer (the control layer) are used to actuate the microvalves, enabling fully controlled manipulation of the liquid in the mirror channel [39].

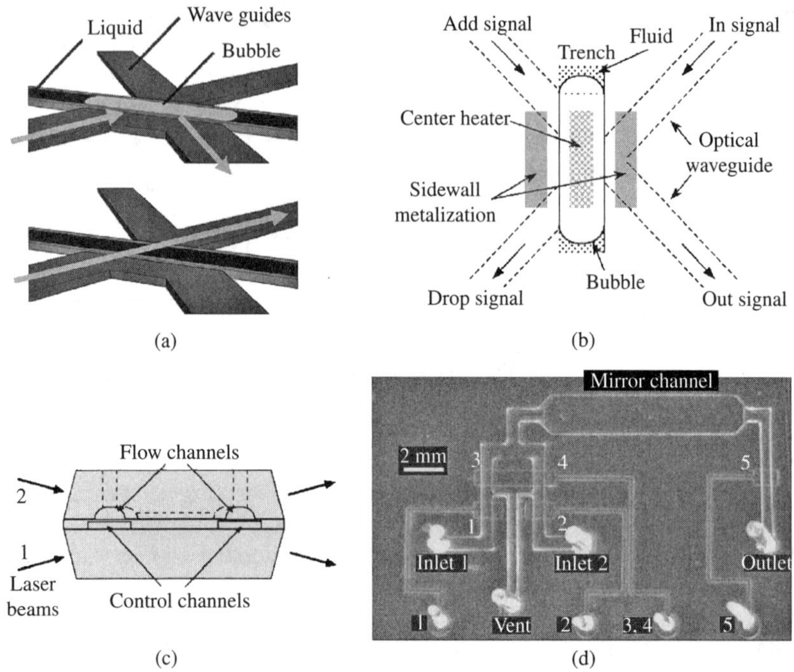

FIGURE 8-1 Three switches are shown: planar bubble switch (a and b), microfluidic 2 × 2 switch (c and d), and diffractive 1 × 4 switch (e and f). (a) Basic component of a bubble switch—cross point between four optical waveguides with a fluidic channel between them. Thermally generated bubble introduced into the cross point reflects the light into a waveguide. (b) Layout of *N* × *N* bubble switch showing a matrix of input-output planar waveguides. (*S. Hengstler, J.J. Uebbing, and P. McGuire, "Laser-activated optical bubble switch element," Proceedings of the 2003 IEEE/LEOS International Conference on Optical MEMS (OMEMS) 2003*), *copyright 2003 IEEE.*) (c) Complete 3-layer elastomer chip with 4 optical quality facets (the arrows on the left show directions of incident laser beams 1 and 2, and the arrows on the right show transmitted and reflected beams). (*Reprinted with permission from K. Campbell, A. Groisman, U. Levy, L. Pang, S. Mookherjea, D. Psaltis, and Y. Fainman, "A microfluidic 2 × 2 optical switch," Appl. Phys. Lett, 85, 6119–6121, 2004. Copyright 2004, American Institute of Physics.*) (d) Microphotograph of the 2 × 2 switch device (control layer inlets are labeled by numbers of the membrane valves they actuate). (*Reprinted with permission from K. Campbell, A. Groisman, U. Levy, L. Pang, S. Mookherjea, D. Psaltis, and Y. Fainman, "A microfluidic 2 × 2 optical switch," Appl. Phys. Lett, 85, 6119–6121, 2004. Copyright 2004, American Institute of Physics.*) (e) Fabricated 1 × 4 switch based on diffraction grating. (*A. Groisman, S. Zamek, K. Campbell, L. Pang, U. Levy, and Y. Fainman, "Optofluidic 1 × 4 switch," Opt. Express, 16 (18), 13499–13508, 2008.*) (f) Concept of operation of 1 × *N* optical switch—the fluid on top of the grating is replaced to redirect the beam into a certain order; the beam is incident from the bottom of the chamber shown on the figure. (*A. Groisman, S. Zamek, K. Campbell, L. Pang, U. Levy, and Y. Fainman, "Optofluidic 1 × 4 switch," Opt. Express, 16 (18), 13499–13508, 2008.*)

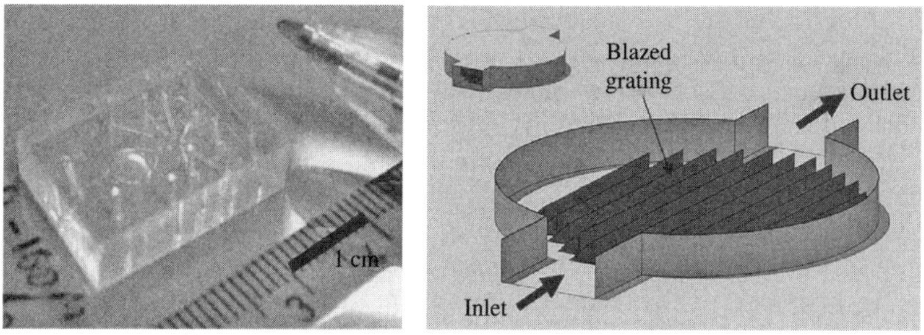

(e) (f)

FIGURE 8-1 (*Continued*)

The first flow layer contains the main functional element of the device—flat mirror channel. It can be filled with either pure water from inlet 1 or salt solution from inlet 2 (see Fig. 8-1d). The salt solution has a refractive index that matches the index of PDMS. The channels connecting the inlets and the outlet with the mirror are completely sealed by the flexible membranes when a pressure is applied to the membranes through channels (control lines) in the second control layer of the chip. The membranes serve as pressure-actuated integrated "push-up" valves [39]. Three of the control lines are connected to the valves on the two inlet channels and the outlet channel (valves 1, 2, and 5). The fourth control line (simultaneously switching two membrane valves 3 and 4) and the vent in the flow layer serve the purpose of purging the dead volumes between the two inlet valves (1 and 2) and the T-junction connecting the inlets with the mirror channel (Fig. 8-1d).

Results of the measurements for the beams 1 and 2 coming from below and above (see Fig. 8-1c) the two orthogonal linear polarization states and the two switching states are summarized in Table 8-1 as values of insertion loss and extinction ratio in decibel (dB).

	Bypass TE, dB		**Exchange TE, dB**		**Bypass TM, dB**		**Exchange TM, dB**	
Laser beam	1	2	1	2	1	2	1	2
Insertion loss	0.78	0.84	1.1	1.1	0.71	0.67	0.83	0.78
Extinction ratio	22.7	24.4	18.2	17.7	17.8	22.5	21.7	16.7

TABLE 8-1 Insertion Loss and Extinction Ratio (in dB) for Transmitted and Reflected Beams in the Two States of the Switch and for Two Inputs at TE and TM

Electrowetting was also employed to actuate liquid in ON/OFF switches [29,30,40]. In switched fluorescence device [29,30], fluorescent oil is immersed in water on top of a waveguide with hydrophobic cladding. The waveguide is used to excite the fluorescent oil with a UV light in the ON state. As voltage is applied to the water, the fluorescent oil layer gets displaced. The new configuration of the water and cladding layers reflects the UV light internally across the entire waveguide area. In this OFF state, no optical path exists that would allow violet light to reach the fluorescent oil layer.

8-1-2 Grating-Based Switches

Diffraction gratings are widely used in optics for beam splitting and spatiotemporal filtering. Beams diffracted by a grating with period Λ travel at different angles given by:

$$\sin\alpha_m = \frac{\lambda}{\Lambda}m \tag{8-1}$$

Here λ is the wavelength of the normally incident beam, and m is an integer, called the diffraction order.

Optofluidics enabled design and fabrication of tunable gratings. Two major families are tunable geometry and tunable refractive index gratings. An example of the first is self-assembled flowing lattices of bubbles [41,42], discussed in detail in Chap 3.

Another family of tunable gratings is based on refractive index tuning. A 1×4 free-space optical switch based on transmission blazed grating was recently reported [43,44] (see Fig. 8-1e and 8-1f). Blazed grating submerged into a liquid exhibits highly efficient diffraction of an incident beam into m'th diffraction order when the fluid satisfies blazed grating condition [45]:

$$m = \frac{l\Delta n}{\lambda} \tag{8-2}$$

Here l is the height of the grating profile, and Δn is the difference between the refractive indices of the fluid and the material used to construct the grating. Introducing a liquid with refractive index that satisfies Eq. (8-2) results in a beam deflection (diffraction) by an angle given by Eq. (8-1).

It should be noticed that for this type of grating the period of the phase modulation of the optical field incident on the grating is modulo Λ when Eq (8-2) is satisfied. For such grating a high diffraction efficiency of order m and consequently low insertion losses are observed [46].

The design of the 1×4 switch was based on salt solutions in water, which provided refractive indices in the range between 1.33 and 1.41 depending on the salt concentration, which allowed switching between four different angles for the given geometry. One of the promises of the suggested design is a low-complexity $N \times N$ switching. For example, only two of these components are required to construct a 4×4 switch, which would otherwise consist of five 2×2 crossbar or sixteen 1×2 switches [47]. The device is fabricated in PDMS, which also suggests easy integration into lab-on-a-chip devices.

8-1-3 Deflectors and Beam Scanners

In many applications it is crucial to have the capability of a continuous scanning of an optical beam. Beam deflectors in which fluids are used as the major functional element were recently demonstrated. Fluid actuation techniques employed for continuous beam scanning are essentially the same as those used for discrete beam switching. Examples include mechanically driven fluidic lens [48], electrowetting-based prism [49], and thermally actuated mirrors [50].

Optical deflector for continuous beam scanning based on electrowetting microprism is shown in Fig. 8-2a. Beam scanning in the

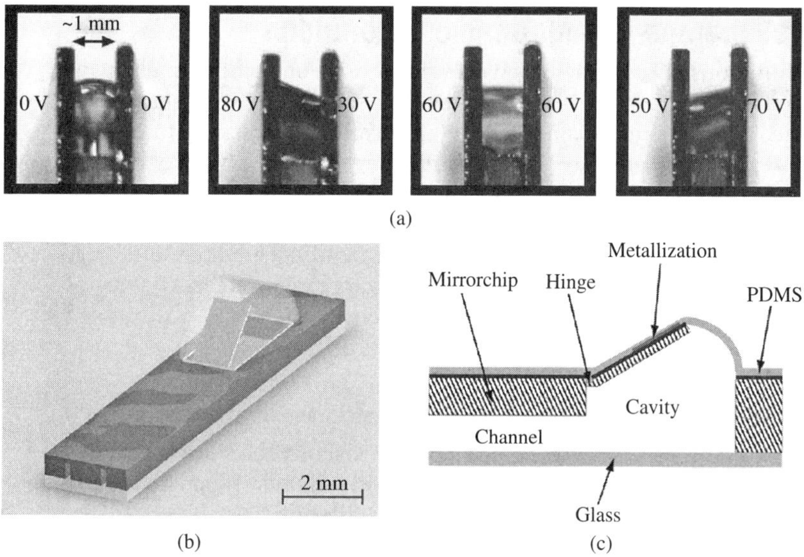

FIGURE 8-2 Tunable optical deflectors: (a) electrowetting-based optical deflector—prism with a variable angle is used to deflect an optical beam. (*N.R. Smith, D.C. Abeysinghe, J.W. Haus, and J. Heikenfeld, "Agile wide-angle beam steering with electrowetting microprisms," Opt. Express, 14 (14), 6557–6563, 2006.*) Tunable micro-mirror; (b) a mirror is embedded into PDMS membrane. (*A. Werber and H. Zappe, "Thermo-pneumatically actuated, membrane-based micro-mirror devices," J. Micromech. Microeng., 16, 2524–2531, 2006. Institute of Physics and IOP Publishing.*) (c) The membrane forms a cavity. When a pressure is applied to inflate the cavity the membrane tilts the mirror. (*A. Werber and H. Zappe, "Thermo-pneumatically actuated, membrane-based micro-mirror devices," J. Micromech. Microeng., 16, 2524–2531, 2006. Institute of Physics and IOP Publishing.*)

range of ±7 degrees with millisecond response time was demonstrated [49]. High steering efficiency, polarization-independent operation, and wide steering range were suggested to be of high interest for laser detection and ranging. According to the author, the same design with higher index fluids would allow further extension up to ±15 degrees of the steering range.

Even wider range of deflection angles was achieved using membrane-based micro-mirror [51]. The micro-mirror is mounted to the membrane and fixed with a thin silicon hinge to the sidewall of the cavity (see Fig. 8-2b and 8-2c). With application of a differential pressure between the cavity and ambient pressure, the PDMS membrane is distended. The hinge limits one degree of freedom, leading to a tilting motion. The micro-mirror, fixed to the membrane, is deflected. By varying the pressure, the tilting angles were varied from 0 to 75 degrees, relative to the substrate surface. Application to medical devices with a variable mirror setup, used for in vivo diagnostics, was suggested [51].

8-2 Membrane-Based Tunable Optofluidics

Numerous optofluidic tunable devices based on soft polymer membranes have recently been reported. In such devices the geometry of the optical element is altered by application of pressure to deform soft polymer elements. In this chapter we review two types of devices based on polymer. We start with a tunable polymer lens and continue the discussion with composite membrane technology. Some review of the mechanical properties of a thin bending membrane is suggested to provide an insight into more complicated mechanics of more general structures, such as composite membranes discussed later in this section.

8-2-1 Mechanics of Pressure Actuated Polymer

Membrane-based devices commonly consist of a chamber with a soft distensible wall, commonly made of a soft polymer. When a pressure is applied to the chamber, the wall bends, altering the geometry of the element. The deflection regime under applied pressure is commonly described in terms of the middle line $u(x, y)$ and thickness $t(x, y)$ of the bending wall (see Fig. 8-3a).

Depending on the design and the materials used, the bending wall can be modeled as a shell or a membrane. The major difference is that a membrane can take bending stresses in addition to tensile and compressive stresses observed in shells. Some guidelines to the model choice can be found in thin plates and shells theory [52,53]. For the completeness of the chapter we review some of the results of the theory.

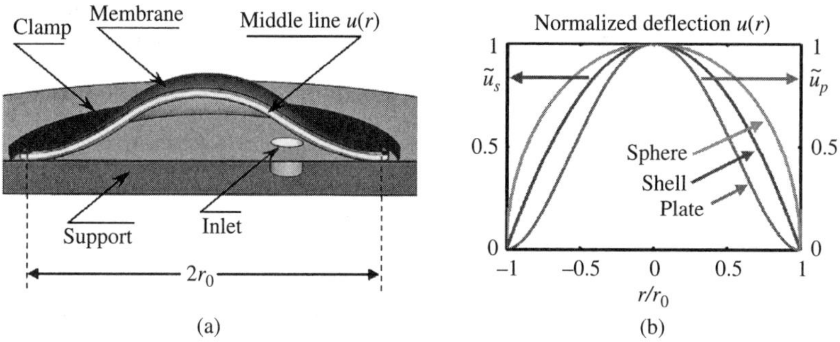

FIGURE 8-3 Typical geometry of a bending wall: (a) the model showing the middle line $u(r)$ of the shell (membrane); (b) the displacement curves of a cross section of $u(r)$; the blue curve shows the normalized displacement $\tilde{u}_s = 4SP^{-1}r_0^{-2}u_s$ of a shell, and the green one shows normalized displacement $\tilde{u}_s = 64DP^{-1}r_0^{-4}u_p$ of a plate (membrane). Spherical surface is shown in red for comparison.

Shell Deflection

As an example we consider large-aperture tunable lenses, introduced in the 1970s [54,55]. These lenses were filled with liquid, and therefore had to retain considerable amounts of liquid and overcome distortions caused by the gravity. To do so, the shell had to have high initial strain. When such a shell is actuated by air or liquid pressure, it obtains a displacement profile $u(x, y)$ that satisfies membrane equilibrium equation [56,57]. This model assumes "large" initial tension and "small" pump pressure P (i.e., the actuation-induced strain is negligible compared with the initial strain), linear response, and no resistance to bending [58]. Using the boundary conditions of a simply supported contour, the displacement of the axisymmetric elastic shell under the applied pressure can be found. For a circular support with a radius r_0 the profile is a paraboloid of revolution:

$$u(r) = \frac{P}{4S}(r_0^2 - r^2) \tag{8-3}$$

Here P and S are the uniform pressure and the isotropic tension per unit length, and $r^2 \equiv x^2 + y^2$. The radius of the curvature of the apex is readily calculated to be $2SP^{-1}$. Assuming thin shell, such curvature produces a lens with focal distance [59]:

$$f \approx 2S(\Delta nP)^{-1} \tag{8-4}$$

where Δn is the difference in the refractive indices between the fluid and the outer medium. The effective focal length of the whole aperture is easily found by applying the best fit of Eq. (8-3) to a spherical surface.

A more general result without the assumption of strong initial strain was provided and the tension per unit length was calculated to be [60]:

$$S^3\left(\frac{\nu-1}{h}\right)+S^2\nu P+CP^2r_0^2=0 \tag{8-5}$$

where h = thickness of the film
ν = Poisson's ratio of the film
$C = 0.04167$, a constant.

Solving this equation for S and substituting S into Eqs. (8-3) and (8-4) allows accurate evaluation of the optical properties of the lens.

Membrane Deflection

A circular membrane with axisymmetric uniform load P with a clamped edge exhibits deflection profile given by [52,61]:

$$u(r)=\frac{P}{64D}\left(r_0^2-r^2\right)^2 \tag{8-6}$$

where $D = Eh^3(1-\nu^2)^{-1}/12$ is the stiffness (flexural rigidity) of the plate
E = modulus of elasticity
ν = Poisson's ratio
h = thickness of the plate.

When no changes in the thickness of the membrane are assumed, the rigidity is constant across the membrane, and the bending profile is explicitly given by Eq. (8-6).

Such shells can be used to construct a liquid-filled or a gas-filled pneumatically actuated lens. For liquid-filled lens, the refractive properties of the filling differ significantly from those of the outer environment. So the optical properties of the lens are primarily defined by the optical path in the liquid, given by $u(r)$. For pneumatically actuated lens, the filling and the outer medium have the same refractive index, so it is solely the optical path $l(r)$ in the membrane that defines the focal length (see Fig. 8-3a). Assuming constant thickness of the membrane and for small deflections, the optical path length is approximately given by:

$$l(r)\approx\frac{1}{2}h\left[1+\left(\frac{\partial u}{\partial r}\right)^2\right]\Delta n \tag{8-7}$$

Here Δn is the difference in the refractive indices between the plate (polymer) and the outer medium. Using the best spherical fit for the function $l(r)$, the radius of curvature of the optical phase and consequently the effective focal length of the lens can be obtained.

8-2-2 Adaptive Optofluidic Lenses

Fast development of optoelectronics in the past decades keeps pushing miniaturization of imaging systems to the new limits. Traditional imaging systems with variable focal length incorporate a pair of lenses, with a variable distance between them. This approach is too cumbersome in many application areas where bulky optics with moving parts are intolerable. A need for the capability of adjusting the focal length without bulky optical components is essential for numerous applications.

A significant step toward miniaturization of adaptive lenses was made in the 1980s with the advent of liquid crystal (LC), which allowed a successful implementation of LC-based lenses [62–66]. Such lenses are based on electro-optic effect, which allows control over the refractive index of the constituent medium. Such lenses suffer from many limitations: first, very strong electromagnetic fields are required to produce a noticeable change in the refractive index with electro-optic effect; second, LC is a birefringent optical medium and so LC lenses are polarization sensitive. Moreover, costly fabrication makes LC lenses impractical in many application areas and limits such lenses to small apertures.

Optofluidic lenses overcame many of the limitations imposed on LC lenses. Liquid-filled and polymer-based optofluidic lenses allow focal tuning from $-\infty$ to $+\infty$, fast response times, easy low-cost fabrication, and a whole range of optical aperture sizes—from meters down to several millimeters. Lenses based on liquid-filled shells or pneumatically deformed polymers that have recently been reported are described throughout the chapter.

Two major types of optofluidic lenses evolved—liquid-filled and pneumatic lenses. Liquid-based lenses commonly offer very wide focal tuning but very limited aperture. Pneumatic lenses, on the contrary, allow large apertures for high performance optics, at the expense of the focal tuning range. So the two types represent tradeoff between aperture size and focal tuning range and span a wide range of application areas including vision devices [67–69], microscopy [70–73], photography [74], optical data storage [75], bioengineering and medicine devices [76,78], biochemical and temperature sensing [79–81], and lab-on-a-chip devices [70,82,83]. These lenses can be driven mechanically [48,77,84–92], chemically [93,94], thermally [95–97], by electrowetting effect [74,98,99], and by radiation pressure [100]. These diverse means of manipulation and control make optofluidic lenses versatile and appealing in a number of application areas mentioned herein.

Liquid-Filled Adaptive Elements

The very old idea of using liquid-based optical lens [54,55,101–103] has been recently revisited for portable adaptive lenses with mechanical actuators [84,104–107]. Micro-lens array integrated into a microfluidic device was demonstrated [51,85]. Trade-offs between focusing and angle of view were analyzed [108]. Fast actuation with a response time of milliseconds was achieved using piezoelectric actuators [109]. Compact varifocal micro-lens with integrated thermal actuator and sensor was implemented [110]; a set of integrated cylindrical lenses with independent biaxial tuning was constructed [69]; variable focus was also achieved by replacing liquids with different refractive indices [111]; chemical actuation of pH-sensitive hydrogel was also employed to design a varifocal lens [80]; hydrodynamical interface between two liquids flowing in a curved microchannel was exploited to construct a variable focus lens [77]; additional designs and actuation techniques have been continuously reported [48,51,84,88,89,112,113].

Lenses mentioned in this section exhibit aspherical surfaces leading to optical aberrations, which were analyzed for a variety of configurations [51,60,85,114–116]. These aberrations may be diminished by using compound structures [117,118], introduction of an aperture [51], arbitrary profile membranes with variable thickness [48], composite membranes [119], and two different lens curvatures [86]. On the other hand, these aberrations can be useful to compensate aberrations of other optical components.

Along with imaging and aberrations control, the liquid-filled lenses were employed for biaxial beam shaping and optical signal processing. A set of two orthogonal cylindrical lenses [69] for beam shaping is reviewed here in greater details. The device consists of three chambers separated by two PDMS membranes and filled with liquids of different refractive indices (see Fig. 8-4a). Differences in the pressures applied to the three chambers result in bending of the membranes. Under the proper design the shape of the bending membrane is nearly cylindrical. Therefore, the shape of the beam becomes highly elliptical with an aspect ratio of approximately 10 (see Fig. 8-4f through 8-4h). Comparison between the beam intensity profiles in the figure suggests that focusing along one direction has no appreciable effect on the other. Therefore, variations of differential pressures are directly translated into changes in the focal distances with practically no cross talk.

Pneumatically Actuated Membrane-Based Lens

Pneumatically actuated optofluidic lenses are lenses in which a polymer with varying shape is used as an optical element for focusing. They are based on the same concept used in liquid-filled lenses—there is a chamber with a soft wall, which bends under applied pressure. In the liquid-filled lens the fluid mimics the behavior of a lens. Pneumatically actuated polymeric lens [71] contains no liquid and the focusing is achieved solely by the differences between the optical path lengths $l(r)$ through the membrane across the aperture (see Sec. 8-2-1). In the liquid-filled lens, it is the bending profile $u(r)$ alone

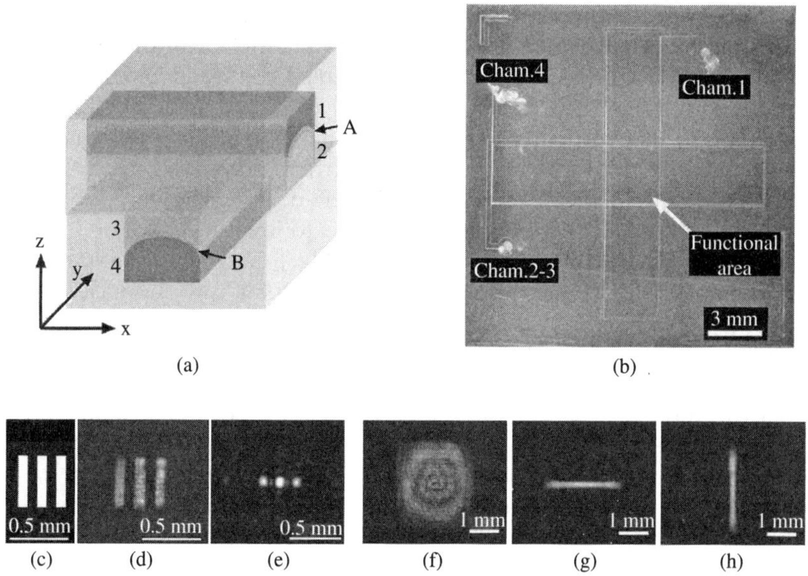

FIGURE 8-4 A set of cylindrical orthogonal lenses [69]: (a) conceptual drawing of the device; (b) measured beam profiles show independent two-axis beam shaping; (c) through (e) demonstration of imaging and optical signal processing: (c) image of a target taken under a microscope; (d) image of the target projected on the CCD by the device; (e) Fourier transform of the target (diffraction pattern) projected on the CCD; (f) through (h) demonstration of beam shaping with a CCD camera placed at $z = 200$ mm behind the device. The figures show patterns of light on the CCD. (*L. Pang, U. Levy, K. Campbell, A. Groisman, and Y. Fainman, "Set of two orthogonal adaptive cylindrical lenses in a monolith elastomer device," Opt. Express, 13, 9003–9013, 2005.*)

(neglecting the thickness of the wall) that defines the focusing properties of the lens (see Fig. 8-3).

Much early work on polymer-based lenses focused on microlenses and lens arrays with constant focus [120,121]. Refractive microlenses with lens diameters of a few to several hundred micrometers are extensively used in optical devices such as detectors and emitters to boost the optical efficiency. They are also likely to be combined with other micro-optical structures in optical sensor arrays, in high-definition display and projection systems, in biomedical systems, and in optical interconnection technology. Many fabrication techniques have been developed [70,94,111,122–141], and micro-lens arrays can now be tailored and fabricated in high volumes.

Tunable polymer-based lenses can be actuated thermally [95,97,142] or pneumatically [71]. For thermally tuned lenses both planar [142] and free space [95] configurations were demonstrated. In the free-space configuration, thermally actuated polymeric lens is composed of a flexible PDMS lens, silicon conducting ring, and silicon heater [96]. The mismatching of the coefficient of thermal expansion and stiffness between PDMS and silicon leads to deformation of polymer lens during heating, so as to further change its focal length. The difficulty to control thermal expansion of a large area limits the aperture to hundreds of micrometers for any practical design. Pneumatically actuated lenses do not bear these limitations and are discussed in further details as follows.

Pneumatically actuated lens is shown in Fig. 8-5a. The membrane is integrated in compound camera lenses that contain two more elements attached to the same mount: a planoconvex glass lens and a diaphragm between the membrane and the lens (see Fig. 8-5a). The mount is sealed by the membrane and the lens, and the pressure of air in it is adjusted through a connector on a side. Application of vacuum to the interior of the mount pulls the membrane inward. The shape of the deformed membrane is modeled as a thin circular plate with clamped edges (see Section 8-2-1). The diaphragm is integrated with the set of

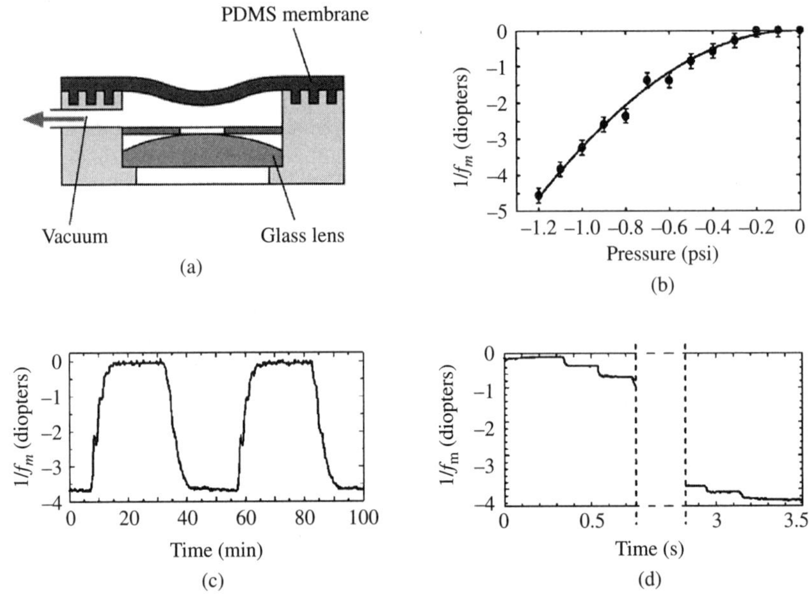

FIGURE 8-5 A set of constant and variable lens [71]: (a) schematic drawing of the device; (b) focusing power measured versus the applied pressure. The plot shows performance of the lens with membrane diameter of 1.2 cm, thickness of 1.66 mm, diaphragm aperture of 3.3 mm, and focal length of 19 mm; (c) variation of the refractive power, of the membrane lenses with time; (d) the pressure fed to the lens is switched from –0.2 to –1.1 psi in 15 steps at a rate of 5 steps/s. (*Reprinted with permission from K. Campbell, Y. Fainman, and Groisman A, "Pneumatically actuated adaptive lenses with millisecond response time," Appl. Phys. Lett., 91 (17), 171111, 2007. Copyright 2007, American Institute of Physics.*)

lenses to reduce spherical aberrations. Such lenses allow construction of large apertures not available in other types of fluidic lenses along with millisecond transition times. Pneumatically actuated lens can be driven at 500 Hz with 4 diopter variation in the refractive power [71]. Focal length variations with millisecond response time are shown in Fig. 8-5c. Considerable focusing power can be obtained at low pressures (see Fig. 8-5b). Therefore, these lenses can be used for fast longitudinal scanning in three-dimensional imaging.

The advantages of such lens as opposed to the liquid-filled lens become obvious. First, the optical performance of a pneumatically actuated lens is not compromised by the bubbles that form in the liquid. Second, apertures much larger than in the other types of fluidic lenses are readily available. Finally, the lens carries no additional mass of the liquid, so the response time is faster, and the impact of mechanical shocks and gravitational distortions is diminished. One drawback is the lower focusing power (i.e., dioptres) than in the other types of fluidic lenses.

8-2-3 Composite Membrane Devices

Thin flexible membranes made of a silicon elastomer PDMS have been intensively used in microfabricated devices to construct pressure-actuated valves [39,143,144], check valves [145,146], and adaptive lenses [71,95,97,142]. The character of deformation of a plain membrane under pressure is defined by the shape of the frame it is attached to, allowing for a very limited tunability. Here we focus on composite membranes with pieces of rigid epoxy grafted inside PDMS. The dimensions and positions of the epoxy pieces are defined with a high precision by UV-lithography, allowing high control of its mode of deformation under applied pressure.

Composite structures in this context are a combination of two or more materials, each of which retains its own elastic properties. Certain combinations of several materials allow deformation to be highly adjustable and easily tailored. This feature was successfully employed to build tunable optical devices [119]. The pattern of epoxy grafted into the membrane defines its mode of deformation under pressure. Planar architecture allows standard easy and precise soft-photolithography fabrication techniques to be used for adaptive optical devices.

It was shown in Sec. 8-1-2 how a blazed grating can be effectively tuned by replacing the liquid on top. Manipulation of liquids with different refractive indices allows tunable beam deflection. Another type of tunable gratings is based on composite membrane technology, which allows grating geometry tuning with mechanical actuation [119]. The membrane's mode of deformation is made highly adjustable. The proposed technique was demonstrated in two types of devices: stretcher and rotator (see Fig. 8-6a, and 8-6b). In the stretcher, the grafted pieces of epoxy focus the pressure-induced extension of the

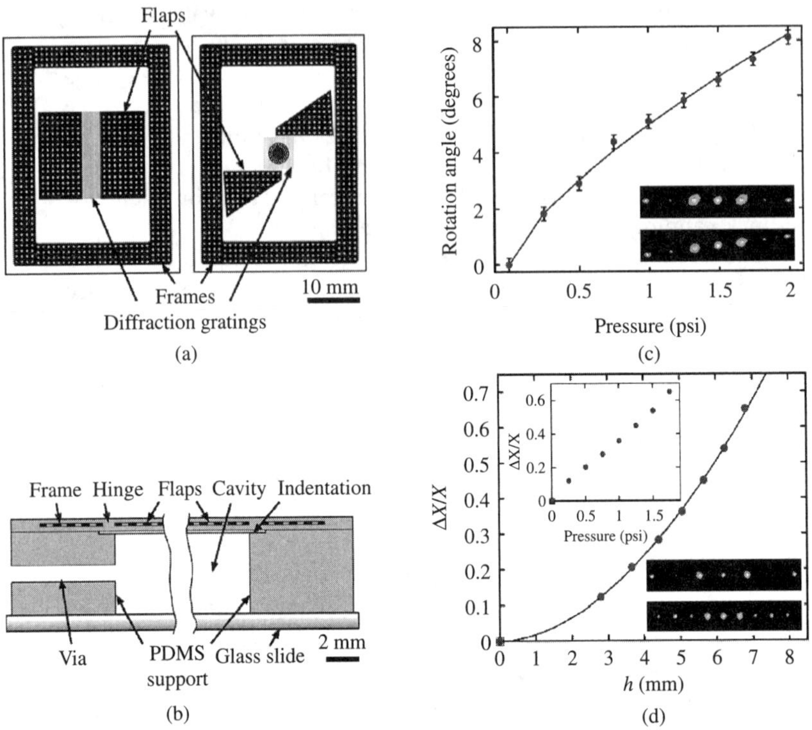

Figure 8-6 Composite membrane devices with stretching and rotating geometries [119]: (a) schematic drawings of the composite membranes in the stretcher (left) and rotator (right); (b) cross section of the device; (c) angle of the in-plane rotation, φ, of the central circle of the composite membrane in the rotator as a function of pressure, P, applied to the device. The continuous line is a power low fit, $\varphi \propto P^{0.72}$; (d) relative extension of the central strips of the stretcher as a function of elevation of the center of the membrane for the stretcher. (inset) Relative extension as a function of pressure, P. (*Reprinted with permission from K. Campbell, U. Levy, Y. Fainman, and A. Groisman, "Pressure-driven devices with lithographically fabricated composite epoxy-elastomer membranes," Appl. Phys. Lett., 89, 154105, 2006. Copyright 2006, American Institute of Physics.*)

membrane to a thin strip of PDMS. In the rotator, the epoxy patterning causes in-plane rotation of a central area of the membrane when pressure is applied.

In the stretcher, the pressure-induced extension provided tunable period grating. Approximately 50% elongation of the grating period was achieved at applied pressure below 1.5 psi. This strong deformation at low pressures in the region of interest is obtained by grafting pieces of epoxy into the membrane. The composite structure transfers most of the deformation onto the thin strip of a soft membrane patterned with a grating profile (see Fig. 8-6d).

In the rotator, the composite structure exhibits in-plane rotation of the grating. Again, epoxy grafted into the membrane decreased membrane's torsion in some areas, while increasing it in the others. Angular rotation of the grating by up to 8 degrees was demonstrated at applied pressure of approximately 2 psi (see Fig. 8-6c).

8-3 Summary

In this chapter we reviewed a number of tunable optofluidic devices. In these devices gases and liquids are used to manipulate and control light in its many forms: planar-guided and free-space, coherent, and broadband. These devices comprise two major parts—the microfluidic device used as an actuator and the optical element that performs the optical tuning function. We provided a detailed review of adaptive and tunable optical elements such as lenses, gratings, scanners, and beam shapers. Integration of these elements with their actuation devices and other optical components allows construction of miniature multifunctional optical devices enabling further integration into micro- and macroscale systems.

Additionally, optofluidic switches use liquids and thus can operate with ultrawide spectral bandwidth and simultaneously be invariant to the state of polarization of optical beams; these properties cannot be achieved with other technologies. $N \times 1$ optical switching based on optofluidic components is made possible and allows significant simplification of $N \times N$ optical switching and interconnections. Both total internal reflection and diffraction phenomena were exploited to construct optofluidic switches.

Pneumatically driven compound lens overcomes pitfalls of traditional liquid-filled tunable lenses. Absence of liquid allows faster operation and improved performance, which is not compromised by bubbles under mechanical stresses and vibrations. Moreover, apertures much larger than in liquid-filled lenses are readily available. These lenses were shown to produce considerable refractive power under comparatively low applied pressures.

Composite membrane technology presented in this chapter is based on soft polymer membranes patterned with rigid epoxy inclusions. These inclusions are designed to obtain the desired mechanical properties of the membrane and allow good control of the deformation profile. As the fabrication relies on planar lithographic techniques, the pattern can be easily produced in high volumes with high accuracy. Although composite membranes were so far exploited for tunable gratings [119], they show much potential for any adaptive and tunable optics where arbitrary geometry is desired.

In this early stage of development of optofluidics as a field of research, every new device opens new perspectives and stimulates research showing improved performance and new applications. The significant progress of micro- and nanofabrication techniques is striving to provide smoother surfaces, smaller form factors, robustness and better repeatability, faster prototyping, and therefore lower costs of optofluidic devices. Commercialization of some of the reviewed technologies shows much promise for optofluidics for adaptation as an evolving new field.

References

1. C. Monat, P. Domachuk, and B.J. Eggleton, "Integrated optofluidics: A new river of light," *Nat. Photonics*, 1, 106–114, 2007.
2. D. Psaltis, S.R. Quake, and C. Yang, "Developing optofluidic technology through the fusion of microfluidics and optics," *Nature*, 442, 381–386, 2006.
3. I.P. Kaminow and E.H. Turner, *Proc. IEEE*, 54, 1374, 1966.
4. T. Motoki, "Low voltage optical modulator using electrooptically induced phase gratings," *Appl. Opt.*, 12, 1472–1476, 1973.
5. R.P. Kenan, C.M. Verber, and Van E. Wood, "Wide-angle electro-optic switch," *Appl. Phys. Lett.*, 24, 428, 1974.
6. R.C. Alferness, "Waveguide electrooptic switch arrays," *Selected Areas in Communications*, 6 (**7**), 1117–1130, 1988.
7. R.W. Dixon, "Photoelastic properties of selected materials and their relevance for applications to acoustic light modulators and scanners," *J. Appl. Phys.* 38, 5149, 1967.
8. D.A. Smith, R.S. Chakravarthy, Z.Y. Bao, J. E. Baran, J.L. Jackel, A. dAlessandro, D.J. Fritz, S.H. Huang, X.Y. Zou, S.M. Hwang, A.E. Willner, and K.D. Li, "Evolution of the acousto-optic wavelength routing switch," *J. Lightwave Technol.*, 14, 1005–1019, 1996.
9. M. Shirasaki, N. Takagi, T. Obokata, and K. Shirai, "Bistable magnetooptic switch using YIG crystal with phase matching films," *IEEE J. Quantum Electron.*, 17 (12), 2498–2499, December 1981.
10. M. Haruna and J. Koyama, "Thermooptic deflection and switching in glass," *Appl. Opt.*, **21**, 3461–3465, 1982.
11. A.A. Zatykin, S.K. Morshnev, and A.V. Frantsesson, "A thermooptic fiber switch," *Kvantovaya Elektronika*, 12 (1), 211–213, 1985.
12. K.E. Petersen, "Micromechanical light modulator array fabricated on silicon [laser display applications]," *Appl. Phys. Lett.*, 31 (8), 521–523, October 1977.
13. M.C. Wu, A. Solgaard, and J.E. Ford, "Optical MEMS for lightwave communication," *J. Lightwave Technol.* 24 (12), 4433–4454, 2006.
14. M.J. Brady, L.V. Gregor, and M. Johnson, U.S. Patent No. 4384761, 1983.
15. J.M. Ginder, J.T. Remillard, and W.H. Weber, U.S. Patent No. 5351319, 1994.
16. M. Saito, M. Takakuwa, and M. Miyagi, *IEICE Trans. Electron.*, E78-C, 1995, 1465.
17. S. Taketomi, *Jpn. J. Appl. Phys.*, 22, 1983, 1137.
18. S. Taketomi, U.S. Patent No. 4812767, 1989.
19. C.Y. Hong, "Optical switch devices using the magnetic fluid thin films," *J. Magn. Magn. Mater.*, 201, 1999, 178–181.
20. G. Da Costa, "All-optical light switch using interaction between low-power light beams in a liquid film," *Opt. Eng.*, 25, 1058, 1986.
21. G. Da Costa and R. Escalona, "Time evolution of the caustics of a laser heated liquid film," *Appl. Opt.*, 29, 1023–1033, 1990.
22. K. Campbell, A. Groisman, U. Levy, L. Pang, S. Mookherjea, D. Psaltis, and Y. Fainman, "A microfluidic 2 × 2 optical switch," *Appl. Phys. Lett*, 85, 6119–6121, 2004.
23. J.L. Jackel and W.J. Tomlinson, "Bistable optical switching using electrochemically generated bubbles," *Opt. Lett.*, 15 (24), 1470, 1990.
24. A. Ware, "New photonic-switching technology for all-optical networks," *Lightwave*, 92–98, 2000.
25. S. Hengstler, J. J. Uebbing, and P. McGuire, "Laser-activated optical bubble switch element," *Proceedings of the 2003 IEEE/LEOS International Conference on Optical MEMS (OMEMS 2003)*, 117–118, August 2003.

26. V.I. Sukhanov, F.L. Vladimirov, and D.A. Monakhov, "How the surface properties of the liquid element affect the response time of a liquid optical deflector," *J. Opt. Technol.*, 71 (3), 2004.
27. J.J. Uebbing, S. Hengstler, D. Schroeder, S. Venkatesh, and R. Haven, "Heat and fluid flow in an optical switch bubble," *J. Microelectrom. Syst.*, 15, 6, 2006.
28. G. Beni and S. Hackwood, "Electro-wetting displays," *Appl. Phys. Lett.*, 38, 207, 1981.
29. J. Heikenfeld and A.J. Steckl, "High-transmission electrowetting light valves," *Appl. Phys. Lett.*, 86 (15), 151121, 2005.
30. J. Heikenfeld and A.J. Steckl, "Intense switchable fluorescence in light wave coupled electrowetting devices," *Appl. Phys. Lett.*, 86 (1), 011105, 2005.
31. L. Hou, N.R. Smith, and J. Heikenfeld, "Electrowetting manipulation of any optical film," *Appl. Phys. Lett.*, 90 (25), 251114, 2007.
32. M. Makihara, M. Sato, F. Shimokawa, and Y. Nishida, "Micromechanical optical switches based on thermocapillary integrated in waveguide substrate," *Lightwave Technol.*, 17(1), 14, 1999.
33. Nam-Trung Nguyen, Tian-Fook Kong, Jun-Hui Goh, and Cassandra Lee-Ngo Low, "A micro optofluidic splitter and switch based on hydrodynamic spreading," *J. Micromech. Microeng.*, 17, 2007, 2169–2174.
34. J.E. Fouquet, S. Venkatesh, M. Troll, D. Chen, H.F. Wong, and P.W. Barth, "A compact, scalable crossconnect switch using total internal reflection due to thermally-generated bubbles," in *IEEE LEOS Annual Meeting*, Orlando, 1998.
35. J.E. Fouquet, "Compact optical cross-connect switch based on total internal reflection in a fluid containing planar lightwave circuit," *OFC*, Baltimore, March 7–10, 2000.
36. S. Hengstler, J.J. Uebbing, and P. McGuire, "Laser-activated optical bubble switch element," *Proceedings of the 29th European Conference on Optical Communication—14th International Conference on Integrated Optics and Optical Fiber Communication (ECOC-IOOC 2003)*, September 2003.
37. T. Sakata, H. Togo, M. Makihara, F. Shimokawa, and K. Kaneko, "Improvement of switching time in a thermocapillarity optical switch," *J. Lightwave Technol.*, 19 (7), 1023–1027, July 2001.
38. S. Venkatesh, J.-W. Son, J.E. Fouquet, R.E. Haven, D. Schroeder, H. Guo, W. Wang, P. Russell, A. Chow, and P.F. Hoffmann, "Recent advances in bubble actuated photonic cross-connect," *Silicon-based and Hybrid Optoelectronics IV*, January 2002.
39. G.H.G.V. Studer, A. Pandolfi, M. Ortiz, W.F. Anderson, and S.R Quake, Scaling properties of a low-actuation pressure microfluidic valve, *J. Appl. Physics*, 95, 393, 2004.
40. R.A. Hayes, B.J. Feenstra, "Video-speed electronic paper based on electrowetting," *Nature*, 425, 383–385, 2003.
41. M. Hashimoto, B. Mayers, P. Garstecki, and G.M. Whitesides, "Flowing lattices of bubbles as tunable, self-assembled diffraction gratings," *SMALL* 2 (11), 1292–1298, 2006.
42. P. Garstecki, I. Gitlin, W. DiLuzio, G.M. Whitesides, E. Kumacheva, and H.A. Stone, "Formation of monodisperse bubbles in a microfluidic flow-focusing device," *Appl. Phys. Lett.*, 2004, 85, 2649.
43. S. Zamek, K. Campbell, L. Pang, A. Groisman, Y. Fainman, "Optofluidic 1 × 4 switch", *CLEO*, CTuR4, 2008.
44. A. Groisman, S. Zamek, K. Campbell, L. Pang, U. Levy, and Y. Fainman, "Optofluidic 1 × 4 switch", *Opt. Express*, 16 (18), 13499–13508, 2008.
45. M. Neviere, "Electromagnetic study of transmission gratings," *Appl. Opt.* 30, 4540–4547, 1991.
46. R. Petit, "Electromagnetic Theory of Gratings," *Topics in Applied Physics*, Springer, 1980.
47. B.E.A. Saleh and M.C. Teich, *Fundamentals of Photonics*, John Wiley and Sons, New York, 1991, Chap. 21, Photonic Switching and Computing.
48. D. Shaw and C-W Lin, "Design and analysis of an asymmetrical liquid-filled lens," *Opt. Eng.*, 46, 123002 (2007).

49. N.R. Smith, D. C. Abeysinghe, J.W. Haus, and J. Heikenfeld, "Agile wide-angle beam steering with electrowetting microprisms," *Opt. Express*, 14, (14), 6557–6563, 2006.
50. L. Truong, E.F. Borra, R. Bergamasco, N. Caron, P. Laird, and A. Ritcey, "Nanoengineered liquid mirrors shaped by thermal fields," *Appl. Opt.*, 44 (9), 2005.
51. A. Werber and H. Zappe, "Tunable microfluidic microlenses," *Appl. Opt.*, 44 (16), 3238–3245, 2005.
52. E. Ventsel and T. Krauthammer, *Thin Plates And Shells—Theory, Analysis and Applications*, Marcel Dekker, New York, 2001.
53. H. Mollmann, *Introduction to the Theory of Thin Shells*, John Willey and Sons, New York, 1981.
54. A.A. Burrows and V.E. Hamilton, "Stereopsis using a large aspheric field lens," *Appl. Opt.*, 13, 4, 1974.
55. V.E. Hamilton, A.H. Lanckton, "Stereopsis using a large liquid-filled aspheric field lens stereoscope," *Photogrammetric Eng. and Remote Sensing*, 39, 3, 294, 1973.
56. S.D. Poisson, *Memoirs of the Academy*, Vol. 8, Academy of Sciences, Paris, 1829.
57. L. Prandtl, "Zur Torsion von prismatischen Stäben," *Phys. Zeits.*, 4, 190, 1903.
58. J.P. Den Hartog , *Advanced Strength of Materials*, Courier Dover Publications, 1987, 11.
59. M. Born and E. Wolf, *Principles of Optics*, 7th ed., Cambridge (UK), Cambridge University Press, 1999.
60. A.H. Rawicz and I. Mikhailenko, "Modeling a variable-focus liquid-filled optical lens," *Appl. Opt.*, 35 (10), 1587–1589, 1996.
61. J.N. Reddy, *Theory and Analysis of Elastic Plates*, Taylor & Francis, 1996.
62. S. Sato, "Liquid-crystal lens-cells with variable focal length," *Jpn. J. Appl. Phys.*, 18, 1679–1684, 1979.
63. D. W. Berreman, US Patent No. 4,190,330 1980.
64. S.T. Kowel, D.S. Cleverly, and P.G. Kornreich,"Focusing by electrical modulation of refraction in a liquid crystal cell," *Appl. Opt.*, 23, 1984, 278.
65. A.F. Naumov, M.Y. Loktev, I.R. Guralnik, and G. Vdovin, "Liquid-crystal adaptive lenses with modal control," *Opt. Lett.*, 23, 992–994, 1998.
66. L.G. Commander, S.E. Day, and D.R. Selviah, "Variable focal length microlenses,"*Opt. Commun.*, 177, 157–170, 2000.
67. von Waldkirch M, Lukowicz P, and Troster G, "Oscillating fluid lens in coherent retinal projection displays for extending depth of focus," *Opt. Comm.*, 253 Issue: (4–6), 407–418, 2005.
68. Varioptic—The Liquid Lens Company, www.varioptic.com.
69. L. Pang, U. Levy, K. Campbell, A. Groisman, and Y. Fainman, "Set of two orthogonal adaptive cylindrical lenses in a monolith elastomer device," *Opt. Express*, 13, 9003–9013, 2005.
70. Y. Gambin, O. Legrand, and S.R. Quake, "Microfabricated rubber microscope using soft solid immersion lenses," *Appl. Phys. Lett.*, 88, 174102, 2006.
71. K. Campbell, Y. Fainman, and Groisman A, "Pneumatically actuated adaptive lenses with millisecond response time," *Appl. Phys. Lett.*, 91 (17), 171111, 2007.
72. P.S. Tsai, B. Migliori, K. Campbell, T.N. Kim, Z. Kam, A. Groisman, and D. Kleinfeld, "Spherical aberration correction in nonlinear microscopy and optical ablation using a transparent deformable membrane," *Appl. Phys. Lett.*, 91, 191102, 2007.
73. F.S. Tsai, Sung Hwan Cho, Yu-Hwa Lo, B.Vasko, and J. Vasko, "Miniaturized universal imaging device using fluidic lens," *Opt. Lett.*, 33 (3), 291–293, 2008.
74. S. Kuiper and B.H.W. Hendriks, "Variable-focus liquid lens for miniature cameras," *Appl. Phys. Lett.*, 85, 1128 (2004).
75. B.H.W. Hendriks, S. Kuiper, M.A.J. Van As, C.A. Renders, and T.W. Tukker, "Electrowetting-based variable-focus lens for miniature systems," *Opt. Rev.*, 12, 3, 255–259, 2005.
76. B.A. Flusberg, E.D. Cocker, W. Piyawattanametha, J.C. Jung, E.L.M. Cheung, and M.J. Schnitzer, "Fiber-optic fluorescence imaging," *Nat. Methods*, 2, 941–950, 2005.

77. Xiaole Mao, John Robert Waldeisen, Bala Krishna Juluri, and Tony Jun Huang, "Hydrodynamically tunable optofluidic cylindrical microlens," *Lab Chip* 7, 1303–1308, 2007.
78. A. Divetia, Tsung-Hsi Hsieh, Jun Zhang, Zhongping Chen, Mark Bachman, and Guann-Pyng Li, "Dynamically focused optical coherence tomography for endoscopic applications," *Appl. Phys. Lett.*, 86, 103902, 2005.
79. L. Dong, Abhishek K. Agarwal, David J. Beebe, and Hongrui Jiang, "Adaptive liquid microlenses activated by stimuli-responsive hydrogels," *Nature* 442, 551–554, 2006.
80. L. Dong and H. Jiang, "pH-adaptive microlenses using pinned liquid-liquid interfaces actuated by pH-responsive hydrogel," *Appl. Phys. Lett.*, 89, 211120, 2006.
81. W.S. Wang, J. Fang, "Variable focusing microlens chip for potential sensing applications," *IEEE Sens. J.*, 7 (1–2), 11–17, 2007.
82. J.C. Roulet, R. Völkel, H.P. Herzig, E. Verpoorte, N.F. de Rooij, and R. Dändliker, "Microlens systems for fluorescence detection in chemical microsystems," *Opt. Eng.*, 40, 814–821, 2001.
83. Jeonggi Seo and Luke P. Lee, "Disposable integrated microfluidics with self-aligned planar microlenses," *Sens. Actuators B Chem.*, 99 (2–3), 615–622, 2004.
84. De-Ying Zhang, Victor Lien, Yevgeny Berdichevsky, Jaehyuck Choi, and Yu-Hwa Lo, "Fluidic adaptive lens with high focal length tenability," *Appl. Phys. Lett.*, 82, 3171, 2003.
85. N. Chronis, G. Liu, K.-H. Jeong, and L. Lee, "Tunable liquid-filled microlens array integrated with microfluidic network," *Opt. Express,* 11, 2370–2378, 2003.
86. K.H. Jeong, G. Liu, N. Chronis, and L. Lee, "Tunable microdoublet lens array," *Opt. Express,* 12 (11) 2494–2500, 2004.
87. J. Chen, W. Wang, J. Fang, and K. Varahramyan, "Variable-focusing microlens with microfluidic chip," *J. Micromech. Microeng.*, 14, 675–680, 2004.
88. H. Ren and S.T. Wu, "Variable-focus liquid lens by changing aperture," *Appl. Phys. Lett.*, 86, 211107, 2005.
89. H. Ren, S.T. Wu, "Variable-focus liquid lens," *Opt. Express*, 15 (10), 5931–5936, 2007.
90. R. Kuwano, T. Tokunaga, Y. Otani, and N. Umeda, "Liquid pressure varifocus lens," *Opt. Rev.*, 12 (5), 405-408, 2005.
91. P.M. Moran, Saman Dharmatilleke, Aik Hau Khaw, Kok Wei Tan, Mei Lin Chan, and Isabel Rodriguez, "Fluidic lenses with variable focal length," *Appl. Phys. Lett.*, 88, 041120, 2006.
92. Sindy K.Y. Tang, Claudiu A. Stan, and George M. Whitesides, "Dynamically reconfigurable liquid-core liquid-cladding lens in a microfluidic channel," *Lab Chip*, 8, 395–401, 2008.
93. C.A. López, C.C. Lee, and A.H. Hirsa, "Electrochemically activated adaptive liquid lens," *Appl. Phys. Lett.*, 87, 134102, 2005.
94. Dorrer C, Prucker O, and Ruhe J, "Swellable surface-attached polymer microlenses with tunable focal length," *Adv. Mater.*, 19 (3), 456, 2007.
95. S.N. Lee, H.W. Tung, W.C. Chen, and W.L. Fang, "Thermal actuated solid tunable lens," *IEEE Photon. Technol. Lett.*, 18, 2191–2193, 2006.
96. S.Y. Lee, W.C. Chen, H.W. Tung, and W. Fang, "Microlens with tunable astigmatism," *IEEE Photon. Technol. Lett.*, 19, 18, 2007.
97. Han Yang, Yong-Hao Han, Xiang-Wei Zhao, Keiji Nagai, and Zhong-Ze Gu, "Thermal responsive microlens arrays," *Appl. Phys. Lett.*, 89, 111121, 2006.
98. Berge, B., "Electrocapillarite et mouillage de films isolants par l'eau", *C. R. Acad. Sci. II,* 317, 157, 1993.
99. Berge, B., and Peseux, J., "Variable focal lens controlled by an external voltage: an application of electrowetting," *Eur. Phys. J. E*, 3, 159–163, 2000.
100. A. Casner and J.-P. Delville, "Adaptive lensing driven by the radiation pressure of a continuous-wave laser wave upon a near-critical liquid-liquid interface," *Opt. Lett.* 26, 1418–1420, 2001.

101. R.L. Gordon, US Patent No. 1269422, June 11, 1918.
102. B.M. Wright, "Improvements in or relating to variable focus lenses," English patent 1209234, March 11,1968.
103. G.C. Knollman, J.L.S. Bellin, and J.L. Weaver, "Variable focus liquid filled hydroacoustic lense," *J. Acoustical Soc. Am.*, 49 (1), 253, 1971.
104. T. Kaneko, T. Ohmi, N. Ohya, N. Kawahara, and T. Hattori, "A new, compact, and quick-response dynamic focusing lens," *Proceedings of the 9th International Conference on Solid-State Sensors and Actuators (Transducers'97)*, Chicago, 16–19 June 1997, 63–66.
105. Si-Hong-Ahn and Yong-Kweon Kim, "Proposal of human eye's crystalline lens-like variable focusing lens," *Broadband Optical Networks and Technologies: An Emerging Reality/Optical MEMS/Smart Pixels/Organic Optics and Optoelectronics, IEEE/LEOS Summer Topical Meetings*, II/89–II/90, 20–24 July 1998.
106. Si-Hong-Ahn and Yong-Kweon Kim, "Proposal of human eye's crystalline lens-like variable focusing lens," *Sens. Actuators A*, 78, 48–53, 1999.
107. Si-Hong-Ahn, Yong-Kweon Kim, "Design and fabrication of variable focusing lens," *Proc. SPIE*, 3515, 270, 1998.
108. M. Agarwal, R A Gunasekaran, P. Coane, and K. Varahramyan, "Polymer-based variable focal length microlens system," *J. Micromech. Microeng.*, 14, 2004, 1665–1673.
109. H. Oku, K. Hashimoto, and M. Ishikawa, "Variable-focus lens with 1-kHz bandwidth," *Opt. Express*, 12, 2138–2149, 2004.
110. W. Wang, Ji Fang, and Kody Varahramyan, "Compact Variable-Focusing Microlens With Integrated Thermal Actuator and Sensor," *IEEE Photon. Technol. Lett.*, 17 (12), 2643–2645, 2005.
111. Kuang-Sheng Hong, JingWang, Alexey Sharonov, Dinesh Chandra, Joanna Aizenberg, and Shu Yang, "Tunable microfluidic optical devices with an integrated microlens array," *J. Micromech. Microeng.*, 16, 1660–1666, 2006.
112. D. Chandra, Shu Yang, and Pei-Chun Lin, "Strain responsive concave and convex microlens arrays," *Appl. Phys. Lett.*, 91, 251912, 2007.
113. S.W. Lee and S.S. Lee, "Focal tunable liquid lens integrated with an electromagnetic actuator," *Appl. Phys. Lett.*, 90, 121129, 2007.
114. N. Sugiura and S. Morita, "Variable-focus liquid-filled optical lens," *Appl. Opt.*, 32 (22), 4181–4186, 1993.
115. Z. Wang, Y. Xu, and Y. Zhao, "Aberration analyses for liquid zooming lenses without moving parts," *Opt. Commun.*, 275 (1), 22–26, 1 July 2007.
116. G. Beadie, M. L. Sandrock, M. J. Wiggins, R. S. Lepkowicz, J. S. Shirk, M. Ponting, Y. Yang, T. Kazmierczak, A. Hiltner, and E. Baer, "Tunable polymer lens," *Opt. Express*, 16 (16), 11847–11857, 2008.
117. H. Feldman, "Nearly spherical acoustic lenses," *J. Acoust. Soc. Am.*, 45, 868, 1969.
118. Y. Tannaka and T. Koshikaw, "Solid-liquid compound hydroacoustic lens of low aberration," *J. Acoust. Soc. Am.*, 53, 2, 590–595, 1973.
119. K. Campbell, U. Levy, Y. Fainman, and A. Groisman, "Pressure-driven devices with lithographically fabricated composite epoxy-elastomer membranes," *Appl. Phys. Lett.*, 89, 154105, 2006.
120. H.P. Herzig, "Micro-optics: elements, systems, and applications," *Refractive Lenslet Array*, ed. M.C. Hutley, 1997.
121. D. Daly, "Applications and fabrication technologies," in *Microlens Array*, Talyer and Francis, London, 2001.
122. D.L. MacFarlane, V. Narayan, J.A. Tatum, W.R. Cox, T. Chen, and D.J. Hayes, "Microjet fabrication of microlens arrays," *IEEE Photon. Technol. Lett.*, 6, 1112–1114, September 1994.
123. P. Heremans, J. Genoe, M. Kuijk, R. Vounckx, and G. Borgh, "Mushroom microlens: optimized microlenses by reflow of multiple layers of photoresist," *IEEE Photon. Technol. Lett.*, 9, 1367–1369, October 1997.
124. P. Nussbaum, I. Philipoussis, A. Husser, and H.P. Herzig, "Simple technique for replication of micro-optical elements," *Opt. Eng.*, 37, 1804–1808, 1998.

125. Z.D. Popovic, R.A. Sprague, and G.A.N. Connell, "Technique for monolithic fabrication of microlens arrays," *Appl. Opt.*, 27, 1281–1284, 1988.
126. S. Biehl, R. Danzebrink, P. Oliveira, and M. A. Aegerter, "Refractive microlens fabrication by ink-jet process," *J. Sol-Gel Sci. Technol.* 13, 177–182, 1998.
127. T. Okamoto, M. Mori, T. Karasawa, S. Hayakawa, I. Seo, and H. Sato, "Ultraviolet-cured polymer microlens arrays," *Appl. Opt.*, 138, 2991–2996, 1999.
128. D.M. Hartmann, O. Kibar, and S.C. Esener, "Characterization of a polymer microlens fabricated by use of the hydrophobic effect," *Opt. Lett.*, 25, 975–977, 2000.
129. C. David, "Fabrication of stair-case profiles with high aspect ratios for blazed diffractive optical elements," *Microelectron. Eng.*, 53, 677–680, 2000.
130. M.-H. Wu and G.M. Whitesides, "Fabrication of two-dimensional arrays of microlenses and their applications in photolithography," *J. Micromech. Microeng.*, 12, 747–758, 2002.
131. W.X. Yu and X.C. Yuan, "UV induced controllable volume growth in hybrid sol-gel glass for fabrication of a refractive microlens by use of a grayscale mask," *Opt. Express*, 11, 2253–2258, 2003.
132. F.T. O'Neill and J.T. Sheridan, "Photoresist reflow method of microlens production Part I: Background and experiments," *Optik*, 113, 391–404, 2002.
133. X.J. Shen, L.W. Pan, and L.W. Lin, "Microplastic embossing process: experimental and theoretical characterizations," *Sens. Actuators A*, 97, 428–433, 2002.
134. M.V. Kunnavakkam, F.M. Houlihan, M. Schlax, J.A. Liddle, P. Kolodner, O. Nalamasu, and J.A Rodgers, "Low-cost, low-loss microlens arrays fabricated by soft-lithography replication process," *Appl. Phys. Lett.*, 82, 1152–1154, 2003.
135. M. Uekawa, H. Sasaki, D. Shimura, K. Kotani, Y. Maeno, and T. Takamori, "Surface-mountable silicon microlens for low-cost laser modules," *IEEE Photon. Technol. Lett.*, 15, 945–947, 2003.
136. Teng-Kai Shin, Jeng-Rong Ho, and J.-W.J. Cheng, "A new approach to polymeric microlens array fabrication using soft replica molding," *IEEE Photon. Technol. Lett.*, 16(9), 2078–2080, September 2004.
137. S.-I. Chang and J.-B. Yoon, "Shape-controlled, high fill-factor microlens arrays fabricated by a 3D diffuser lithography and plastic replication method," *Opt. Express*, 12 (25), 6366–6371, 2004.
138. T. Bourouina, T. Masuzawa, and H. Fujita, "The MEMSNAS process: Microloading effect for micromachining 3-D structures of nearly all shapes," *J. Microelectromech. Syst.*, 13, 190–199, 2004.
139. A. Llobera, A.R. Wilke, D.W. Johnson, and S. Buttgenbach, "Polymer microlenses with modified micromolding in capillaries (MIMIC) technology," *IEEE Photon. Technol. Lett.*, 17, 2628–2630, 2005.
140. J.B. Orhan, V.K. Parashar, A. Sayah, and M.A.M. Gijs, "Fabrication and characterization of three-dimensional microlens arrays in sol-gel glass," *J. Microelectromech. Syst.*, 15, 1159–1164, 2006.
141. T.H. Lin, H. Yang, and C.K. Chao, "Concave microlens array mold fabrication in photoresist using UV proximity printing," *Microsystem Technologies*, 13, 11, 2007, *Dans Symposium on Design, Test, Integration and Packaging of MEMS/MOEMS–DTIP* Stresa, Lago Maggiore, Italie, 2006.
142. A.L. Glebov, L.D. Huang, S. Aoki, M. Lee, and K. Yokouchi, "Planar hybrid polymer-silica microlenses with tunable beamwidth and focal length," *IEEE Photon. Technol. Lett.*, 16, 1107–1109 (2004).
143. M. A. Unger, H. P. Chou, T. Thorsen, A. Scherer, and S. R. Quake, "Monolithic Microfabricated Valves and Pumps by Multilayer Soft Lithography," *Science*, 288, 113, 2000.
144. W.H. Grover, A.M. Skelley, C.N. Liu, E.T. Lagally, and R.A. Mathies, "Monolithic membrane valves and diaphragm pumps for practical large-scale integration into glass microfluidic devices," *Sens. Actuators B - Chem.*, 89, 315, 2003.

145. N.L. Jeon, D.T. Chiu, C.J. Wargo, H.K. Wu, I.S. Choi, J.R. Anderson, and G.M. Whitesides, "Design and fabrication of integrated passive valves and pumps for flexible polymer three-dimensional microfluidic system," *Biomed. Microdevices*, 4, 117, 2002.
146. M.L. Adams, M.L. Johnston, A. Scherer, and S.R. Quake, "Polydimethylsiloxane based microfluidic diode," *J. Micromech. Microeng.*, 15, 1517, 2005.

CHAPTER 9

Bio-Inspired Fluidic Lenses for Imaging and Integrated Optics

Frank S. Tsai, Jessica Godin, Yu-Hwa Lo

Electrical and Computer Engineering, Jacobs School of Engineering, University of California at San Diego

Sung Hwan Cho

Materials Science and Engineering Program, Jacobs School of Engineering, University of California at San Diego

Wen Qiao

Electrical and Computer Engineering, Jacobs School of Engineering, University of California at San Diego

One can find a natural connection between optics and fluidics in animal vision. Animal eyes, both compound eyes and camera-like eyes, are great examples of using fluidic media to form images. Although beyond the scope of this chapter, it is interesting to ponder why nature selects lenses of tunable shape for vision accommodation, rather than varying the spacing between the lens and retina. The concept similar to the latter approach (i.e., change of spacing between lenses to change the focal length of an optical system) has been used by all human-designed optical systems until recently.

Starting from a simple epithelial cell with photodetection capabilities, animal eyes have evolved into highly compact, efficient, and diversified vision systems. For the best interest of the species, animal vision has been specialized either into a telephoto system (e.g., bird's eye) or into a wide angle system (e.g., fish eye). In spite of the diversity of animal vision, images of high resolution and sensitivity have been obtained in numerous animal eyes. For example, with merely two lenses, of which one is a tunable lens, human eyes can achieve a resolution as high as one arc minute in the fovea area.

Although the anatomy of animal eyes varies widely, the optics in most eyes are simple yet highly effective. This is in sharp contrast with human-made imaging systems, which are far more complicated and bulkier than animal eyes. The most marked difference in optics between animal eyes and human-made optics is that the former achieves focusing by change of lens shape, while the latter achieves focusing by varying the distances between fixed-shaped lenses. The human ciliary muscles can achieve an accommodation range of 10 D (corresponding to a focal range from 10 cm to infinity) with less than 8 g of force and a motion no greater than 0.1 mm. We are not aware of any existing human-made systems capable of the same performance with such limited force and travel.

By changing the curvature of the lens, animals can have a wide tuning range in an extremely compact design. For instance, a young person can achieve a tuning range of up to 14 D, producing a focal range from around 8 cm to infinity. Some water birds have a tuning range as wide as 50 D. Many evidences in animal vision show that changing lens shape is an effective and economical way to change the focal length of an optical system. This offers a particularly attractive tuning mechanism for miniature cameras in laptop computers, cellular phones, and other handheld devices where both the image quality and the form factor of the cameras are of primary concerns. Our exploration of fluidic lens optics is motivated by its significant potential in commercial applications as well as the intrinsic elegance found in animal vision.

This chapter covers the following subjects: fundamentals of fluidic lenses (Sec. 9-1), fluidic lens imaging systems (Sec. 9-2), fluidic intraocular lenses (IOLs) for implanted IOL (Sec. 9-3), and two extended areas from the core technology: fluid-filled tunable molding techniques (Sec. 9-4) and photonic integrated circuits using fluidic optics (Sec. 9-5).

In Sec. 9-1, we give a detailed discussion on the fabrication process and the characteristics of the devices, which provide readers the fundamentals of fluidic lenses. Similar to the biological lenses in most animal eyes, which are generally of aspherical shapes, fluidic lenses can also obtain aspherical shapes to compensate for aberrations and to be most space efficient. Understanding the actual lens profile and developing the ability to control the profile are essential for achieving high-performance fluidic lens systems.

The main focus of this chapter is to demonstrate the unique functionality and superb performance of bio-inspired fluidic lens systems.

Among these systems are miniaturized universal imagers integrating the functions of cameras and microscopes, the world's smallest optical zoom lens, and the design of next-generation surgical cameras for minimally invasive surgery. To show the potential of the fluidic lens as a paradigm-shifting technology for future optical systems, examples are discussed in Sec. 9-2.

The potential application of the fluidic lens as an IOL after cataract removal is described in Sec. 9-3. This section demonstrates how the fluidic lens inspired by the optics in nature can restore human vision. Because of the length limit, we have skipped many medical-related discussions involving biocompatibility and surgical procedures for lens implantation. Instead, we focus on the potentially achievable performance of the fluidic IOL. Readers interested in the medical discussions may find additional information in the references.

Another interesting subject is to extend the fluidic lens technology into a tunable molding process to fabricate aspherical lenses of designed properties. Aspherical lenses are indispensable for compact, high-performance imaging systems, and can be found in almost all human-made devices as well as in animal eyes. Creating the mold master for aspherical lenses, however, is a tedious, difficult, and expensive task. In Sec. 9-4, we demonstrate that our fluidic tunable molding method enables us to achieve aspherical lenses of designed shapes for fast prototyping and design verification, a major contribution in the general field of optics.

Finally, in Sec. 9-5, we extend the fluidic lens technology to two-dimensional, in-plane optics as a new platform for polymer-based integrated optics that may find broad applications in biosensing. Here we discuss not only in-plane fluidic lenses but also a large family of optical elements including beam stops, prisms, waveguides, and the like that can be readily integrated with microfluidics to form lab-on-a-chip devices. We use one of the most important biomedical instruments, flow cytometer or fluorescence-activated cell sorter (FACS), as an example to demonstrate the potential benefits of fluidic photonic integrated circuits (PICs).

9-1 Bio-Inspired Fluidic Lens: Structures and Operations

There are two general light bending mechanisms for lenses: index of refraction gradient and lens curvature. These mechanisms are also utilized in fluidic lenses. Over the past decade, there have been numerous studies on fluidic lenses using these two mechanisms. A brief discussion of various fluidic lenses is given next.

9-1-1 Graded-Index-Tunable Fluidic Lens

One way to form an index gradient is to use liquid crystal (LC). The molecular structure of liquid crystal is either elongated (rodlike) or flat (disklike) [1]. Because of its structure, liquid crystal exhibits a birefringent behavior that the incoming light may experience a different index of refraction depending on its state of polarization. By varying the strength of the electric field in the direction normal to the propagation of light, the liquid crystal molecules can be oriented to create an index gradient radially, thus (de)focusing the light passing the device [2]. Several groups have utilized liquid crystals to create fluidic lenses. One example of the liquid crystal fluidic lens is shown in Fig. 9-1.

One approach to control liquid crystals is by designing the electrodes such that an axially symmetrical inhomogeneous electric field is created in the liquid crystal layer [3]. This has been done by sectioning the electrodes and applying different voltages [4–6] or by placing curved glasses into the fluidic lens to change the spacing between the electrodes [7–9]. Another approach to create liquid crystal fluidic lens is by mixing liquid crystals with polymers. When the polymer polymerizes, it will lock the liquid crystal in its original orientation, keeping it from moving when electric field is applied. This causes the lens to have a varying index of refraction throughout the entire lens and create the lensing effect [10–12].

There are several applications for the liquid crystal fluidic lens. For instance, it has been applied to micro-lenses [13,14], confocal microscopy [15], and disc storage systems [16,17].

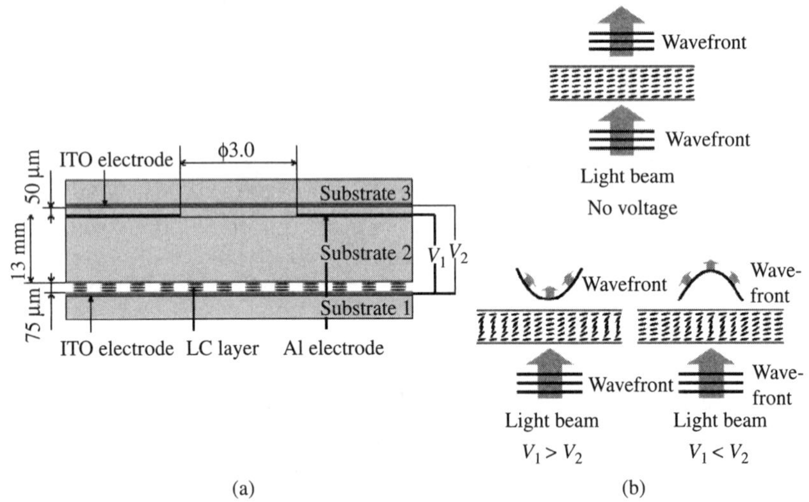

FIGURE 9-1 Liquid crystal (LC) tunable liquid lens. (a) A nematic LC is sandwiched between glass substrates 1 and 2. A transparent indium tin oxide (ITO) film is sputtered on substrate 1 and an aluminum (Al) film is coated on substrate 2. There is a circular hole in the center of the Al electrode. Above the Al layer, there is another ITO electrode sputtered on substrate 3. By applying different voltages to the electrodes, focusing and diverging lens can be formed, as shown in (b). (*B. Wang, M. Ye, and S. Sato, "Liquid crystal lens with focal length variable from negative to positive values," Photonics Technology Letters, IEEE, copyright 2006 IEEE.*)

9-1-2 Curvature-Tunable Fluidic Lens

In contrast to fluidic lens that tunes its index or refraction gradient, the other major type of fluidic lens changes its curvature. To change the curvature, different physical mechanisms can be used. Among them are electrowetting effect, dielectrophoretic effect, surface tension, and the structure of fluid-filled chamber covered by an elastomer membrane, to name a few.

The effect of electrowetting is manifested by the change of contact angle of a liquid droplet by an applied electric field. Lippman-Young equation is the governing equation:

$$\cos(\theta) = \cos(\theta_0) + \frac{\varepsilon\varepsilon_0}{2\gamma_{LG} d} V^2$$

where θ = the contact angle
θ_0 = initial contact angle with zero voltage applied
ε = dielectric constant of the dielectric layer
ε_0 = permittivity of vacuum
γ_{LG} = interfacial energy between the droplet and the ambient
d = thickness of the dielectric layer
V = voltage applied between the substrate and the droplet.

When the contact angle of the droplet changes, the curvature of the lens changes [18–23]. An example of such structure is shown in Fig. 9-2.

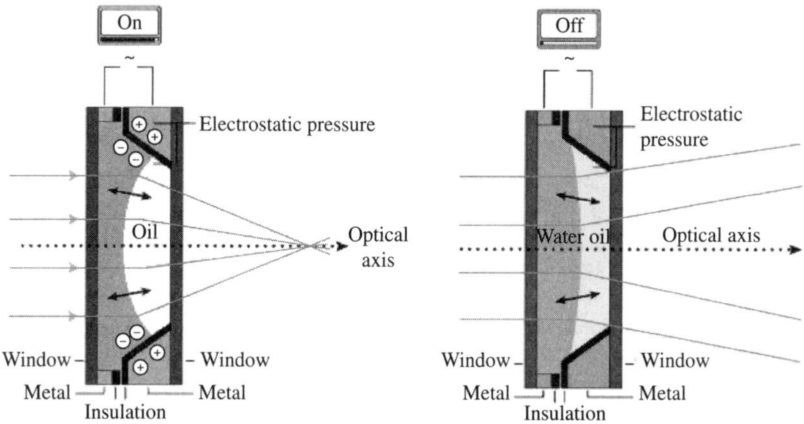

FIGURE 9-2 Electrowetting liquid lens. First, a water drop is deposited on a substrate made of metal, covered by a thin insulating layer. The voltage applied to the substrate modifies the contact angle of the liquid drop. The liquid lens uses two isodensity liquids, one is an insulator while the other is a conductor. The variation of voltage leads to a change of curvature of the liquid-liquid interface, which in turn leads to a change of the focal length of the lens. (*http://www.varioptic.com/en/index.php.*)

Dielectrophoresis presents an alternative mechanism to control the curvature of fluidic lens. Because of the effect of dielectrophoresis, a dielectric molecule experiences a net force in an electrical field gradient. Reference 24 utilizes this phenomenon to control the curvature of a liquid crystal droplet. Liquid crystals are dielectric molecules, and electrical field gradients can apply forces to the dielectric molecules to change the curvature of a droplet. References 25 and 26 report another liquid lens structure consisting of two types of dielectric materials. By manipulating the electric field with patterned electrodes, the dielectric liquid can change its curvature, as shown in Fig. 9-3, for an example.

Among many fluidic lenses, we believe the most attractive design is to use an optically clear elastic membrane to constrain the fluid in a lens chamber. The structure of the fluidic lens is shown in Fig. 9-4. The lens power is determined by the lens curvature and the refractive index difference between air and the optical fluid. The deformable elastic membrane is used to constrain the optical fluid and to produce the desired lens profile under a pressure difference between the lens chamber and the ambient. When optical fluid is injected into the lens chamber to create a positive pressure, the elastomer membrane produces a convex shape for a positive lens. Conversely, when optical fluid is withdrawn from the lens chamber into a reservoir, a negative pressure is formed to produce a concave lens. Such design offers flexibilities and characteristics (i.e., tuning power) matched by no other

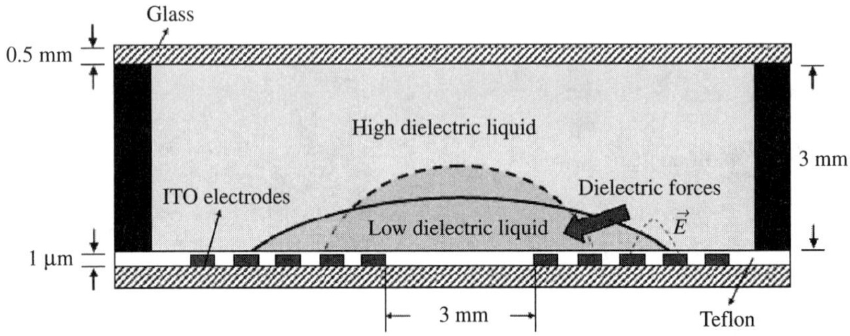

FIGURE 9-3 Dielectric fluidic lens. The liquid lens consists of a 15 μL (liquid) droplet with a low dielectric constant and a sealing liquid with a high dielectric constant. The bottom diameter of the droplet was 7 mm when no voltage was applied. The two liquids were injected inside a 3-mm-thick PMMA (polymethylmethacrylate) chamber that was sealed between two ITO glass substrates. The concentric ITO electrods on the bottom glass substrate were coated with 1-μm-thick Teflon® to reduce friction between the droplet and the glass substrate. As the voltage was applied, a dielectric force arose on the droplet due to the difference in the dielectric constant between the two liquids. The dielectric force shrunk the droplet, increasing the droplet's contact angle and shortening the focal length of the liquid lens. (*C. C. Cheng and J. A. Yeh, "Dielectrically actuated liquid lens," Optics Express, vol. 15, pp. 7140–7145, 2007.*)

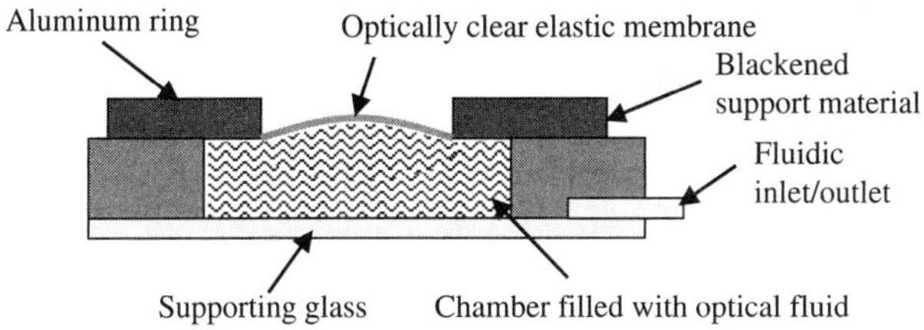

FIGURE 9-4 Membraned fluidic lens. The main components are (1) optically clear elastomer membrane and (2) lens chamber filled with optical fluid. Optical fluids having an index of refraction from 1.40 and 1.67 can be chosen for different applications and design requirements. Fluid can be moved into or out of the chamber by an actuator, producing a concave or convex lens. Fluidic lens with this structure provides superior mechanical stability, scalability, thermal stability, and tuning range unmatched by other liquid lenses.

fluidic lenses. As long as the fluid does not interact with the elastomer membrane, one has the freedom to select the optical fluid of desired index and dispersion properties for optimal system performance. Another merit of the design is that the lens profile is determined solely by the mechanical properties of the membrane, independent of the lens fluid. Hence, one can engineer the lens profile by adjusting the processing and material structure of the membrane. Different lens configurations, including plano-convex, biconvex, plano-concave, biconcave, positive meniscus, negative meniscus, and the like, have been demonstrated [27]. Various lens actuation mechanisms have been investigated, including piezoelectric actuator, thermal actuator [28], pneumatic actuator [29], aperture size actuator [30,31], electromagnetic actuator [32], fibrous actuator [33], pH-responsive hydrogel actuator [34], and so on.

To assess the pros and cons of various liquid lens technologies, we consider the following figures of merit: tuning range, scalability, and mechanical and thermal stability. Tuning range determines the functionality and performance. A small tuning range yields a limited focal range for auto-focusing system, but creates great difficulties for compact optical zoom systems. Scalability refers to the range of optical clear aperture (or lens diameter) a technology can produce. High image quality is often achieved with a larger clear aperture of the lens. Finally, to live up to the expectation of being a paradigm-shifting solution in the field of optics, fluidic lens needs to be mechanically robust, thermally stable, and low cost to manufacture. We have found that fluid-filled, elastomer-membrane lenses perform well in all the above areas. Such fluidic lenses can produce a record high tuning range of 200 D, can change the lens type between convex and concave structures, can scale to a relatively large clear aperture of over 10 mm, and have superb mechanical robustness and reliability. Therefore, we will focus on this fluidic lens technology in the remaining part of this chapter.

9-1-3 Fluidic Lens Fabrication

The process to fabricate a membrane-based bio-inspired fluidic lens can be divided into tsections: (1) membrane fabrication and (2) fluid chamber fabrication. To fabricate the membrane, we first mix, degas, and spin-coat prepolymer polydimethylsiloxane (PDMS) onto a chlorotrimethylsilane-coated silicon wafer. The purpose of chlorotrimethylsilane treatment is to avoid PDMS/Si cross-links that cause difficulties in separating PDMS from the Si handle wafer. The PDMS-coated wafer is kept in a 65°C oven for 40 min to cure the PDMS. The fully cured PDMS membrane, having a typical thickness of 50 to 200 μm is then separated from the Si handle wafer and mounted onto a precision-machined aluminum ring. The donut-shaped aluminum ring has the flatness, circularity, and concentricity that meet the requirements for an optical system. The permanent bonding of the PDMS membrane and the Al ring is performed immediately after the UV-ozone surface treatment. This membrane/aluminum ring is then bonded or clamped onto the lens chamber. In the final step, the lens chamber is vacuum-filled with optical fluid. More information on the fluidic lens fabrication process can be found in several references [27,35–39].

9-1-4 Lens Profile Analysis

Although a lens is usually characterized by its focal length or radius of curvature, high-quality imaging systems cannot be designed without the accurate knowledge of the detailed lens profile when the lens is not perfectly spherical. When the surface of a lens is not spherical, aspheric terms have to be added in the mathematical representation of the lens profile. The most frequently used mathematical model for aspherical lenses is the elliptical equation. Two parameters, curvature and conic constant, are needed to uniquely define an elliptical equation. Curvature is the reciprocal of the radius at the vertex of the lens and conic constant specifies the nature of the elliptical equation, being elliptical, parabolic, or hyperbolic. Equation (9-1) is the general expression of an elliptical equation:

$$z = \frac{c \cdot r^2}{1 + \sqrt{1 - (k+1)c^2 r^2}} \tag{9-1}$$

where c = radius of curvature at the vertex of the lens
r = distance from the center of the lens
k = conic constant.

When the conic constant (k) is zero, the equation describes a spherical lens. When $k = -1$, the shape is a parabola. When $k > 0$, the shape is an oblate ellipsoid. The shape difference has mechanical and optical implications. Mechanically, when $k > 0$, more fluid is needed to create the same curvature than a spherical lens ($k = 0$). Conversely, when $k < 0$, less fluid is needed to achieve the same curvature. Since curvature, in a first-order approximation, determines the lens power, the sign and value of conic factor means different requirements for the actuators that have to move a certain amount of fluid into and out of the lens chamber to vary the focal length. Optically, different conic factors mean different geometric aberrations. Hence, the value of conic factor affects how the optical system is designed, and the ability of controlling the conic factor is important to achieve optimal system performance. In this sense, the IOL in human eye has a highly desirable aspherical shape for lowered spherical aberration and high tuning efficiency for the actuator (i.e., ciliary muscles).

To better understand the lens profile, the elastic membrane is modeled using the COMSOL Multiphysics simulation software. The analysis is done by solving partial differential equations (PDEs) using the finite element method (FEM). The most suitable mechanical model for rubbers where PDMS belongs to is the hyperelastic model where the stress-strain relation is specified by a strain-energy density function [40]. The strain-energy function (W) is a function of the deformation gradient tensor (e.g., $W_{hyp} = W(\mathbf{F})$ where $\mathbf{F}$ is the deformation gradient tensor). For isotropic material, the strain-energy function becomes a function of the principal invariants of the right Cauchy-Green deformation tensor, for example, $W_{hyp} = W(I_1, I_2, I_3)$ [41,42]. The right Cauchy-Green deformation tensor is defined as $\mathbf{C} = \mathbf{F}^T\mathbf{F}$ and the principal invariants are

$$I_1 = trace(\mathbf{C})$$
$$I_2 = \frac{1}{2}\{trace((\mathbf{C}))^2 - trace(\mathbf{C}^2)\}$$
$$I_3 = \det(\mathbf{C})$$

We use Mooney-Rivlin constitutive equations to model the incompressible isotropic elastomer (i.e., PDMS). The strain energy function is defined in COMSOL as follows:

$$W_{hyp} = C_{10}(\bar{I}_1 - 3) + C_{01}(\bar{I}_2 - 3) + \frac{1}{2}K(J_{el} - 1)^2 \tag{9-2}$$

where C_{10}, C_{01}, and K are material properties for PDMS membrane. $\bar{I}_1 = I_1 \cdot J^{-2/3}$ and $\bar{I}_2 = I_2 \cdot J^{-4/3}$ where I_1 and I_2 are the first and second invariant of the left Cauchy-Green tensor and J is the determinant of the deformation gradient tensor, $J = \det(\mathbf{F})$

$$J_{el} = \frac{J}{(1+\varepsilon_{th})^3}$$

where ε_{th} is thermal strain. Basically, J describes the volume change and J_{el} is the elastic part of volume change. Note that $\overline{I}_1$ and $\overline{I}_2$ remain constant under volume change.

One major improvement in membrane processing is the application of prestretch where the membrane is in tensile stress before being bonded to the Al ring. Prestretch of the membrane not only prevents wrinkling at low lens power (nearly flat surface) but also greatly suppresses the effect of gravity. In addition, adjusting the amount of prestretch allows us to adjust the value of conic factor [43]. The effect of prestretch is simulated in COMSOL. Figure 9-5a shows the simulated profile under different fluid pressure. After COMSOL Multiphysics analysis, simulated lens profiles are fitted to the elliptical equations. The relationship between the conic factor and the radius of curvature is shown in Fig. 9-5b. The rapid increase in the value of conic constant has an insignificant effect in real applications since it happens when the lens power becomes very low.

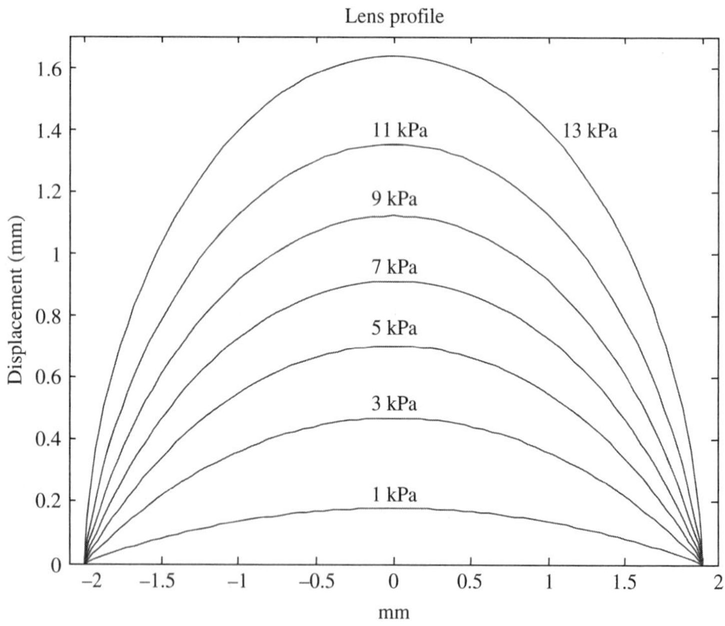

FIGURE 9-5 (a) Lens profile simulated from COMSOL Multiphysics (b) Conic constant vs. curvature from simulation. (*F. S. Tsai, S. H. Cho, W. Qiao, N. H. Kim, and Y. H. Lo, "Miniaturized unified imaging system using bio-inspired fluidic lens," Proceedings of SPIE, vol 7061, 70610N, 2008.*)

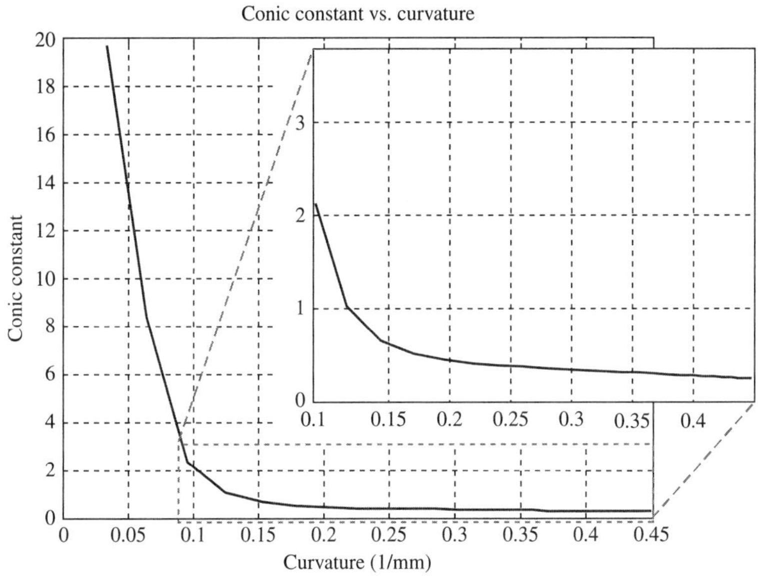

FIGURE 9-5 (*Continued*)

9-2 Fluidic Lens for Imaging

It is helpful to briefly introduce how an optical system is designed. Each surface of the lens in a system provides a certain amount of converging (positive) or diverging (negative) power to the light beam. Associated with the bending of the light by each lens surface is a certain amount of aberration determined by the lens material and profile. These aberrations, if not corrected, degrade the image quality. According to the nature of aberrations, the effects on the final image may be described as spherical aberration, coma, stigmatism, distortion, axial color shift, lateral color shift, and the like. To achieve high-quality images, optical designers use different lens surfaces to cancel these aberrations. Readers interested to know how optical systems are designed may find useful information in [Refs. 44 to 50].

Although the general methodology of optical design using fluidic lenses is not different from that of a conventional system, one needs to keep in mind that bio-inspired fluidic lens has an unprecedented tuning range. If this property is effectively utilized, optical systems of unique functions can be obtained. In this section, we will demonstrate that bio-inspired fluidic lens can produce the world's first miniature universal imager which integrates two systems, camera and microscope, that are considered entirely incompatible in conventional optics. The bio-inspired universal imager not only possesses such unique functions but also has an extremely short (<1.2 cm) total track length. Similarly, the world's shortest (< 1.5 cm) 3× optical zoom lens can also be achieved with fluidic lenses.

9-2-1 Auto-Focusing Miniaturized Universal Imager

Fluidic lens has the inherent capability of changing its focal distance over a wide range. However, a single lens, being tunable or not, is not sufficient for producing high-quality image on a solid-state image sensor. As mentioned previously, one needs to introduce additional surfaces to compensate for the aberrations. One major aberration that can be easily understood is field curvature, which is the deviation of the focal point from the image plane as the ray incident angle increases. This can be seen in Fig. 9-6.

In human and most mammals' eyes, the retina is curved to accommodate for the large field curvature [51]. However, current image sensors, both complementary metal-oxide-semiconductor (CMOS) and charged-coupled device (CCD), are flat. The limitations imposed by current image sensor technologies require additional lenses to correct the effect of field curvature. This leads to a hybrid optical system consisting of fluidic lenses and fixed lenses. The former produce tunability and the latter correct aberrations such as field curvature. One example of such fluidic/fixed lens system is shown in Fig. 9-7. In this design, the fluidic lens is placed in front of a Cooke Triplet consisting of three fixed lenses. When the fluidic lens has zero power, the system is optimized at an object distance of 10 m. The simulated image height is 4.4 mm (the diagonal of the CMOS image sensor), corresponding to a 1/4 in optical format. By varying the power of the fluidic lens, we show that such an optical system can form images of objects at infinity and as close as 2 cm from the lens, as a result of the wide tuning range of the fluidic lens. When the object is at infinity, the simulated system has a resolution of 205 line-pairs/mm (diffraction-limited) on the axis, 148 line-pairs/mm at 50% field, and 95 line-pairs/mm at 100% field. For an object distance of 2 cm, the resolution is reduced to 56 line-pairs/mm on the axis, 88 line-pairs/mm at 50% field, and

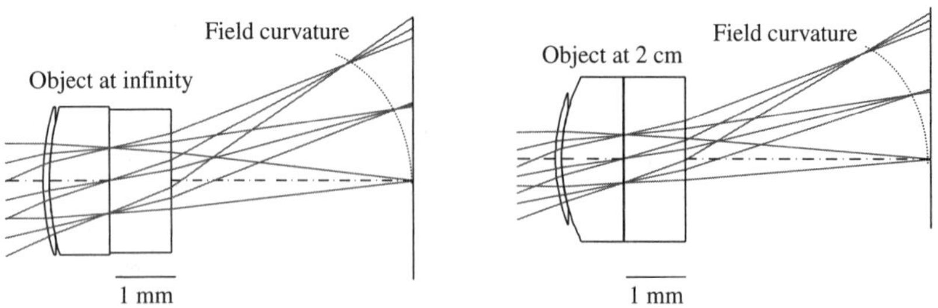

FIGURE 9-6 Ray tracing of a single lens imaging system. Any single lens imaging system is riddled with aberrations. One easy-to-spot aberration is the field curvature. One can easily observe that the focus moves away from the image plane as the ray incident angle increases.

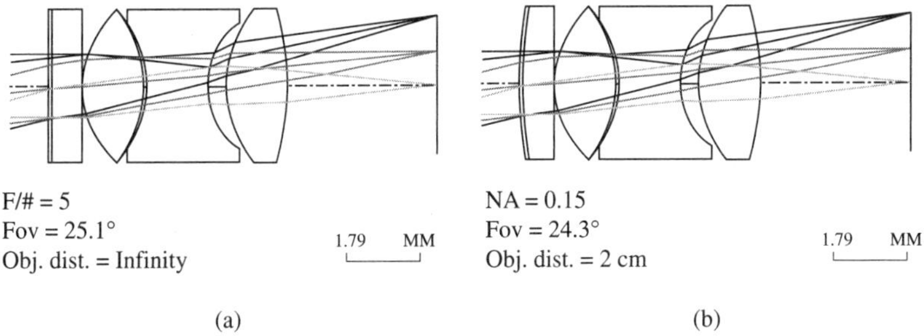

FIGURE 9-7 Lens View of the miniaturized unified imager consisting of 3 fixed lenses and 1 fluidic lens. The fluidic lens has different shapes for objects at (a) infinity and at (b) 2 cm from the lens. Compared to Fig. 2.2.1, field curvature and other aberrations are greatly suppressed. (*F. S. Tsai, S. H. Cho, Y.-H. Lo, B. Vasko, and J. Vasko, "Miniaturzied universal imaging device using fluidic lens," Optics Letters, vol. 33, pp. 291–293, 2008.*) (See also color insert.)

48 line-pairs/mm at 100% field. Resolution is measured at a modulation transfer function (MTF) of 0.3 considering C-d-F (656.3-578.6-486.1 nm) wavelengths.

A fluidic lens was fabricated and integrated with the fixed lenses mentioned above. The total track length of the optical system is less than 12 mm. Figure 9-8a and b show the photographs for objects at 15 m distance and as close as 1 cm distance. The camera's capability of revealing details of an object invisible to human eyes is demonstrated in Fig. 9-9. As we move the object from 14 to 1 cm, the device can clearly resolve the fabric of a business card. The most striking feature is that the imager can also function as a microscope. Such an example is shown in Fig. 9-10 where lines as narrow as 2.5 μm on the Edmund Optics 1951 Air Force Target can be resolved.

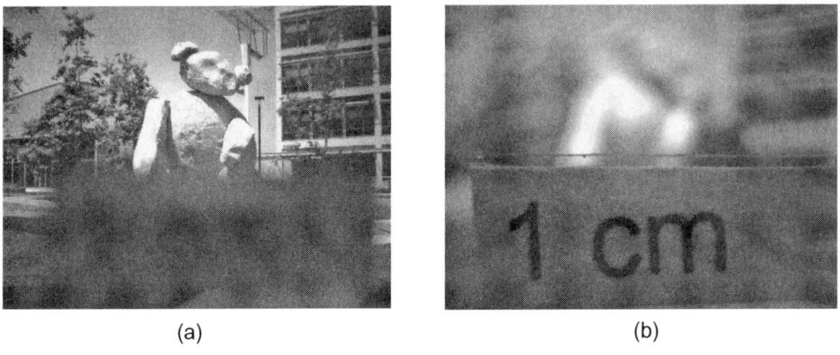

FIGURE 9-8 Demonstration of ultra-wide focusing range of the miniaturized unified imager. The device is stone bear 15 away paper mounted slide at 1 cm from device focused at (a) the meters away, and (b) to a glass the device. (*F. S. Tsai, S. H. Cho, Y.-H. Lo, B. Vasko, and J. Vasko, "Miniaturzied universal imaging device using fluidic lens," Optics Letters, vol. 33, pp. 291–293, 2008.*)

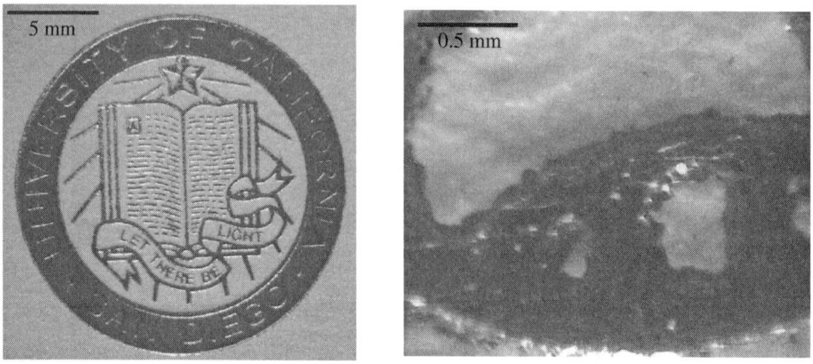

(a) (b)

FIGURE 9-9 (a) UCSD Emblem on a business card placed at 14 cm away from the imaging system. (b) The same object placed at 1 cm from the device. (*F. S. Tsai, S. H. Cho, Y.-H. Lo, B. Vasko, and J. Vasko, "Miniaturzied universal imaging device using fluidic lens," Optics Letters, vol. 33, pp. 291–293, 2008.*)

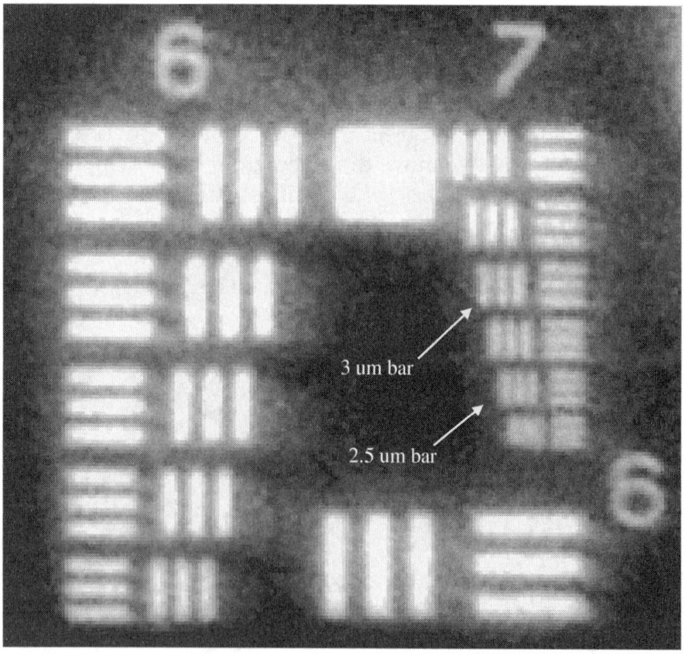

Edmund optics 1951 USAF target

FIGURE 9-10 With fluidic lens, the device can resolve 2.5 μm features, showing that the unified imager can function as a camera and a microscope. (*F. S. Tsai, S. H. Cho, Y.-H. Lo, B. Vasko, and J. Vasko, "Miniaturzied universal imaging device using fluidic lens," Optics Letters, vol. 33, pp. 291–293, 2008.*)

9-2-2 Fluidic Zoom Lens

Animal eyes have an elegant way to produce accommodation. However, no animal eyes possess the capability of optical zoom where the magnification of the system can be changed without varying the object distance. Optical zoom is an essential feature for most modern imaging systems including digital still cameras and camcorders. In this section, we demonstrate that the bio-inspired fluidic lens can provide compact optical zoom systems with high image quality.

Optical zoom refers to the change of magnification or image size without varying the object distance. Equivalently, one may think of optical zoom as varying the field of view. For a given image sensor, a larger image means a smaller field of view and a smaller image means a larger field of view. Optically, the field of view has a nearly proportional relation with the effective focal length (EFL) of the optical system, and the zoom ratio of an optical system is defined as the ratio between the maximum EFL and the minimum EFL [44]. In a greatly simplified model, conventional zoom lenses comprise two movable lenses. In such a two-lens system, the total power of the system is related to the power of each lens in the following way: $\Phi_T = \Phi_1 + \Phi_2 - d \cdot \Phi_1 \cdot \Phi_2$ where Φ_1 and Φ_2 are the power of individual lenses and d is the distance between the lenses. The only way for a traditional optical system to change the EFL is to change the distance between the lenses. Moving lenses along the optical axis is a complicated process, making today's zoom lenses bulky, expensive, and hard to manufacture. Optical zoom can be achieved without varying the lens distance provided fluidic lenses are used. Considering a similarly simplified optical system comprising only two fluidic lenses, one can change the EFL and the field of view by varying the lens shapes instead of the lens spacing. A longer EFL or narrower field-of-view is achieved if the front lens has a convex shape and the rear lens has a concave shape. Conversely, a shorter EFL or larger field of view is achieved if the front lens has a concave shape and the rear lens has a convex shape. The elimination of axial lens movements greatly simplifies the fabrication of zoom lens system, and the large tuning range of fluidic lenses reduces the system's total track length.

One example of 1/4-in format fluidic zoom lens system is shown in Fig. 9-11. In Fig. 9-11a, the front lens is tuned to a convex lens and the back lens is tuned to a concave lens. Such a configuration is equivalent to a telephoto system with a diagonal full field of view (FFoV) of 13°. By changing the curvature of the fluidic lenses, as shown in Fig. 9-11b, a reversed telephoto system consisting of a concave front lens and a convex back lens can be created. The FFoV becomes 60°. Ignoring aberrations, the field of view of an optical system is determined by its effective focal length (EFL) and the diagonal of the image sensor (i.e., optical format). The EFL of each configuration is 14.2 and 4.1 mm, respectively. This corresponds to greater than 3× optical zoom.

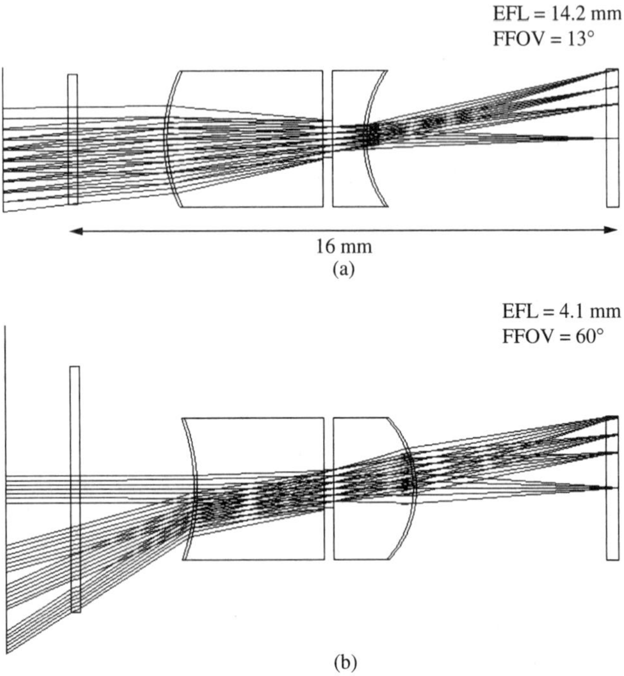

Figure 9-11 Miniaturized zoom lens consisting of only two fluidic lenses. Switching between (a) a telephoto (small field of view) and (b) a reversed telephoto (large field of view) configuration, zoom effect can be achieved. The effective focal length (EFL) and the diagonal full field of view (FFoV) are shown on the upper right of the lens plot. The system achieves more than 3× optical zoom at a total track length of 12 mm. (*F. S. Tsai, S. H. Cho, W. Qiao, N. H. Kim, and Y. H. Lo, "Miniaturized unified imaging system using bio-inspired fluidic lens," Proceedings of SPIE, vol. 7061, 70610N, 2008.*)

To evaluate the performance of the optical system, we used the Geometric Bitmap Image Analysis option in ZEMAX ray-tracing software. An image of the real-sized pancreas model is placed at 15 cm distance from the fluidic lens for both zoom-in and zoom-out configurations. Analysis results are shown in Fig. 9-12.

9-2-3 Application Example: Surgical Camera

In this section, we demonstrate that the bio-inspired fluidic lens can provide compact optical zoom systems with high image quality. We use fluidic lenses to design a laparoscopic camera for minimally invasive surgery (MIS), taking advantage of its small form factor and its ability to achieve optical zoom without moving parts.

Invented some 15 years ago, laparoscopy is a method of surgery in which small incisions are made on the abdominal wall for surgical tools to go through. Small incisions ensure better and faster recovery

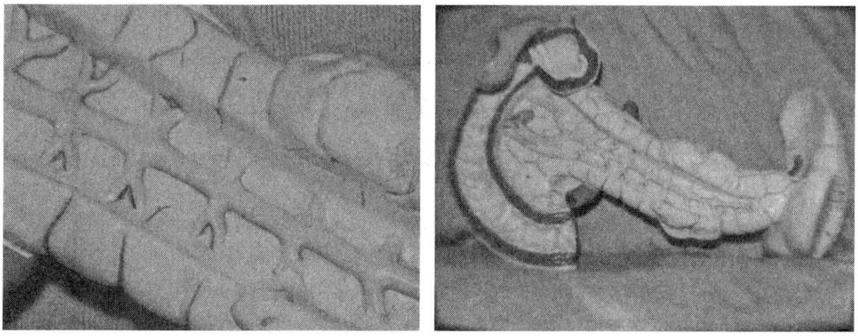

(a) (b)

Figure 9-12 Using ZEMAX ray-tracing simulation, we perform Geometric Bitmap Image Analysis on the images of a real sized pancreas model placed at 15 cm from the fluidic zoom lens. (a) Zoom-in image. (b) Zoom-out image. The blurry effect at the corners of (b) is mainly attributed to astigmatism.

for the patient. One important recent progress in this area is Natural Orifice Translumenal Endoscopic Surgery (NOTES) where all surgical tools enter the human body through a natural orifice (e.g., mouth, vagina, bladder, or colon). The incision is usually made internally, at the stomach or at the vagina, thus eliminating any abdominal incision. All methods of minimally invasive surgery (MIS) produce less chance of infection, less pain, faster recovery, and better overall surgical outcomes [52]. The main obstacle for NOTES and the overall MIS procedures is the limited vision provided by surgical cameras. Today's laparoscopic cameras produce neither auto-focusing nor optical zoom, and are attached to a long tube with image relay optics. The laparoscopic camera has to be operated by a delegate person and is unable to provide surgeons with vision comparable to the natural human vision in open surgery. This prompts us to apply fluidic lenses to surgical cameras to achieve superb auto-focusing and optical zoom capabilities in a small form factor. The camera can be made small enough to be inserted through a small incision or natural orifice and mounted on the CO_2 inflated abdominal wall of the patient.

One noticeable feature from Fig. 9-12 is that most organs are strongly biased toward red color. Pictures inside human body contain very weak blue component in general. Furthermore, if the camera has the capability to operate at near-infrared (NIR) wavelengths (e.g., 850 nm), this will enable surgeons to more clearly observe blood veins through fat. Thus surgeons can obtain vision even superior to human vision and thereby have a lower probability of surgical mistakes. NIR vision can be obtained using light emitting diode (LED) light sources and silicon CMOS or CCD sensors. To take advantage of the NIR images, an NIR fluidic zoom lens system producing high (HDTV) quality images is designed. The lens system is shown in Fig. 9-13.

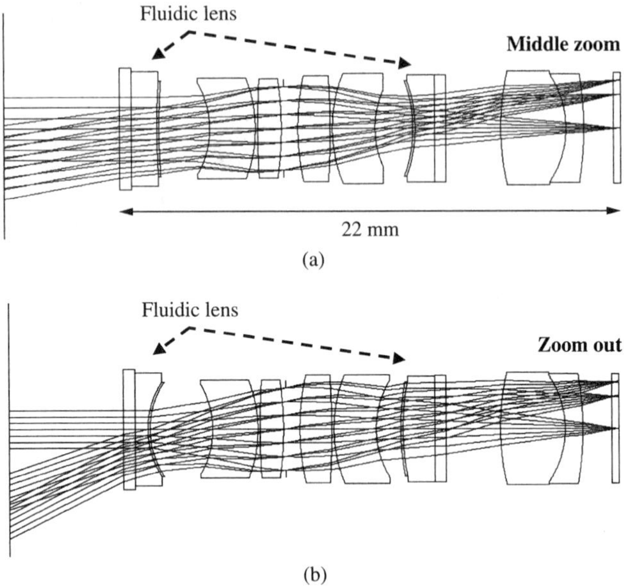

FIGURE 9-13 Improved surgical camera design. Two fluidic lenses are used to create a telephoto and reversed telephoto system. Fixed lenses are added between the fluidic lenses to reduce aberrations. The total track length is 22 mm and the maximal lens diameter is less than 8 mm, suitable for camera insertion. (*F. S. Tsai, S. H. Cho, W. Qiao, N. H. Kim, and Y. H. Lo, "Miniaturized unified imaging system using bio-inspired fluidic lens," Proceedings of SPIE, vol. 7061, 70610N, 2008.*)

Similar to the design in Fig. 9-11, two fluidic lenses are used to provide optical zoom and auto-focusing. Additional fixed lenses are introduced between the fluidic lenses to correct astigmatism and other aberrations. The curvatures of the fluidic lenses change in a similar fashion to the previous design to achieve telephoto (zoom-in) and reversed telephoto (zoom-out) characteristics. However, by placing fixed lenses between the fluidic lenses, aberrations are greatly reduced. In Fig. 9-13, four glass lenses, two in front of the aperture stop and two behind the aperture stop, are inserted to form a quasi-symmetric design that tends to produce higher quality images. As the fluidic lenses change their shapes and the system becomes increasingly asymmetric, image quality starts to degrade. The optimization program of the ray-tracing software makes a graceful degradation of the image quality from its optimal position to assure that HDTV quality images are obtained over the entire optical zoom range.

To evaluate the improved design of 3× optical-zoom miniature medical cameras, ZEMAX Geometric Bitmap Image Analysis is performed at an object distance of 15 cm for both zoom-in and

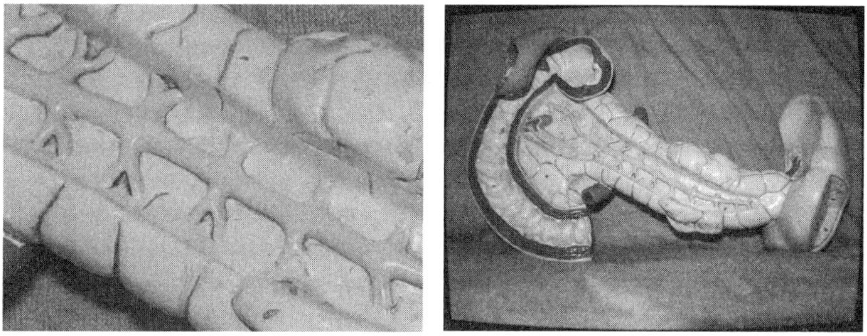

(a) (b)

FIGURE 9-14 ZEMAX Geometric Bitmap Image Analysis on the images produced from the improved design in Fig. 2.2.8. (a) Zoom-in image. (b) Zoom-out image, corresponding to Fig. 2.2.8 (b). In both cases, image quality is superior to that in Fig. 2.2.7.

zoom-out images. The analysis result is shown in Fig. 9-14. Comparing Figs. 9-12 and 9-14, we demonstrate that the quality has been greatly improved.

9-2-4 Summary

Fluidic lens brings new opportunities to optical imaging systems. By integrating a fluidic lens with fixed lenses, compact universal imagers with camera and microscope functions can be formed. Furthermore, with fluidic lenses, optical zoom without moving lenses is demonstrated.

9-3 Bio-Inspired Intraocular Lens—Restoration of Human Vision

As discussed previously, in a constrained space, bio-inspired fluidic lens can achieve much greater accommodation than solid lenses. Since fluidic lenses are inspired by animal vision, it is natural to apply the devices to restore (or enhance) human vision. In this section we report the preliminary investigation of the feasibility of using fluidic lenses to replace the crystalline intraocular lenses (IOLs) [53]. Because of the inherent difficulty in quantitative assessment of the performance of fluidic lenses for human vision, we have set up a scaled-up eye model using a CMOS sensor to simulate the retina near the fovea.

A crystalline lens in human eye is a transparent, lentil-shaped tunable lens. The lens not only provides one-third of the eye's total power, but also changes the focal length of an eye continuously from the change of the surface curvature. Accommodation, which is the ability to focus far and near, is made possible by the crystalline lens. For patients with cataract disease, a crystalline lens becomes cloudy and finally obstructs the transmission of light. The most effective and common treatment is to remove the lens from the capsular bag and replace it with a synthetic lens, called IOL [54,55]. There are several different designs of IOLs, but the simplest design is to use a fixed lens with one or multiple focal distances [56,57]. Although the patients regain their vision after surgical replacement, they lose most of the accommodation capability. Due to the inconveniences caused by fixed focus IOLs, efforts have been made to restore accommodation in human vision. The majority of commercially available accommodating IOLs achieve accommodation by axial linear motion. When the IOL is moved closer or further away from the retina via mechanical coupling to the ciliary muscle [58–60], accommodation is achieved. This is essentially how conventional optical systems achieve auto-focusing, a mechanism that has been proved to be less efficient than animal eyes. Indeed, theoretical analysis shows that 1-mm travel of a 20-D single-optic IOL can only yield an accommodation of ~1.2 D. Even with a dual-optic IOL (with 32-D anterior-moving lens and −12-D correcting posterior lens), the same travel only yield an accommodation of ~2.2 D [61,62]. These results are significantly worse than the 7.0-D average tuning range of young adults [63]. The large performance gap between human crystalline lens and artificially made IOL is due to the fundamental fact that optic-shift is a much less efficient mechanism than change of lens shape.

Capsular bag refilling with gel can be considered as one type of fluidic lens to restore ocular accommodation. It is achieved by filling the capsule with an injectable malleable material while preserving capsule integrity. Though many experimental studies have already been done [64–66], several problems remain to be solved (e.g., achieving emmetropia in the relaxed state, adequate accommodative response upon zonuler relaxation [67]). In this section, we explore the use of bio-inspired fluidic lens as IOL to restore the clarity and accommodation of human vision.

9-3-1 Optical Simulation of Eye Model

A pseudoaphakic eye based on Liou and Brennan's eye model was adopted to theoretically investigate the optical performance of fluidic IOL [68]. In a pseudoaphakic eye, the native crystalline lens is replaced with an implanted synthetic lens because of cataract or other eye diseases. The corneal lens has radii of curvature of 7.77 mm for the anterior (front) surface and 6.40 mm for the posterior (rear) surface. The pseudoaphakic eye has an axial length of 23.95 mm. The refractive index of the corneal lens is 1.376 at the wavelength of 555 nm, which is the peak of the photopic curve. The pupil is modeled as a circular aperture stop on the front surface of the crystalline lens with its center offset by 0.5 mm nasally from the optical axis. The angle between the optical axis and the visual axis is 5 [69]. The refractive indices of aqueous and vitreous humors are 1.336. To simulate the effects of fluidic

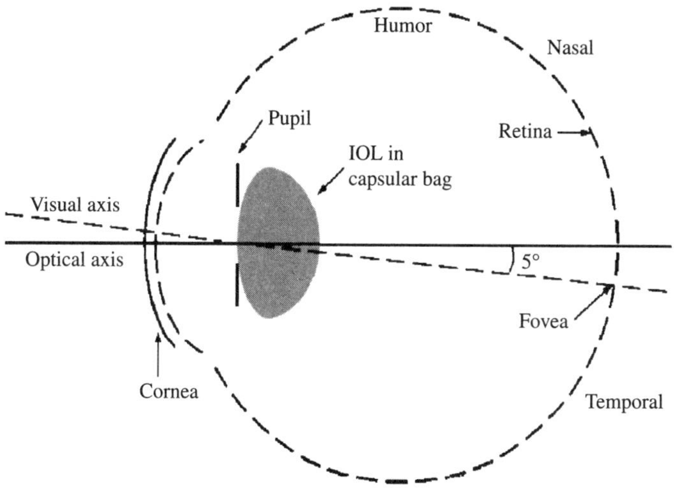

Figure 9-15 Schematic of a pseudoaphakic eye based on Liou and Brennan's model eye (top view). (*W. Qiao, F. Tsai, S. H. Cho, and Y.-H. Lo, "Fluidic intraocular lens with a large accommodation range," IEEE Photonic Technology Letters, copyright (year) IEEE.*)

IOL, the lens in Liou and Brennan's eye model is replaced by the fluidic lens (Fig. 9-15). A ray-tracing program (Code V) is used to calculate the resolution, distortion, and other relevant properties. The fluidic IOL is optimized to achieve best optical performance on the fovea, which is located 5 temporally.

One of the most effective approaches to evaluate the performance of an optical system is to calculate its modulation-transfer-function (MTF) curves. The averages of sagittal and tangential MTF curves for the model eye with fluidic accommodative IOL are shown in Figs. 9-16 and 9-17. The on-axis (i.e., at fovea) (MTF) curves, which are focused at infinity and 25 cm respectively, are compared with Liou and Brennan's eye model (Fig. 9-16). The off-axis MTFs (5 object angle from the fovea) in both horizontal and vertical planes are presented as well (Fig. 9-17). All the MTF curves are simulated at wavelengths of 475, 555, and 625 nm with a weighting factor ratio of 1:2:1 and with an object distance of infinity. The results show that fluidic IOL in pseudoaphakic eye can produce optical performance comparable to human eye.

9-3-2 Experimental Results

The fluidic IOL consists of a polydimethylsiloxane (PDMS) elastomer membrane, a fluid-containing lens chamber, and a flat supporting substrate. The lens chamber is filled with silicone oil. To experimentally evaluate the optical performance and to record images, a scaled-up eye model is constructed (Fig. 9-18) to simulate the eye optics and a 2 million pixel CMOS sensor is used to simulate the retina. The output of the

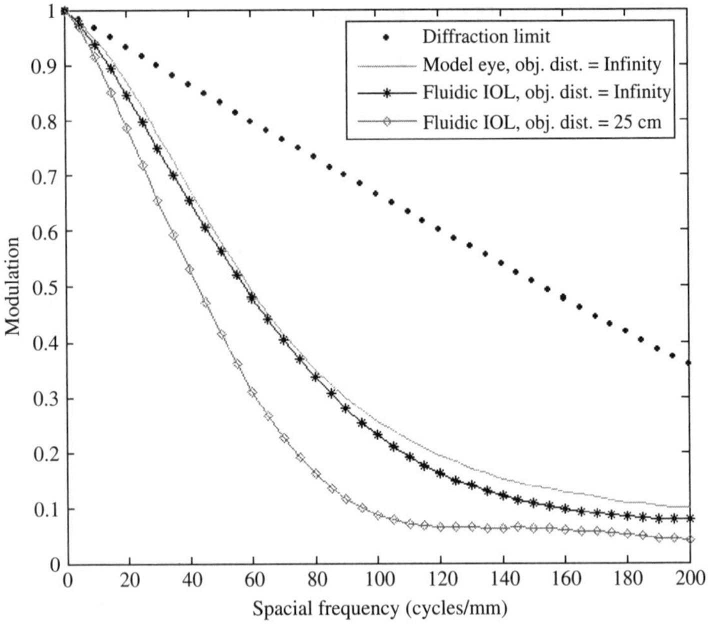

FIGURE 9-16 On-axis polychromatic MTFs of fluidic IOL with pseudoaphakic eye for a 3-mm pupil. Also included in the figure is the result (blue) from Liou and Brennan model eye with fixed lenses focused at infinity.

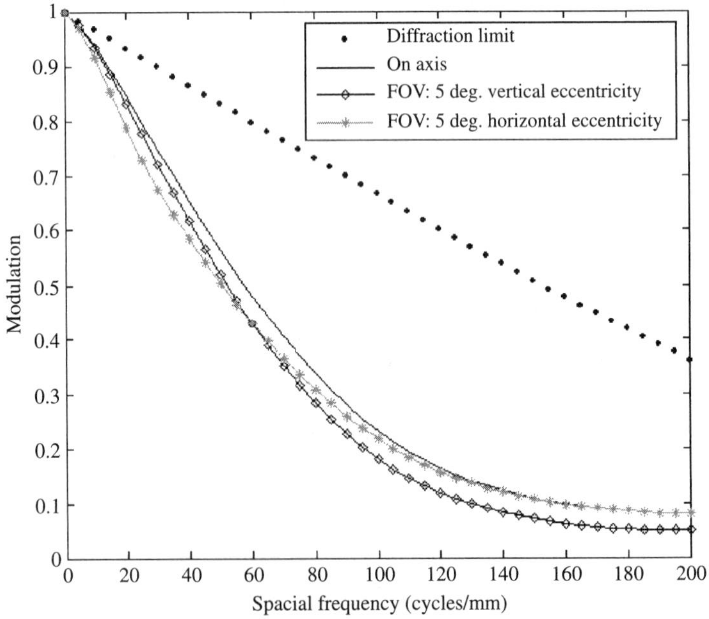

FIGURE 9-17 On-axis (at fovea) and off-axis (5-degree from fovea) MTFs of fluidic IOL with pseudoaphakic eye.

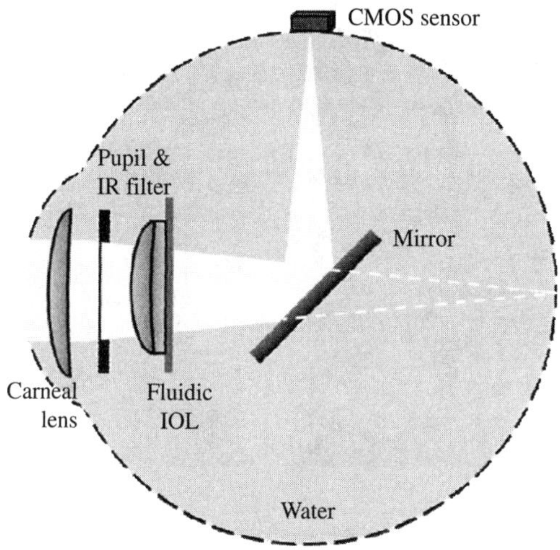

FIGURE 9-18 Schematic of experimental pseudoaphakic eye with a plano-convex fluidic IOL. (*W. Qiao, F. Tsai, S. H. Cho, and Y.-H. Lo, "Fluidic intraocular lens with a large accommodation range," IEEE Photonic Technology Letters, copyright (year) IEEE.*)

CMOS sensor is displayed on the computer screen and digitally recorded for visualization and further analysis. Behind the plano-convex corneal lens is a biconvex lens, which is later on replaced with the fluidic lens to evaluate the image quality and range of accommodation. A 5-mm-diameter pupil and an infrared filter are inserted between the cornea and the lens. The experimental eye model is filled with water. A 45-mirror deflects the light path toward the CMOS image sensor, which is placed right above water to acquire data at the equivalent position of the fovea. The dimension of the pixel is 3.18 μm. The near point of the experimental eye model is defined to be 45 cm. At this distance, the eye model shows a resolution of 0.07 mm, corresponding to the same resolution as the near point of 25 cm for emmetropia eye with normal visual acuity (i.e., one arc minute). After finishing the optical quality evaluation of the eye model, the fixed biconvex lens is replaced with a plano-convex fluidic IOL. When changing the curvature of the fluidic IOL, a well-focused image can be formed from one half of the near distance (22.5 cm) to infinity. Since one half of the near point in the experiment is equivalent to 12.5 cm for emmetropia eye, the experimental eye model can achieve a subject accommodation range of 8 D. Taking pseudo-accommodation (i.e., accommodation caused by the influence of depth of field) into consideration, the real tuning range of effective focal length is a bit smaller than subjective accommodation range.

Figures 9-19 and 9-20 are taken with the aforementioned pseudoaphakic eye model using a plano-convex fluidic IOL. Figure 9-19a

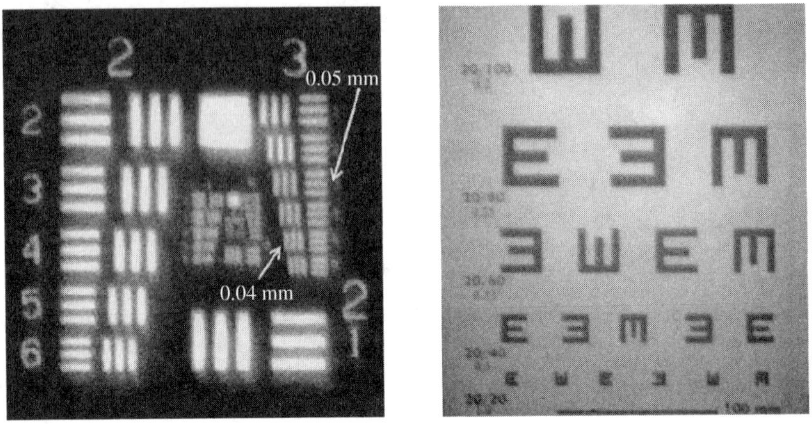

(a) (b)

FIGURE 9-19 Pictures taken by pseudoaphakic eye with a fluidic IOL. (a) 1951 USAF target (Edmund Optics) is placed at 25 cm away. Features as small as 0.04 mm can be resolved by our device. (b) Resized eye chart at a distance of 3.7 m. It shows a visual acuity better than 20/20. (*W. Qiao, F. Tsai, S. H. Cho, and Y.-H. Lo, "Fluidic intraocular lens with a large accommodation range," IEEE Photonic Technology Letters, copyright (year) IEEE.*)

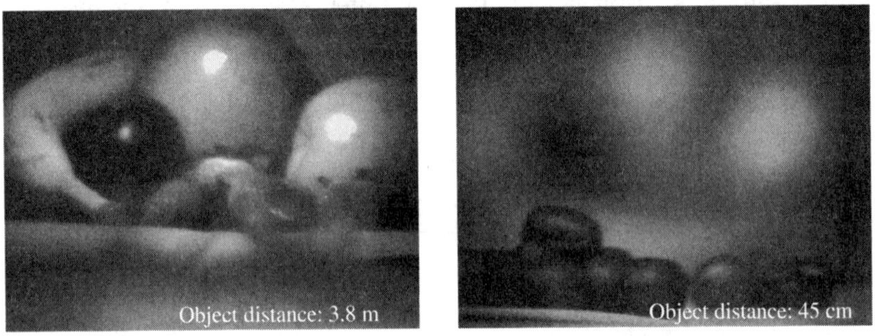

(a) (b)

FIGURE 9-20 (a) Image of still fruits 3.8 m away from the pseudoaphakic eye. (b) Image of green beans 45 cm away from the pseudoaphakic eye. (*W. Qiao, F. Tsai, S. H. Cho, and Y.-H. Lo, "Fluidic intraocular lens with a large accommodation range," IEEE Photonic Technology Letters, copyright (year) IEEE.*)

shows the image of a 1951 U.S. Air Force (USAF) target obtained from the eye model with a fluidic IOL. A resolution of 0.04 mm at an object distance of 25 cm has been achieved. This is about twice the resolution of an emmetropia eye, which can resolve 0.04 mm features from a distance of 13 cm. The eye chart shown in Fig. 9-19b has been resized following the calibration chart and placed at 3.7 m from the corneal lens. The result demonstrates that the pseudoaphakic eye model with fluidic IOL has a visual acuity better than 20/20 at 3.7 m. 20/20 is the visual acuity needed to discriminate two points separated by 1 arc minute and considered to be normal vision. Figure 9-20 shows fruits and green beans at 3.8 m and 45 cm, respectively. The results demonstrate a 4-D tuning range of fluidic IOL.

9-3-3 Mechanical Modeling of Fluidic Intraocular Lens

Movement of the ciliary muscle from contracted/accommodated state to the relaxed/unaccommodated state increases tension in the zonules and changes the capsular bag into a thinner and longer shape. The state-of-art technique to actuate the IOL is by utilizing the shape change of the capsular bag when the human eye accommodates. In order to explore the pressure distribution of fluidic IOL, the mechanism of membrane is simulated using finite element analysis. A nonlinear hyperelastic model is applied to the PDMS membrane. Assuming the fluidic IOL is rotational symmetric, one can simplify the computation using an axisymmetric model [70]. The simulated IOL profile under given fluidic pressure is then fitted to an elliptical equation [Eq. (9-1)] characterized by the radius of curvature and conic factor [71]. The conic factor accounts for any aspherical profile of the fluidic IOL to the first order. Figure 9-21 shows the relationship between actuating force and accommodation range. Based on the material properties of PDMS

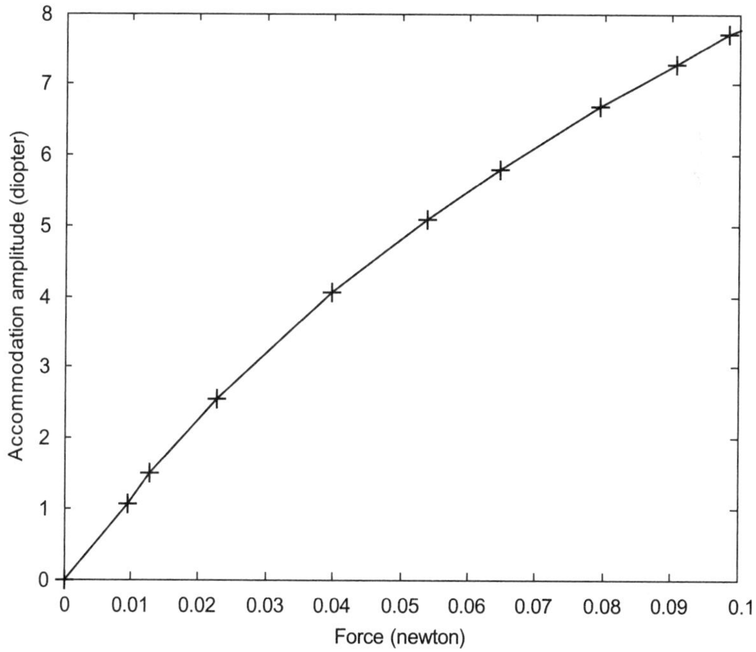

FIGURE 9-21 The accommodation amplitude as a function of driving force. (*W. Qiao, F. Tsai, S. H. Cho, and Y.-H. Lo, "Fluidic intraocular lens with a large accommodation range," IEEE Photonic Technology Letters, copyright (year) IEEE.*)

(Gelest 1.41), the force needed for driving the fluidic IOL to achieve an accommodation range of 7 D is approximately 0.085 N. Given that ciliary muscles in human eyes generally produce approximately 0.08 N of net force during accommodation [72–74], the elastic properties of the fluidic IOL match well with the muscular force. If desired, the required driving force can be further reduced via engineering of the mechanical properties of PDMS membrane.

9-3-4 Summary

Unlike optic-shift IOL having very low accommodation range, the fluidic IOL demonstrated in this section possesses an accommodation range comparable to or even wider than a young human eye and can produce good image quality. The result also indicates that the required force to achieve the desired accommodation range is comparable to the force generated from ciliary muscles of human eyes. We hereby conclude that fluidic IOLs hold promise to fully restore the acuity and accommodation range of vision and could potentially produce superhuman vision.

9-4 Liquid Molding Technique—Prototyping of Aspherical Lenses

Most optical systems are designed using a ray-tracing software equipped with efficient routines that optimize the profile of each lens surface according to user-defined merit functions. Before volume production, prototypes are produced first to verify the design. These prototypes often reveal design faults and potential issues not conceived in the original design. The use of aspherical lenses has become increasingly popular because, compared with systems solely made of spherical lenses, systems containing aspherical lenses produce superior performance and are more compact. Although plastic aspherical lenses can be made cost-effectively in very high-volume production, small-volume fabrication of aspherical lenses for prototypes is time consuming and costly [75–77]. In this section, we will show that when a fluidic lens is used as a tunable mold, we can form aspherical lenses of designed shape and property promptly and cost-effectively, and thus greatly shorten the design cycle for optical systems.

9-4-1 Tunable Liquid-Filled Molding Technology

A simple yet accurate process to fabricate aspherical polymer lenses using a tunable liquid-filled molding technology has been reported [78]. The authors modified a fluidic lens [35–39] into a tunable liquid-filled mold and used it to fabricate a variety of focus-fixed polymer lenses of designed aspherical profiles. A layer of 100- to 300-μm-thick optically transparent elastomer membrane is formed in the same way as described in Sec. 2-1-3. Stretching the membrane before mounting it to the aluminum fluid chamber prevents wrinkles. The amount of prestretch is defined as

$$\frac{\Delta R}{R_0} \times 100(\%)$$

where ΔR is the radial displacement by the applied tensile stress and R_0 is the initial radius in the relaxed state. Fluid is then injected into the membrane-mounted chamber till the membrane is deformed to the desired shape. The amount of fluid injection is controlled by a syringe pump and the fluid pressure is monitored by a pressure sensor. To fabricate lenses using the inflated (or deflated) membrane as a mold, uncured PDMS is poured onto the membrane. After curing, the profile of the membrane of the liquid-filled mold is transferred to the cured PDMS. Demolding the cured lens from the membrane is rather easy because cross-linking does not occur between the second PDMS and the already-cured membrane. As a result, focus-fixed lenses of complementary surface profiles to the liquid-filled mold are produced. The fabrication process of a focus-fixed plano-convex PDMS lens out of a concave fluid-filled mold is illustrated in Fig. 9-22.

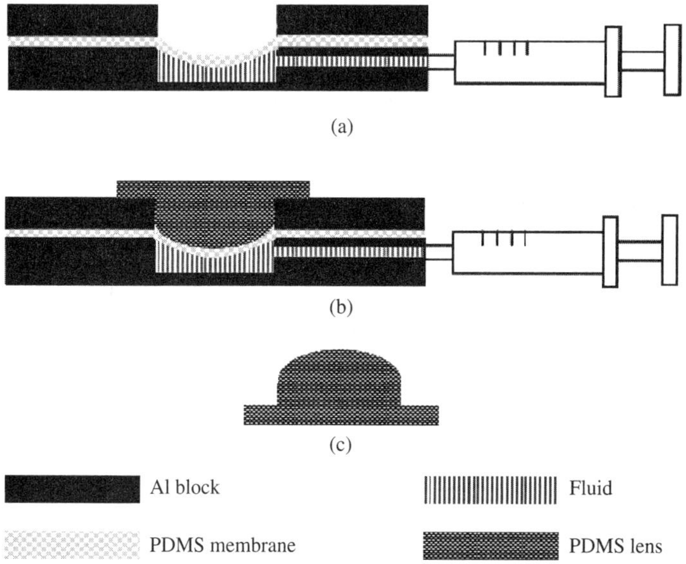

FIGURE 9-22 Process flow of molding a convex lens: (a) achieve the desired curvature by controlling the fluid volume and/or pressure; (b) pour PDMS prepolymer onto the membrane and cure PDMS; and (c) demold PDMS lens from the liquid-filled mold. (*Adapted from Sung Hwan Cho, Frank S. Tsai, Robert Vasko, Jeff Vasko, and Yu-Hwa Lo, "Fluid-filled tunable mold for polymer lens," IEEE Conference on Lasers and Electro-Optics, 2008 and 2008 Conference on Quantum Electronics and Laser Science, Copyright 2008 IEEE.*)

When the mold is overfilled with fluid to produce a convex profile, plano-concave lenses can be molded. After one lens is molded, the liquid-filled mold can be tuned to another shape.

The roughness of the surface of the molded lens is measured using an optical interferometer (Veeco NT 1100). The root-mean-square (RMS) surface roughness of the molded PDMS lenses is less than 6 nm. The surface quality is comparable to commercial lenses and adequate for most optical systems although more demanding optical systems may require an RMS roughness of less than 2 nm [75].

As discussed earlier, radius of curvature alone is not sufficient to describe the profile of the molded lenses. The lens profile can be more precisely represented by an aspherical function [Eq. (9-1)]. In order to fabricate aspherical lenses of designed profiles using the liquid-filled mold, both the curvature and the conic factor need to be precisely controlled. A simple calculation shows that the surface profile of an infinitely thin elastic membrane under fluidic pressure is parabolic, corresponding to a conic factor of −1 [79]. However, in reality, the elastic membrane has a finite thickness and shows nonlinear hyper-elastic behavior. Figure 9-23 shows that a wide range of conic factors can be obtained from different process and material parameters, such as membrane thickness and amount of prestretch, suggesting that a large family of focus-fixed aspherical lenses can be produced from a tunable liquid-filled mold.

Figure 9-23 represents various combinations of curvature and conic factor of the liquid-filled tunable mold [80]. Different extents of prestretch, for example, result in various combinations of conic factors and curvatures. In this specific case, only positive conic factors are obtained. Negative conic factors can be obtained with different process parameters since the conic factor approaches −1 (e.g., parabola) for infinitely thin elastic membrane [79]. Sophisticated membrane processing techniques other than thickness and prestretch offer possibilities for even more complicated lens profiles. These techniques include varying elastic constants and membrane thickness along the radial axis.

9-4-2 Summary

We have discussed a new way of fabricating aspherical polymer lenses using tunable liquid-filled molds. The surface roughness of the molded lenses is less than 6 nm, which is adequate for most optical systems. The curvature and the conic factor, two key parameters to define an aspherical lens, can be controlled by changing process parameters such as the applied pressure, the percentage of prestretch, the membrane thickness, and the mechanical constants of the membrane. This unique capability enables us to prototype an optical system in a cost-effective and time-efficient manner.

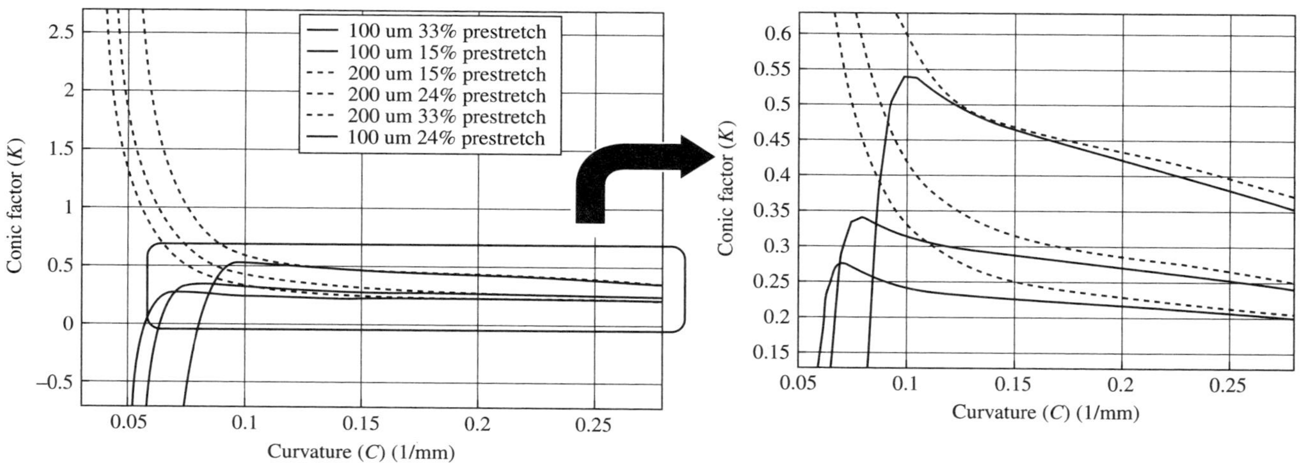

FIGURE 9-23 Plot of curvature (C) vs. conic factor (K) of the aspherical PDMS lenses. The ratio of pre-stretch and the thickness of the membrane affect both the curvature and the conic factor of the PDMS lenses.

9-5 Fluidic Lens for Lab-on-a-Chip and Micro-Total-Analysis Systems

So far we have discussed the applications of fluidic lenses in imaging optics. In this section, we introduce another class of fluidic lens applications. Here fluidic lenses are reduced to two-dimensional devices to bend light beams that are confined to the plane of the substrate. Great benefits can be derived from the ability to reduce a three-dimensional optical system to a fully functional, miniature two-dimensional system. The goal of such miniature two-dimensional systems is to preserve the original functionality and to add new functionality while reducing system size, cost, and complexity by orders of magnitude. In-plane, two-dimensional non-image-forming optical systems are useful for a number of applications, including optical telecommunications [81] and lab-on-a-chip devices. In such systems, there is a need to relay light from position A to position B in a specific fashion: for example, by collimating the beam for a longer travel distance, by shrinking or expanding the beam to a specific diameter, or perhaps by changing the numerical aperture of the system. A two-dimensional miniaturized nonimaging system offers the benefits of simplified light manipulation and drastically reduced path length in a robust, integrated fashion. Such systems can be employed in lab-on-a-chip devices for fluorescence detection assays, light scatter measurements, or absorption measurements.

Miniaturized two-dimensional (in-plane) optical systems generally consist of small optical elements created in the plane of the substrate. Such systems will typically require at least two materials to create the refractive index contrast needed for refraction (Snell's law). Many systems, especially some of the pioneering work in two-dimensional lenses for miniaturized optical chips, employ some air-material interface to satisfy this need [81–85]. Light exiting a rectangular waveguide end facet with outward curvature into air will be focused. Similarly, a small pocket of air, tenths of microliters in volume or less, with inwardly curved sidewalls can also serve to focus an incident light beam.

For many materials, the index contrast in such an air-material system can be quite high, yielding strong lenses but unacceptably high reflective and scattering losses. A more moderate refractive index contrast will be desired, leading to the need to fill the tiny above-mentioned spaces with some higher-index material. Fluidic lenses and other optical elements are thus often created [81,86–88], as the tiny nature of these spaces readily lends itself to capillary filling, and a fluid-filled space circumvents the problem of unintentional interfacial air gaps forming between two solid materials. Fluids can further offer the capability of lens tuning [89–92] and can help reduce light scatter by effectively reducing sidewall roughness [90], as discussed briefly in the following paragraphs.

The simplest nonimaging opto-fluidic lens takes the form of a modified waveguide. The end facet(s) will be curved to yield a focusing or defocusing effect. By controlling the curvature, the refractive index difference, and the dispersion of the substrate and the optical fluid, the effective focal length can be tailored to suit the needs of the system. Glebov et al. have created arrays of such curved waveguide facets in conjunction with index-matching fluid for use in dense wavelength division multiplexing (DWDM) networks, where arrayed beam collimation and focusing is required [81].

While such curved-facet waveguides provide the ability to add some light beam manipulation into planar optical designs, their capabilities can be rather limited, compromising the performance of many miniaturized systems relative to their macro counterparts. The lens positions and lens power have been restricted by the location and dimensions of the waveguide ends. A more fully functional miniaturized analog to a benchtop optical system would include "freestanding" lenses that can be any aperture size or any curvature needed, and can further be positioned at any location needed to create the desired system.

The "freestanding" fluid-filled two-dimensional lens is an analog to the air-pocket lens, allowing freely positionable lenses with the further ability to choose the refractive index difference and the lens dispersion qualities. These benefits, offered by the use of fluid, allow lens power and chromatic dispersion to be chosen by design. Filling is readily accomplished via capillary action or with the aid of a vacuum chamber. Two-dimensional fluid-filled optics, such as prisms, y-couplers, and lensed waveguide facets, were demonstrated by Kou et al. [87] in 2004. Godin et al. demonstrated freestanding fluid-filled lenses in 2006 [86,93,94], allowing for systems comprising multiple lenses to be realized. Such lenses can readily be designed with highly custom profiles, such as aspheric or parabolic curvatures. This provides a significant advantage over bulk systems, where creating such unique curvatures becomes costly and labor intensive.

While the use of lens fluid helps to reduce the refractive index difference and thereby reduce reflective losses, the need for optically smooth sidewalls nevertheless remains. For mold-replicated devices, smooth sidewalls start with a smooth mold. Smoother molds can be obtained by using improved etching techniques, such as cryogenic etching, or improved photolithography processes (when creating polymer-based molds). Etched molds allow for post-etch smoothing processes, such as anisotropic wet etching [82]. To keep improving the performance of two-dimensional optical systems, increasing attention needs to be paid to sidewall smoothness.

Two-dimensional optics lack the ability to contain and control light in one dimension, generally out of the plane of the chip or device. For very short optical paths, vertical losses may not be catastrophic for some applications. For applications such as optical computing or low-level light collection, these losses badly degrade performance. This is especially problematic when creating optical systems from such "freestanding" lenses mentioned above, where the total travel distance can become quite large relative to the vertical beam expansion. Slab waveguiding has been employed in various material systems to combat this issue [83,84,86]. The necessary layered structure can pose some physical robustness issues in terms of the interfaces. For multimaterial systems, some bonding is necessary [83,92,95]. The "interfaceless" all-PDMS system developed by Lien et al. [95,96] uses cure-bonding of PDMS waveguides to the PDMS device body, along with UV/ozone bonding of the two device body layers, to create a layered system without physical interfaces, allowing for the creation of a robust, slab-waveguided device.

While fluid-filled lenses are practical and sensible for simple and complete space filling as well as ready control of optical properties, they are unfortunately somewhat less practical for commercialization. Some fluids may be absorbed into the surrounding solid material. One simple solution is often overlooked: the use of curable polymers as a lens fluid, similar to the way waveguides have been created [88,95]. The precured polymer allows the designer to choose the optical properties, maintains the ease and completeness of filling (no air gaps), but then is heated or photo-cured into a solid. In a polymer body, a polymer-filled lens will bond to the sidewalls, preventing the formation of air gaps due to thermal cycling or physical stresses. Thus by utilizing uncured polymer as a lens fluid, all of the benefits of fluid use can be maintained while sidestepping its pitfalls.

Custom-shaped, replica-molded optics can take a variety of forms beyond lenses. Not only can this technology be used to create spherical, parabolic, or aspheric lens profiles, but other elements such as Fresnel lenses [97], prisms [87,98], reflectors [99], and apertures (or beam stops) [85,90,100] can also be fabricated, allowing for a complete optical-system-on-a-chip to be molded. These optical elements can be seamlessly integrated with microfluidic channels and associated technologies, creating a truly optofluidic device. A frequent test bed for the marriage of these technologies is the flow cytometer, a research and biomedical device commonly used for blood tests as well as for cell and bacteria studies [101,102]. The basic operation involves illuminating cells as they flow one by one past a laser beam and then detecting light scatter and/or fluorescence signals from each individual cell, allowing sample statistics to be acquired. This device relies heavily on optical detection; indeed, adequate system performance cannot be obtained with a simple fiber-and-channel configuration [103]. From the optics standpoint, signal resolution relies on providing a small, uniform source of illumination and on collecting light from a well-defined area by minimizing background illumination levels and collection of background signal. Lenses, apertures, and beam blocks are commonly found in benchtop cytometers.

These devices are generally made up of bulk optics and can easily take up an entire tabletop, though some smaller versions do exist.

Miniaturization of the flow cytometer would allow for point-of-care testing, testing in rural areas lacking infrastructure for a permanent clinic, or in-the-field testing of samples ranging from blood to water sources. Much of the work referred to in this section involves some form of lab-on-a-chip approach to flow cytometry. Over the years, these systems have grown in complexity from simple fiber-against-channel approaches to approaches integrating lenses and other optical elements to create a sophisticated system on a chip [83,84,86,93,104]. As the field of on-chip optics progresses, more optical elements are added to the "toolbox" for creating lab-on-a-chip devices, enabling higher quality systems to be created. Figure 9-24 shows a prototype lab-on-a-chip cytometer integrating fibers, waveguides, lenses of various shapes and sizes, and beam blocks. Figure 9-25 shows a parabolic lens. Figure 9-26 shows a beam block,

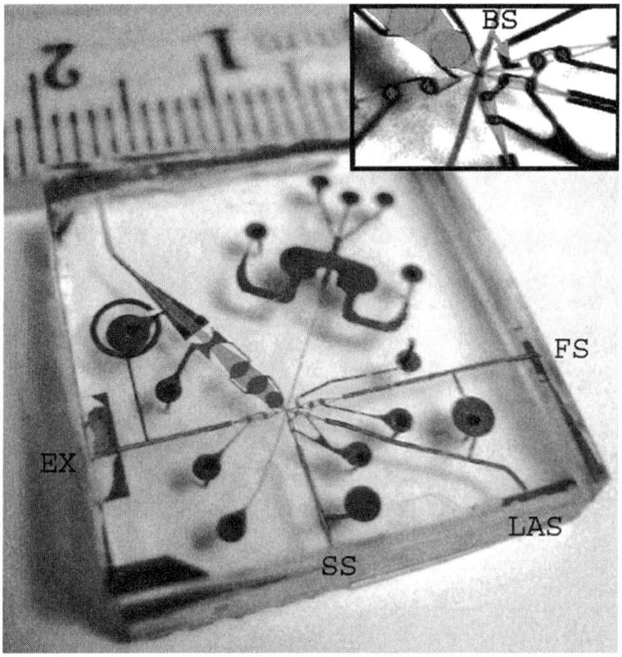

FIGURE 9-24 A prototype lab-on-a-chip cytometer, consisting of an illumination line, two scatter collection lines (all highlighted in blue), and a large numerical aperture fluorescence collection line (highlighted in red). This compact, highly integrated device is manufactured by simple molding techniques. (*J. Godin, C.-H. Chen, S. H. Cho, W. Qiao, F. Tsai, and Y.-H. Lo, "Microfluidics and photonics for Bio-System-on-a-Chip: A review of advancements in technology towards a microfluidic flow cytometry chip," Journal of Biophotonics 2008. Copyright Wiley-VCH Verlag GmbH & Co. KGaA. Reproduced with permission.*)

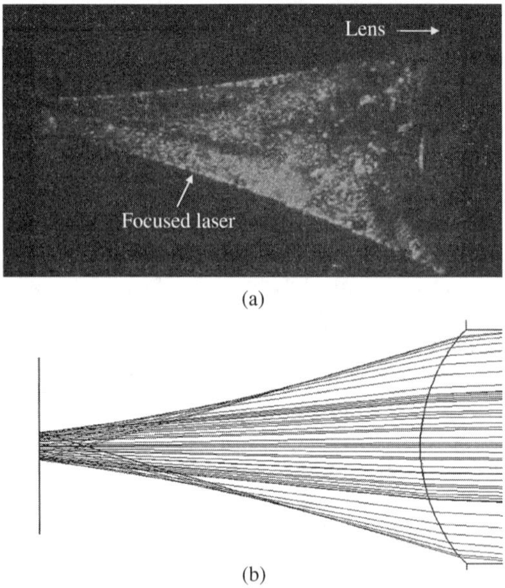

Figure 9-25 A planar spherical lens. Laser light comes from the right-hand side and is focused. Good agreement between experiment and simulation is achieved. Nanoparticles are incorporated as scattering centers to help visualize the light focusing effect. (*Reprinted with permission from J. Godin, V. Lien, and Y. H. Lo, "Demonstration of two-dimensional fluidic lens for integration into microfluidic flow cytometers," Applied Physics Letters, vol. 89, p. 061106, 2006. Copyright 2006, American Institute of Physics.*)

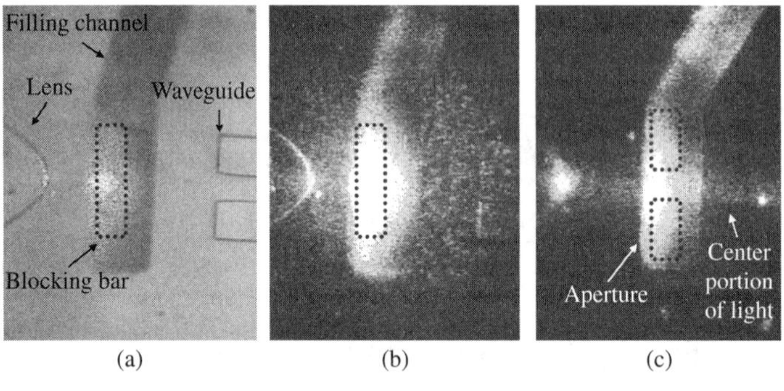

Figure 9-26 (a) Structure of a blocking bar between a lens and an exit waveguide. A filling channel is located below the plane of light travel enters from the top of the image. The blocking bar is above the filling channel and is indicated by the dotted box. (b) The blocking bar demonstrates the light-blocking capabilities that can be employed in on-chip devices. Light blocking is critical for reducing stray light collection and improving signal resolution. (c) An aperture in an on-chip device. The apertures is denoted in the red box. The filling channel connects the two sides of the aperture from below the plane of the slab waveguide. The aperture blocks light from the sides, but allows the central portion to travel through. (*S. H. Cho, J. Godin, C. H. Chen, F. S. Tsai, and Y. H. Lo, "Microfluidic photonic integrated circuits," Proceedings of SPIE, 7135, 71350M, 2008.*)

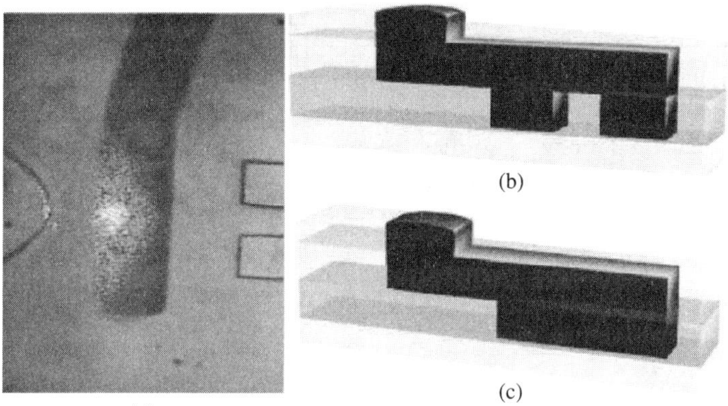

(a) (b) (c)

FIGURE 9-27 (a) Structure of blocking elements. A bar sits in the plane of light travel, between a lens (left) and a waveguide (right). The bar blocks much of the light from entering the waveguide, seen visually by scttaering upward from the bar. Both apertures (b) and blocking bars (c) can be made in this fashion. Each is filled by a filling channel, located outside of the plane of the slab waveguide (the darker blue region in the images) [100]

the performance of which is better understood by comparing to Fig. 9-27, showing an aperture allowing light to pass through the central region while blocking some light from the sides and from below. By integrating microfluidic channels not just with fiber optics, but with lenses, apertures, prisms, and other high-quality optical elements, a miniaturized cytometer can be devised that may start to rival the performance of at least the basic flow cytometer, and one day perhaps even the most advanced cytometers. With these same tools, the miniaturization and mass production of many optical systems can be realized.

9-6 Summary

Through millions of years, animal eyes have evolved into highly efficient optical designs that, in many critical areas, are superior to today's human-made optical devices. Fluidic lens represents a major step toward bio-inspired imaging systems. Among numerous methods of producing fluidic lenses, we believe fluid-filled elastomer membrane lenses possess the tunability, scalability, and reliability suitable for many applications. The new fluidic lens changes the concept of optical design and could shift the paradigm of future optical systems, in favor of more compact, versatile, and functional devices found in ever-increasing number of consumer and industrial products. Furthermore, with the wide-tuning range, fluidic lens can be used to replace intraocular lens to restore and improve human vision. The maturing technology of membrane processing also makes possible for an innovative tunable molding technology to allow aspheric lenses to be made easily and fast. Stemming from three-dimensional imaging optics, two-dimensional nonimaging fluidic lenses can be integrated into next-generation lab-on-a-chip devices and micro-total-analysis chips such that advanced optical interrogation systems can be integrated with microfluidic devices on a single chip.

References

1. A. Yariv and P. Yeh, *Photonics—Optical Electronics in Modern Communications*, Oxford University Press, New York, 2007.
2. K. Iizuka, *Elements of Photonics*, vol. I, Wiley-Interscience, 2002.
3. S. Sato, "Applications of liquid crystals to variable-focusing lenses," *Optical Review*, vol. 6, pp. 471–485, 1999.
4. M. Ye, B. Wang, and S. Sato, "Liquid-crystal lens with a focal length that is variable in a wide range," *Applied Optics*, vol. 43, pp. 6407–6412, 2004.
5. M. Ye, B. Wang, and S. Sato, "Liquid crystal lens with focus movable in focal plane," *Optics Communications*, vol. 259, pp. 710–722, 2006.
6. B. Wang, M. Ye, and S. Sato, "Liquid crystal lens with focal length variable from negative to positive values," *Photonics Technology Letters, IEEE*, vol. 18, pp. 79–81, 2006.
7. H. Ren and S. T. Wu, "Adaptive liquid crystal lens with large focal length tunability," *Optics Express*, New York, vol. 14, pp. 11292–11298, 2006.
8. Y. H. Fan, H. Ren, X. Liang, H. Wang, and S. T. Wu, "Liquid crystal microlens arrays with switchable positive and negative focal lengths," *IEEE/OSA Journal of Display Technology*, New York, vol. 1, pp. 151–156, 2005.
9. B. Wang, M. Ye, and S. Sato, "Lens of electrically controllable focal length made by a glass lens and liquid-crystal layers," *Applied Optics*, vol. 43, pp. 3420–3425, 2004.
10. V. V. Presnyakov and T. V. Galstian, "Electrically tunable polymer stabilized liquid-crystal lens," *Journal of Applied Physics*, vol. 97, pp. 103101, 2005.
11. V. Presnyakov and T. Galstian, "Polymer-stabilized liquid crystal lens for electro-optical zoom," *Proceedings of SPIE*, vol. 5577, p. 861, 2004.
12. V. Presnyakov, K. Asatryan, T. Galstian, and A. Tork, "Polymer-stabilized liquid crystal for tunable microlens applications," *Optics Express*, vol. 10, pp. 865–870, 2002.
13. H. Ren, Y. H. Fan, S. Gauza, and S.-T. Wu, "Tunable microlens arrays using polymer network liquid crystal," *Optics Communications*, vol. 230, pp. 267–271, 2004.
14. L. G. Commander, S. E. Day, and D. R. Selviah, "Variable focal length micro-lenses," *Optics Communications*, vol. 177, pp. 157–170, 2000.
15. A. M. R. L. Yang, E. M. McCabe, L. A. Dunbar, and T. Scharf, "Confocal microscopy using variable-focal-length microlenses and an optical fiber bundle," *Applied Optics*, vol. 44, pp. 5928–5936, 2005.
16. J. Knittel, H. Richter, M. Hain, S. Somalingam, and T. Tschudi, "Liquid crystal lens for spherical aberration compensation in a Blu-ray disc system," *Science, Measurement and Technology, IEE Proceedings*, vol. 152, pp. 15–18, 2005.
17. M. Hain, R. Gl kner, S. Bhattacharya, D. Dias, S. Stankovic, and T. Tschudi, "Fast switching liquid crystal lenses for a dual focus digital versatile disc pickup," *Optics Communications*, vol. 188, pp. 291–299, 2001.
18. B. Berge and J. Peseux, "Variable focal lens controlled by an external voltage: An application of electrowetting," *The European Physical Journal E-Soft Matter*, vol. 3, pp. 159–163, 2000.

19. S. Kuiper and B. H. W. Hendriks, "Variable-focus liquid lens for miniature cameras," *Applied Physics Letters*, vol. 85, pp. 1128–1130, 2004.
20. J. Park, C.-X. Liu, and J.-W. Choi, "A planar liquid lens design based on electrowetting," presented at Sensors, *2007 IEEE*, 2007.
21. F. Krogmann, R. Shaik, W. Monch, and H. Zappe, "Repositionable liquid micro-lenses with variable focal length," *Micro Electro Mechanical Systems, 2007, MEMS, IEEE 20th International Conference on*, pp. 707–710, 2007.
22. S. Grilli, L. Miccio, V. Vespini, A. Finizio, S. De Nicola, and P. Ferraro, "Liquid micro-lens array activated by selective electrowetting on lithium niobate substrates," *Optics Express*, vol. 16, pp. 8084–8093, 2008.
23. http://www.varioptic.com/en/index.php.
24. C. C. Cheng, C. A. Chang, and J. A. Yeh, "Variable focus dielectric liquid droplet lens," *Optics Express*, vol. 14, pp. 4101–4106, 2006.
25. C. C. Cheng and J. A. Yeh, "Dielectrically actuated liquid lens," *Optics Express*, vol. 15, pp. 7140–7145, 2007.
26. H. Ren and S. T. Wu, "Tunable-focus liquid microlens array using dielectrophoretic effect," *Optics Express*, vol. 16, pp. 2646–2652, 2008.
27. D. Zhang, N. Justis, and L. Y.-H., "Fluidic adaptive lens of transformable lens type," *Applied Physics Letters*, vol. 84, pp. 4194–4196, 2004.
28. W. Wang and J. Fang, "Design, fabrication and testing of a micromachined integrated tunable microlens," *Journal of Micromechanics and Microengineering*, vol. 16, pp. 1221–1226, 2006.
29. K. Campbell, Y. Fainman, and A. Groisman, "Pneumatically actuated adaptive lenses with millisecond response time," *Applied Physics Letters*, vol. 91, p. 171111, 2007.
30. H. Ren and S. T. Wu, "Variable-focus liquid lens by changing aperture," *Applied Physics Letters*, vol. 86, p. 211107, 2005.
31. H. Ren, D. Fox, P. A. Anderson, B. Wu, and S. T. Wu, "Tunable-focus liquid lens controlled using a servo motor," *Optics Express*, vol. 14, pp. 8031–8036, 2006.
32. S. W. Lee and S. S. Lee, "Focal tunable liquid lens integrated with an electromagnetic actuator," *Applied Physics Letters*, vol. 90, p. 121129, 2007.
33. R. Kuwano, M. Y., T. Tokunaga, and Y. Otani, "Liquid pressure varifocus lens using a fibrous actuator," *Proceedings of SPIE*, vol. 6374, pp. 1–4, 2006.
34. L. Dong and H. Jiang, "pH-adaptive microlenses using pinned liquid-liquid interfaces actuated by pH-responsive hydrogel," *Applied Physics Letters*, vol. 89, 2006.
35. F. S. Tsai, S. H. Cho, Y.-H. Lo, B. Vasko, and J. Vasko, "Miniaturzied universal imaging device using fluidic lens," *Optics Letters*, vol. 33, pp. 291–293, 2008.
36. D. Zhang, N. Justis, and Y.-H. Lo, "Integrated fluidic adaptive zoom lens," *Optics Letters*, vol. 29, pp. 2855–2857, 2004.
37. D. Zhang, N. Justis, and Y.-H. Lo, "Integrated fluidic lenses and optic systems," *IEEE Journal of Selected Topics Quantum Electronics*, vol. 11, pp. 97–106, 2005.
38. D. Zhang, V. Lien, Y. Berdichevsky, J. Choi, and Y.-H. Lo, "Fluidic adaptive lens with high focal length tunability," *Applied Physics Letters*, vol. 82, pp. 3171–3172, 2003.
39. D. Y. Zhang, N. Justis, and Y. H. Lo, "Fluidic adaptive zoom lens with high zoom ratio and widely tunable field of view," *Optics Communications*, vol. 249, pp. 175–182, 2005.
40. A. F. Bower, "Applied mechanics of solids," 2008.
41. Y. C. Fung, *A First Course in Continuum Mechanics,* 3d ed. Englewood Cliffs, New Jersey, Prentice Hall, 1994.
42. A. Shabana, *Computational Continuum Mechanics*, Cambridge University Press, New York, 2008.
43. F. S. Tsai, S. H. Cho, W. Qiao, N. H. Kim, and Y. H. Lo, "Miniaturized unified imaging system using bio-inspired fluidic lens," *Proceedings of SPIE*, 2008.
44. W. J. Smith, *Modern Lens Design*, McGraw-Hill Professional, USA, 2005.
45. R. Kingslake, *Lens Design Fundamentals*, Academic Press, USA, 1978.

46. R. R. Shannon, *The Art and Science of Optical Design*, Cambridge University Press, 1997.
47. W. J. Smith, *Modern Optical Engineering*, 4th, McGraw-Hill Professional, USA, 2007.
48. M. Laikin, *Lens Design*, CRC Press, USA, 2007.
49. J. M. Geary, *Introduction to Lens Design: With Practical Zemax Examples*, Willmann-Bell, USA, 2002.
50. G. H. Smith, *Practical Computer-Aided Lens Design*, Willmann-Bell Richmond, Va, Virginia, 1998.
51. M. F. Land and D. E. Nilsson, *Animal Eyes*, Oxford University Press, Oxford, 2002.
52. http://www.noscar.org/faq.php.
53. W. Qiao, F. Tsai, S.-H. Cho, and Y.-H. Lo, "Fluidic intraocular lens with a large accommodation range," *IEEE Photonic Technology Letters*, vol. 21(5), March 1,2009.
54. A. Glasser, "Accommodation: mechanism and measurement," *Ophthalmology clinics of North America*, vol. 19, pp. 1–12, v, 2006.
55. J. F. Doane and R. T. Jackson, "Accommodative intraocular lenses: considerations on use, function and design," *Current Opinion in Ophthalmology*, vol. 18, pp. 318–324, 2007.
56. J. Tabernero, P. Piers, and P. Artal, "Intraocular lens to correct corneal coma," *Optics Letters*, vol. 32, pp. 406–408, 2007.
57. M. Packer, I. H. Fine, R. S. Hoffman, and P. A. Piers, "Improved functional vision with a modified prolate intraocular lens," *Journal of Cataract and Refractive Surgery*, vol. 30, pp. 986–992, 2004.
58. J. S. Cumming, D. M. Colvard, S. J. Dell, J. Doane, I. H. Fine, R. S. Hoffman, M. Packer, and S. G. Slade, "Clinical evaluation of the crystalens AT-45 accommodating intraocular lens: results of the U.S. Food and Drug Administration clinical trial," *Journal of Cataract and Refractive Surgery*, vol. 32, pp. 812–825, 2006.
59. A. Galand, "Performance of the 1CU accommodating IOL," *Journal of Cataract and Refractive Surgery*, vol. 32, p. 3, author reply 3–4, 2006.
60. M. W. Matthews, H. C. Eggleston, and G. E. Hilmas, "Development of a repeatedly adjustable intraocular lens," *Journal of Cataract and Refractive Surgery*, vol. 29, pp. 2204–2210, 2003.
61. S. D. McLeod, L. G. Vargas, V. Portney, and A. Ting, "Synchrony dual-optic accommodating intraocular lens. Part 1: optical and biomechanical principles and design considerations," *Journal of Cataract and Refractive Surgery*, vol. 33, pp. 37–46, 2007.
62. S. D. McLeod, V. Portney, and A. Ting, "A dual optic accommodating foldable intraocular lens," *The British Journal of Ophthalmology*, vol. 87, pp. 1083–5, 2003.
63. J. E. Wold, A. Hu, S. Chen, and A. Glasser, "Subjective and objective measurement of human accommodative amplitude," *Journal of Cataract and Refractive Surgery*, vol. 29, pp. 1878–88, 2003.
64. J. Kessler, "Experiments in refilling the lens," *Archives of Ophthalmology*, vol. 71, pp. 412–417, 1964.
65. O. Nishi, K. Nishi, Y. Nishi, and S. Chang, "Capsular bag refilling using a new accommodating intraocular lens," *Journal of Cataract and Refractive Surgery*, vol. 34, pp. 302–309, 2008.
66. O. Nishi and K. Nishi, "Accommodation amplitude after lens refilling with injectable silicone by sealing the capsule with a plug in primates," *Archives of Ophthalmology*, vol. 116, pp. 1358–1361, 1998.
67. R. Menapace, O. Findl, K. Kriechbaum, and C. Leydolt-Koeppl, "Accommodating intraocular lenses: a critical review of present and future concepts," *Graefe's Archive for Clinical and Experimental Ophthalmology = Albrecht von Graefes Archiv fur klinische und experimentelle Ophthalmologie*, vol. 245, pp. 473–489, 2007.

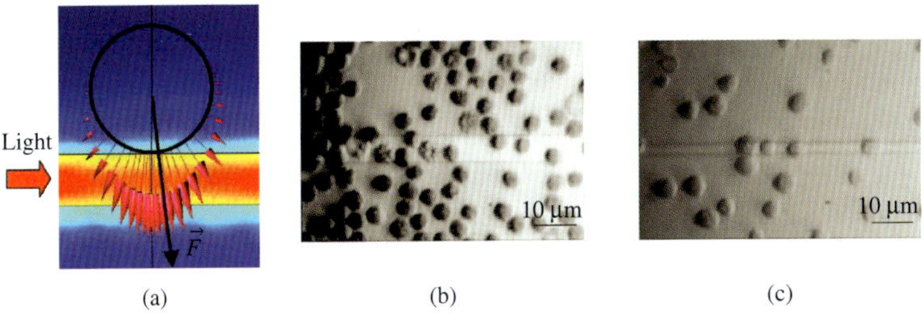

Figure 5-3 Optofluidic transport of biological species. (a) Finite element simulation of optical field in a channel waveguide and forces acting on a glass particle. (b) Image of radiation pressure transport of red blood cells. (c) Yeast cells. *(S. Gaugiran, S. Getin, J. M. Fedeli, G. Colas, A. Fuchs, F. Chatelain, and J. Derouard, "Optical manipulation of microparticles and cells on silicon nitride waveguides," Optics Express, 13(18), (2005), 6956–6963.)*

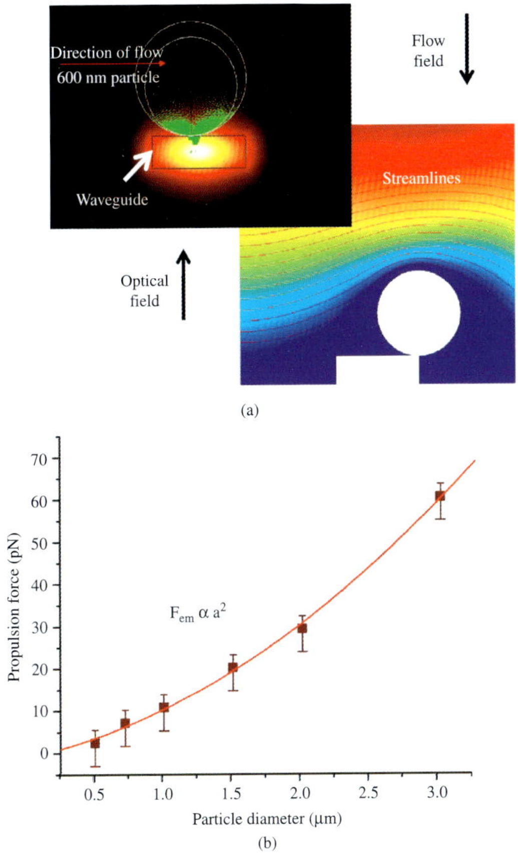

Figure 5-7 Forces on a particle trapped on a waveguide. (a) Cross section of guided mode in the waveguide and streamlines for a particle trapped on the waveguide and subject to a crossflow. (b) Propulsion force computed on particles trapped on the waveguide.

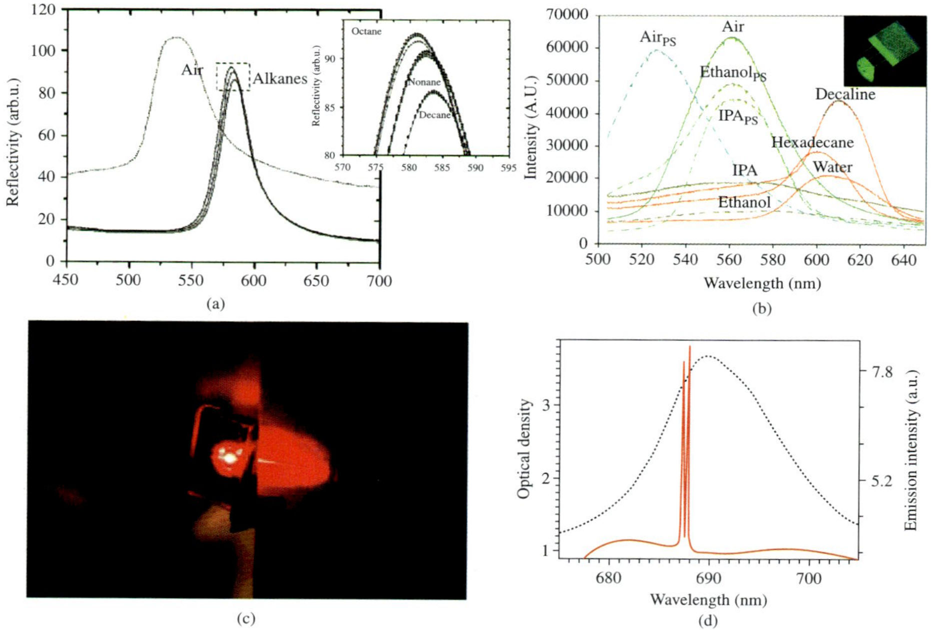

FIGURE 6-4 (a) The reflectance spectra showing optical chromatography results collected from the capillary with built-in colloidal crystal which is infiltrated with different types of alkanes (*U. Kamp, V. Kitaev, G. von Freymann, G. A. Ozin, and S. A. Mabury, "Colloidal Crystal Capillary Columns-Towards Optical Chromatography," Advanced Materials, 17, (2005), 438–443. Copyright Wiley-VCH Verlag GmbH & Co. KGaA. Reproduced with permission*). (b) The reflectance spectra from hybrid colloidal crystals composed of silica and PS particles. Here, the colloidal crystals were prepared in the centrifugal microfluidic chips. Solid and dotted lines represent the reflectance spectra of silica and PS colloidal crystals, respectively. (*S.-K. Lee, G-R. Yi, and S.-M. Yang, "High-speed fabrication of patterned colloidal photonic structures in centrifugal microfluidic chips," Lab on a Chip, 6, (2006), 1171–1177. Reproduced by permission of the Royal society of chemistry.*) (c) Optical image of laser emission at colloidal crystals immersed in dye solution. (d) Optical density (dotted line) of colloidal crystals and laser emission spectra at the band edge of lower wavelength. (*M. N. Shkunov, Z. V. Vardeny, M. C. DeLong, R. C. Polson, A. A. Zakhidov, and R. H. Baughman, "Tunable, gap-state lasing in switchable directions for opal photonic crystals," Advanced Functional Materials, 12, (2002), 21–26. Copyright Wiley-VCH Verlag GmbH & Co. KGaA. Reproduced with permission.*)

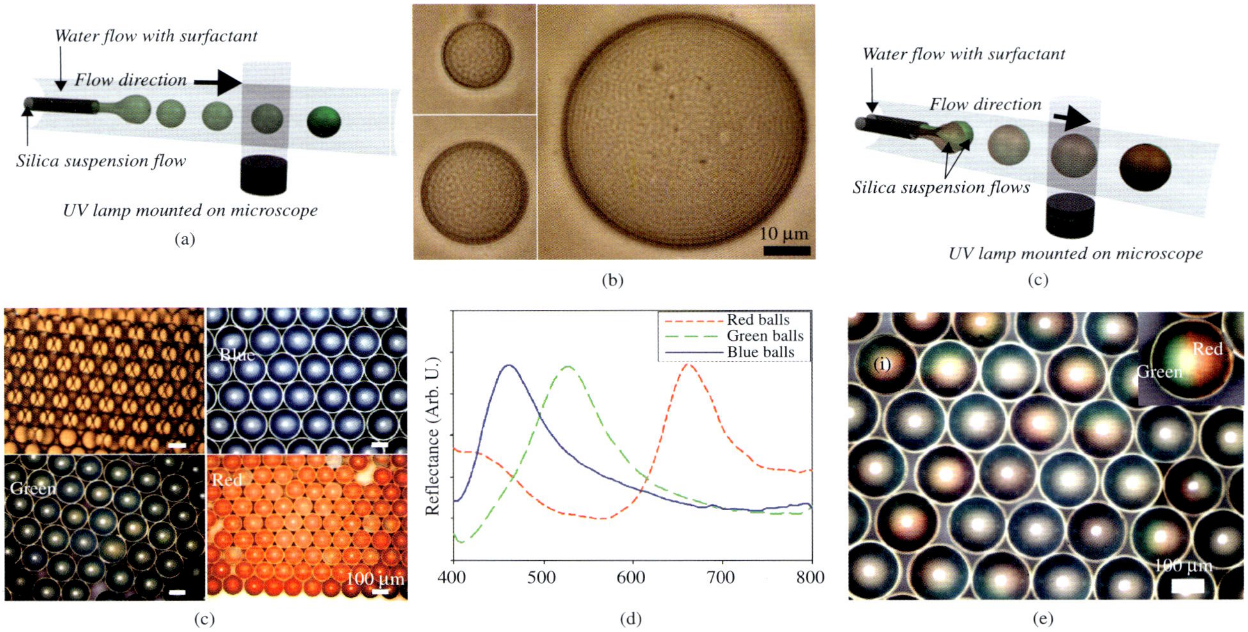

FIGURE 6-5 (a) Schematic illustration of optofluidic device for creating the photonic balls. (b) Optical microscope images of cross sections of emulsion droplets which contain 1-µm silica particles at 33% volume fraction. Layered structure from the outermost layers is observed. (c) Optical microscope image of monodisperse emulsion droplets and blue, green, and red photonic balls. (d) Normalized reflectance spectra from the blue, green, red photonic balls in (c). (e) Schematic of optofluidic device with paired inner capillaries for creating the photonic Janus balls. (f) Optical microscope image of photonic Janus balls which have green and red faces of a single photonic ball. (*S.-H. Kim, S.-J. Jeon, G.-R. Yi, C.-J. Heo, J. H. Choi, and S.-M. Yang, "Optofluidic assembly of colloidal photonic crystals with controlled sizes, shapes and structures," Advanced Materials, 20, (2008), 1649–1655. Copyright Wiley-VCH Verlag GmbH & Co. KGaA. Reproduced with permission.*)

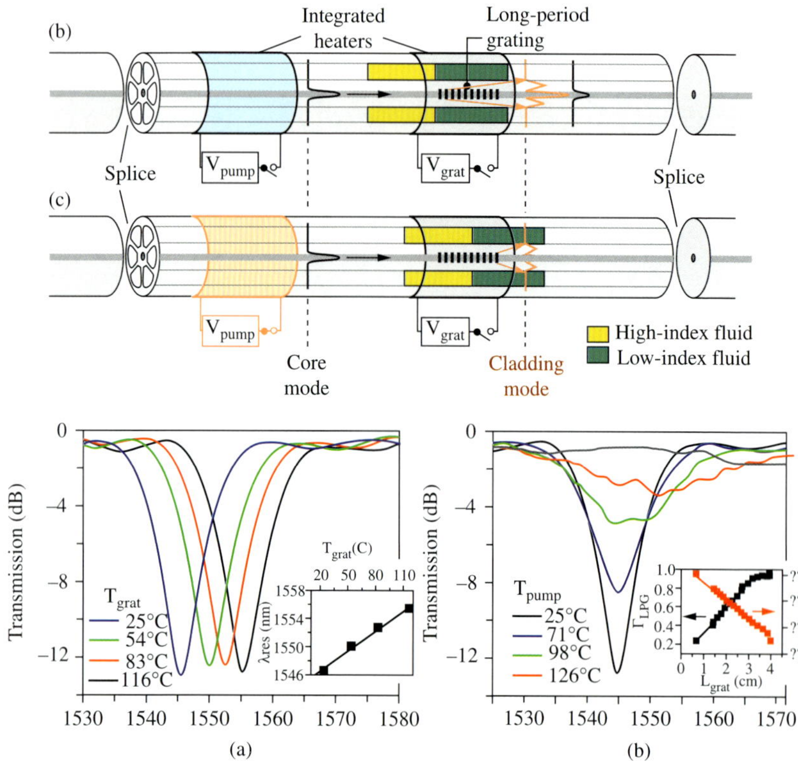

Figure 7-8 (top) A schematic of the fluid-filled grapefruit fiber with dual heaters to control the fluid position with respect to the LPG and a heater to tune the LPG directly. (bottom) Spectrum tunability as a function of both heaters' temperature. (*Reprinted with permission from P. Mach, M. Dolinski, K. W. Baldwin, et al., "Tunable microfluidic optical fiber," Appl. Phys. Lett., 80, 4294–4296 (2002). Copyright 2002, American Institute of Physics.*)

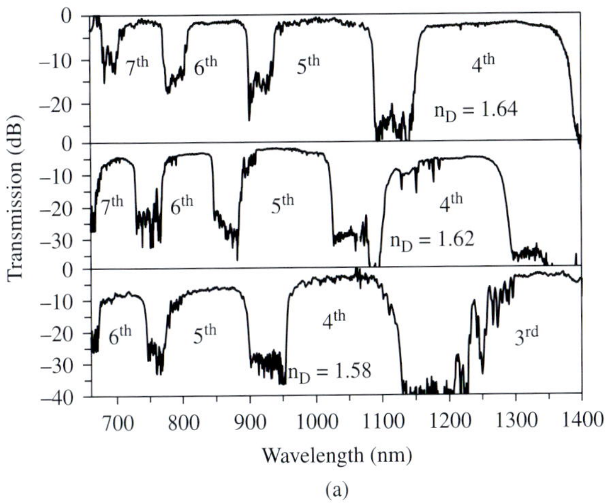

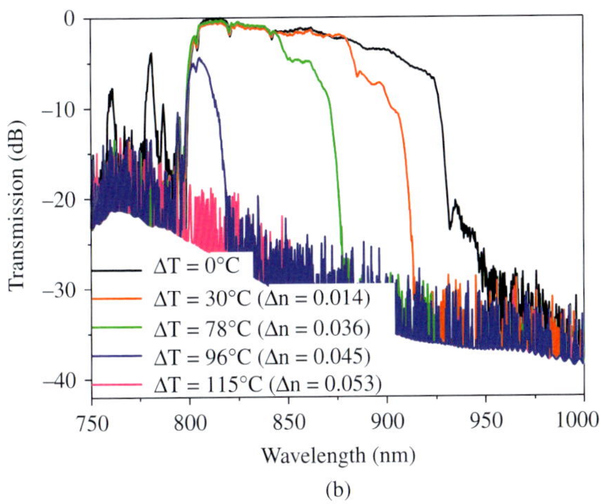

Figure 7-20 (a) Measured transmission spectra through fluidic PBGFs ($d = 3.5$ μm, $\Lambda = 7.7$ μm, single defect core, 4 rings of holes) filled with index fluids with $n_D = 1.64$, 162, and 1.58, where the fiber lengths are 10, 50, and 38 cm, respectively. Ordinal numbers indicate bandgap order, where the 1st order would be the lowest frequency fundamental bandgap. (b) Demonstration of dynamic filtering actuated by a thermal gradient, where a 25 mm section of a 12.4 cm long fluidic PBGF is heated ($d = 1.7$ μm, $\Lambda = 3.2$ μm, $n_D = 1.65$); legend indicates the average temperature difference across the fiber and corresponding index gradient in the fluid. (*P. Steinvurzel, B. J. Eggleton, C. M. de Sterke, et al., "Continuously tunable bandpass filtering using high-index inclusion microstructured optical fiber," Electron. Lett., Copyright 2005 IEEE.*)

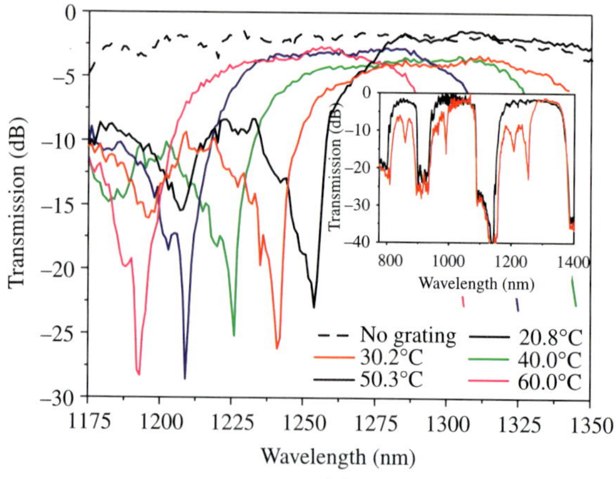

(a)

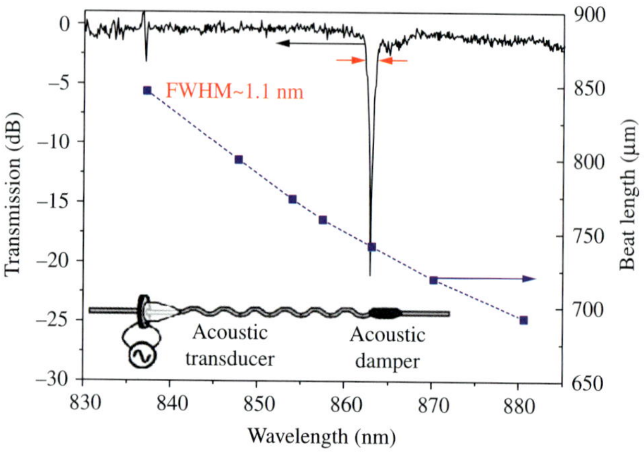

(b)

FIGURE 7-21 (a) Inset: transmission spectrum through a fluidic PBGF with (red) and without (black) a mechanical stress-induced LPG with a period of 660 μm. (*P. Steinvurzel, E. D. Moore, E. C. Mägi, et al., "Long period grating resonances in photonic bandgap fiber," Opt. Express, 14, 3007–3014 (2006).*) Main figure shows temperature tuning of the LPG resonance at 1250 nm [116]. (b) Spectrum of an acousto-optic LPG in a similar fiber for an acoustic wavelength of 740 μm (left axis). Squares indicate the optical beat length as a function of wavelength (right axis), where the large slope of this curve (proportional to the group index difference between the coupled modes) is what makes the resonance narrow. (*D.-I. Yeom, P. Steinvurzel, B. J. Eggleton, et al., "Tunable acoustic gratings in solid-core photonic bandgap fiber," Opt. Express, 15, 3513–3518 (2007).*)

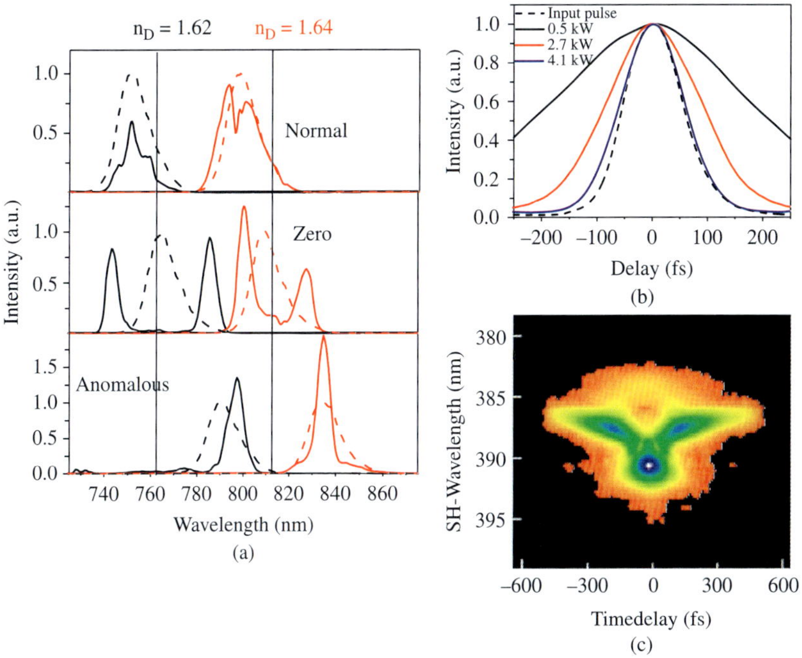

FIGURE 7-23 (a) Measured spectra of 70–100 fs pulses showing the effect of index scaling on waveguide dispersion, where ~0.02 index change shifts the dispersion profile by ~50 nm. Dashed lines correspond to input pulse spectrum; solid lines to spectrum after 40–60 cm length of PBGF with nonlinear parameter γ~16.2 (kW·m)$^{-1}$; black = n_D 1.62, 4.1 kW peak power; red = n_D 1.64, 3.7 kW peak power; center pulse wavelengths are as indicated and vertical lines indicate dispersion zero. Top spectra show SPM induced broadening, middle show dispersive wave radiation and soliton recoil, and bottom show soliton propagation with Raman self-frequency shift (see Ref. 126). (b) Autocorrelation time traces showing soliton formation with increasing peak power for n_D = 1.62 fiber at 780 nm. (*A. Fuerbach, P. Steinvurzel, J. A. Bolger, et al., "Nonlinear propagation effects in anti-resonant high-index inclusion photonic crystal fibers," Opt. Lett., 30, 830–832 (2005).*) (c) Measured-time and frequency-resolved spectrograph of pulse propagation at the dispersion zero, where the short wavelength band elongated along the time axis corresponds to the dispersive waves and the long wavelength band compressed along the time axis is the soliton. (*A. Fuerbach, P. Steinvurzel, J. A. Bolger, et al., "Nonlinear pulse propagation at zero dispersion wavelength in anti-resonant photonic crystal fibers," Opt. Express, 13, 2977–2987 (2005).*)

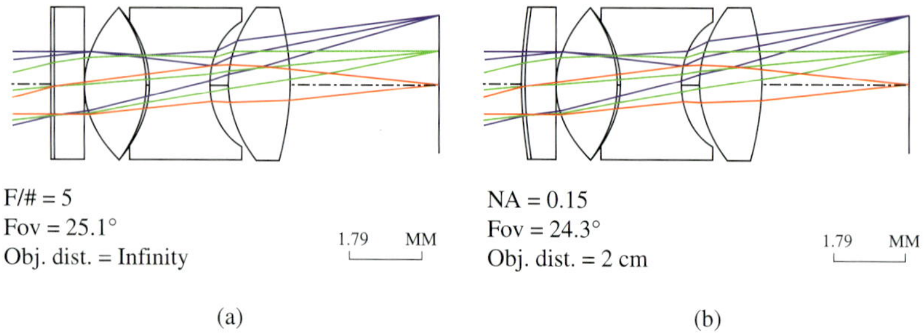

Figure 9-7 Lens View of the miniaturized unified imager consisting of 3 fixed lenses and 1 fluidic lens. The fluidic lens has different shapes for objects at (a) infinity and at (b) 2 cm from the lens. Compared to Fig. 2.2.1, field curvature and other aberrations are greatly suppressed. (*F. S. Tsai, S. H. Cho, Y.-H. Lo, B. Vasko, and J. Vasko, "Miniaturzied universal imaging device using fluidic lens," Optics Letters, vol. 33, pp. 291–293, 2008.*)

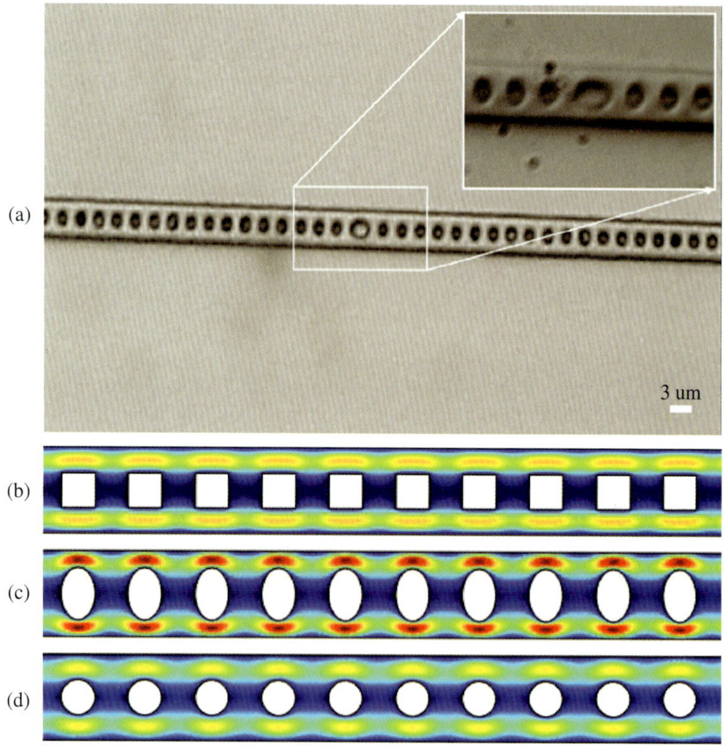

Figure 10-8 Panel (a) shows an optical micrograph of the DFB laser fabricated by Li and Psaltis [*Z. Li and D. Psaltis, "Optofluidic Distributed Feedback Dye Lasers," IEEE J. Top. Quant. Electron. 13(2), 185–193 (2007)*]. Panels (b) through (d) show two-dimensional laminar flow profiles calculated from Stokes' equation with the aid of a finite-element method. The three cases are for the same flow-rate and dark regions correspond to a vanishing flow velocity.

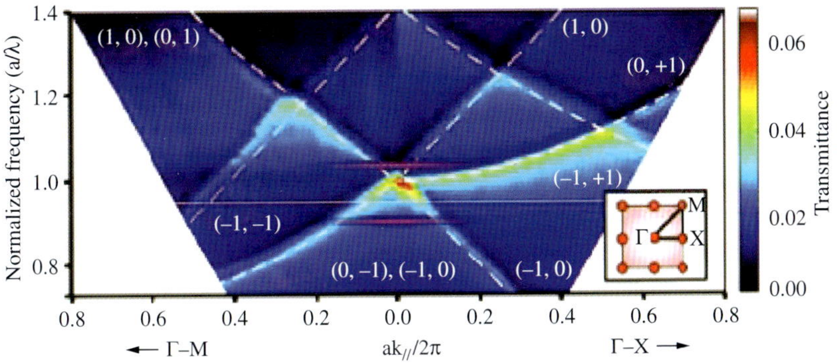

FIGURE 14-9 Transmittance through a 2D array of 350-nm holes in a 100-nm-thick aluminum film on GaAs substrate. To cover the domain of normalized frequencies a/λ shown, data from two samples with hole periods $a = 1.4\cdot\mu m$ and $a = 1.6\cdot\mu m$ are combined; the horizontal white line at $a/\lambda \sim 0.95$ indicates the boundary between the two data sets. Dashed white lines represent the momentum-matching condition [Eq. (14-3)]. The two pink ovals mark the angular and spectral range of short pulses in a separate experiment described later in this section (Fig. 14-11). (*Reprinted with permission from R. Rokitski, K. A. Tetz, and Y. Fainman, "Propagation of femtosecond surface plasmon polariton pulses on the surface of a nanostructured metallic film: space-time complex amplitude characterization," Phys. Rev. Lett., 95, 17, 177401, 2005. Copyright 2005 by the American Physical Society.*)

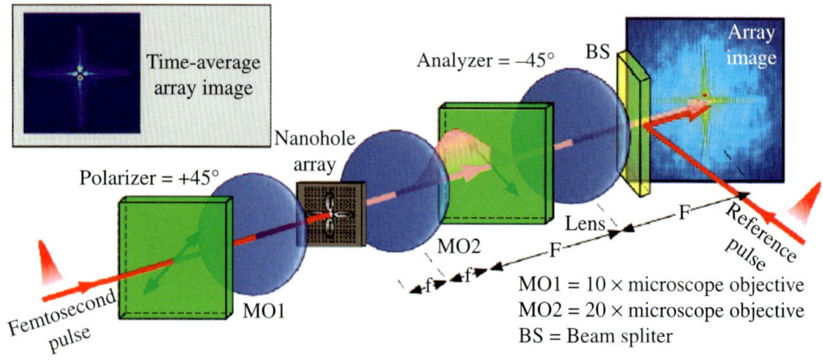

FIGURE 14-11 Schematic diagram of SPP imaging setup. A ~200-fs pulse excites SPP at the center of the nanohole array. As the SPP propagate across the sample, they reradiate part of their energy; by imaging this radiation into a CCD camera, the SPP paths can be visualized as seen in the inset. If the image is mixed with a delayed reference pulse, an interference pattern is observed, such as one shown in false color in the image plane. By varying the delay and noting the location of interference fringes, the progress of the SPP across the sample plane over time can be mapped out. The orthogonally oriented polarizer-analyzer pair suppresses directly transmitted light as explained in the text. (*Reprinted with permission from R. Rokitski, K. A. Tetz, and Y. Fainman, "Propagation of femtosecond surface plasmon polariton pulses on the surface of a nanostructured metallic film: space-time complex amplitude characterization," Phys. Rev. Lett., 95, 17, 177401,2005. Copyright 2005 by the American Physical Society.*)

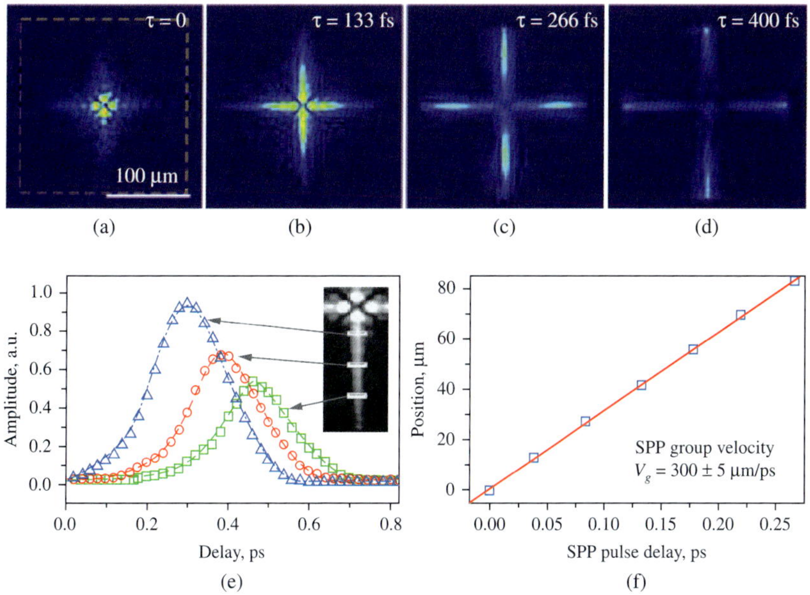

FIGURE 14-12 Temporal evolution of scattered femtosecond SPP pulse. (a through d) Spatial amplitude of the femtosecond SPP pulse (shown in false color) scattered from nanohole array with $a = 1.6\ \mu\text{m}$ hole period, measured for several time delays between SPP and reference pulses. (e) Normalized temporal cross-correlation of scattered SPP and reference pulses, measured at three spatial locations along the SPP propagation path. (f) Delay of the SPP pulse peak measured at different positions along the propagation path gives SPP pulse group velocity $\sim 300 \pm 5\ \mu\text{m/ps}$. Dashed line in (a) outlines dimensions of the nanohole array. (*Reprinted with permission from R. Rokitski, K. A. Tetz, and Y. Fainman, "Propagation of femtosecond surface plasmon polariton pulses on the surface of a nanostructured metallic film: space-time complex amplitude characterization," Phys. Rev. Lett., 95, 17, 177401,2005. Copyright 2005 by the American Physical Society.*)

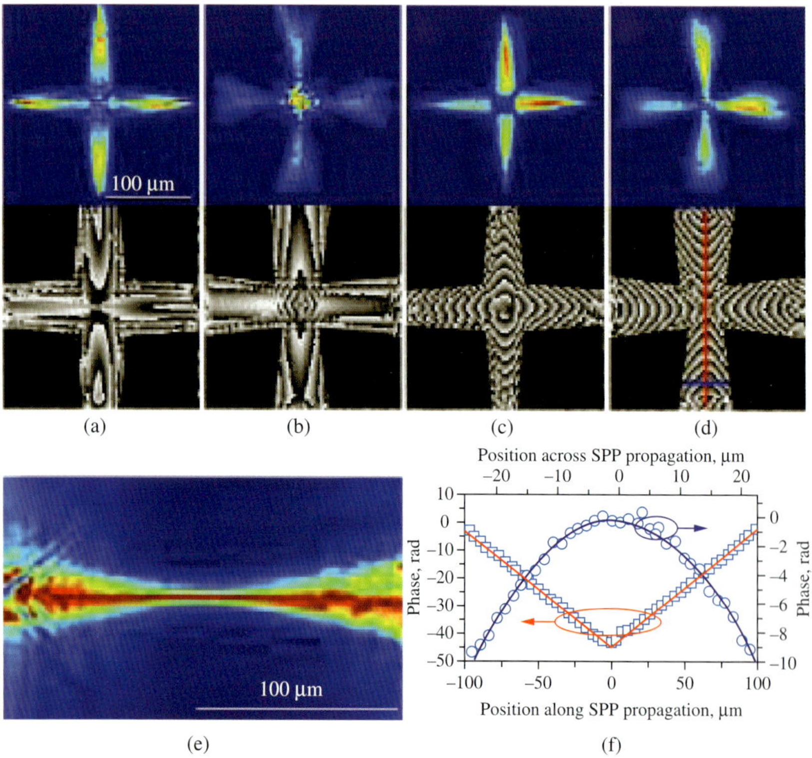

FIGURE 14-13 Snapshots of scattered SPP field over nanohole arrays with hole-to-hole distance $a = 1.6$ μm (a,b) and $a = 1.4$ μm (c,d), excited by spatially converging (a,c) and diverging (b,d) Gaussian pulses. In-plane convergence and divergence of SPP is clearly seen. (e) In-plane SPP focal waist formation, measured in nanohole array with 1.6 μm period. Amplitude is renormalized in the SPP propagation direction (left to right) to compensate for attenuation. (f) Unwrapped longitudinal (open squares) and transverse (open circles) cross sections of the spatial phase distribution from (d) with linear and quadratic least squares–fitting plots. Linear phase corresponds to 6.0° phase-matching angle for (−1,0) and (0,−1) SPP modes, while the quadratic spatial phase of the SPP field is acquired from the quadratic spatial phase of the incident optical field. (*Reprinted with permission from R. Rokitski, K. A. Tetz, and Y. Fainman, "Propagation of femtosecond surface plasmon polariton pulses on the surface of a nanostructured metallic film: space-time complex amplitude characterization," Phys. Rev. Lett., 95, 17, 177401,2005. Copyright 2005 by the American Physical Society.*)

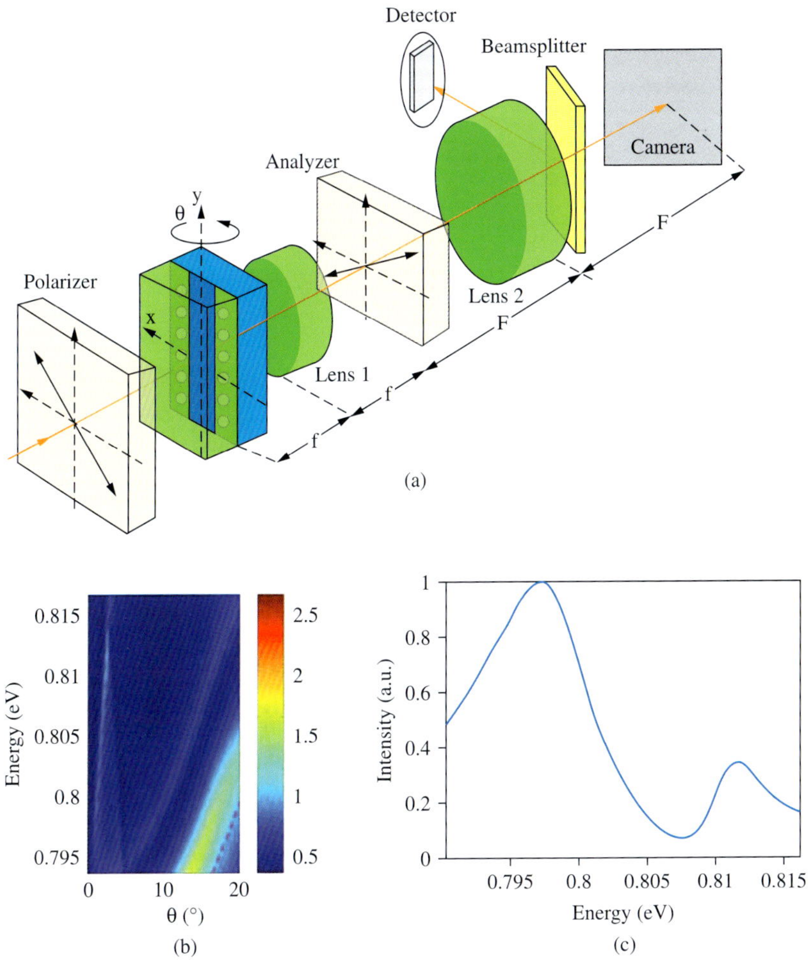

FIGURE 14-15 Observation of SPP mode splitting: (a) Experimental setup; (b) angle- and frequency-resolved transmittance data; (c) a slice through (b) at $\theta = 18°$, showing the low-energy *bright* and high-energy *dark* modes. *Reprinted with permission from L. Pang, K. A. Tetz, and Y. Fainman, "Observation of the splitting of degenerate surface plasmon polariton modes in a two-dimensional metallic nanohole array," Appl.Phys. Lett., 90, 111103, 2007. Copyright 2007, American Institute of Physics.*)

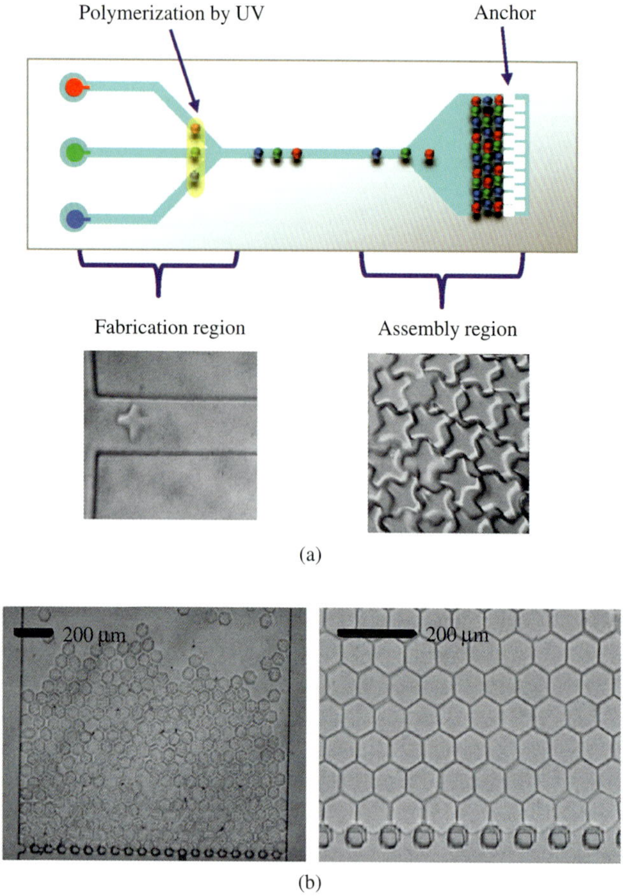

Figure 17-6 OFML system as a platform for self-assembled scalable systems. (a) A microfluidic channel can have a fabrication region and an assembly region, so that the particles continuously generated at a fixed area are assembled simultaneously by the flow. (b) Fluidic self-assembly of hexagonal microstructures in a microfluidic channel.

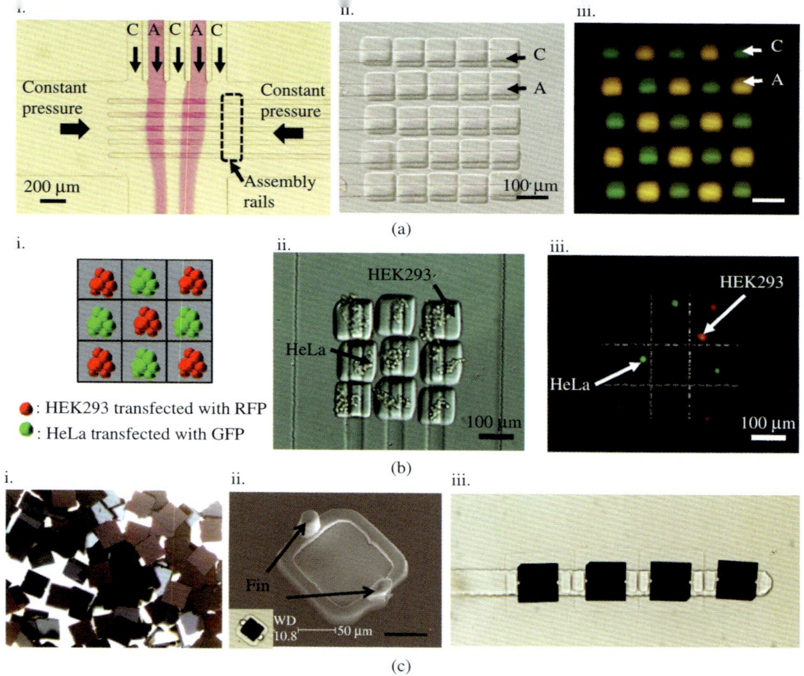

Figure 17-10 Applications of railed mcirofluidics. (a) Heterogeneous self-assembly of microstructures. Two-dimensional heterogeneous assembly process (left). 5×5 matrix assembly of squared microstructures (center). Corresponding fluorescent microscopic image displaying a checkerboard pattern of two different colors (right). (b) Heterogeneous cell patterning. 3×3 microscale hydrogel matrix (left), 3×3 patterns with two different living cells (center). Corresponding fluorescent image (right). (c) Microchips polymer packaging. Externally fabricated silicon microchips (left). SEM image of a silicon microchip in the polymer package (center). Packaged silicon chips are fluidically placed with desired uniform spacing (right). (*Reprinted by permission from Macmillan Publishers Ltd: Nature Materials S. E. Chung, W. Park, S. Shin, S. A. Lee, and S. Kwon, "Guided and fluidic-self-assemlby of microstructures using railed microfluidic channels," Nature Materials, 7, 2008, 581–587, copyright 2008.*)

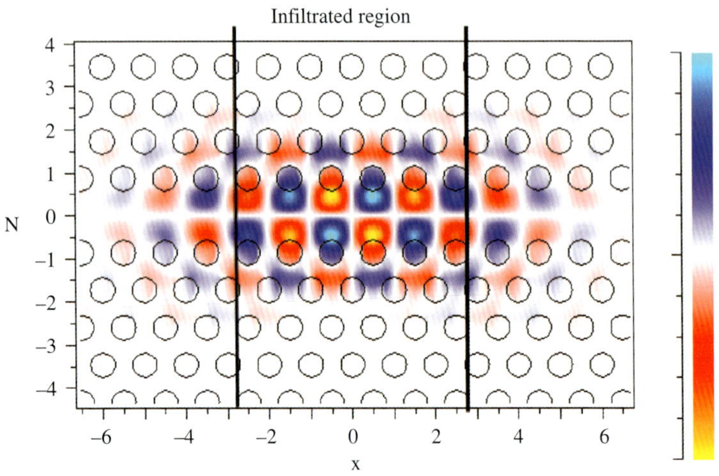

Figure 18-6 The calculated cross section of the major electric field component amplitude, E_x, of the resonant mode in the lower mode-gap. Circles indicate the holes; the infiltrated region is denoted by the vertical lines.

68. H. L. Liou and N. A. Brennan, "Anatomically accurate, finite model eye for optical modeling," *Journal of the Optical Society of America A-Optics Image Science and Vision*, vol. 14, pp. 1684–1695, 1997.
69. A. N. Simonov, G. Vdovin, and M. C. Rombach, "Cubic optical elements for an accommodative intraocular lens," *Optics Express*, vol. 14, pp. 7757–7775, 2006.
70. R. A. Schachar and A. J. Bax, "Mechanism of human accommodation as analyzed by nonlinear finite element analysis," *Annals of Ophthalmology*, vol. 33, pp. 103–112, 2001.
71. M. Dubbelman, G. L. Van der Heijde, and H. A. Weeber, "Change in shape of the aging human crystalline lens with accommodation," *Vision Research*, vol. 45, pp. 117–132, 2005.
72. C. H. M. Chien, T. Huang, and R. A. Schachar, "Analysis of human crystalline lens accommodation," *Journal of Biomechanics*, vol. 39, pp. 672–680, 2006.
73. E. A. Hermans, M. Dubbelman, G. L. van der Heijde, and R. M. Heethaar, "Estimating the external force acting on the human eye lens during accommodation by finite element modelling," *Vision Research*, vol. 46, pp. 3642–3650, 2006.
74. H. J. Burd, S. J. Judge, and J. A. Cross, "Numerical modelling of the accommodating lens," *Vision Research*, vol. 42, pp. 2235–2251, 2002.
75. J. Deegan, W. Hurley, B. Bundschuh, and K. Walsh, "Precision glass molding technical brief," 2007.
76. Y. Aono, M. Negishi, and J. Takano, "Development of large aperture aspherical lens with glass molding," *Proceedings of SPIE*, vol. 4231, 2000.
77. Y.-C. Tung, M. Zhang, C.-T. Lin, K. Kurabayashi, and S. J. Skerlos, "PDMS-based opto-fluidic micro flow cytometer with two-color, multiangle fluorescence detection capability using PIN photodiodes," *Sensors and Actuators B: Chemical*, vol. 98, pp. 356–367, 2004.
78. S. H. Cho, F. S. Tsai, R. Vasko, J. Vasko, and Y.-H. Lo, "Fluid-filled tunable mold for polymer lenses," presented at *Optical Society of America-CLEO/QELS Conference*, San Jose, 2008.
79. N. Sugiura and S. Morita, "Variable-focus liquid-filled optical lens," *Appl. Opt*, vol. 32, pp. 4181–4186, 1993.
80. S. H. Cho, F. S. Tsai, W. Qiao, N.-H. Kim, and Y.-H. Lo, "Fabrication of aspherical polymer lenses using tunable liquid-filled mold," University of California at San Diego, 2008.
81. A. L. Glebov, L. Huang, S. Aoki, M. G. Lee, and K. Yokouchi, "Two-dimensional microlens arrays in silica-on-silicon planar lightwave circuit technology," *Journal of Microlithography, Microfabrication, and Microsystems*, vol. 2, p. 309, 2003.
82. J. Seo and L. P. Lee, "Fluorescence amplification by self-aligned integrated microfluidic optical systems," *TRANSDUCERS, Solid-State Sensors, Actuators and Microsystems, 12th International Conference on, 2003*, vol. 2, 2003.
83. Z. Wang, J. El-Ali, M. Engelund, T. Gotsæd, I. R. Perch-Nielsen, K. B. Mogensen, D. Snakenborg, J. P. Kutter, and A. Wolff, "Measurements of scattered light on a microchip flow cytometer with integrated polymer based optical elements," *Lab on a Chip*, vol. 4, pp. 372–377, 2004.
84. S. Camou, H. Fujita, and T. Fujii, "PDMS 2D optical lens integrated with microfluidic channels: principle and characterization," *Lab on a Chip*, vol. 3, pp. 40–45, 2003.
85. K. W. Ro, K. Lim, B. C. Shim, and J. H. Hahn, "Integrated light collimating system for extended optical-path-length absorbance detection in microchip-based capillary electrophoresis," *Analytical Chemistry*, vol. 77, pp. 5160–5166, 2005.
86. J. Godin, C.-H. Chen, S. H. Cho, W. Qiao, F. Tsai, and Y.-H. Lo, "Microfluidics and photonics for Bio-System-on-a-Chip: A review of advancements in technology towards a microfluidic flow cytometry chip," *Journal of Biophotonics*, vol. 5, pp. 355-76, 2008.

87. Q. Kou, I. Yesilyurt, V. Studer, M. Belotti, E. Cambril, and Y. Chen, "On-chip optical components and microfluidic systems," *Microelectronic Engineering*, vol. 73, pp. 876–880, 2004.
88. C. L. Bliss, J. N. McMullin, and C. J. Backhouse, "Rapid fabrication of a microfluidic device with integrated optical waveguides for DNA fragment analysis," *Lab on a Chip*, vol. 7, pp. 1280–1287, 2007.
89. S. K. Y. Tang, B. T. Mayers, D. V. Vezenov, and G. M. Whitesides, "Optical waveguiding using thermal gradients across homogeneous liquids in microfluidic channels," *Applied Physics Letters*, vol. 88, p. 061112, 2006.
90. S. K. Y. Tang, C. A. Stan, and G. M. Whitesides, "Dynamically reconfigurable liquid-core liquid-cladding lens in a microfluidic channel," *Lab on a Chip*, vol. 8, pp. 395–401, 2008.
91. L. Dong and H. Jiang, "Tunable and movable liquid microlens in situ fabricated within microfluidic channels," *Applied Physics Letters*, vol. 91, p. 041109, 2007.
92. X. Mao, J. R. Waldeisen, B. K. Juluri, and T. J. Huang, "Hydrodynamically tunable optofluidic cylindrical microlens," *Lab on a Chip*, vol. 7, pp. 1303–1308, 2007.
93. J. Godin, V. Lien, and Y. H. Lo, "Integrated fluidic photonics for multiparameter in-plane detection in microfluidic flow cytometry," *Lasers & Electro-Optics Society, IEEE*, pp. 605–606, 2006.
94. J. Godin, V. Lien, and Y. H. Lo, "Demonstration of two-dimensional fluidic lens for integration into microfluidic flow cytometers," *Applied Physics Letters*, vol. 89, p. 061106, 2006.
95. V. Lien, Y. Berdichevsky, and Y.-H. Lo, "A prealigned process of integrating optical waveguides with microfluidic devices," *Photonics Technology Letters, IEEE*, vol. 16, pp. 1525–1527, 2004.
96. R. Mazurczyk, J. Vieillard, A. Bouchard, B. Hannes, and S. Krawczyk, "A novel concept of the integrated fluorescence detection system and its application in a lab-on-a-chip microdevice," *Sensors and Actuators B: Chemical*, vol. 118, pp. 11–19, 2006.
97. S. Balslev, B. Bilenberg, O. Geschke, A. M. Jorgensen, A. Kristensen, J. P. Kutter, K. B. Mogensen, and D. Snakenborg, "Fully integrated optical system for lab-on-a-chip applications," *Micro Electro Mechanical Systems (MEMS), 2004, 17th IEEE International Conference on*, pp. 89–92, 2004.
98. A. Llobera, R. Wilke, and S. Büttgenbach, "Poly (dimethylsiloxane) hollow Abbe prism with microlenses for detection based on absorption and refractive index shift," *Lab on a Chip*, vol. 4, pp. 24–27, 2004.
99. A. Llobera, S. Demming, R. Wilke, and S. Büttgenbach, "Multiple internal reflection poly (dimethylsiloxane) systems for optical sensing," *Lab on a Chip*, vol. 7, pp. 1560–1566, 2007.
100. S. H. Cho, J. Godin, C. H. Chen, F. S. Tsai, and Y. H. Lo, "Microfluidic photonic integrated circuits," *Proceedings of SPIE*, vol. 7135, pp. 71350M, 2008.
101. H. M. Shapiro, *Practical flow cytometry*, 4th ed. Hoboken, NY, Wiley-Liss, 2003.
102. A. L. Givan, *Flow Cytometry: First Principles*, 2d ed. New York, Wiley-Liss, 2001.
103. H. M. Shapiro, "Trends and developments in flow cytometry instrumentation," *Clinical Flow Cytometry, Annals of the New York Academy of Sciences*, vol. 677, pp. 155–166, 1993.
104. D. Huh, W. Gu, Y. Kamotani, J. B. Grotberg, and S. Takayama, "Microfluidics for flow cytometric analysis of cells and particles," *Physiol. Meas*, vol. 26, pp. R73–R98, 2005.

CHAPTER 10

Optofluidic Dye Lasers

Anders Kristensen

Department of Micro and Nanotechnology, Technical University of Denmark

N. Asger Mortensen

Department of Photonics Engineering, Technical University of Denmark

10-1 Introduction

Lab-on-a-chip applications call for a reconsideration of light sources. The light must be either coupled into, or generated on the chip itself. Application of external light sources has the advantage that a large variety of light sources are readily available, and often at reasonable price. However, in any product involving integrated optics, the coupling of light onto microchips represents a major challenge, being both labor-intensive and costly. On-chip light generation offers an alternative route. In lab-on-a-chip microsystems, light sources, or even lasers, can be integrated in a simple way by infiltrating dedicated microfluidic components with liquid light emitters, for example, liquid laser dye solutions (see Fig. 10-1). With this optofluidic approach the on-chip light sources are basically added without adding further steps in the fabrication procedure [1]. Furthermore, there is by virtue no alignment issues with the light sources, as all optical components are defined in the same lithography process, and thereby benefit from the high placement accuracy of modern microlithography.

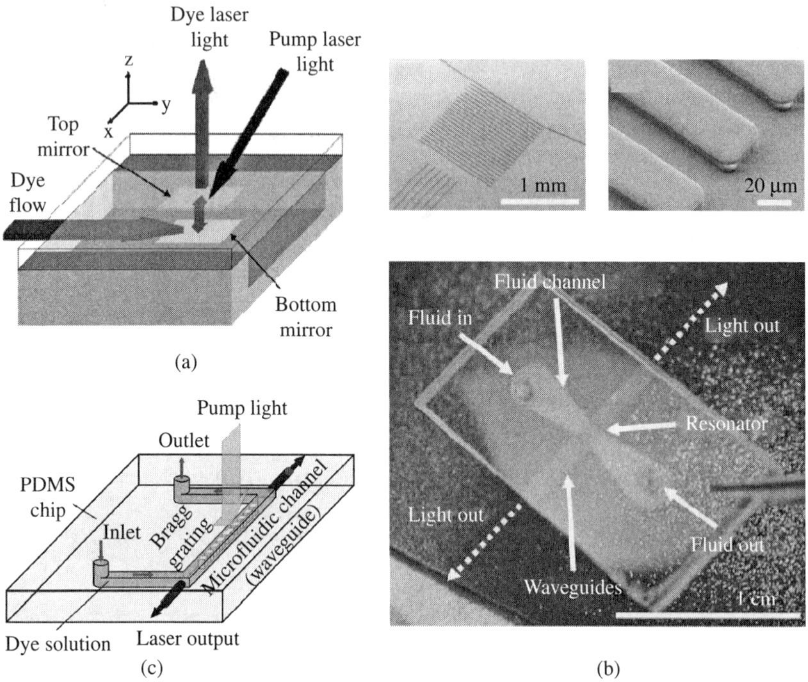

FIGURE 10-1 Three examples of microfabricated optofluidic dye lasers. Panel (a) shows an outline of a vertically emitting Fabry–Perot cavity laser [*B. Helbo, A. Kristensen, and A. Menon, "A micro-cavity fluidic dye laser," J. Micromech. Microeng. 13(2), 307–311 (2003)*]. Panel (b) shows the chip layout zoom-ins on the distributed-feedback optical resonator. In this device, the laser emits laterally, in the plane of the device, and the emission is coupled directly into integrated waveguides [*S. Balslev, A. M. Jorgensen, B. Bilenberg, K. B. Mogensen, D. Snakenborg, O. Geschke, J. P. Kutter, and A. Kristensen, "Lab-on-a-chip with integrated optical transducers," Lab Chip 6(2), 213–217 (2006) Reproduced by permission of the Royal Society of Chemistry*]. Panel (c) shows a laterally emitting optofluidic distributed-feedback laser where the laser resonator is imbedded in a liquid-core waveguide [*Z. Li, Z. Zhang, T. Emery, A. Scherer, and D. Psaltis, "Single mode optofluidic distributed feedback dye laser," Opt. Express 14(2), 696–701 (2006)*].

In this chapter we will discuss the main challenges and opportunities for realizing miniaturized optofluidic dye lasers:

- Design and performance of optofluidic laser resonators
- Strategies for frequency tuning
- Dye replenishment to compensate for bleaching

The discussion is focused on the involved physics rather than entering detailed technical discussions of various device implementations. The current state of the art for optofluidic laser device implementations is well covered in a series of review articles [2–4]. For the completeness we begin with a brief and pedestrian introduction to laser physics and dye lasers.

10-2 Laser Basics

Today, lasers are used in a large variety of daily life applications including telecommunication, audio and video disk players, machining, surgery, dentistry, positioning, and sensing. These applications rely on the key properties of laser light: monochromaticity and coherence, as opposed to other light sources such as Edison's glow bulb and light-emitting diodes (LED) which emit noncoherent photons in a wide band of frequencies. The word "laser" is an acronym for *light amplification by stimulated emission of radiation.*

The laser can simply be described as a feedback coupled amplifier for photons, as illustrated in Fig. 10-2a, which to first order oscillates at a resonance frequency determined by the spectral properties of the feedback and gain medium. This is quite analogous to

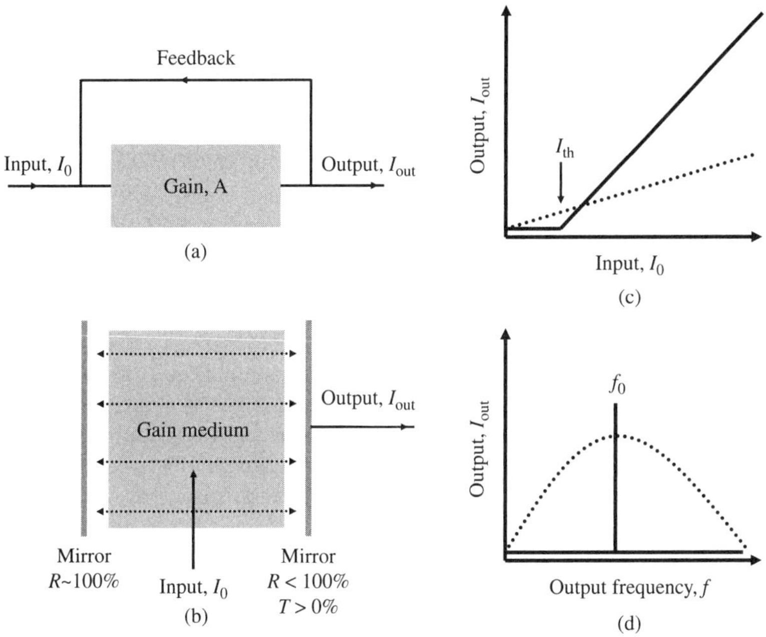

FIGURE 10-2 The laser can be described as a feedback coupled amplifier for photons, see panel (a). The amplification provided by a gain medium inserted in an optical resonator, for example, formed by two mirrors, see panel (b). Panel (c) outlines the input-output characteristics of a laser (solid line) with a characteristic threshold, I_{th} for onset of lasing, opposed to the behavior in the absence of lasing, for example, florescence (dotted line). Panel (d) shows the corresponding characteristic narrow-line emission spectrum from a laser (solid line), compared to the broad emission spectrum of traditional light source.

feedback in audio systems, for example, when the output from the loudspeakers is picked up in the microphone and fed back to the amplifier. Both the amplitude and frequency of these oscillations are very sensitive to the feedback conditions, for example, by modifying the amplification or changing the microphone position. In lasers, the input power, I_0, can be provided by different means, but most commonly the energy is supplied by an electrical current or by absorption of incident photons. These types of lasers are referred to as electrically or optically pumped devices, respectively. The optical feedback in lasers is generally provided by classical optical components like mirrors and gratings (see Fig. 10-2b), while the amplification relies on quantum physics and the statical properties of photons. The optical gain and amplification in the laser medium is obtained through the process of stimulated emission of photons, where existing photons stimulate emission of further "identical" photons having the same frequency and phase properties. Through the optical feedback in combination with the emission spectrum of the gain medium, a narrow band photon population with a common phase is built up in the optical resonator. A fraction of these photons are extracted from the cavity, for example, through a minute transmittance through one of the mirrors, to form the monochromatic and coherent laser emission. For a thorough introduction to laser physics, we refer to Svelto's classical textbook [5].

How to recognize a laser? Figure 10-2c and 10-2d illustrate the main characteristics of laser output (solid lines) as compared to conventional light sources (dotted lines). Spectrally, the laser manifests itself by emitting one color at a well-defined frequency f_0. This is referred to as monochromaticity, as opposed to multicolor or spectrally broad emission from conventional sources. The output power I_{out} of a laser exhibits a characteristic, sudden increase I_{out}, when the input power I_0 reaches a threshold, I_{th}.

The threshold, I_{th}, that marks the onset of lasing is reached when the round-trip gain experienced by the photons overcomes the round-trip cavity losses, leading to a positive net feedback, and hence oscillations. The slope $\partial I_{out}/\partial I_0$ above the lasing threshold, called the slope efficiency, describes the efficiency by which pump energy is converted into laser output energy.

10-3 Dye Lasers

Dye lasers are optically pumped devices using organic dye molecules as gain medium. When optically pumped, the organic dye molecule can provide optical gain over a broad frequency range, whereby the laser emission frequency is determined by the spectral properties of the laser cavity feedback. A range of different dye molecules with optical gain spanning the full spectrum from ultraviolet over the visible to near infrared frequencies are commercially available. Since the

1970s, liquid dye has been widely used for frequency tunable and high-power lasers in the visible frequency range, as reviewed in the book of Schäfer [6].

The optofluidic dye lasers we are concerned with here employ a liquid gain medium which consists of organic dye molecules in a suitable solvent. The organic dye rhodamine 6G is commonly applied for lasers emitting in the yellow (vacuum wavelength 570–600 nm), and has been used in most realizations of optofluidic lasers. Figure 10-3a shows (part of) the energy level diagram for the rhodamine 6G molecule. The electron states form bands of molecular rotation/vibration states (S_0, S_1, and T_1) separated by energy gaps. The spectrum exhibits a singlet-triplet system (S_i and T_i), where the lasing transition is between the ground (S_0) and first excited S_1 singlet bands, of frequency given by $hf = E_B - E_C$, with h being the Planck constant. Energy is provided to the system by absorption of pump photons to excite molecules from the S_0 rotation/vibration ground state to excited rotation/vibration states in the S_1 band. The transition times τ for the different transitions are shown on the figure. The triplet state is meta-stable with a decay time, τ_{Ph} of microseconds. Decay from S_1 into the triplet band T_1 is detrimental for laser operation, since triplet-triplet absorption has a spectral overlap with the singlet-singlet fluorescence, used for lasing. This is illustrated in Fig. 10-3b. If a significant population occupies triplet bands, these molecules will not only be unavailable for lasing, but will also enhance cavity losses through triplet-triplet absorption. Triplet band population is minimized by lowering the molecule-pump radiation interaction time below the S_1 to T_1 transition time $\tau_{TS} = 160$ ns. Despite this requirement, continuous wave (CW) lasing can be obtained in conventional macroscopic dye lasers, where interaction times typically less than 10 ns are achieved by flushing the liquid laser dye solution through the optical cavity in a jet steam of velocity in the range of 5 m/s.

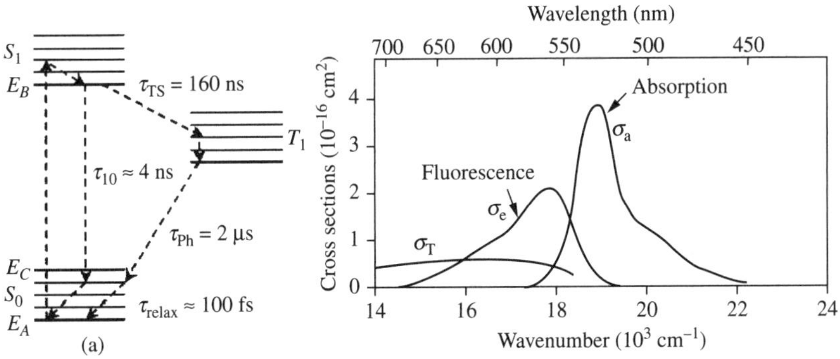

FIGURE 10-3 Panel (a) shows a simplified diagram for rhodamine 6G. Panel (b) shows the cross sections for singlet-singlet absorption, fluorescence, and triplet-triplet absorption for rhodamine 6G dissolved in ethanol [*O. Svelto, Principles of Lasers, 4th ed. (Springer, Heidelberg, 1998)*].

10-4 From Macro to Micro

In microfluidic devices, the 100 mm/s flow velocities would imply excessive pressure gradients, due to the large flow resistance. The hydraulic resistance of a tube of circular cross section is given by [7]

$$R_{\mathrm{hyd}} = \frac{8}{\pi}\eta L \frac{1}{a^4} \tag{10-1}$$

where a = inner radius
L = length
η = viscosity of the liquid flowing through the tube

The Hagen–Poiseuille law

$$\Delta p = R_{\mathrm{hyd}} Q \tag{10-2}$$

can be used to calculate the pressure gradient $-\Delta p/L$ for the flow rate $Q = \pi a^2 v_0$. With a pressure drop of 1 atm over a 10-mm-long microfluidic channel of inner radius $a = 10\ \mu\mathrm{m}$, a maximum average flow velocity of $v_0 = 100$ mm/s can be achieved.

The low flow velocities attainable in microfluidic devices have hindered CW operation of optofluidic dye lasers. Instead, triplet excitation is minimized by pulsing the optical pump radiation, with a pulse length typically below 10 ns.

The limitations on flow velocity are a first illustration that optofluidic dye lasers involve more than straightforward miniaturization. In the rest of the chapter we will discuss three main challenges and their potential solutions.

10-5 Laser Resonators

An optofluidic laser is basically a microfluidic channel with an embedded optical resonator, as illustrated in Fig. 10-1. The first optofluidic laser [8], (see Fig. 10-1a), was a vertically emitting device, where a Fabry–Perot optical resonator was embedded in a 10-μm-high microfluidic channel by placing thin-film gold mirrors in the floor and ceiling of the channel. The mode-spacing of the Fabry–Perot resonator is determined by the condition for standing waves:

$$k_N\, nL = N\pi,\ N = 1, 2, 3, \ldots \text{ (Fabry–Perot)} \tag{10-3}$$

where $k = 2\pi/\lambda$ is the vacuum wavenumber
n = refractive index of the liquid in the cavity

In the actual device, the μm scale mirror separation yields mode spacings ~10 nm in the relevant wavelength range, 570 to 600 nm.

For integration it would be an advantage to have a lateral emission of laser radiation. A laterally emitting Fabry-Perot optofluidic laser was realized in a hybrid approach, where two cleaved, end-face metallized optical fibers were aligned face-to-face in a microfluidic channel in a polymer device [9].

Figure 10-1b shows a laterally emitting distributed feedback (DFB) optofluidic laser. The DFB resonator is formed by placing a periodic array of polymer walls inside the microfluidic channel. Light traversing the microfluidic channel at the resonator will experience a periodic refractive index modulation. For simplicity we treat this as a one-dimensional problem by considering a periodic stack of liquid and polymer layers with the refractive index varying between n_1 = 1.36 in the liquid dye and n_2 = 1.59 in the polymer. Commensurability between the period L of the structure and the wavelength of light will give rise to resonances and standing waves according to the Bragg condition for the free-space wavenumber:

$$k_N\,(n_1L_1 + n_2L_2) = N\pi,\ N = 1, 2, 3,\ldots \text{ (Bragg)} \tag{10-4}$$

with $L = L_1 + L_2$, where L_1 is the width of the fluidic channels and L_2 is the width of the polymer walls. At resonance, a standing wave is formed along the periodic structure. Partial reflection of light at the liquid polymer interfaces yields a distributed optical feedback across the entire structure, as opposed to feedback between the two discrete mirrors in the Fabry–Perot cavity. The optofluidic DFB laser device in Fig. 10-1b has a mode spacing of a few nanometers and operates in a high Bragg order.

For simplicity we could consider to realize a "symmetric" structure $L_1 = L_2 = L/2$, of the same materials as the device in Fig. 10-1b: n_1 = 1.36 and n_2 = 1.59. To build a first-order optofluidic DFB laser N = 1 emitting at a vacuum wavelength λ around 600 nm, a grating period L below 200 nm is required.

In the one-dimensional treatment previously, we have neglected the finite transverse dimensions, for example, the finite height of the microfluidic channel. If we consider the full three-dimensional structure, we note that the liquid dye has a lower refractive index n_{dye} = 1.36 than the surrounding polymer n_{polymer} = 1.59, that is, the light is not confined in the liquid by total internal reflection.

This issue is addressed in the device shown in Fig. 10-1c [10], where light confinement in the liquid dye is enabled by choosing a dye solvent of high refractive index (mixture of ethanol and ethylene glycol n = 1,409) in combination with a lower refractive index polymer (PDMS n = 1.406) for the device. The low index contrast between liquid core and polymer cladding allows single mode waveguiding at wavelength λ ~600 nm even when the channel cross section (i.e., waveguide core) is as large as 2 μm × 5 μm. Truly single-mode laser oscillation is obtained by inserting a longitudinal optical DFB resonator of sufficiently large mode spacing, that only one mode frequency falls within the gain spectrum of the laser dye. The mode spacing, or free spectral range of the DFB resonator is given by

$$\mathrm{FSR} = \lambda_{N-1} - \lambda_N = \frac{\lambda_N}{N-1},\ N = 1, 2, 3, \ldots\ \text{(Bragg)} \qquad (10\text{-}5)$$

With a gain spectrum width of around 100 nm, this calls for DFB orders of $N \leq 15$.

The light confinement strongly reduces the losses in the laser resonator, and thereby implicitly reduces the pumping threshold for lasing. The light confinement in a Bragg grating also depends on the reflection order, N. For high-order modes the optical field samples high- and low-refractive index regions equally, whereas lower reflection order occupies the high-index regions to a larger extend. This implies that higher-order modes in general are less localized in the plane of the device, and thereby more lossy. For this reason it is more advantageous to employ low-order DFB modes [11]. Considering the Bragg condition, Eq. (10-4), which implies $\lambda_N = 2(n_1L_1 + n_2L_2)/N$, reduction of Bragg order N requires smaller dimensions of the Bragg structure. Current nanofabrication methods have allowed realization of third-order nanofluidic DFB lasers with a period of $L \simeq 600$ nm defined by electron beam lithography in a 300-nm-thick polymer layer [11,12].

Closed-loop waveguiding structures can form optical ring resonators, where resonant modes are determined by the condition of constructive interference

$$\kappa(k_N)R = N,\ N = 1, 2, 3, \ldots\ \text{(Ring)} \qquad (10\text{-}6)$$

Here, $\kappa(\omega)$ is waveguide wave vector [the inverse relation $\omega(\kappa) = ck(\kappa)$ is often referred to as the *waveguide dispersion relation*], R is the radius of the ring resonator, and N is again a positive integer. Figure 10-4 shows an example of an optofluidic laser where the walls of the glass capillary are used to form a ring resonator. The light guided in the capillary walls has a small evanescent tail into the hollow part of the capillary. Infiltrating the capillary with liquid dye ensures a small but sufficient overlap of the modes with the gain material. The design is thus conceptually different from the previously mentioned designs where the gain medium to a larger degree occupies the cavity rather than being situated in the close vicinity of the cavity.

Microdroplet-based optofluidic lasers have close similarities with the ring resonator–based lasers. Here, modes are only confined at the exterior boundary of the droplet, while there is no interior boundary

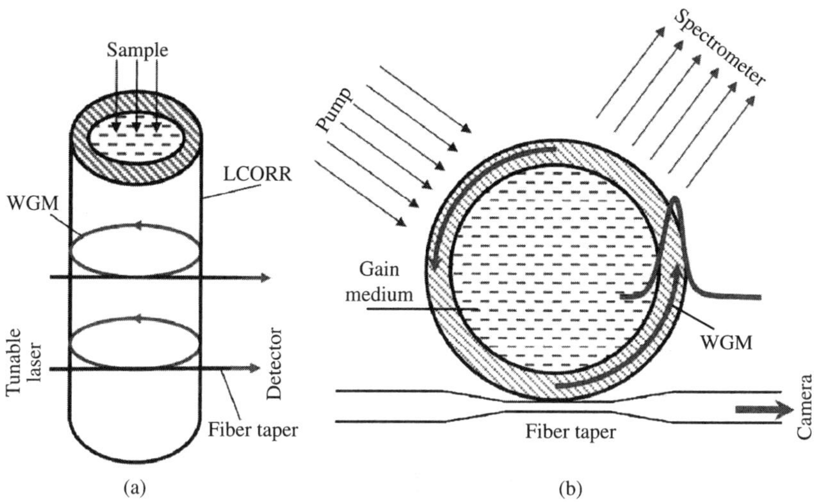

FIGURE 10-4 Outline of an optofluidic ring laser, with optical feedback provided by a liquid-core optical ring resonator (LCORR). (*Reprinted with permission from [S. I. Shopova, H. Zhou, X. Fan, and P. Zhang, "Optofluidic ring resonator based dye laser," Appl. Phys. Lett. 90(22), 221,101 (2007). Copyright 2007, American Institute of Physics*). In this configuration light is guided in a capillary glass wall and amplification is provided through evanescent coupling to the liquid laser dye inside the capillary tube.

as it is the case with the inner boundary of the waveguide. The resonances are known as whispering-gallery modes which form by total internal reflection at the surface of the droplet. The spectrum is very similar to that of the ring resonator and the following spectrum can be derived:

$$k_N nR \simeq N, \quad N = 1, 2, 3, \ldots \text{ (Whispering gallery)} \qquad (10\text{-}7)$$

Here, R is the radius of the droplet and n the refractive index of the liquid forming the droplet. Lasing in droplets has been reported by a large number of papers considering freely falling droplets. More recently also levitated and suspended droplets have been studied as illustrated in Fig. 10-5.

10-6 Tunable Lasers

As discussed previously, the laser dye can provide optical gain over a broad frequency range, whereby the laser emission frequency is determined by the spectral properties of the laser cavity feedback. The lasing frequency of dye lasers can therefore be continuously tuned by changing the resonance frequency of the laser resonator. Most macroscopic

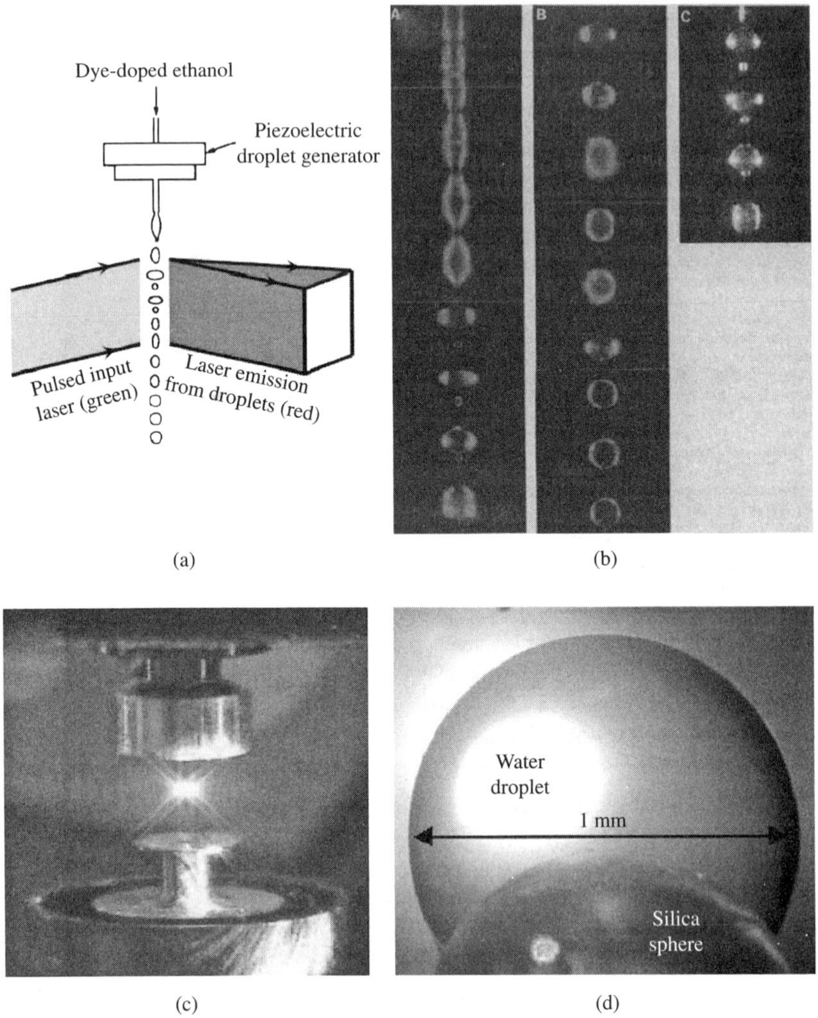

FIGURE 10-5 Examples of optofluidic droplet dye lasers and resonators. Panels (a) and (b) show free-falling droplets [*From S. X. Qian, J. B. Snow, H. M. Tzeng, and R. K. Chang, "Lasing droplets—highlighting the liquid-air interface by laser-emission," Science 231(4737), 486–488 (1986)*]. *Reprinted from AAAS with permission.* panel (c) shows an ultrasonically levitated droplet [*H. Azzouz, L. Alkhafadiji, S. Balslev, J. Johansson, N. A. Mortensen, S. Nilsson, and A. Kristensen, "Levitated droplet dye laser,"* Opt. Express *14(10), 4374–4379 (2006)*], and panel (d) shows a droplet trapped on an engineered substrate with silica micro spheres [*M. Hossein-Zadeh and K. J. Vahala, "Fiber-taper coupling to Whispering-Gallery modes of fluidic resonators embedded in a liquid medium,"* Opt. Express *14(22), 10,800–10,810 (2006)*].

dye lasers are tuned by mechanically altering the feedback conditions, for example, by adjusting the angle of incidence on a grating, which acts as wavelength selective element in the laser resonator.

The free spectral range of the optical resonator determines the size of the tuning range. If the FSR, or mode-spacing, is larger than the gain-width of the dye, the same mode of the optical resonator can ideally be frequency tuned across the full gain region of the laser dye without other modes entering the gain region. If the FSR is smaller than the width of the gain region, two or more optical resonator modes overlap with the gain region, and the laser oscillations will be able to jump from one mode to another. This is called mode hopping, and will result in abrupt jumps in lasing frequency when the laser is frequency tuned. Considering the optical feedback schemes and resonance conditions, [Eqs. (10-3), (10-4), and (10-6)] the miniaturization of optofluidic dye lasers becomes an advantage. For the optofluidic lasers discussed in this chapter, the resonance frequency is controlled by the optical pathlength. The output frequency of the optofluidic dye lasers can therefore be tuned in two ways: mechanically to alter geometric pathlength(s) or by fluidics to modify refractive indices of fluids.

The microfluidic platform offers the ability to precisely control a liquid medium and thereby control optical properties, such as refractive index and optical gain media concentration. These parameters can be altered in real time through, for example, microfluidic mixing of liquids in a precise and predictable fashion [13].

The refractive index of the liquid laser dye can be tuned by applying different solvents and mixtures thereof. Optofluidic dye lasers have been operated with the laser dye rhodamine 6G dissolved in different solvents, listed in Table 10-1. Continuous tuning of the refractive index is obtained by applying mixtures of solvents. As an example, mixtures of refractive index between 1.43 and 1.538 can be obtained by mixing ethylene glycol and benzyl alcohol, where the

Solvent	Refractive Index
Methanol	1.33
Ethanol	1.33
Ethylene glycol	1.43
DMSO	1.478
Benzyl alcohol	1.538

Table 10-1 Solvents Applied with Rhodamine 6G Laser Dye for Microfluidic Refractive Index Tuning of Optofluidic Dye Lasers

refractive index of the mixture is determined by the experimentally verified linear extrapolation between the two pure solvents [12]. In these experiments, the emission wavelength of a third-order DFB laser (with estimated FSR of 291 nm at the third-order Bragg reflectance) is tuned between 580.60 nm and 587.35 nm by changing the refractive index of the dye solution from 1.43 to 1.485.

The efficiency of microfluidic or refractive index tuning is limited by the spatial overlap between the optical mode and

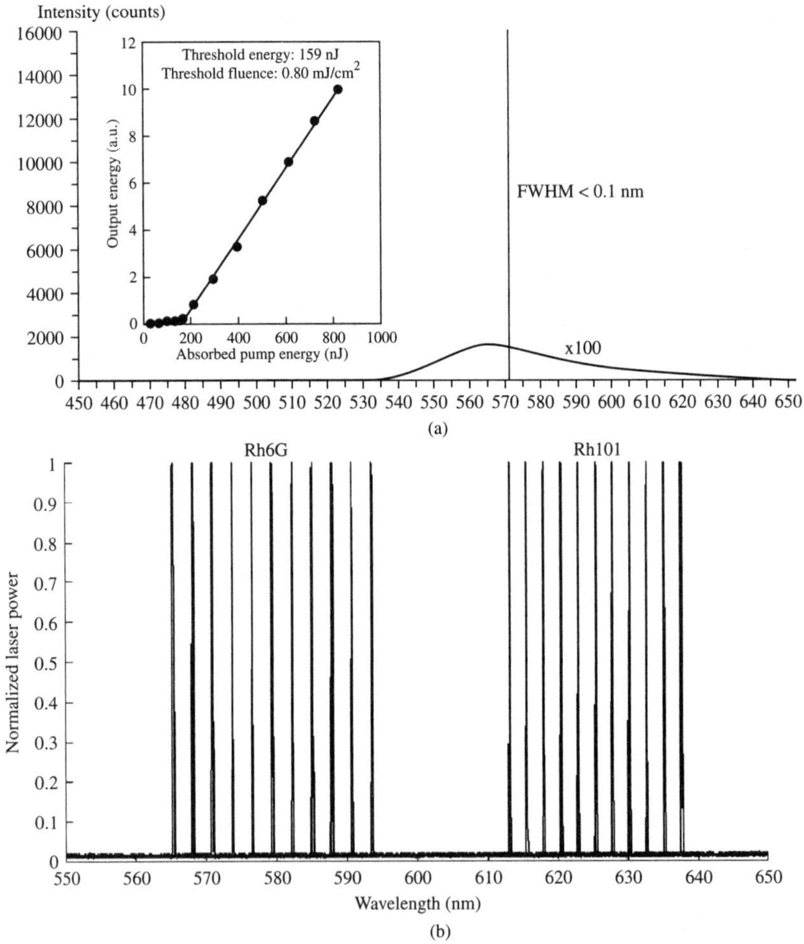

FIGURE 10-6 Panel (a) shows a lasing spectrum and input-output characteristic (inset) from the liquid-core waveguide DFB laser illustrated in Fig. 10-1c. Panel (b) shows tuning curves for the laser obtained with two different laser dyes, rhodamine 6G and rhodamine 101 [*Z. Li and D. Psaltis, "Optofluidic Distributed Feedback Dye Lasers," IEEE J. Top. Quant. Electron. 13(2), 185–193 (2007)*].

the liquid which is refractive index–tuned. In general this can be expressed by [14]

$$\frac{\Delta\lambda}{\lambda} = \frac{\Delta n_1}{n_1} \times f_1 + \frac{\Delta n_2}{n_2} \times f_2 \text{ (Refractive index tuning)} \tag{10-8}$$

where f represents the device geometry–specific tuning efficiency and $f_1 + f_2 = 1$. For Bragg resonators where refractive index n_1 is tuned, f_1 can be estimated to

$$f_1 = \frac{n_1 L_1}{n_1 L_1 + n_2 L_2} \leq 1 \text{ (Bragg)} \tag{10-9}$$

The laser in Fig. 10-1c is fabricated in an elastomeric material, and the geometric length of the resonator can be tuned mechanically by stretching the device [15]. Assuming the induced strain is uniform across the resonator, that is,

$$\frac{\Delta L_1}{L_1} = \frac{\Delta L_2}{L_2} = \frac{\Delta L}{L} \tag{10-10}$$

$$\frac{\Delta\lambda}{\lambda} = \frac{\Delta L}{L} \text{ (Strain tuning)} \tag{10-11}$$

The applied elastomer, PDMS, has a low Young's modulus, typically less than 1000 kPa and can take large strain without plastic deformation. Using this approach, a DFB laser with a grating period, $L = 3$ μm, operating on the fifteenth-order Bragg reflection, is wavelength tuned over 29 nm using a single dye mixture (see Fig. 10-6).

10-7 Dye Bleaching

The organic laser dyes gradually degrade when exposed to visible and ultraviolet radiation. This is referred to as *dye bleaching*. Over time the gain medium will thus turn less active. Typically, the problem of dye bleaching is compensated by a continuous convective flow of liquid-dissolved dye molecules, thus compensating the bleaching dynamics caused by the pump radiation. The required, convective dye-replenishing flow has been achieved by external fluid-handling apparatus. As an alternative to ordinary mechanical pumps, one often relies on syringe pumps for lab-on-a-chip applications. Fabrication of on-chip micro-fluidic pumps has also been pursued [15,16]. More recently, capillary effects have also been used to generate a convective flow without the need for any complicated pumping schemes. Finally, considering the microfluidic platform, optofluidic lasers and other devices may potentially be operated for days by diffusion without the need for a convective flow. Below we give a general account for the physics related to dye replenishment in optofluidic dye lasers. The central hypothesis will be that to lowest order the gain will scale linearly with the concentration C of unbleached dye molecules.

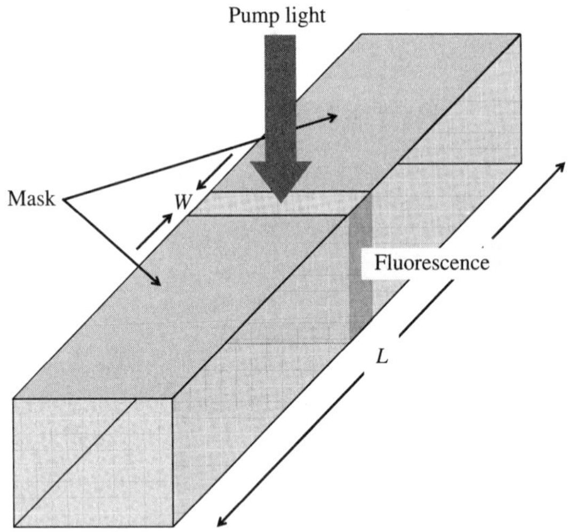

FIGURE 10-7 Experimental setup for the one-dimensional diffusion experiments in (*Reprinted with permission from M. Gersborg-Hansen, S. Balslev, N. A. Mortensen, and A. Kristensen, "Bleaching and diffusion dynamics in optofluidic dye lasers," Appl. Phys. Lett. 90(14), 143,501 (2007). Copyright 2007, American Institute of Physics*). Closed channel containing a liquid solution of dye molecules. The dye molecules are optically pumped by a pulsed, frequency-doubled Nd:YAG laser through a slit of width w covering the sample.

Consider the device illustrated in Fig. 10-7, which for simplicity can be modeled as a one-dimensional problem with the concentration of unbleached dye molecules C being a function of both the spatial coordinate x as well as the time t. The diffusion-convection dynamics is governed by the one-dimensional diffusion-convection equation:

$$\frac{\partial}{\partial t}C(x,t) = D\frac{\partial^2}{\partial x^2}C(x,t) - v\frac{\partial}{\partial x}C(x,t) - \Gamma_b(x,t)C(x,t) \qquad (10\text{-}12)$$

The first term on the right-hand side is the diffusion term with D being the diffusion constant. The second term is the convection term with v being the liquid flow velocity in the x direction. Finally, the third term is the source term, which here accounts for the dye bleaching, with Γ_b being the bleaching rate.

In Fig. 10-7, the dye molecules are optically pumped, and thus bleached, in a narrow segment of width w. In the context of Eq. (10-12) we thus assume that Γ_b takes a constant time-averaged value in the pumped region, while it is zero outside the pumped region. Ideally the concentration of unbleached dye molecules in the optically pumped region $C(x = 0, t)$ should be kept constant at a high level, preferable the initial concentration level $C_0 = C(x = 0, t \to 0)$. This can be achieved by means of a convective flow, that is, a sufficient high flow velocity. The corresponding convection rate is given by $\Gamma_c = v/w$. The condition that

$$\Gamma_c >> \Gamma_b \text{ (convective replenishment condition)} \tag{10-13}$$

is amply satisfied with the $v \simeq 5$ m/s jet flow in conventional dye lasers, and likewise, convective flow is an efficient dye replenishment mechanism in optofluidic dye laser. Making a similar analysis of the diffusion term we arrive at a diffusion rate given by

$$\Gamma_d = \frac{D}{w^2} \tag{10-14}$$

In microfluidics it is a key observation that while the diffusion constant is scale invariant, that is, D does not depend on the size of the device, the *diffusion rate* Γ_d increases as $1/w^2$ when w goes to zero [17]. Thus, a steady state can equally be achieved by diffusive driven molecule exchange with a large reservoir, or an *ideal reservoir* where $\partial C/\partial t = 0$. The condition for this is that

$$\Gamma_d >> \Gamma_b \text{ (diffusive replenishment condition)} \tag{10-15}$$

Equation (10-12) as well as experimental studies [18] indicate that diffusion alone may be sufficient to replenish bleached dye in a miniaturized dye laser under typical optical pumping levels and repetition rates.

As another example where dye replenishment is achieved through a combination of convection and diffusion is shown in Fig. 10-8. The figure shows a finite element calculation of the laminar flow profile in the laser device in Fig 10-1c [10]. In this device the laminar flow profile, and hence also the convective dye replenishment is spatially very inhomogeneous. The flow simulations in Fig. 10-8 reveal that convective flow only occurs off-center in the microfluidic channels, while stagnant fluid volumes ($v \sim 0$) are present in between the polymer posts in the center of the channel. In the stagnant regions the dye replenishment must instead rely on dye molecule diffusion between the stagnant volume and convective flow regions. In this context the convective flow regions act as ideal reservoirs. Using the previously estimated diffusion constant for rhodamine 6G in ethylene glycol, $D \sim 1.5 \times 10^{-11}$ m^2/s, and a typical width $w \sim 1$ µm of the stagnant regions, we arrive at a characteristic diffusion rate $\Gamma_d = D/w^2 \sim 15\ s^{-1}$. This is larger than typical repetition rates of the pulsed pump radiation, thus ensuring an efficient diffusive dye replenishment in the stagnant regions, allowing for a steady laser output.

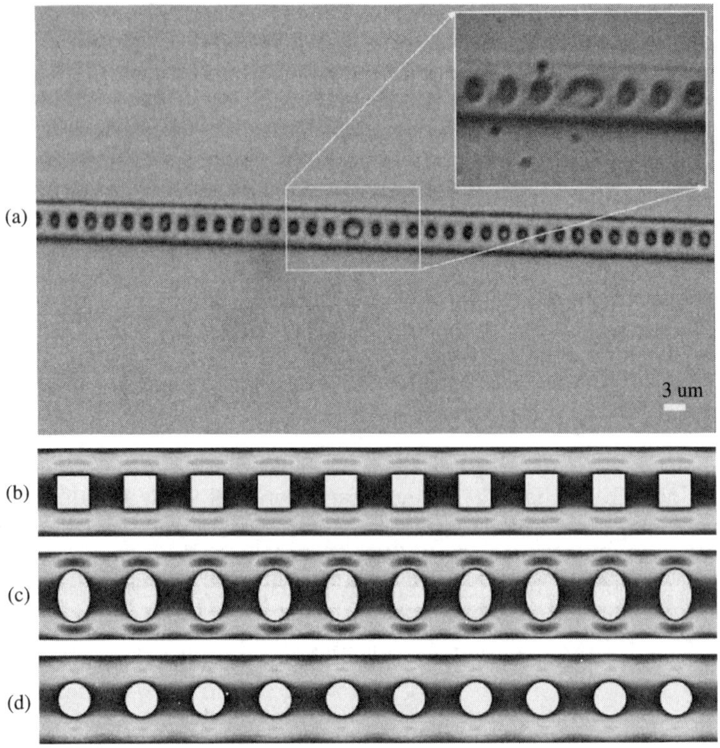

FIGURE 10-8 Panel (a) shows an optical micrograph of the DFB laser fabricated by Li and Psaltis [*Z. Li and D. Psaltis, "Optofluidic Distributed Feedback Dye Lasers," IEEE J. Top. Quant. Electron. 13(2), 185–193 (2007)*]. Panels (b) through (d) show two-dimensional laminar flow profiles calculated from Stokes' equation with the aid of a finite-element method. The three cases are for the same flow-rate and dark regions correspond to a vanishing flow velocity. (Also see color insert.)

10-8 Summary

Optofluidic dye lasers represent a conceptually simple and flexible approach for integration of single mode and frequency tunable laser light sources, which can span the entire range from ultraviolet over visible to near-infrared. The optofluidic dye laser devices are simply customized microfluidic components, which can be added to a lab-on-a-chip microsystem without additional process steps. The microfluidic platform implies both challenges and opportunities. Multiple, single-color light sources can easily be integrated on a chip, where the on-chip generated light is coupled directly into integrated waveguides. Although output power levels are inherently very low, a wide range of sensing applications can be envisaged, either by applying the generated light in integrated optics, or by using the on-chip, microfluidic laser as an intracavity sensor itself.

Among the major conceptual challenges discussed in the chapter are (a) the design and performance of high-quality optical resonators, which can be realized by patterning a thin dielectric film, (b) frequency tuning schemes for the miniaturized laser devices, and (c) strategies to overcome dye bleaching.

Optofluidic dye lasers represents an active research field and although optofluidic dye lasers have not yet been developed for true applications, their size, integration and functionality holds promise for applications within lab-on-a-chip technology.

From a more fundamental point of view, miniaturized lasers and optofluidic lasers in particular are interesting since they pose new challenges and physics not encountered in macroscopic laser realizations. In particular, low mode-volume high-Q resonators may dramatically enhance the feedback and consequently lower the optical threshold power where gain outbalances cavity losses. So far, Fabry–Perot, DFB and ring resonators have been applied to realize optofluidic dye lasers. Photonic crystals offer rich opportunities for further development of the field, exploiting band-edge lasing and other types of dispersion engineering. By pushing laser cavities to yet higher Q factors, the lasing threshold approaches zero asymptotically. This is often referred to as zero-threshold lasing. While the quest for zero-threshold lasing may seem somewhat academic we foresee that low-threshold lasing will find applications in sensing applications where a low-power pump source can be used to power a low-threshold laser cavity employed in an intracavity sensing setup where minute chemical changes will perturb the onset of lasing and/or shift the lasing wavelength.

References

1. S. Balslev, A. M. Jorgensen, B. Bilenberg, K. B. Mogensen, D. Snakenborg, O. Geschke, J. P. Kutter, and A. Kristensen, "Lab-on-a-chip with integrated optical transducers," *Lab Chip* 6(2), 213–217 (2006).
2. D. Psaltis, S. R. Quake, and C. H. Yang, "Developing optofluidic technology through the fusion of microfluidics and optics," *Nature* 442(7101), 381–386 (2006).
3. C. Monat, P. Domachuk, and B. J. Eggleton, "Integrated optofluidics: a new river of light," *Nat. Photon.* 1(2), 106–114 (2007).
4. Z. Li and D. Psaltis, "Optofluidic dye lasers," *Microfluid. Nanofluid.* 4(1–2), 145 (2008).
5. O. Svelto, *Principles of Lasers*, 4th ed. (Springer, Heidelberg, 1998).
6. F. P. Schäfer, ed., *Dye Lasers*, 3rd ed. (Springer, Berlin, 1990).
7. H. Bruus, *Theoretical Microfluidics*, (Oxford Master Series in Physics, Oxford, 2008).
8. B. Helbo, A. Kristensen, and A. Menon, "A micro-cavity fluidic dye laser," *J. Micromech. Microeng.* 13(2), 307–311 (2003).
9. J. C. Galas, C. Peroz, Q. Kou, and Y. Chen, "Microfluidic dye laser intracavity absorption," *Appl. Phys. Lett.* 89(22), 224,101 (2006).

10. Z. Li, Z. Zhang, T. Emery, A. Scherer, and D. Psaltis, "Single mode optofluidic distributed feedback dye laser," Opt. Express 14(2), 696–701 (2006).
11. M. Gersborg-Hansen and A. Kristensen, "Optofluidic third order distributed feedback dye laser," *Appl. Phys. Lett.* 89(10), 103,518 (2006).
12. M. Gersborg-Hansen and A. Kristensen, "Tunability of optofluidic distributed feedback dye lasers," *Opt. Express* 15(1), 137–142 (2007).
13. R. Daw and J. Finkelstein, "Lab on a chip," *Nature* 442(7101), 367–367 (2006).
14. N. A. Mortensen, S. Xiao, and J. Pedersen, "Liquid-infiltrated photonic crystals—enhanced light-matter interactions for lab-on-a-chip applications," *Microfluid. Nanofluid.* 4(1-2), 117 (2008).
15. Z. Li and D. Psaltis, "Optofluidic Distributed Feedback Dye Lasers," *IEEE J. Top. Quant. Electron.* 13(2), 185–193 (2007).
16. J. C. Galas, J. Torres, M. Belotti, Q. Kou, and Y. Chen, "Microfluidic tunable dye laser with integrated mixer and ring resonator," *Appl. Phys. Lett.* 86(26), 264,101 (2005).
17. D. Janasek, J. Franzke, and A. Manz, "Scaling and the design of miniaturized chemical-analysis systems," *Nature* 442(7101), 374–380 (2006).
18. M. Gersborg-Hansen, S. Balslev, N. A. Mortensen, and A. Kristensen, "Bleaching and diffusion dynamics in optofluidic dye lasers," *Appl. Phys. Lett.* 90(14), 143,501 (2007).
19. S. I. Shopova, H. Zhou, X. Fan, and P. Zhang, "Optofluidic ring resonator based dye laser," *Appl. Phys. Lett.* 90(22), 221,101 (2007).
20. S. X. Qian, J. B. Snow, H. M. Tzeng, and R. K. Chang, "Lasing droplets—highlighting the liquid-air interface by laser-emission," *Science* 231(4737), 486–488 (1986).
21. H. Azzouz, L. Alkhafadiji, S. Balslev, J. Johansson, N. A. Mortensen, S. Nilsson, and A. Kristensen, "Levitated droplet dye laser," *Opt. Express* 14(10), 4374–4379 (2006).
22. M. Hossein-Zadeh and K. J. Vahala, "Fiber-taper coupling to Whispering-Gallery modes of fluidic resonators embedded in a liquid medium," *Opt. Express* 14(22), 10,800–10,810 (2006).

CHAPTER 11

Optofluidic Microscope

Xiquan Cui and Changhuei Yang

Department of Electrical Engineering and Bioengineering, California Institute of Technology, Pasadena, California

11-1 Introduction

Optical microscopy pervades almost all aspects of modern bioscience researches and clinical procedures. However, the fundamental microscope design has undergone little change since its invention in the 1600s. A typical microscope still consists of an objective, space for relaying the image, and an eyepiece or an imaging lens to project a magnified image onto a person's retina or a camera. The focus of modern microscopy research and development has predominantly been on adding more imaging functionalities to the microscope. Through the efforts of researchers over the years, phase imaging ability, fluorescence imaging ability, and other sophisticated techniques have dramatically broadened the information-gathering capability of the microscope. Yet, with the development of higher-quality and broader-capability microscopes, the sophistication and price tag of microscopes have also steadily crept up in tandem. These microscope systems will likely remain important workhorses in the foreseeable future; yet, they are also rapidly becoming limiting factors in bioscience and clinical applications by reason of their relatively low throughput, high cost, and large space requirements [1]. The number of microscopes in a typical bioscience laboratory is strongly constrained by the cost and size. An increase in the number of microscopes per laboratory by a factor of hundreds or thousands, via a dramatic microscope cost and size reduction, will lead to significant efficiency enhancement. In addition, cheap and disposable microscopes that can fit easily on a person's fingertip can also dramatically improve the quality of clinical care by forming the imaging engine of cheap point-of-care analysis units and by cutting down on contamination risks (by being disposable).

The field of optofluidics offers us an opportunity to redesign the conventional microscopy from the ground up. The optofluidic microscope (OFM) developed by our group capitalizes on the ease of transporting cells and microorganisms via microfluidic flow and the high-quality optical sensor grid that are readily available at remarkably low cost [2].

This chapter is divided into four sections. In Sec. 11-2, we will introduce the OFM's operating principle. In Sec. 11-3, we will summarize the experimental implementation and evaluation of the first OFM prototypes. In Sec. 11-4, we will discuss some of the applications that the OFM is well suited to make an impact.

11-2 Operating Principle

The OFM principle is best explained by recounting the phenomenon that inspired the idea—the "floater" phenomenon that most of us occasionally observe when looking at a clear patch of sky. Floaters are caused by debris in the vitreous humor that drifts close to the retina. Under uniform illumination, they cast sharp shadows onto the retina and "appear in our perception." The clarity of floaters is a direct function of their proximity to the retina; the closer they are, the sharper the shadows cast. Despite the fact that floaters are tiny, we often see them with excellent detail. It is also interesting to note that our ability to see these tiny objects is not influenced by our eye glasses or the intrinsic lenses in our eyes (if in doubt, try putting on or off a pair of glasses the next time you see floaters). This observation points to the fact a direct projection imaging strategy (basis of the floater phenomenon) is capable of rendering high-resolution images as long as (1) we can place the target close to the sensor grid, and (2) the sensor grid pixels are small.

The direct projection imaging strategy has previously been used by other groups [3]; however, the quality of the images is less than satisfactory for microscopy applications as the image resolution is bounded by the size of the sensor pixel. Since the typical pixel size of a commercial CCD or CMOS sensor is larger than 3 μm (getting down to smaller pixel size is difficult from a semiconductor fabrication point of view), the resolution achievable is much poorer than the resolution achieved with a conventional microscope.

It is difficult to imagine that a single-time-point direct projection imaging strategy for collecting images at resolution better than the sensor pixel size exists. However, if we permit ourselves to exploit the time dimension during the image-acquisition process, it is possible to develop viable high-resolution direct projection imaging strategies in which resolution and sensor pixel size are independent. To begin, consider the following sensing platform—a sensor grid that is coated with a thin metal layer and that has a line of small apertures that are etched onto the metal layer. Each aperture should be situated at the center of each sensor pixel. The sensor pixel will then be sensitive only to light transmitted through the aperture. By placing a target object on top of the grid, we can then obtain a sparsely sampled image of the object (Fig. 11-1a). We can "fill in" the image by raster-scanning the object over the grid (or equivalently, raster-scanning the grid under the object) and compositing the time-varying transmissions through the apertures appropriately (Fig. 11-1b). We can see that in this case, the resolution is fundamentally determined by the aperture size and not the pixel size. Therefore, by choosing the appropriate aperture size, we can achieve high resolution.

This imaging strategy can be simplified by tilting the aperture grid slightly and replacing the raster-scan pattern with a single linear

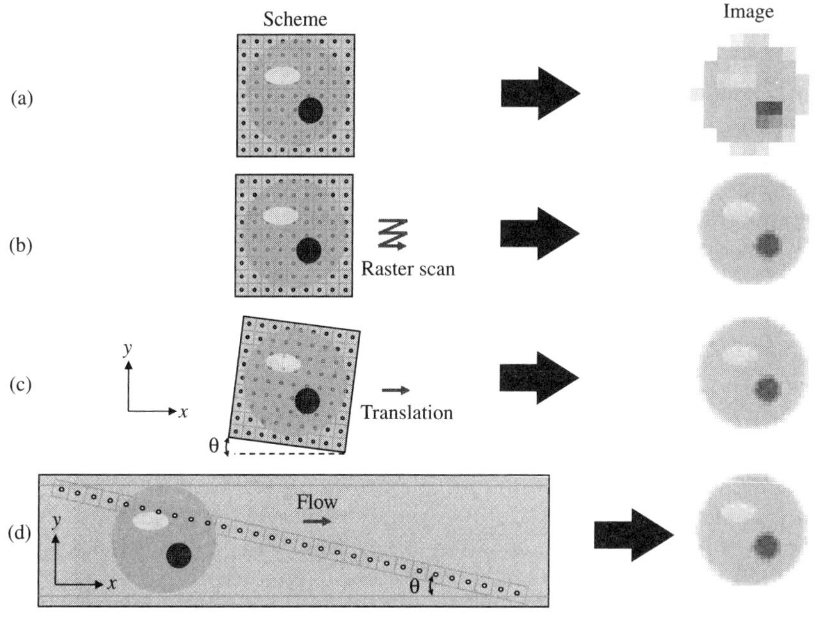

FIGURE 11-1 Comparison of direct projection imaging strategies. (a) By placing the specimen on a grid of apertures, we can obtain a sparsely sampled image of the object. (b) We can "fill in" the image by raster-scanning the object over the grid (or equivalently, raster-scanning the grid under the object) and compositing the time-varying transmissions through the apertures appropriately. (c) This imaging strategy can be simplified by tilting the aperture grid slightly and replacing the raster-scan pattern with a single linear translation of the object across the grid. (d) This design can be further simplified by replacing the tilted 2D aperture grid with a long tilted 1D aperture array. This scheme is the basis for the optofluidic microscopy method. (*X. Cui, L. M. Lee, X. Heng, W. Zhong, P. W. Sternberg, D. Psaltis, and C. Yang, "Lensless high-resolution on-chip optofluidic microscopes for Caenorhabditis elegans and cell imaging," Proceedings of the National Academy of Sciences of the United States of America, vol. 105 (31), pp. 10670–10675, 2008. Copyright (2008) National Academy of Sciences, USA.*)

translation of the object across the grid (Fig. 11-1c). As the object passes across each hole, the time-varying transmission represents a line scan across the object. By choosing the small angle between the grid orientation and the translation direction, we can ensure that the object is fully scanned by the apertures. This design can be further simplified by replacing the tilted 2D aperture grid with a long tilted 1D aperture array (Fig. 11-1d). This imaging strategy is the basis of the optofluidic microscopy (OFM) method. The OFM method shares a lot of similarities with near-field scanning optical microscopy methods. In fact, the OFM aperture array can be interpreted as a series of NSOM apertures. Whereas NSOM sensors are generally raster-scanned over the target objects, the OFM approach uses object translation to accomplish scanning. This is a significant advantage for objects that are suspended in fluids as we can apply microfluidic technology to implement flow controls in a compact and cost-effective fashion.

In terms of implementation, our current typical OFM prototype consists of a metal-coated sensor with apertures etched onto the metal layer as the base layer. The top layer consists of a transparent structure containing a carefully aligned microfluidic channel for sample delivery and scanning. An illumination source situated above the device completes the design. To perform imaging, we flow the targets through the channel and electronically acquire line scans. The image composition processing is minimal and simply involves compiling the line scans appropriately.

11-3 Prototype Evaluations

11-3-1 *Caenorhabditis elegans* Imaging

Our on-chip OFM prototype (Fig. 11-2a and 11-2b) utilizes the above-mentioned core design with one change—two parallel OFM arrays are implemented (Fig. 11-2c). We choose to use two parallel OFM arrays for two reasons. First, by measuring the time difference between when the target object first passes across each array, we can determine the flow speed of the object by dividing the distance separation between the arrays by that time difference. Knowledge of the speed is important for the correct computation of the delay and the correct matching of the collected line scans to generate OFM images. Second, significant differences in the two acquired images will indicate object shape changes, flow speed variations, and/or object rotations during the data-acquisition process. Accurate OFM imaging requires the absence of these variations, and therefore, discrepancy in the images is a good criterion for rejecting that image pair. In our experiments, we reject image pairs when their correlation is less than 50%. During our initial experiments, approximately 50% of the samples were rejected based on this criterion.

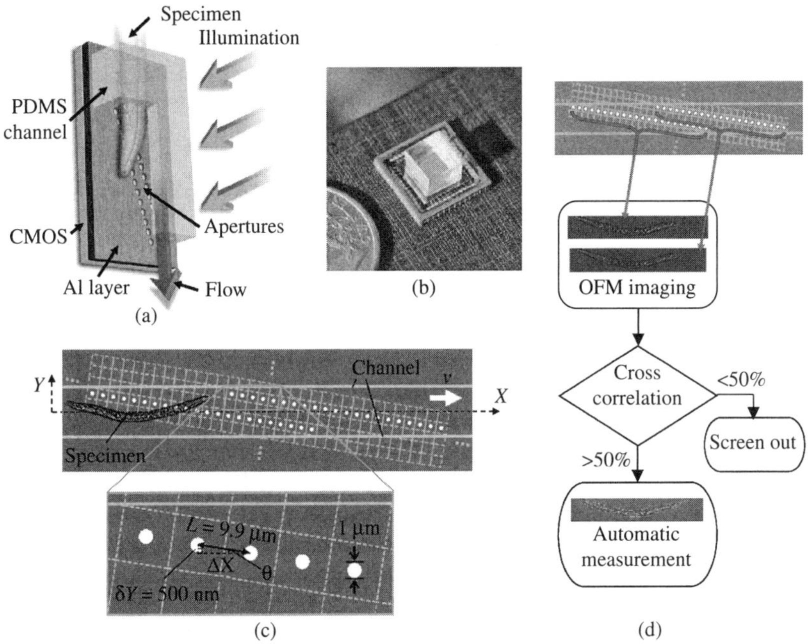

FIGURE 11-2 OFM prototype. (a) Upright operation mode. (b) The actual device compared with a U.S. quarter. (c) Schematic of the device (top view). The OFM apertures (circles) are defined on the Al (gray)-coated 2D CMOS image sensor (dashed grid) and span across the whole microfluidic channel (lines). (d) Flow diagram of the automated OFM operation. Two OFM images of the same *C. elegans* are acquired by the two OFM arrays, respectively (arrows). If the image correlation is greater then 50%, the image pair is rejected. Otherwise, the area and the length of the worms are measured automatically by evaluating the contour (dashed line) and the midline (dashed line). (*X. Cui, L. M. Lee, X. Heng, W. Zhong, P. W. Sternberg, D. Psaltis, and C. Yang, "Lensless high-resolution on-chip optofluidic microscopes for Caenorhabditis elegans and cell imaging," Proceedings of the National Academy of Sciences of the United States of America, vol. 105 (31), pp. 10670–10675, 2008. Copyright (2008) National Academy of Sciences, USA.*)

The OFM prototype was fabricated on a commercial CMOS imaging sensor (Micron MT9V403C12STM). The CMOS pixel size was 9.9 μm × 9.9 μm. The data-acquisition program was written to only read out signals from the two lines of apertures; this allowed us to increase the line readout rate (f) to 1000 frames per second for each OFM array. We planarized the surface of the sensor die with a layer of SU8 resin (2 μm thick), and then coated it with a 300-nm-thick layer of Al to mask the sensor from light. The two lines of apertures (1-μm diameter) were milled by a focused ion beam (FIB) machine (Nova 200, FEI Company). Each line consisted of approximately 100 apertures spaced apart by the width of a sensor pixel. We spin-coated a 0.2-μm-thick PMMA layer onto the Al film to protect the OFM apertures. Finally, we bonded an optically transparent polydimethylsiloxane (PDMS) microfluidic chip containing a channel (width = 50 μm, height = 15 μm) on top of the sensor chip. The system was illuminated with a halogen lamp (~20 mW/cm^2—approximately equal to the intensity of sunlight).

The microfluidic channel was designed with a smooth funnel at both ends. The channel was oriented at a small angle $\theta = 0.05$ rad with respect to the aperture arrays, which ensured that approximately 100 apertures in each row spanned the channel. Oxygen plasma was used to make the inner surface of the PDMS microfluidic channel hydrophilic. Prior to use, we additionally flushed the channel with a PEG solution (10% concentration) to reduce potential sample adhesion to the channel walls. We chose to operate the completed system in the upright mode (Fig. 11-2a), so that gravity can drive the flow and eliminate the need for bulky external pressure pumps. When the specimen solution (newly hatched *C. elegans* at a number concentration of ~20 per μL) was injected into the top funnel, the solution would wet the microfluidic channel and the specimens would be pulled continuously into the microfluidic channel by gravity. To prevent excessive nematode wiggle motions, we immobilized them by subjecting them to a 70°C heat bath for 3 min. The maximum observed throughput was approximately five worms per minute. However, the nematode flow speed v in the channel was fairly uniform and was approximately 500-μm/s. Imaging of each nematode required approximately 2.5 s.

The OFM sampling scheme effectively establishes a virtual sensing grid. Unlike the physical sensing grid in CCD and CMOS image sensors, the pixel density of the OFM virtual sensing grid can be adjusted by changing the number of apertures spanning the channel, the flow speed of the target objects, and the pixel readout rate. For our prototype, the grid spacing along the Y direction equals $\delta Y = L\sin\theta = 0.5$ μm, and the grid spacing along the X direction equals $\delta X = v/f = 0.5$ μm. We note that pixel density is distinct from system's resolution. In the case of the OFM, the pixel density is not limited by the aperture size. Higher pixel density is helpful as it allows us to oversample the object and prevent undesirable aliasing artifacts from appearing in the images.

Figure 11-3a shows a pair of OFM images acquired by the two OFM arrays from the same wild-type *C. elegans* L1 larva. The image correlation between them is 56%. Consistent internal structures are found in both OFM images. For comparison, Fig. 11-3b shows an image collected from a similar nematode that was placed directly onto an unprocessed CMOS sensor (note that the pixel size is 9.9 μm × 9.9 μm); the nematode was barely distinguishable in this poor-resolution direct projection image. Figure 11-3c shows a conventional microscope image of a similar larva acquired through a 20× Olympus objective lens (650-nm resolution for 555-nm wavelength under Sparrow's criterion) [4]. Similar internal structures of *C. elegans* appear in both the microscope and the OFM

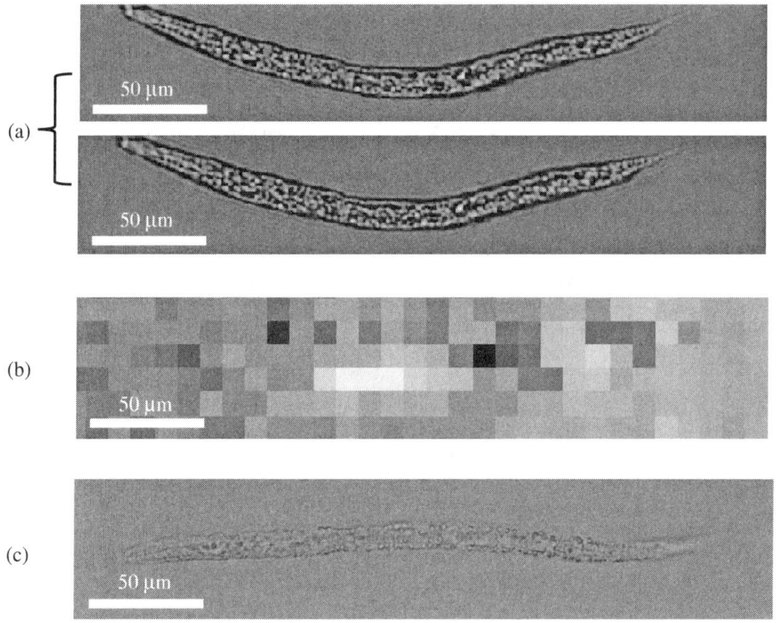

FIGURE 11-3 Images of wild-type *C. elegans* L1 larvae. (a) Duplicate OFM images acquired by the two OFM arrays for the same *C. elegans*. (b) Direct projection image on a CMOS sensor with 9.9-µm pixel size. (c) Conventional microscope image acquired with a 20× objective. (*X. Cui, L. M. Lee, X. Heng, W. Zhong, P. W. Sternberg, D. Psaltis, and C. Yang, "Lensless high-resolution on-chip optofluidic microscopes for Caenorhabditis elegans and cell imaging," Proceedings of the National Academy of Sciences of the United States of America, vol. 105 (31), pp. 10670–10675, 2008. Copyright (2008) National Academy of Sciences, USA.*)

images. It confirms that the OFM can render images with comparable quality to a conventional microscope of similar resolution.

For the OFM, the plane closest to the aperture array is rendered with the best resolution. The resolution degrades as the plane moves away from the apertures. The point spread function (PSF) associated with each OFM aperture is a good measure for characterizing the OFM. We characterized the PSF of our prototype by laterally scanning a NSOM (Alpha-SNOM, WITec Gmbh) tip across an aperture on the prototype at various heights (H) from the aperture and measuring the optical signal detected by the underlying CMOS sensor pixel (Fig. 11-4a). The NSOM tip was less than 100 nm in diameter and can be approximated as a point source. The inset of Fig. 11-4b shows representative OFM PSF plots at $H = 0.1$, 1.5, and 2.5 µm. The width of the PSF broadens as a function of point source height. We can quantify the height-dependent resolution of our prototype by the PSF's width. Figure 11-4b shows the resolution (Sparrow's criterion) [5] as a function of H. From the plot,

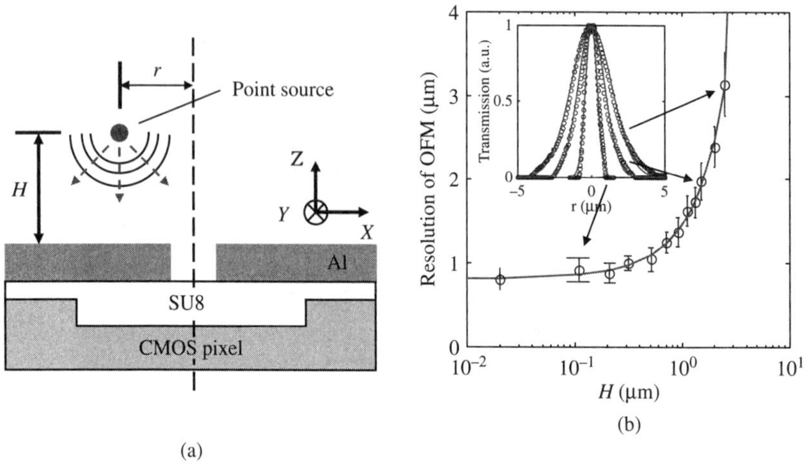

FIGURE 11-4 Resolution of the OFM prototype. (a) Schematic of the PSF measurement. (b) Resolution of the prototype at various heights H above a 1-μm-diameter aperture under Sparrow's criterion. (Inset) Representative OFM PSF plots at $H = 0.1$, 1.5, and 2.5 μm for the 1-μm-diameter aperture. (*X. Cui, L. M. Lee, X. Heng, W. Zhong, P. W. Sternberg, D. Psaltis, and C. Yang, "Lensless high-resolution on-chip optofluidic microscopes for Caenorhabditis elegans and cell imaging," Proceedings of the National Academy of Sciences of the United States of America, vol. 105 (31), pp. 10670–10675, 2008. Copyright (2008) National Academy of Sciences, USA.*)

we can see that the resolution limit of our prototype was 0.9 μm and the resolution degraded to 3 μm at $H = 2.5$ μm. We additionally find that the degradation of resolution with respect to H is approximately linear. The result is consistent with the findings in our more detailed study on aperture-based imaging systems in Ref. 6.

As an illustration of the OFM's utility for bioscience applications, we further implemented a demonstration experiment where we used the OFM prototype to image and analyze the body size phenotypes of *C. elegans* L1 larvae from three different genotypes—wild-type, *sma-3*, and *dpy-7*. The two mutant strains are expected to differ in size from the wild-type strain. This study is motivated by the extensive use of phenotype characterization, especially morphology, in the genetic studies of *C. elegans*, other microorganisms, and cells. Quantitative analysis of phenotype characteristics can also provide insights in the study of the effects of environmental stress, nutrition and drugs on growth, reproduction, and life span of *C. elegans*. Often, the variation in physical dimensions amongst individual animals within the same strain can mask the subtle phenotype difference between different strains. As such, to perform meaningful phenotype characterization, measurements on a large sample size of nematode are generally needed in order to quantify the mean

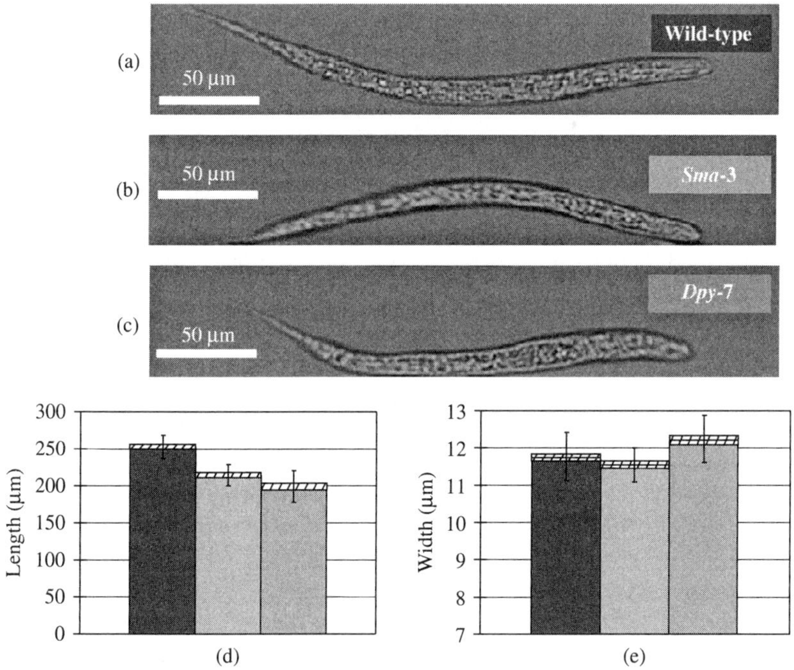

FIGURE 11-5 Phenotype characterization of *C. elegans* L1 larvae (a through c) Typical OFM images of wild-type, *sma-3*, and *dpy-7* worms, respectively. (d and e) The length (d) and effective width (e) of wild-type, *sma-3*, and *dpy-7* worms, respectively. The columns represent the mean values in the population; the hatched areas correspond to the confidence intervals of the mean values; and the error bars are the standard deviations indicating the variation between individuals in the population. Twenty-five worms were evaluated for each phenotype. (*X. Cui, L. M. Lee, X. Heng, W. Zhong, P. W. Sternberg, D. Psaltis, and C. Yang, "Lensless high-resolution on-chip optofluidic microscopes for Caenorhabditis elegans and cell imaging," Proceedings of the National Academy of Sciences of the United States of America, vol. 105 (31), pp. 10670–10675, 2008. Copyright (2008) National Academy of Sciences, USA.*)

length (or any other phenotype characteristics) with a sufficiently narrow confidence interval. Cheap, automated and compact microscopes can significantly improve this otherwise labor-intensive process.

To compare body sizes of the three *C. elegans* strains, we imaged 25 animals of each strain. Figure 11-5a through 11-5c shows typical OFM images of wild-type, *sma-3,* and *dpy-7* worms, respectively. The images show that the *sma-3* worm is smaller and thinner than the wild-type worm, and the *dpy-7* worm is fatter and shorter than the wild-type worm. These observations are consistent with observations made with a conventional microscope.

Since OFM images are naturally digitalized, we can perform large volume and automatic quantitative information extraction by computer assisted postprocessing. We developed a MATLAB program to trace the contour of the *C. elegans* and determine the area and length of the *C. elegans* in batches (Fig. 11-2d). From those two quantities, we then computed an effective width for each nematode by dividing the area by the length. In Fig. 11-5d and 11-5e, the columns represent the mean length and width of the three *C. elegans* strains; the hatched areas correspond to the confidence intervals of our mean length and width estimates. The standard deviations (error bars) of the measurement indicate the variation between individuals within the strain.

11-3-2 Cell Imaging

The imaging of cells with the OFM method requires a different flow mechanism. This is because a pressure-based microfluidic flow has a parabolic velocity profile (Poiseuille flow) that arises from the nonslip boundary condition on the channel side-walls. Objects flowing under such condition tend to rotate and tumble due to the torque they receive from the nonuniform fluid push. While *C. elegans* simply do not have the space to rotate in the microfluidic channel, ellipsoidal/spherical cells do not have such constraints.

Fortunately, we found that the use of dc (direct current) electrokinetics provides a simple and direct way to control the motion of biological cells in the on-chip OFM system as to suppress rotation and to allow a constant translational motion in the microfluidic channel. This method is simple to implement—apply an electric field along the channel by introducing a potential difference between the two ends of the microfluidic channel. We typically apply approximately 25-V difference along a 3-mm-long channel. By varying the potential difference, we can easily alter the speed of the objects.

There are three mechanisms involved. First, the electric field causes the translation of the electric double layer at the channel walls (electrosmosis). This in turn drags the entire fluid column uniformly through the channel. Second, a cell would typically carry a net electric charge and the interaction of this charge with the electric field will likewise actuate the cells (electrophoresis). Third, the electric field will induce a dipole moment on a cell. Alternately, the heterogeneous distribution of electric charge on a cell can also create a natural dipole. The interaction of the dipole with the electric field will cause the cell to orientate itself in the channel to minimize the associated electric potential energy (electro-orientation).

Using this method, we were able to control cell motions well and achieve good-quality OFM imaging of cells. Figure 11-6 shows comparison images acquired by the OFM and a conventional microscope.

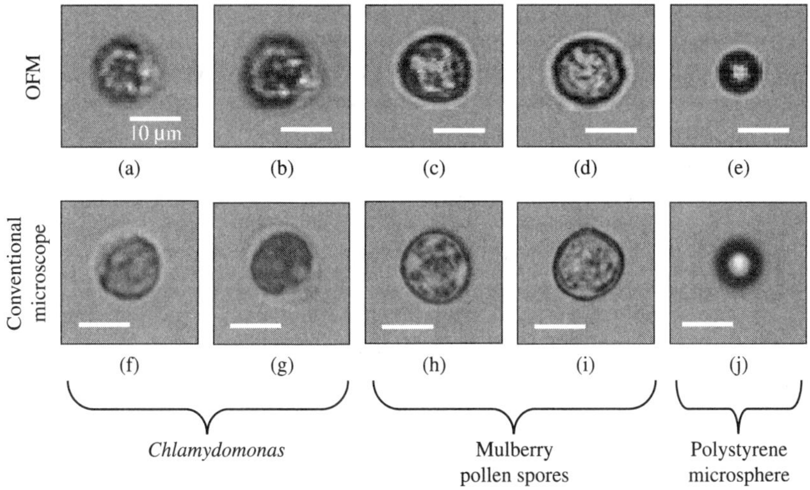

FIGURE 11-6 Cell and microsphere images. (a through e) Images taken from the on-chip OFM driven by dc electrokinetics of *Chlamydomonas* (a and b), mulberry pollen (c and d), and a 10-µm polystyrene microsphere (e). (f through j) Images taken from a conventional light-transmission microscope with a 20× objective of *Chlamydomonas* (f and g), mulberry pollen (h and i), and a 10-µm polystyrene microsphere (j). (Scale bars: 10 µm.) (*X. Cui, L. M. Lee, X. Heng, W. Zhong, P. W. Sternberg, D. Psaltis, and C. Yang, "Lensless high-resolution on-chip optofluidic microscopes for Caenorhabditis elegans and cell imaging," Proceedings of the National Academy of Sciences of the United States of America, vol. 105 (31), pp. 10670–10675, 2008. Copyright (2008) National Academy of Sciences, USA.*)

11-4 Potential Applications

The OFM design is very compact and easily implemented via semiconductor fabrication. It is possible to create tens or hundreds of such devices on a single chip. This advantage is particularly significant in the context of high-throughput bioscience imaging and analysis where large numbers of objects need to be examined, for example, screening and harvesting stem cells or circulating tumor cells from a solution of other cells. We believe that, with the OFM method, we can now begin to consider furnishing bioscientists and clinicians with large numbers of microscopes that they can operate in parallel.

Another application that the OFM may be useful for is blood parasite analysis in developing countries. Malaria and sleeping sickness are two such diseases that the OFM usage can have a significant impact on. For example, the gold standard for malaria diagnosis remains the examination of a blood smear through a high-magnification microscope. A low-cost, compact and rugged (no lenses to break) OFM-based analysis system (the size of an iPod) that uses sunlight as the illumination source can replace the conventional microscope in such applications.

Yet another potential application is the use of the OFM for image-based flow cytometry. Image-based flow cytometry for white blood cell typing and counting can potentially complement existing commercial cytometer units by providing additional cell characterizations for identification purposes. Specifically, the indirect characterization of cell types by conventional flow cytometers by two parameters (forward and side scattering) is intrinsically less accurate than a histopathology analysis where the cells are imaged and distinguished via morphology. These systems are also susceptible to artifact errors from platelet aggregation and nucleated red blood cells. Finally, the relative size and maturity of specific white blood cell populations, which are not measured by these systems, are important parameters for the detection of certain diseases, such as leukemia. Due to its imaging nature, we do not expect the OFM to ever achieve flow cytometer's throughput. We see the methods as complementary. The OFM can provide an accurate differentiation with samples that flow cytometry has difficulty with.

References

1. M. Oheim, "High-throughput microscopy must re-invent the microscope rather than speed up its functions," *British Journal of Pharmacology,* vol. **152**, p. 1, 2007.
2. X. Cui, L. M. Lee, X. Heng, W. Zhong, P. W. Sternberg, D. Psaltis, and C. Yang, "Lensless high-resolution on-chip optofluidic microscopes for Caenorhabditis elegans and cell imaging," *Proceedings of the National Academy of Sciences of the United States of America,* vol. **105**, pp. 10670– 10675, 2008.
3. D. Lange, C. W. Storment, C. A. Conley, and G. T. A. Kovacs, "A microfluidic shadow imaging system for the study of the nematode Caenorhabditis elegans in space," *Sensors and Actuators B-Chemical,* vol. **107**, pp. 904–914, 2005.
4. Airy Patterns and Resolution Criteria, Olympus Inc., http://www.olympusconfocal.com/java/resolution3d/index.html.
5. A. W. Jones and J. Bland-Hawthorn, "Towards a general definition for spectroscopic resolution," *ASP Conference Series,* vol. **77**, pp. 503–507, 1995.
6. X. Heng, X. Cui, D. W. Knapp, J. Wu, Z. Yaqoob, E. J. McDowell, D. Psaltis, and C. Yang, "Characterization of light collection through a subwavelength aperture from a point source," *Optics Express,* vol. **14**, pp. 10410–10425, 2006.

CHAPTER 12
Optofluidic Resonators

Dominik G. Rabus
Baskin School of Engineering, University of California, Santa Cruz

Optofluidic resonators are a new class of devices that have emerged recently with the implementation of resonant optical structures in fluidic channels. There are several reasons why this resonator-fluidic merger has gained increasing resonance, the first one being the focus on sensors and especially biosensors where it is inevitable to analyze substances in liquids. The other reason is the advent of optofluidic light sources where integrated resonators are needed to provide feedback and thus enable higher-output powers as it is known from conventional integrated light sources.

12-1 Optofluidic Resonators

This chapter is dedicated to provide an overview on optofluidic resonators, especially photonic crystal, Bragg grating, ring resonators, and Fabry-Perot resonators. Suitable fabrication techniques will be briefly explained and highlighted. Examples on demonstrated resonator devices will conclude this chapter.

12-1-1 Resonators

Optofluidic resonators realized so far in different material systems are, for example, photonic crystals [1], Bragg gratings [2], ring resonators [3], and Fabry-Perot resonators [4]. In order to design and fabricate devices, the basic theories of selected resonator structures are given in this section.

Photonic crystals have emerged recently and have found numerous applications like the use of optofluidic photonic crystal fiber. The potential of photonic crystals was first realized in 1987 by Eli Yablonovitch.

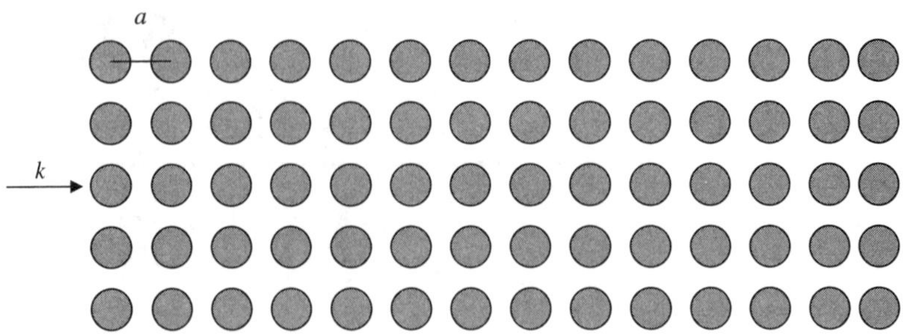

FIGURE 12-1 Example of a photonic crystal structure.

The basic principle behind photonic crystals is a regular, defined pattern of structures as shown in Fig. 12-1, to form a so-called crystal structure. The structures in our example pillars are separated by the distance *a*. A wavefront in the form of a planewave with vector *k* propagates through the crystal:

$$\vec{E}, \vec{H} \sim e^{i(\vec{k}\vec{x}-\omega t)} \tag{12-1}$$

$$\left|\vec{k}\right| = \frac{\omega}{c} = \frac{2\pi}{\lambda} \tag{12-2}$$

Beams propagate through the photonic crystal for most wavelengths without scattering as scattering cancels coherently. Only some wavelengths, that are a multiple of twice the pillar distance *a*, will not be able to pass through the photonic crystal. This bandwith is defined as the photonic bandgap. Numerous examples exist in literature that explain in detail the behavior of optical waves propagating through a photonic crystal device; therefore, this section is limited to the basic explanation.

The other important resonator suitable for integration in optofluidic devices is the Bragg grating. Bragg gratings are not only used in optical fibers, but also used especially as laser resonators. A Bragg grating is realized by a periodic or aperiodic perturbation of the effective refractive index of a wave guiding layer. This perturbation is periodic over a certain length which depends on the type of grating to be fabricated. The period is of the order of hundreds of nanometers. This leads to the reflection of light for a specific bandwidth of wavelengths. The reflected wavelengths satisfy the so-called Bragg condition. The lasing wavelength of a Bragg grating based laser for example is given by

$$m\lambda_m = 2n_{\text{eff}}\Lambda \tag{12-3}$$

where λ_m = *m*th-order resonant wavelength
n_{eff} = effective index of the guided mode
Λ = grating period

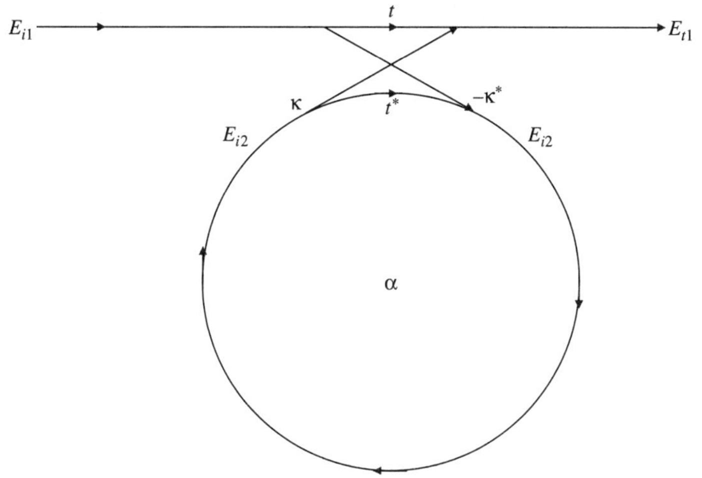

FIGURE 12-2 Model of a single ring resonator with one waveguide.

One of the emerging and promising new optofluidic devices are ring resonator configurations. The basic configuration (Fig. 12-2), which consists of unidirectional coupling between a ring resonator with radius r and a waveguide, is described briefly in the following paragraph, which is adapted from Ref. 5.

Defining that a single unidirectional mode of the resonator is excited, the coupling is lossless, single polarization is considered, none of the waveguide segments and coupler elements couple waves of different polarization, and the various kinds of losses occurring along the propagation of light in the ring resonator filter are incorporated in the attenuation constant, the interaction can be described by the matrix relation:

$$\begin{pmatrix} E_{t1} \\ E_{t2} \end{pmatrix} = \begin{pmatrix} t & \kappa \\ -\kappa^* & t^* \end{pmatrix} \begin{pmatrix} E_{i1} \\ E_{i2} \end{pmatrix} \tag{12-4}$$

The complex mode amplitudes E are normalized, so that their squared magnitude corresponds to the modal power. The coupler parameters t and κ depend on the specific coupling mechanism used. The $*$ denotes the conjugated complex value of t and κ, respectively.

The matrix is symmetric because the networks under consideration are reciprocal. Therefore,

$$\left|\kappa^2\right| + \left|t^2\right| = 1 \tag{12-5}$$

In order to further simplify the model, E_{i1} is chosen to be equal to 1. Then the round trip in the ring is given by

$$E_{i2} = \alpha \cdot e^{j\theta} E_{t2} \tag{12-6}$$

where α is the loss coefficient of the ring (zero loss: $\alpha = 1$) and $\theta = \omega L/c$, L being the circumference of the ring, which is given by $L = 2\pi r$, r being the radius of the ring measured from the center of the ring to the center of the waveguide, c the phase velocity of the ring mode ($c = c_0/n_{\text{eff}}$), and the fixed angular frequency $\omega = kc_0$; c_0 refers to the vacuum speed of light. The vacuum wavenumber k is related to the wavelength λ through $k = 2\pi/\lambda$. Using the vacuum wavenumber, the effective refractive index n_{eff} can be introduced easily into the ring coupling relations by

$$\beta = k \cdot n_{\text{eff}} = \frac{2\pi \cdot n_{\text{eff}}}{\lambda} \tag{12-7}$$

where β is the propagation constant. This leads to

$$\theta = \frac{\omega L}{c} = \frac{kc_0 L}{c} = k \cdot n_{\text{eff}} \cdot 2\pi r = \frac{2\pi \cdot n_{\text{eff}} \cdot 2\pi r}{\lambda} = 4\pi^2 n_{\text{eff}} \frac{r}{\lambda} \tag{12-8}$$

From Eqs. (12-4) and (12-6) we obtain

$$E_{t1} = \frac{-\alpha + t \cdot e^{-j\theta}}{-\alpha t^* + e^{-j\theta}} \tag{12-9}$$

$$E_{i2} = \frac{-\alpha \kappa^*}{-\alpha t^* + e^{-j\theta}} \tag{12-10}$$

$$E_{t2} = \frac{-\kappa^*}{1 - \alpha t^* e^{j\theta}} \tag{12-11}$$

This leads to the transmission power P_{t1} in the output waveguide, which is

$$P_{t1} = \left|E_{t1}\right|^2 \frac{\alpha^2 + |t|^2 - 2\alpha|t|\cos(\theta + \varphi_t)}{1 + \alpha^2|t|^2 - 2\alpha|t|\cos(\theta + \varphi_t)} \tag{12-12}$$

where $t = |t|\exp(j\varphi_t)$, $|t|$ representing the coupling losses and φ_t the phase of the coupler.

The circulating power P_{i2} in the ring is given by

$$P_{i2} = |E_{i2}|^2 = \frac{\alpha^2(1 - |t|^2)}{1 + \alpha^2\,|t|^2 - 2\alpha\,|t|\cos(\theta + \varphi_t)} \tag{12-13}$$

On resonance, $(\theta + \varphi_t) = 2\pi m$, where m is an integer, the following is obtained:

$$P_{t1} = |E_{t1}|^2 = \frac{(\alpha^2 - |t|)^2}{(1 - \alpha\,|t|)^2} \tag{12-14}$$

and

$$P_{i2} = |E_{i2}|^2 = \frac{\alpha^2(1-|t|^2)}{(1+\alpha|t|)^2} \tag{12-15}$$

A special case happens when $\alpha = |t|$ in Eqs. (12-14), when the internal losses are equal to the coupling losses. The transmitted power becomes zero. This is known in literature as critical coupling, which is due to destructive interference.

In using the Eqs. (12-4) and (12-15), it is possible to get a good idea of the behavior of a simplified basic ring resonator filter configuration consisting of only one waveguide and one ring.

Similar to the aforementioned ring resonator is the Fabry-Perot resonator, which is described in the following section briefly. The Fabry-Perot resonator consists of two parallel reflecting surfaces. If a light wave hits one of these reflecting surfaces, new light waves are generated at this specific surface (see Fig. 12-3)—one reflecting wave and one transmitting wave. The phase difference of these two light waves differs depending on the optical path length and the way reflection occurred.

If we consider an incident light wave with amplitude E_0 representing the direction of the inserted light into the resonator, then θ is the entrance angle of the light waves that are reflected in the resonator. The incident light wave has the vacuum wavelength λ_0 and the effective refractive index between the plates is n. For simplification, the electric field vector is considered to be linearly polarized with respect to the vertical and parallel incident planes. In order to describe the mathematical behavior of the light waves, the parameters of Fig. 12-3 are used. The reflection and transmission coefficients of the incident wave traveling from left to right will be defined as positive

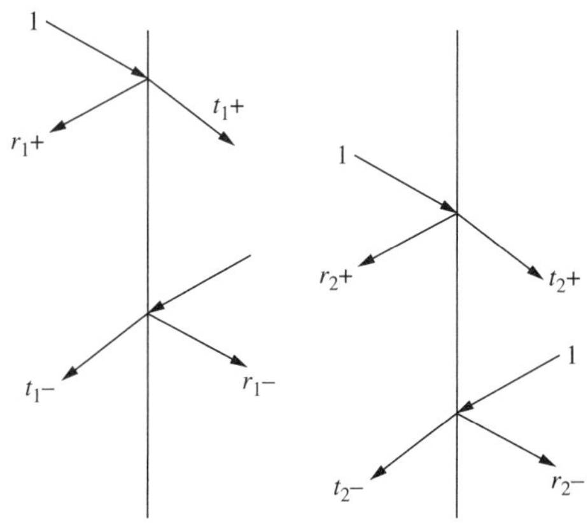

FIGURE 12-3 Fabry-Perot resonator transmission of light waves.

waves, and those traveling from right to left are defined as negative waves. The coefficients are complex numbers. The back-and-forth traveling waves in the resonator generate a phase difference that is calculated to be

$$\varphi = \frac{2\pi(2nd\cos(\theta))}{\lambda_0} \tag{12-16}$$

Using the principle of superposition, the amplitude for a wave traveling from left to right through the resonator after m passes is given by

$$\begin{aligned} E_t(m) &= t_1^+ t_2^+ \left\{1 + r_1^- r_2^+ e^{i\varphi} + \cdots + \left(r_1^- r_2^+\right)^{m-1} e^{i(m-1)\varphi}\right\} \\ &= \frac{t_1^+ t_2^+ \left[1 - \left(r_1^- r_2^+\right)^m e^{im\varphi}\right]}{1 - r_1^- r_2^+ e^{i\varphi}} \end{aligned} \tag{12-17}$$

For an infinite number of reflections $m \to \infty$ and $r_1^- r_2^+ < 1$:

$$E_t \to E_t(\infty) = \frac{t_1^+ t_2^+}{1 - r_1^- r_2^+ e^{i\varphi}} \tag{12-18}$$

The resulting transmitted intensity is given by

$$I_t = E_t E_t^* = \frac{\left|t_1^+ t_2^+\right|^2}{1 + \left|r_1^- r_2^+\right|^2 - 2\left|r_1^- r_2^+\right| \cos\psi} \tag{12-19}$$

with $\Psi = \varphi + \varepsilon$; ε is a correction factor for the phase difference occurring during the reflection.

$$\varepsilon = \arg r_1^- + \arg r_2^+ \tag{12-20}$$

If the surfaces of the Fabry-Perot resonator are made out of the same dielectric layers, the coefficients r and t can be considered as being real numbers. Then for a single reflecting surface:

$$t^+ t^- = T;\ r^+ = -r^-;\ (r^+)^2 = (r^-)^2 = R;\ R + T = 1 \tag{12-21}$$

R and T are coefficents for the intensity of the reflection and the transmission of the surface. Using Eq. (12-19) in Eq. (12-21) and $\varepsilon = 0$ and $t_2^+ = t_1^-; r_2^+ = r_1^-$:

$$I_T = \frac{T^2}{1+R^2-2R\cos\varphi} = \frac{T^2}{(1-R)^2+4R\sin^2\left(\frac{\varphi}{2}\right)}$$

$$= \frac{T^2}{(1-R)^2}\left(\frac{1}{1+\left[\frac{4R}{(1-R)^2}\right]\sin^2\left(\frac{\varphi}{2}\right)}\right)$$

$$= \left[\frac{T}{1-R}\right]^2\left[1+K\sin^2\left(\frac{\varphi}{2}\right)\right]^{-1} = \left[\frac{T}{1-R}\right]^2 \mathrm{A}(\varphi) \qquad (12\text{-}22)$$

With $K = 4R/(1-R)^2$, $\mathrm{A}(\varphi)$ is defined as the Airy function. The amplitude of the resulting electric field vector of the back-reflected light waves $E_r(m)$ is obtained using again the principle of superposition for m reflected waves:

$$E_r(m) = r_1^+ + t_1^+t_1^-r_2^+e^{i\varphi}\left\{1+r_1^-r_2^+e^{i\varphi}+\cdots+\left(r_1^-r_2^+\right)^{m-2}e^{i(m-2)\varphi}\right\}$$

$$= r_1^+ + \frac{t_1^+t_1^-r_2^+e^{i\varphi}\left(1-r_1^-r_2^+e^{i(m-1)\varphi}\right)}{1-r_1^-r_2^+e^{i\varphi}} \qquad (12\text{-}23)$$

For an infinite number of reflections $m \to \infty$

$$E_r \to E_r(\infty) = r_1^+ + \frac{t_1^+t_1^-r_2^+e^{i\varphi}}{1-r_1^-r_2^+e^{i\varphi}} \qquad (12\text{-}24)$$

Considering two identical dielectric surfaces, using Eqs. (12-21) and (12-24):

$$E_r = r_1^+ - \frac{t_1^+t_1^-r_1^+e^{i\varphi}}{1-r_1^{+2}e^{i\varphi}} = \sqrt{R}\,\frac{(1-\mathrm{R}e^{i\varphi}-Te^{i\varphi})}{1-\mathrm{R}e^{i\varphi}} = \sqrt{R}\,\frac{1-e^{i\varphi}}{1-\mathrm{R}e^{i\varphi}} \qquad (12\text{-}25)$$

which leads to

$$I_R = E_rE_r^* = R\frac{(2-2\cos\varphi)}{1+R^2-2R\cos\varphi} = \frac{4R\sin^2\left(\frac{\varphi}{2}\right)}{(1-R)^2+4R\sin^2\left(\frac{\varphi}{2}\right)}$$

$$= K\sin^2\left(\frac{\varphi}{2}\right)\left[1+K\sin^2\left(\frac{\varphi}{2}\right)\right]^{-1} \qquad (12\text{-}26)$$

If there is no absorption and no scattering of the light waves at the two reflecting surfaces, the sum of the intensities of the transmitted and reflected light must be equal to 1. Using Eqs. (12-22) and (12-26):

$$I_T + I_R = \frac{\dfrac{T^2}{(1-R)^2} + K\sin^2\left(\dfrac{\varphi}{2}\right)}{1 + K\sin^2\left(\dfrac{\varphi}{2}\right)} \xrightarrow{R+T=1} 1 \text{ QED} \tag{12-27}$$

Other characteristics describing a Fabry-Perot resonator are similar to a ring resonator. The free spectral range (FSR) is given by

$$\text{FSR} = \frac{\lambda_m}{m+1} = \frac{\lambda^2}{2nd} \qquad \lambda_m = \frac{2nd}{m} \tag{12-28}$$

The FWHM is defined as the full width at half maximum and is the same as in the case of the ring resonator. The quality factor is given in terms of the finesse (F). The finesse of a resonator gives information about the quality of the reflecting surfaces and the spectral resolution of a Fabry-Perot resonator. The finesse is given by

$$F = \frac{\text{FSR}}{\text{FWHM}} \tag{12-29}$$

The reflection finesse is given by

$$F_R = \frac{\pi\sqrt{R}}{1-R} = \left(\frac{\pi}{2}\right)\sqrt{K} \tag{12-30}$$

In the ideal case, the finesse is identical with the refection finesse. In practical cases another finesse is present, the so-called surface finesse F_S. The relation between all of them is given by

$$\frac{1}{F^2} = \frac{1}{F_R^2} + \frac{1}{F_S^2} \tag{12-31}$$

The maximum intensity of a single peak is in the ideal case equal to the intensity of the incident wave I_0. Due to absorption (A) and/or scattering, this intensity will be weakened and is given by

$$I_{\max} = I_0\left[1 - \frac{A}{A+T}\right]^2 = I_0\frac{T^2}{(1-R)^2} \text{ due to the fact that } R+T+A=1 \tag{12-32}$$

The minimum intensity is given by

$$I_{min} = I_0 \frac{\left[1 - \frac{A}{A+T}\right]^2}{1+K} = I_0 \frac{\left[1 - \frac{A}{A+T}\right]^2}{\left[\frac{(1+R)^2}{(1-R)^2}\right]} = I_0 \frac{T^2}{(1+R)^2} \tag{12-33}$$

The achievable contrast between maximum and minimum intensity inside the Fabry-Perot resonator is given by

$$\frac{I_{max}}{I_{min}} = \frac{(1+R)^2}{(1-R)^2} \tag{12-34}$$

The contrast depends on the quality of the surfaces. The theoretical contrast would be infinite at $R = 1$. Figure 12-4 shows the transmission spectrum for different reflectivity.

These are the basic equations for describing a Fabry-Perot resonator. In optofluidic devices the focus lies on detuning the wavelength of a resonator. There are three possibilities in doing so: either n, θ, or d needs to be changed. Optofluidic devices preferably change the refractive index inside the resonator.

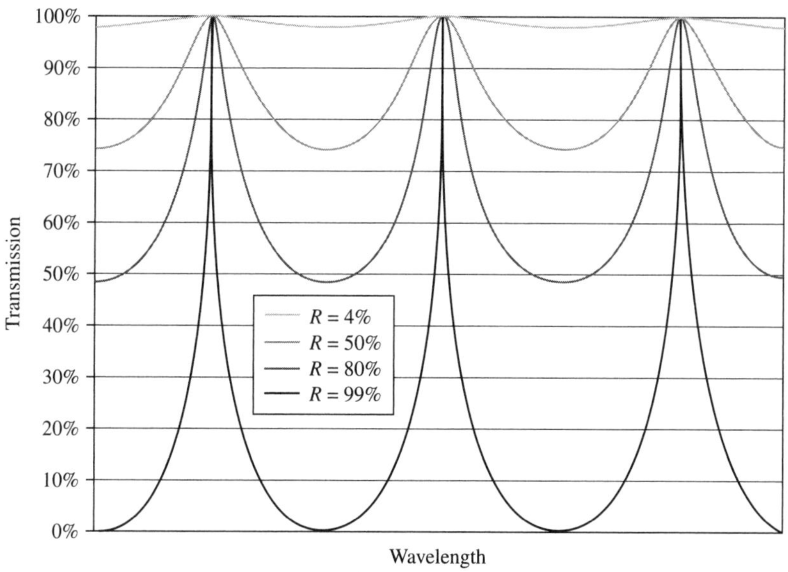

FIGURE 12-4 Characteristic transmission of a Fabry-Perot resonator with different facet reflectivity.

12-1-2 Fabrication Methods

The emergence of optofluidic devices is largely enabled by the recent advances in microfabrication, microfluidics, and polymer processing technologies [6,7]. The methods of choice are micromachining, soft lithography, and embossing techniques, which enable the fabrication of micron-scale fluidic channels in silicon, glass, polymer, and elastomer materials.

Polymers have been accepted as the material of choice for the integration of photonic integrated circuits and fluidic devices, mainly due to their increasing performance, rapid processibility, capability for precise tailoring of their optical properties, and their comparatively low cost. Another important aspect is the biocompatibility of polymer materials and the fact that these materials are already in use in many bio- and nonbio laboratories, which increases the acceptance of polymer based optofluidic devices. This advantage requires the improvement of fabrication technologies as well as the development of application-specific tailored materials. As stated before, polymer optical waveguides have been fabricated by various techniques, such as dry etching, UV curing, and soft lithography replica-molding, and embossing.

In recent years, hot embossing of microcomponents has become a routinely used replication technology for thermoplastic polymers. Low flow rates and slow molding speeds ensure that even the smallest details in the nanometer range are replicated perfectly. Hot embossing is particularly suited for structuring planar plates and foils, as only a small amount of plastic has to be molded. In contrast to injection-molding, the polymer flows a very short distance from the foil into the microstructure during hot embossing. As a result, very little stress is induced into the polymer and the molded parts are well suited to optical applications, such as waveguides and lenses [8]. The setup of the hot embossing machine is relatively simple. Setup times are short as the mold-insert and the polymer are easily exchanged. Nickel shims of only a few hundred micrometers can be used for replication without major effort. The electroplating process for such shims takes much less time than for more compact tools, as the electroplating time increases linearly with shim thickness. Therefore, tools can be manufactured from an existing photomask design within several days. A photograph of a nickel tool and an embossed substrate is shown in Fig. 12-5.

In order to integrate optics and fluidics, it would be advantageous to use a similar technology to create an optofluidic device. The deep UV technique [9] is one method of combining optics, which includes waveguides and light sources [10] and fluidics.

Two types of polymers have been investigated: PMMA (Hesa@ Glas, a homopolymer from Notz-Plastic, Switzerland) and alicyclic methacrylate copolymers which were obtained from Hitachi Chemical

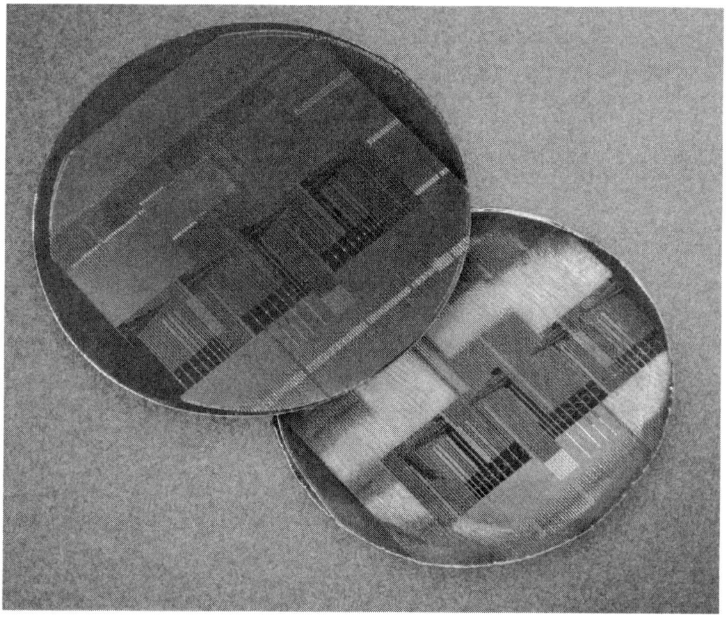

FIGURE 12-5 Nickel shim with photonic structures (top left) and replicated part.

as OPTOREZ-series (OZ-series). For deep UV (DUV) modification, a commercial UV-exposure system is used, a mask aligner EVG620 having a DUV lamp combined with a cold mirror with reflectance in the wavelength range of 200–240 nm in the exposure system.

Using the DUV process, it is possible to fabricate fluidic channels and reservoirs. There are several possibilities of realizing fluidic channels with this method. PMMA can be spin coated onto a glass wafer and then be exposed and developed. PMMA is used in this case like a conventional photoresist. The DUV fabrication method can also be used to realize a Ni-shim, which can then be used for hot embossing of fluidic channels. Another possibility is to directly expose a PMMA bulk substrate and develop the exposed regions. The penetration depth of the DUV light is only a few micrometers which defines the maximum height of the channels. The advantage of using spin-coated substrates is a defined height structure for realizing a flat and smooth bottom of the channels. A photograph of an unsealed T-junction is shown in Fig. 12-6. The fluidic channels have a width of approximately 5 μm.

As the fluidic channels are fabricated in PMMA, it takes only another DUV-aligned exposure to integrate the waveguides (Fig. 12-7).

In a next step cover plates are heat sealed onto the fluidic channel. DUV flood exposure is applied to both the substrate containing

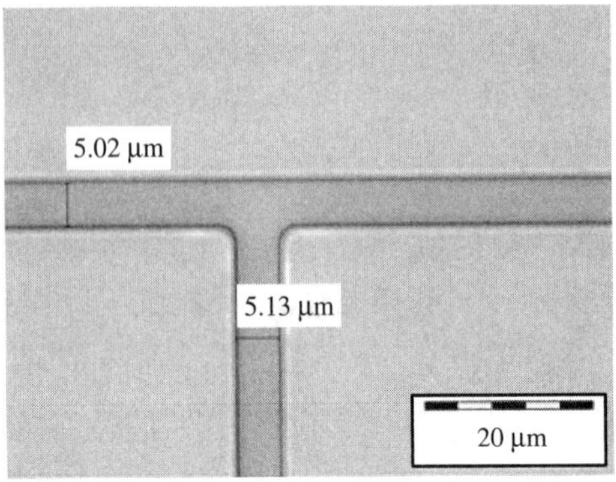

FIGURE 12-6 Unsealed fluidic channels in PMMA.

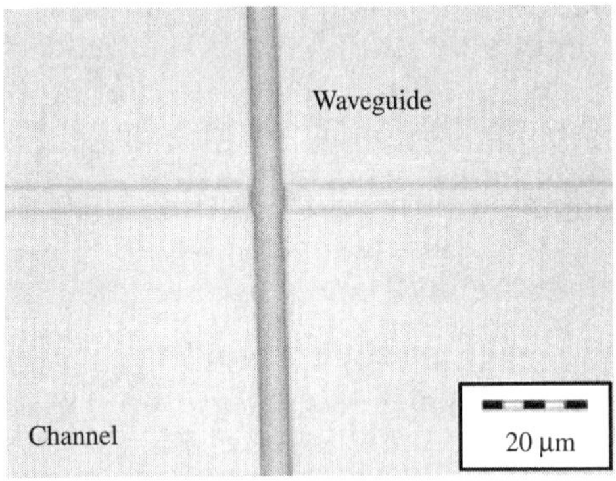

FIGURE 12-7 Fluidic channel and waveguide integrated on the same chip.

fluidic channel and waveguides and the cover plates. DUV radiation lowers the glass transition temperature, which is used advantageously in this bonding process making welding at lower temperatures possible.

12-1-3 Optofluidic Resonator Devices

Optofluidic resonator devices have been fabricated and characterized recently with great success. Optofluidic resonators are mainly used to realize sensor and laser architectures due to their ability to enhance the light traveling in these resonant structures. Several examples of optofluidic devices based on photonic crystals, Bragg gratings, ring resonators, and Fabry-Perot resonators will be given in this section.

To begin with, an example of a fabricated photonic crystal structure is presented (Fig. 12-8). Mandal and Erickson [1] demonstrate in this paper a new biosensor platform, which is referred to as nanoscale

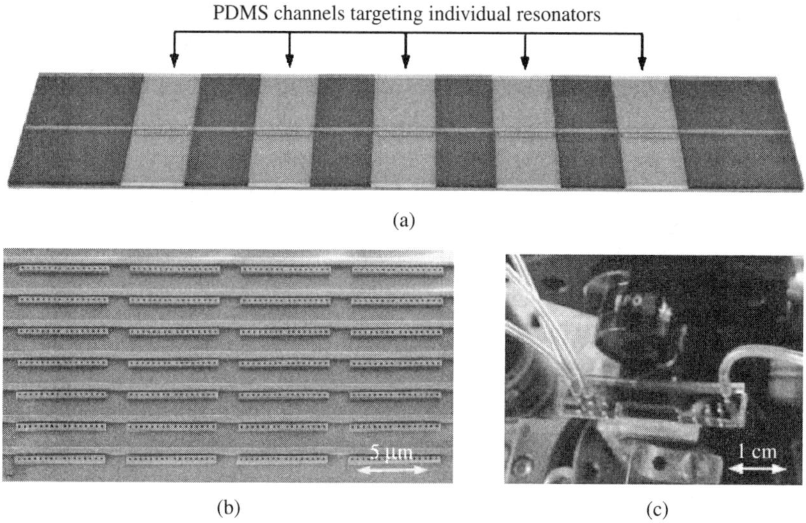

Figure 12-8 (a) Three-dimensional schematic showing a PDMS channel running across the side resonator. This channel allows the fluidic targeting of individual sensing sites. (b) SEM of a NOSA device. It illustrates how this architecture is capable of two-dimensional multiplexing, thus affording a large degree of parallelism. (c) Actual NOSA chip with an aligned PDMS fluidic layer on top. (*S. Mandal and D. Erickson, "Nanoscale optofluidic sensor arrays," Opt. Express, 16, 1623–1631, 2008.*)

optofluidic sensor arrays. These devices comprise of a waveguide with a series of evanescently coupled "side resonators." A change in the refractive index of the near-field region surrounding the optical cavity results in a shift in the resonant wavelength. The sensitivity of the device is characterized. The results suggest a bulk refractive index resolution of 7×10^{-5} which translates to a mass limit of detection of approximately 35 ag. Q factors of the devices were demonstrated to be approximately 3000.

A so-called photonic bandgap-edge optofluidic biosensor is demonstrated theoretically in Xiao and Mortensen [11]. It is shown that the simulated structures are strongly sensitive to the refractive index of the liquid, which is used to tune the dispersion of the photonic crystal. The calculated high sensitivity makes such devices interesting for biochemical sensing applications.

Photonic crystals have been fabricated in several material systems, and a logical consequence now is to use these devices in combination with integrated fluidics to create optofluidic photonic crystal sensor devices. Hence these kinds of devices will be seen more often in the future development of optofluidic sensors.

The other type of resonator structure, which was briefly introduced in the previous section, was the Bragg reflector. This type of resonator is mainly used in optofluidic lasers, which are also highlighted in this book. Therefore only a limited number of examples are given here. Light sources are essential for future optofluidic lab-on-chip devices in order to measure on chip and eliminate the need to couple light in to the device. Gersborg-Hansen and Kristensen [2] demonstrate a polymer-based optofluidic third-order Bragg grating–distributed feedback dye laser. The device relies on light confinement in a nanostructured polymer film where the individual resonator elements (nanofluidic channels and polymer walls) are of subwavelength dimensions. The resonator consists of an array of nanofluidic channels forming a third-order DFB Bragg grating resonator.

Another Bragg grating–based mechanically tunable optofluidic-distributed feedback dye laser presented by Li and coworkers [12] with a similar configuration as shown in Fig. 12-9 (ring resonator) except that a Bragg grating is used instead of a ring resonator as the feedback element. The optical feedback is realized by a phase-shifted higher-order Bragg grating embedded in the liquid core of a single-mode buried channel waveguide. The DFB laser is fabricated in PDMS. Due to the soft elastomeric nature of PDMS, the authors were able to tune the laser frequency mechanically by stretching the grating period. This mechanism is only limited by the gain bandwidth. A tuning range of nearly 60 nm is demonstrated from a single-dye laser chip by combining two common dye molecules—rhodamine 6G and rhodamine 101. Single-mode operation was maintained with less than 0.1-nm linewidth.

One of the thriving optofluidic devices is the ring resonator. Ring resonators are ideal for integration, as no facets or gratings are needed to provide optical feedback and resonance enhancement.

One of the first optofluidic ring resonators (OFRRs) is used for creating a dye laser on a monolithic polydimethylsiloxane (PDMS) chip [13]. A laser threshold of 9.2 nJ is obtained with a single-mode liquid-core waveguide-based microring cavity. The schematic of the realized device and a photograph of the fabricated devices are shown in the Fig. 12-9.

Ring resonators are versatile devices as is described in detail in Ref. 5. Besides integrated laser sources, ring resonator-based sensors have been developed. The advantage is the achieved resonance-enhancement and hence an achievable lower detection limit. A novel sensor architecture based on a liquid-core optical ring resonator (LCORR) in which a fused silica capillary is utilized to carry the aqueous sample and to act as the ring resonator is demonstrated by White and coworkers [14]. The device uses whispering-gallery modes as the sensing mechanism. The wall thickness of the LCORR is controlled to a few micrometers to expose the whispering-gallery mode to the aqueous core. Optical characterization with a water-ethanol mixture shows that the spectral sensitivity of the LCORR sensor is approximately 2.6 nm per refractive index unit.

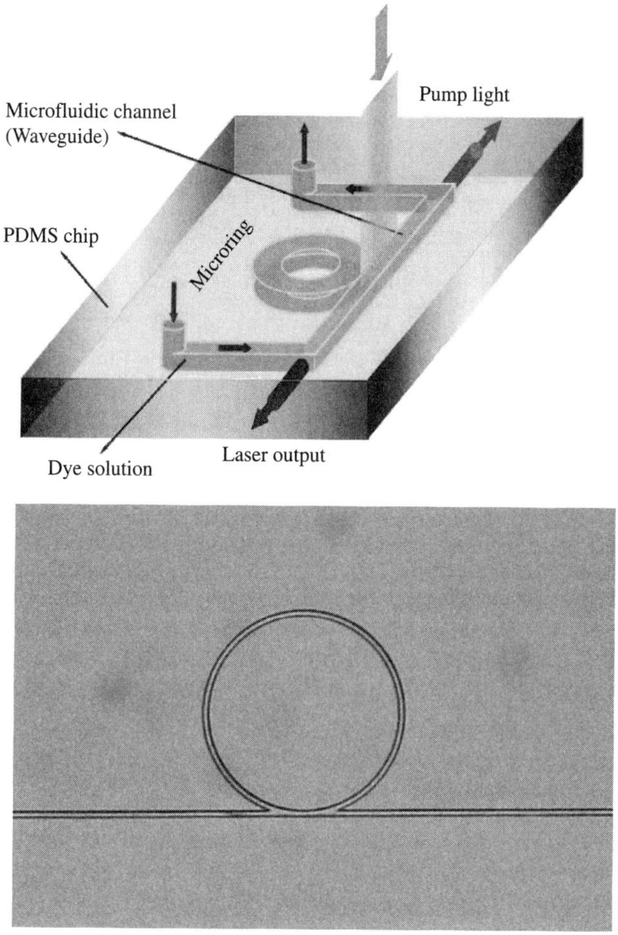

FIGURE 12-9 Optical micrograph of an optofluidic microring resonator in PDMS. (*Z. Li, Z. Zhang, A. Scherer, and D. Psaltis, "Optofluidic microring dye laser," IEEE LEOS Summer Topicals, Copyright 2007 IEEE.*)

The same principle is used by the group to demonstrate biomolecule sensing [15] and label-free viral detection [16] with what the group calls an OFRR. The setup of the sensor is shown in Fig. 12-10.

In the presented label-free viral detection experiment, filamentous bacteriophage M13 is used as a safe model system. Virus samples are flowed through the OFRR, whose surface is coated with M13-specific antibodies. The sensor performance is studied by monitoring in real time the virus and antibody interaction. It is shown that the OFRR can detect M13 with high specificity and sensitivity. The detection limit is approximately 2.3×103 pfu mL^{-1} and the detection dynamic range spans seven orders of magnitude.

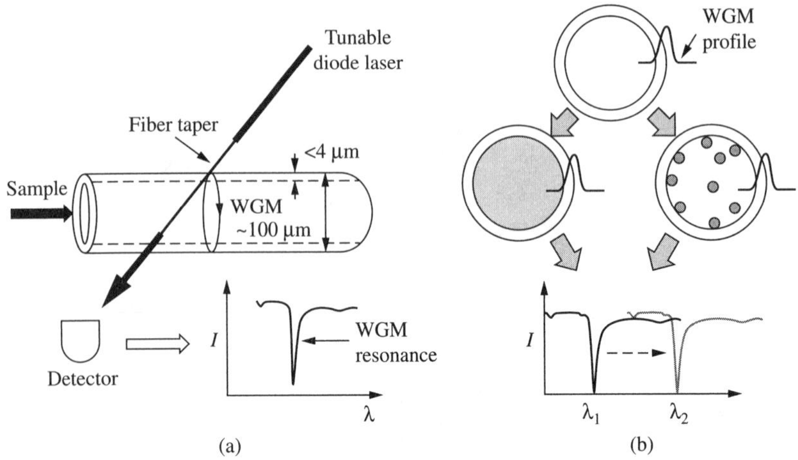

FIGURE 12-10 (a) Overall configuration for the OFRR. (b) The cross section of an OFRR. The WGM circulates along the OFRR circumference. Its evanescent field extends beyond the interior surface of the OFRR and interacts with the analytes in the core. The WGM resonant wavelength shifts in response to the bulk solution change or the binding of molecules to the OFRR interior surface.

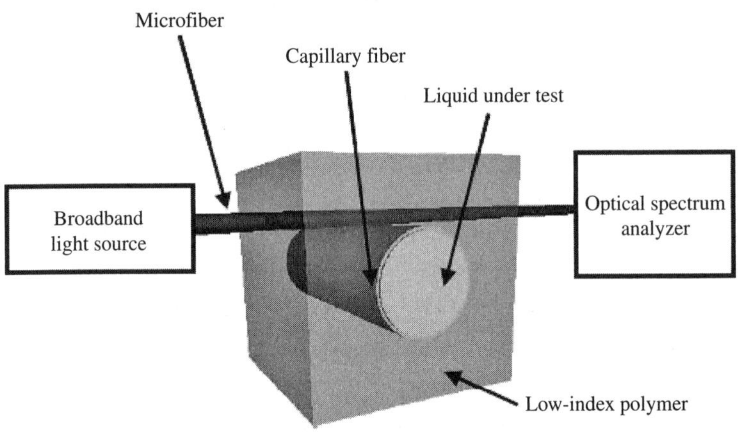

FIGURE 12-11 Schematic of a liquid-core optical ring-resonator sensor. (*M. Sumetsky, R. S. Windeler, Y. Dulashko, and X. Fan, "Optical liquid ring resonator sensor," Opt. Express, 15, 14376–14381, 2007.*)

Another ring resonator sensor based on whispering-gallery modes is demonstrated by Sumetsky and coworkers [17]. The schematic of the device is shown in Fig. 12-11. Sensing is based on measurement of the transmission spectrum of the whispering-gallery modes, which are excited across the liquid channel by an optical microfiber imbedded into the polymer. The achieved sensitivity is 800 nm/RIU.

The development of biosensors is highly sought after and a promising technology for biosensing is surface-enhanced Raman spectroscopy (SERS). The combination of SERS and OFRRs is demonstrated by White and coworkers [18], using again the aforementioned whispering-gallery-mode-based ring resonator platform. In SERS, the Raman signal enhancement is enabled by metal nanostructures, for example, silver or gold nanoparticles. A detection limit of 400 pM is achieved in using this configuration.

Ring resonators as demonstrated by several examples are advantageous in creating novel optofluidic devices and hold several opportunities in the future.

Similar to the transmission characteristics of ring resonators are Fabry-Perot resonators. Fabry-Perot resonators have been around for quite some time and have numerous applications in photonics. The use of Fabry-Perot resonators in combination with microfluidics is only a consequent step. One such device is presented by Shao and coworkers [4]. A microfluidic Fabry-Perot cavity in a Pyrex glass substrate is demonstrated. The structure of the devices is shown in Fig. 12-12. The microfluidic channels are etched in HF solution on a

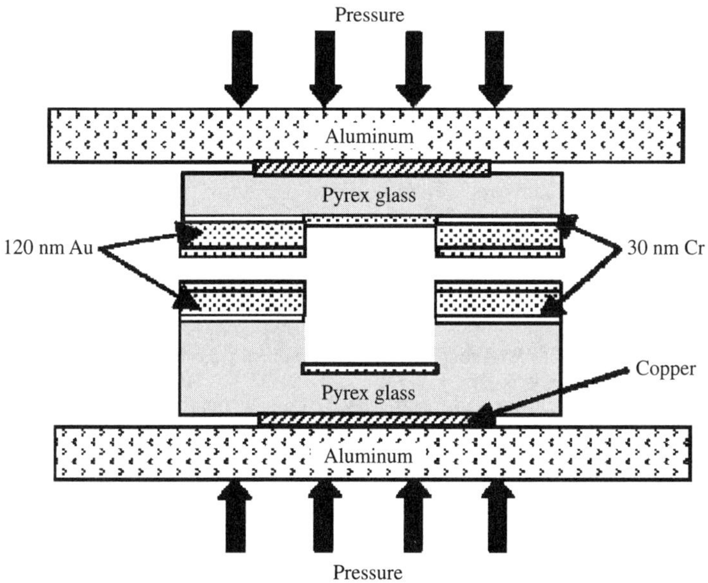

FIGURE 12-12 Schematic diagram of thermal compressive gold-gold diffusion bonding. (*H. Shao, D. Kumar, S. A. Feld, and K. L. Lear, "Fabrication of a Fabry-Pérot cavity in a microfluidic channel using thermocompressive gold bonding of glass substrates," J. Microelectromech. Syst., Copyright 2005 IEEE.*)

glass substrate using a Cr/Au/photoresist etching mask resulting in a channel-bottom roughness of 1.309 nm. An effective thermocompressive gold-gold bonding technique is used to bond the photolithographically etched glass substrates inside a 350°C oven in a 10^3 torr vacuum. Pressure is applied to the glass pieces by using two aluminum blocks with intermediate copper sheets. This method takes advantage of using Cr/Au layers both as a wet etching mask and as intermediate bonding layers, requiring only one lithography step for the entire process. The device has been used in Ref. 19 for optofluidic intracavity spectroscopy to measure single cells.

Biological cells have also been studied by the same group using vertical cavity laser with an incorporated microfluidic channel [20].

Fabry-Perot resonators are a useful class of devices whose principle is well known and, due to the availability of different fabrication technologies, a potential candidate for optofluidic devices. The essential difference to the previously described ring resonator devices is the need for parallel-aligned reflecting surfaces, which are not needed in the case of ring resonators and are a major drawback for integrated optofluidic Fabry-Perot devices. Here the advantage lies in standalone devices as demonstrated for spectroscopy purposes.

12-2 Summary

The use of different important resonator principles has been highlighted and the theoretical background is briefly explained. Several state-of-the-art fabrication methods have been described, which enable the realization of these novel optofluidic resonator devices. The advent of optofluidic resonator devices has already begun and promising devices have already been demonstrated for integrated laser sources and biosensors. This is one step toward all optofluidic integrated sensor platforms (OISPs).

References

1. S. Mandal and D. Erickson, "Nanoscale optofluidic sensor arrays," *Opt. Express*, **16**, 1623–1631, 2008.
2. M. Gersborg-Hansen and A. Kristensen, "Tunable optofluidic third order DFB dye laser," *OSA* 1-55752-834-9.
3. S. I. Shopova, H. Zhou, and X. Fan, "Optofluidic ring resonator based dye laser," *Appl. Phys. Lett.*, **90**, 221101, 2007.
4. H. Shao, D. Kumar, S. A. Feld, and K. L. Lear, "Fabrication of a Fabry-Pérot cavity in a microfluidic channel using thermocompressive gold bonding of glass substrates," *J. Microelectromech. Syst.*, **14** (4), 756–762, August 2005.
5. D. G. Rabus, *Integrated Ring Resonators—The Compendium*, Springer, Berlin, Heidelberg, New York 2007.
6. Gina S. Fiorini and Daniel T. Chiu, "Disposable microfluidic devices: fabrication, function, and application," *BioTechniques*, **38** (3), 429–446, March 2005.
7. Y. Xia and G. M. Whitesides, "Soft lithography," *Angew. Chem.*, Int. Ed., **7** (5), 550–575, December 1998.

8. M. Heckele and W. K. Schomburg, "Review on micro molding of thermoplastic polymers," *J. Micromech. Microeng.*, 14, R1–R14, 2004.
9. D. G. Rabus, M. Bruendel, Y. Ichihashi, A. Welle, R. A. Seger, M. Isaacson, "A bio-fluidic-photonic platform based on deep UV modification of polymers," *IEEE J. Select. Topics Quantum Electron.*, **13**, 214–222.
10. M. Bruendel, Y. Ichihashi, J. Mohr, M. Punke, D. G. Rabus, M. Worgull, V. Saile, "Photonic integrated circuits fabricated by deep UV and hot embossing," *IEEE LEOS Summer Topicals*, Paper TuB2.6, 2007.
11. S. Xiao and N. A. Mortensen, "Proposal of highly sensitive optofluidic sensors based on dispersive photonic crystal waveguides," *J. Opt. A: Pure Appl. Opt.*, **9**, S463–S467, 2007.
12. Zhenyu Li, Zhaoyu Zhang, Axel Scherer, and Demetri Psaltis, "Mechanically tunable optofluidic distributed feedback dye laser," *Opt. Express*, **14** (22), 10494, Oct. 30, 2006.
13. Z. Li, Z. Zhang, A. Scherer, and D. Psaltis, "Optofluidic microring dye laser," *IEEE LEOS Summer Topicals*, 2007.
14. Ian M. White, Hesam Oveys, and Xudong Fan, "Liquid-core optical ring-resonator sensors," *Opt. Lett.*, **31** (9), 1319–1321, May 1, 2006.
15. H. Zhu, I. M. White, J. D. Suter, P. S. Dale, and X. Fan, "Analysis of biomolecule detection with optofluidic ring resonator sensors," *Opt. Express*, **15**, 9139–914, 2007.
16. H. Zhu, I. M. White, J. D. Suter, M. Zourobb, and X. Fan, "Opto-fluidic microring resonator for sensitive label-free viral detection," *Analyst*, **133**, 356–360, 2008.
17. M. Sumetsky, R. S. Windeler, Y. Dulashko, and X. Fan, "Optical liquid ring resonator sensor," *Opt. Express*, **15**, 14376–14381, 2007.
18. I. M. White, J. Gohring, and X. Fan, "SERS-based detection in an optofluidic ring resonator platform," *Opt. Express*, **15**, 17433–17442, 2007.
19. Hua Shao, Dhiraj Kumar, and Kevin L. Lear, "Single-cell detection using optofluidic intracavity spectroscopy," *IEEE Sens. J.*, **6** (6), 1543–1550, December 2006.
20. D. Kumar, H. Shao, and K. L. Lear, "Vertical cavity laser and passive Fabry-Perot interferometer based microfluidic biosensors," *Laser Applilcations to Chemical, Security and Environmental Analysis, Paper TuD3*, 2006.

CHAPTER 13

High-*Q* Resonant Cavity Biosensors

Andrea Armani

Mork Family Department of Chemical Engineering and Materials Science, University of Southern California, Los Angeles, California

13-1 Overview of Resonant Microcavities

Resonators and oscillators are prevalent throughout science and engineering, with examples found in mechanics (mechanical springs), electronics (capacitors and inductors), acoustics (tuning forks), and optics (photonic crystals and microcavities). The defining characteristic of a resonant device is its ability to store large amounts of energy built up from a considerably weaker input. In optics, this translates to the storing and building up intense optical fields. As the quality factor (Q) of the cavity increases, the length of time that light can be confined within the cavity also increases (linearly). Therefore, the intensity of the stored energy also increases.

13-1-1 Introduction to Optical Resonant Devices

Optical microresonators can be broadly classified into two categories depending on how they confine light: those that rely on total internal reflection (TIR) and those that rely on Bragg reflection for optical confinement.

Examples of TIR microresonators include microspheres [1,2], microdisks [3–6], microtoroids [7], and microrings [8–11]. The size of a TIR microresonator is limited by the TIR condition, or the index difference between the guiding region and the cladding, similar to optical fiber. Index-guided resonators are typically easier to fabricate than Bragg resonators. For example, in the case of the highest-Q resonant cavity, the microtoroid resonator, the fabrication process requires only photolithography and two etching steps [7]. These devices are also easier to couple light in and out of, because the design of phase-matched couplers is better understood. Moreover, this type of resonator has demonstrated Q factors in excess of 100 million [7]. Ultrahigh-Q factors offer opportunities to explore numerous fundamental aspects of optics, such as parametric effects [12], optomechanical coupling [13], and light-atom coupling [14], which are inaccessible with lower-Q devices because the circulating intensity of the light is lower.

The second category of microresonators relies on Bragg reflection to confine light. Examples of Bragg resonators include quarter-wave-shifted distributed feedback, photonic crystal, Bragg annular, and onion cavities [15–18]. This type of resonator can possess significantly smaller physical sizes than index-guided resonators since they are not limited by total internal reflection. However, they have the inverse problems of those mentioned previously (difficult to couple light into, complex fabrication process, etc.).

The focus of the present chapter is on TIR microresonators. These devices are also known as whispering gallery mode microcavities, named after the Whispering Gallery chamber at St. Paul's Cathedral, London.

Microresonator Essentials

There are three parameters which are often used in the characterization of a resonant cavity: the quality factor (Q) and the free spectral range (FSR). Occasionally, the Q will be expressed in terms of the finesse of the cavity. Depending on the application, often the mode volume and the circulating intensity will often be cited as well.

The free spectral range (FSR) expresses both the optical path inside the resonant cavity and gives the frequency/wavelength spacing between sequential resonant frequencies. The FSR is defined as

$$\Delta\omega_{\mathrm{FSR}} \equiv \Omega_{m+1} - \Omega_m \qquad (13\text{-}1)$$

where Ω_{m+1} and Ω_m are consecutive resonance orders. The resonance condition is satisfied whenever

$$\beta_m L_{\mathrm{RT}} + \phi_o = 2m\pi \qquad (13\text{-}2)$$

where $\beta_m = \Omega_m n_{\mathrm{eff}}/c$ is the propagation constant, L_{RT} is the round-trip length of the resonator, and ϕ_o is any additional phase that the light may accumulate in a round-trip. As expected, the FSR is dependent on the refractive index, the geometrical properties of the cavity and the testing frequency. Effective refractive index n_{eff} is the effective index of the resonator (dependent on the refractive index of the resonant cavity material and the environment), and c is the speed of light. Therefore, substituting into Eq. (13-1), we obtain

$$\Delta\omega_{FSR} = \frac{2\pi c}{n_g L_{RT}} \tag{13-3}$$

where n_g is the group index. Depending on the size of the resonant cavity, the FSR can range from gigahertz to terahertz.

The second property most commonly used to characterize a resonant cavity is the quality factor or Q. The Q factor describes the losses of the resonator and is defined as

$$Q \equiv \Omega \times \frac{\text{field energy stored}}{\text{power dissipated}} \tag{13-4}$$

where Ω is the resonance frequency of the resonator. This general expression can be more precisely defined by assuming that U is the field energy stored, α_{RT} is the fractional loss per round-trip in the resonator, and τ_{RT} is the round-trip time. Substituting these expressions into Eq. (13-4), the power dissipated by the resonator is

$$\text{Power dissipated} = \frac{\alpha_{RT} U}{\tau_{RT}} \tag{13-5}$$

Substituting this into Eq. (13-4) yields

$$Q = \Omega \frac{\tau_{RT}}{\alpha_{RT}} \tag{13-6}$$

From this expression, it is obvious that the Q factor is inversely related to the losses of the cavity, and it is therefore possible to improve the Q factor by minimizing these losses. An alternative expression for Q is to explicitly list these loss mechanisms according to the simple formula [2]:

$$Q_{tot}^{-1} = Q_{mat}^{-1} + Q_{ss}^{-1} + Q_{rad}^{-1} + Q_{coup}^{-1} + Q_{cont}^{-1} \tag{13-7}$$

where Q_{mat} is material loss, Q_{ss} is surface scattering loss, Q_{rad} is whispering gallery loss, Q_{coup} is coupling loss, and Q_{cont} is contamination loss [1]. To maximize the quality factor or the sensitivity of the cavity, all of the loss mechanisms must be minimized; a few comments are in order concerning each of these mechanisms.

Whispering gallery resonators always experience a certain amount of tunnel-leakage of the radiation from the confined mode. This leakage or radiation loss increases as the diameter of the cavity is reduced (scaled by the wavelength). It also depends on the refractive index contrast between the resonator material and the surrounding medium. Whispering gallery loss Q_{rad} is therefore readily controlled by appropriate selection of the resonator diameter, the material refractive index, and also the operational wavelength. Contamination loss can likewise be made small by fabrication and testing in a sufficiently clean environment. It should be noted that one component of contamination is the intentionally introduced biomolecules themselves. In the case of resonant-wavelength-shift detection, it is therefore important to understand the impact of the biomolecules on the optical Q factor. Alternatively, if variation in optical loss (or Q factor) is to be used as the detection mechanism, then it will be desirable that the molecules produce a discernable contribution to optical loss.

The need to couple optical power both to and from the resonator implies that a necessary component of loss is associated with the waveguide used to achieve this intentional coupling. In addition to the desirable coupling, the waveguide, itself, can create unintended parasitic loss. Coupling loss Q_{coup} contains both of these components.

Radiation, parasitic-coupling, and contamination can be controlled so as to not limit the intrinsic Q factor (i.e., Q factor in the absence of desirable waveguide-coupling-induced loss). This leaves material and surface-scattering loss contributions as the dominant contributors that limit the intrinsic Q factor. Surface scattering must be controlled through detailed attention to microfabrication and/or applying special techniques that smooth the dielectric boundary to lower scattering.

Of resonator materials that have, so far, been studied in the context of biodetection, silica resonators provide the lowest material losses and hence highest Q_{mat}. Silica is also a common dielectric in many wafer-based processing methods and hence has a practical value even beyond its low material loss. Beyond silica, semiconductors such as silicon have been used to attain Q factors of more than 500,000 [5] and polymer-ring resonators have also achieved Q values of more than 100,000 [19–21]. However, these Q factors are several orders of magnitude lower than for the silica ultrahigh-Q microcavities, such as microtoroids [7] or microspheres [2], which have attained Q values ranging from 500 million to 10 billion.

Another important consideration in material selection is operational wavelength. Because of the importance of operation in an aqueous bath for detection of biomolecules, most sensing experiments are performed in the visible, where the loss of water is low [22]. In this regard, silica, owing to its very-broad, low-loss spectral window extending from the ultraviolet in the infrared, is an excellent material choice.

An alternate expression for Q, if the lineshape is lorentzian, is simply:

$$Q = \Omega\tau_L = \frac{\Omega}{\Delta\omega} = \frac{\lambda}{\Delta\lambda} \tag{13-8}$$

where $\Delta\omega$ and $\Delta\lambda$ are the full-width half-max in frequency and wavelength domain of the lineshape, λ is the resonance wavelength, and τ_L is the lifetime of the photon in the cavity. This equation is particularly useful for whispering gallery mode resonant cavities as $\Delta\omega$, $\Delta\lambda$, and λ are experimentally measurable parameters.

As stated previously, instead of expression the Q of a cavity, often the finesse (F) is given. This originally arose from the resolving power of a Fabry-Perot etalon and is the ratio of the FSR to the FWHM ($\Delta\lambda$) of the resonance:

$$F = \frac{\Delta\omega_{\mathrm{FSR}}}{\Delta\omega} = \frac{\Delta\lambda_{\mathrm{FSR}}}{\Delta\lambda} \tag{13-9}$$

F can be viewed as a metric that combines FSR and Q, in the case of the lorentzian lineshape.

Applications of Microresonators

In addition to telecommunications, whispering gallery mode optical microcavities have numerous applications [8,23]. As a result of the high circulating optical fields, very low threshold lasers have been demonstrated using rare earth dopants, nanocrystals, and laser dyes [24–27]. Additionally, these devices have been used to study nonlinear optical effects, such as second and third harmonic generation, Raman lasing, and four-wave mixing [13,26,28–30]. More recently, planar devices have shown optical-induced mechanical behavior and cooling effects, as well as frequency comb generations [13,28]. These devices have been used to study quantum optic effects, such as quantum entanglement and cavity quantum electrodynamics [31–33]. Finally, resonant cavities have also been used in the biosensing field, studying protein folding, cell membrane structure, and single-molecule detection [34–36].

13-1-2 Whispering Gallery Mode Devices

Optical microcavities can be fabricated from numerous materials and in many different geometries [37,38]. A nonexhaustive overview of some of these methods will be given in this section along with a brief description on common resonant cavity characterization techniques.

Fabrication Techniques and Geometry

As a result of the different loss mechanism (material, surface roughness), historically there have been two regimes of Q factor: high-Q and ultrahigh-Q. The dividing line between high-Q and ultrahigh-Q has been somewhere between 10^5 and 10^8, which is a very large range, resulting in some confusion. Recently, as more devices have emerged, this dividing line has become even more blurred as this intermediate region is becoming more crowded. This is shown in Table 13-1, which summarizes the most prominent devices geometries, some common

	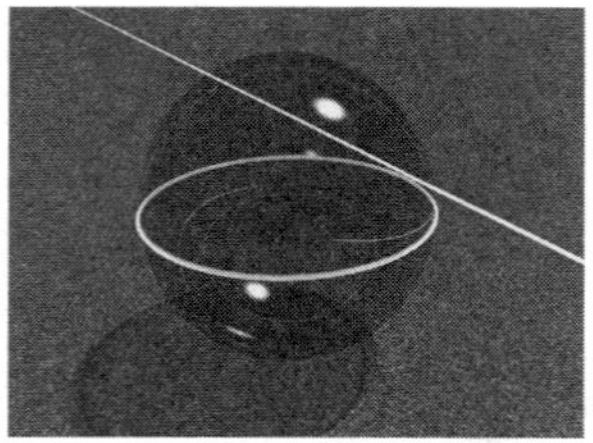		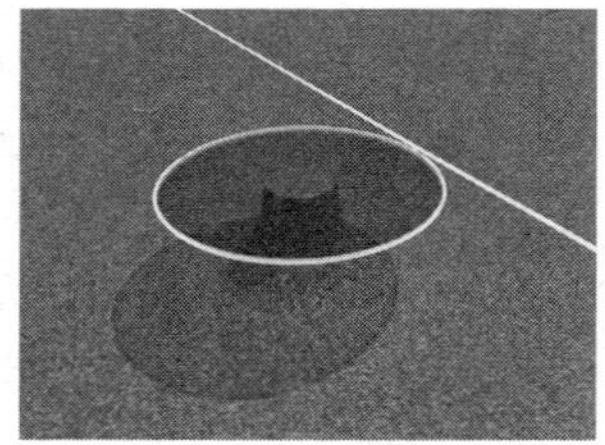	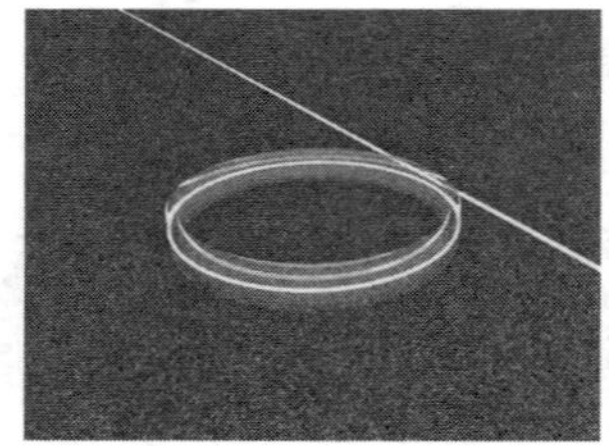
Geometry	Microsphere [2]	Microtoroid [7]	Microdisk [5,7,39]	Microring [8,10,21]
Possible material	Silica	Silica	Silica, silicon, silicon nitride, AlGaAs	Polymeric materials, silicon
Quality factor	$>10^9$	$>10^8$	$\sim10^5$–10^7	$\sim10^3$–10^5
Fabrication method	Reflow	Lithography/ reflow	Lithography	Lithography/ molding

Table 13-1 A Summary of the Most Commonly Used Whispering Gallery Mode Resonant Cavity Geometries, along with the Materials Used to Fabricate Them and the Corresponding Quality Factors. The Images Were Generated in Povray, a Ray-Tracing Program, and Are Indicative of the General Structure of the Different Geometries. The Whispering Gallery Mode Is Highlighted in White and Light Is Coupled into the Device Using a Waveguide Such as a Tapered Fiber

materials used to fabricate the microcavities, the fabrication method, and the Q factor.

From a cursory glance at the table, it quickly becomes evident that structures, which have been reflowed, have higher Q factors. The reflow process entails using a CO_2 or a flame to melt the surface of the resonant cavity, thereby removing any imperfections. This method reduces the surface roughness loss and creates a device whose Q is limited by either material losses or radiation losses. Currently there are two types of silica ultrahigh-Q microcavities: the microtoroid [7], and the microsphere [1,2]. Microspheres are fabricated serially by heating the tip of an optical fiber while microtoroids are fabricated in large arrays using photolithographic techniques.

Lower-Q (Q < 100,000) resonant devices are typically fabricated using e-beam lithography or soft lithography, and have already been integrated with waveguides and a plethora of other optical components [8,20,40,41]. However, as a result of the surface roughness induced by the lithography, the Q factors have been limited to 100,000. One reason for their continued success despite the lower Q factors is their ability to integrate a waveguide on-chip, creating a complete optical package. Using the waveguide-resonant cavity as a fundamental building block, optical systems like add-drop filters, buffers, and laser have been constructed.

Bridging the gap between the lower-Q and higher-Q devices are the wedge-shaped devices. The higher-Q factor results from the whispering gallery mode being forced toward the interior of the device, away from the lithographically rough surface. Therefore, while the Q is still limited by surface roughness, the effect is minimized. The microdisk devices fall into this category. These wedge-shaped devices form the third (new) regime of optical microcavities.

Waveguide Coupling Methods

There are, to date, four different methods of coupling light to and from resonant cavities, assuming a waveguide has not already been integrated, as in the case of the microring resonator. They are (a) prism coupling [42], (b) half block coupler [43], (c) angle polished fiber couplers [44], and (d) fiber taper coupler [32,45]. All of these methods are based on the evanescent coupling of light between a waveguide and the resonant cavity. Figure 13-1 illustrates these methods used in coupling to a silica microsphere resonant cavity. In all cases, it is necessary to bring the resonator into close proximity to the coupling device to allow for efficient coupling. The precise distance or gap between the coupler and the resonant cavity is dependent on the testing wavelength.

The highest-efficiency coupling (i.e., minimal parasitic loss) to date has been obtained using fiber tapers [32]. These devices are also inherently fiber compatible and hence provide a convenient means of

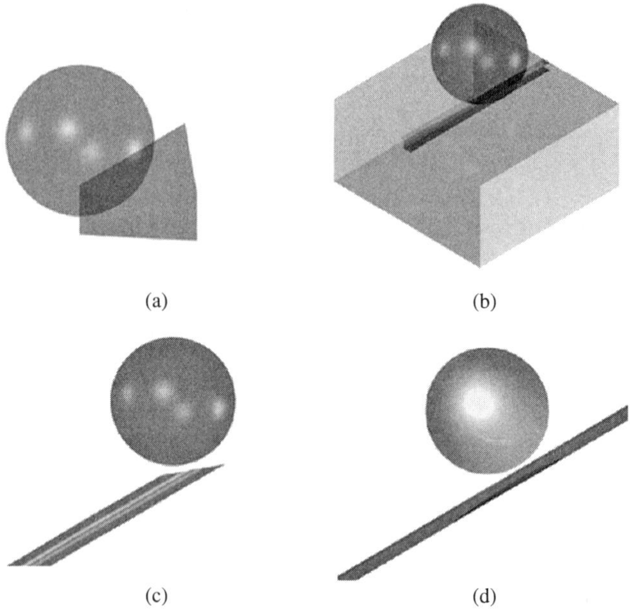

Figure 13-1 Renderings of four commonly used coupling devices illustrated in the context of a silica microsphere resonator. All devices evanescently couple light into the resonant cavity; however, the efficiency or contributions of parasitic loss vary significantly between the different techniques. (a) Prism coupler, (b) half block coupler, (c) angle polished fiber coupler, and (d) fiber taper coupler.

"pigtailing" to a fiber-coupled laser and directly coupling output power to a photodetector.

Tapered optical fiber waveguides are fabricated by heating the center of an optical fiber with an oxyhydric flame and pulling at both ends until the waist diameter is smaller than the operating wavelength, thus creating an evanescent region [32,46]. An alternative method for fabricating tapered optical fibers is the "flame-brush" technique [47]. In this method, the flame is repeatedly brushed across the fiber as it is pulled from both ends. The total length of tapered optical fibers fabricated using the flamebrush technique is longer than those fabricated using a fixed flame. However, both techniques have demonstrated low loss-tapered optical fibers function from the visible through the near-IR [32,46–48].

Experimental Characterization Techniques

There are several methods that can be used to characterize an optical cavity. The most common is spectral characterization in which a tunable laser is used to scan over a series of wavelengths to determine both the resonant frequency of the cavity and the resonant linewidth.

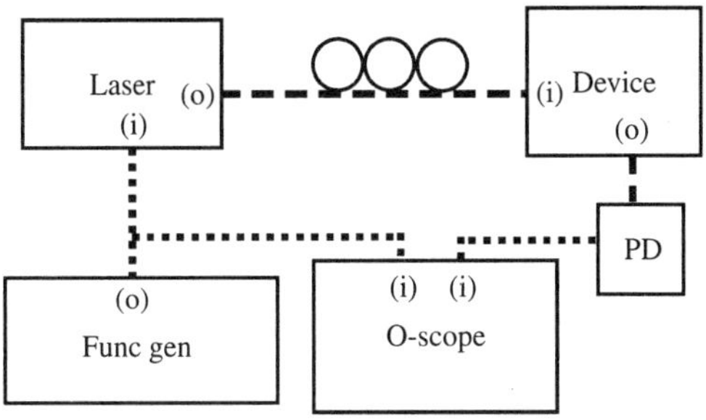

FIGURE 13-2 Schematic of the spectral measurement setup. Light from a narrow linewidth, CW, mode-hope-free, tunable laser (Laser) passes through a polarization controller and is coupled in and out of the optical microresonator (device) using optical fiber. From the resonator, the optical signal is detected using a high-speed photodiode (PD) on an oscilloscope (O-scope). If the Q factor is sufficiently high ($> 10^6$), it will be necessary to fine-scan the laser in order to accurately measure the linewidth; this requires an additional level of control over the laser using a function generator. Note, all optical signals are indicated in dashed lines and electrical signals are shown as dotted lines.

Depending on the Q of the cavity, this process can be very complex and is outlined in the schematic in Fig. 13-2.

As shown in Fig. 13-2, light from a narrow line-width tunable laser is coupled into the device using standard, single-mode optical fiber, and the transmitted light is collected using a single-mode fiber coupled to a high-speed (GHz) photodiode, which is connected to an oscilloscope. From there, the broadband spectra (or free spectral range) and the Q of the cavity can be determined. Additional information, such as the loss of the device, can be found by sending a fraction of the power to a power meter.

A few notes should be made at this point about the subtleties of Q measurements. The spectrum should always be taken in the undercoupled regime to minimize coupling-induced losses; however, assuming the waveguide was very low loss, it is possible to use a coupling model to extract the intrinsic Q of the device by varying the coupling and using a simple model [45]. It is also important to monitor the lineshape to ensure that the linewidth being measured is accurate, and is not being distorted by the presence of nonlinear effects.

13-2 Biosensing with Optical Microcavities

Whispering gallery mode resonators are a member of the larger group of label-free optical sensors, which includes surface plasmon resonance sensors and waveguide sensors among others. It is the interaction of the whispering gallery mode and the environment/molecule which results in detection. In biosensing, as is also true in numerous other applications of microcavities, long confinement times (or high-Q

factors) are beneficial. This property has been used to boost sensitivity in biological detection using optical microresonators and to improve signal-to-noise ratio [21,35,36,49–54]. Using this improved sensitivity, single-molecule detection has been demonstrated. Previously, this sensitivity level was only possible using labeled methods, such as surface-enhanced Raman spectroscopy (SERS) or total internal reflection fluorescence microscopy (TIRF).

13-2-1 Resonant Cavity–Detection Mechanisms

When a molecule binds on the surface of the resonant cavity, it triggers a cascade of events, all of which can be used as the detection signal. However, typically, in any given experiment, one proves to be the more reliable signal, and therefore it is monitored and recorded. The most commonly used detection mechanism is resonant frequency shift; however, Q change has also been demonstrated. While monitoring, a change in transmission is also possible, so this method is less commonly used in any detection scheme because of the unreliability of the method. While very few external signals can result in resonant frequency shifts, numerous external triggers can cause a transmission change, including laser noise. Therefore, only the first two will be discussed here.

Resonant Wavelength Shift

The resonant frequency of the cavity is like the cavity's signature. It is dependent on all of the inherent properties of the cavity (material, geometry, etc.) in addition to the testing conditions (operating wavelength, environment, etc.). Therefore, any time that a molecule binds to the surface of the cavity, it will act as a perturbation to this signature. There are two mechanisms that can be used to induce a resonant frequency shift. The first is based upon a polarizability change; the second is based upon a thermal change. Which mechanism is used in detection depends on the input power or the intensity of the circulating field.

Detection based on the polarizability of the molecule scales with the polarizability, the interaction area (surface area of the molecule), and the testing wavelength. Similarly, thermo-optic detection scales with the circulating intensity, interaction area, testing wavelength, and absorption cross section.

Q Factor Change

As seen in Sec. 13-1, there are many variables that affect the Q of the cavity. While this is typically viewed negatively, from the viewpoint of a sensor, anything that is sensitive to its environment can be used as a sensing modality. Therefore, as long as the resonant cavity is operating in or near the Q_{mat} (material limited Q) regime, by monitoring the Q, it is possible to determine if the environment around the microcavity has changed.

13-2-2 Optimization for Detection

Because the Q factor plays such a pivotal role in determining the sensitivity, it is necessary to maintain the Q throughout the experiment. While microtoroid quality factors in excess of 100 million are relatively easy to obtain for operation in air, in water it becomes more challenging due to the –OH overtones of the water molecule which increase the absorption, especially in the near-IR. Aqueous operation also has the side effect of increasing the radiation loss from the resonators at a given diameter. This occurs because the refractive index contrast between silica and water is lower than for silica-to-air operation. Nonetheless, operation in water is essential to keep biological species in their native state in order to maintain activity. To determine the ideal operational wavelength and diameter (from the point of view of Q optimization), finite element modeling of the system has been performed and experiments have verified the theoretical predictions [48]. These experiments will be briefly reviewed in the next section, as they experimentally validate several of the previously discussed loss mechanisms. Additionally, they provided the foundation for future biosensing research using optical microcavities.

Q Factor Optimization

As explained earlier, the radiation loss component (which scales as $Q^{-1} \propto e^{-D/\lambda}$) is dependent on diameter while the material absorption loss is strongly dependent on wavelength [2]. For a small enough microcavity diameter, the radiation loss is dominant, while at larger diameters, the material absorption loss (in the case of aqueous operation, this is typically the water loss) is dominant. There are thus two regimes of loss controlled by the diameter, with the transition diameter between these regimes determined by the operational wavelength.

Water and water-based solutions are the primary fluids used in biological detection experiments. However, deuterium oxide (D_2O) was particularly useful for comparison to water. D_2O (heavy water) and H_2O (water) have nearly identical refractive indices, but from 680 nm through 1550 nm, the absorption of D_2O is less than H_2O [22]. Therefore, while the Q_{rad} would be nearly identical (at a given diameter) for operation in these fluids, the Q_{mat} should diverge as diameter is increased. However, by proper optimization of both operational wavelength and microtoroid size, it is possible to recover the ultra-high-quality factors that are possible for air operation.

To demonstrate the sensitivity of Q factor to both the diameter and operational wavelength, ultrahigh-Q silica microtoroid resonators were fabricated over a wide range (50–250 μm) of major diameters using the previously outlined process [7]. Experiments were performed in both H_2O and D_2O [48]. The D_2O was purchased from Aldrich. Measurements of the resonator quality factor and analysis of the modal structure were performed at three wavelength bands

(680, 1300, and 1500 nm) using the process described in the previous section. Light was coupled into the resonant cavities using tapered optical fiber waveguides.

The testing chamber was formed from a single glass-slide spacer and a glass cover slip. By using this type of chamber, it was possible to first couple in air, using both top- and side-view cameras, and then couple in water using the top-view camera. A "liquid" gap between the toroid and the taper was maintained when determining the quality factor in either H_2O or D_2O in order to maintain constant coupling between the microtoroid resonator and the taper waveguide. The quality factor of the microtoroid resonator was first determined in air to ensure that it was above the theoretical limit for a given toroid diameter (once immersed in liquid) [48]. The highest Q resonance was located by first scanning over a wide wavelength range or performing a broadband scan; an example is shown in Fig. 13-3a.

The intrinsic Q factors measured in the 680-nm band for both water and heavy water plotted versus toroid major diameter are presented in Fig. 13-3b (circles and triangles, respectively) [48]. The model-predicted values are also shown in the plot (dotted and dashed lines). Q factors trend to larger values with increasing toroid size. This behavior is in good agreement with the predictions of the model and results from decreasing radiation loss. The maximum quality factors achieved at the time of the publication of the corresponding paper [48] were 2.3×10^8 in H_2O and 1.3×10^8 in D_2O. These values are notable as they represent the highest Q factors reported to date for operation in an aqueous environment. The highest aqueous Q factor reported previously was approximately 10^6 in a silica microsphere [50]. Measurements beyond Q factors of 500 million were not possible in this experiment due to laser linewidth stability. In principle, however, larger toroid diameters should exhibit quality factors as high as 1×10^9 in water and 1×10^{10} in D_2O.

For comparison, the same measurements were also taken in the near-IR (1300 and 1550 nm). Both the radiation-loss-limited regimes and the absorption-loss-limited regimes are clearly visible in these plots (Fig. 13-3c and 13-3d). Within these wavelength bands, D_2O has a lower optical absorption and hence exhibits an absorption-limited Q plateau that is significantly higher than for H_2O. The highest quality factors achieved in water at 1300 and 1550 nm were 8×10^5 and 7×10^4, respectively. In D_2O, the highest quality factors achieved at 1300 and 1550 nm were 2×10^7 and 2.8×10^6, respectively.

Surface Functionalization

In addition to the sensitivity that the high-Q provides, it is also necessary for a detector to be specific. Specificity is gained through surface functionalization of the optical whispering gallery using various biological or chemical molecules.

There are numerous well-developed surface functionalization methods. In particular, silica resonant cavities benefit from the wealth

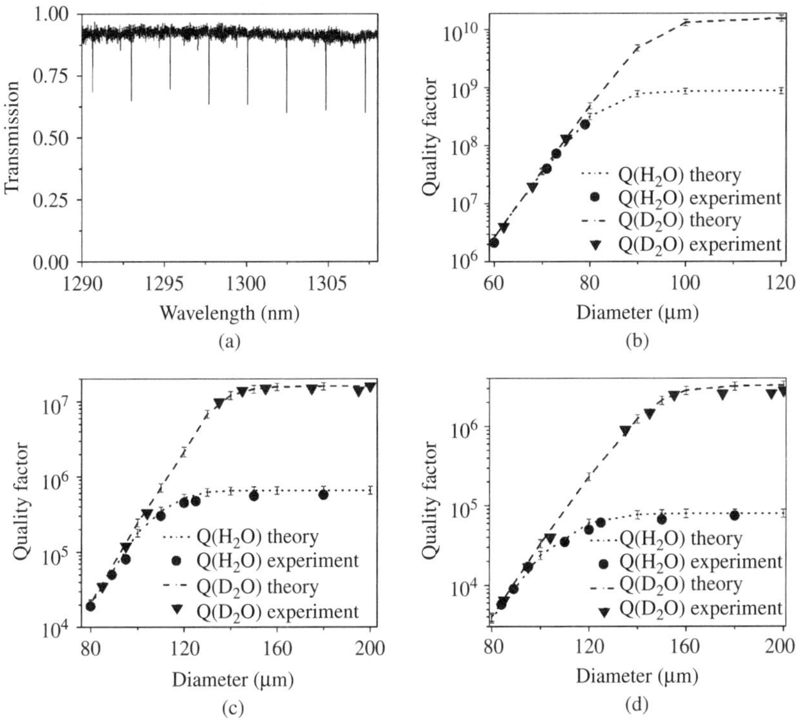

FIGURE 13-3 Transmission spectra and quality factors of the resonator in an aqueous environment. (a) Transmission spectra of a microtoroid resonator in D_2O at 1300 nm. The resonator is highly under-coupled in the spectrum presented. (b) Quality factors measured and predicted in the 680-nm band plotted versus toroid major diameter. *Q* increases with major diameter over the range of diameters wherein radiation loss is the dominant loss mechanism. It then plateaus at values set by absorption of the aqueous environment. Above 5×10^8 data taking is unreliable due to laser-linewidth stability limitations. The maximum quality factor measured in H_2O was 2.3×10^8 and in D_2O was 1.3×10^8. (c) Quality factors measured and predicted in the 1300-nm band. In H_2O, the maximum quality factor measured is 8×10^5. By changing to D_2O, the maximum quality factor increased to 2×10^7. (d) Quality factors measured and predicted in the 1550-nm band. In H_2O, the maximum quality factor measured is 7×10^4, while by changing to D_2O, the maximum quality factor increased to 2.8×10^6. (*Reprinted with permission from A. M. Armani, D. K. Armani, B. Min, and K. J. Vahala, "Ultra-high-Q microcavity operation in H_2O and D_2O," Appl Phys Lett, 87, 151118 (2005). Copyright 2005, American Institute of Physics.*)

of knowledge previously developed for microscopy. These techniques include biotin-streptavidin attachment [55,56], antibody-antigen [57], APTES [58], and silanization [56] to name a few.

The most commonly used of these techniques are either biotin or antibody surface functionalization. These two protocols are highly desirable from a biological standpoint. The biotin-streptavidin dissociation constant is extremely small, indicating a very strong attachment. In fact, the biotin-steptavidin interaction is the strongest noncovalent bond currently known. This protocol is often used in immobilizing proteins because of its stability across a wide temperature and pH range. However, this technique requires labeling either the target molecule with biotin or performing a sandwich assay.

If either of these two techniques are not feasible, or if a higher degree of specificity is preferred and stability is not a concern, then antibodies are preferred. Antibody-antigen immobilizations are found commonly in medical diagnostic assays, such as enzyme-linked immunoassays or ELISA protocols [57]. The specificity of the antibody-antigen interaction allows for array screenings for many antigens simultaneously.

However, it is important to note that not all antibodies are equal in their degree of specificity or affinity. Researchers often use polyclonal antibodies because they are simple and relatively inexpensive to generate in large quantities in a short period of time. However, these antibodies often will have multiple binding sites for the antigen, each with a different binding affinity. This can be improved by appropriate screening methods, but these are costly and reduce the quantity of the antibody available for experiments. In contrast, monoclonal antibodies have a single, homogenous binding site, typically with a very high affinity for the antigen. However, these antibodies are very costly and time-consuming to make.

13-2-3 Experimental Examples of Detection

While there are numerous demonstrations of using resonant cavities as biosensors, the next two examples are meant to simply give a flavor of the diversity of these experiments. Other chapters demonstrate high concentration biodetection using liquid-core whispering gallery mode resonators; therefore, in this chapter, the examples are restricted to either chemical detection or single-molecule detection.

Heavy Water Detection

The change in Q factor that occurs when the microtoroid is immersed in either water or heavy water suggests a method to measure the concentration of heavy water in water [49]. To have the highest sensitivity, the difference between the Q in water and Q in heavy water must be maximized. Of the wavelengths studied in the previous section, the sensitivity was greatest at 1300 nm.

To demonstrate this effect, a simple testing procedure was used: (1) immerse the microtoroid in 100% D_2O, (2) gradually increase the concentration of H_2O in D_2O until 100% H_2O is reached, and (3) return the concentration of D_2O to 100% [49]. The microtoroid diameter was chosen such that the quality factor (in H_2O and D_2O) was liquid-limited [48]. The Q is determined, as before, by monitoring the linewidth and extinction of a particular optical mode. Further details on the measurement are contained in Ref. 49.

Two series of measurements were performed. In the first series of measurements, the solutions were prepared in 10% increments (10% H_2O in D_2O, 20% H_2O in D_2O, etc.), starting with 100% D_2O. The quality factor was measured for a given concentration. The chamber was then flushed 5 times with the next concentration solution and the quality factor was determined again. Figure 13-4 shows a series of Q factor measurements taken in this manner. As expected, when the concentration of D_2O was reduced, the quality factor decreased. The theoretical values for each concentration were calculated and are indicated by the dashed line. This Q decrease was reversible, and by increasing the D_2O concentration, the quality factor is recovered as can be seen in the sequential runs.

To determine the detection sensitivity, larger dilutions of D_2O in H_2O were prepared, ranging from 0.01 to 1×10^{-9}%. As can be seen in the inset to Fig. 13-4, there is a strong signal at 0.001% D_2O in H_2O and a small, yet detectable, shift occurs with the 0.0001% D_2O solution [49].

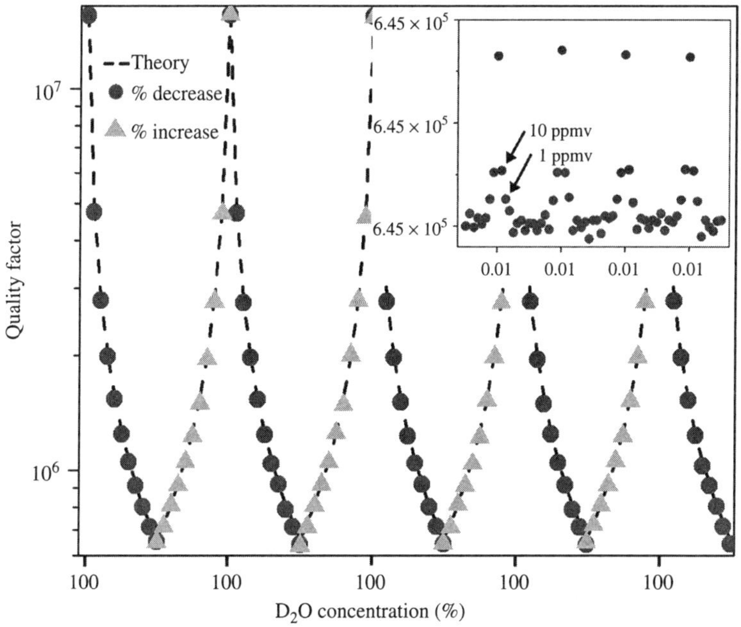

FIGURE 13-4 Ultrasensitive detection of heavy water. The quality factor is decreased (circles) and recovered (triangles) as the D_2O and H_2O are exchanged repeatedly in 10% concentration increments. The measurement is cycled several times showing that the measurement is reversible. (Inset) Starting with 100% H_2O, the concentration of D_2O was gradually increased using low-concentration solutions ranging from 1×10^{-9} to 0.01%. The minimum detectable change in Q was at 0.0001% (1 part per million per volume (ppmv)), indicated by arrows. (*A. M. Armani and K. J. Vahala, "Heavy water detection using ultrahigh-Q microcavities," Optics Letters, 31, 1896–1898, 2006.*)

These values are not believed to reflect the fundamental limit of the detection sensitivity of this device since no attempt was made to reduce operational sources of noise.

Based on the difference in optical absorption between H_2O and D_2O, the ultrahigh-Q microcavity has demonstrated the ability to detect 0.0001% (1 ppmv) of D_2O in H_2O. This form of detection illustrates a mode of operation in which Q factor is directly varied by a substance.

Allergen Detection

The present set of experiments are unique in that they directly compare detection results obtained using a microcavity with those obtained using a fluorescent measurement technique [59]. ELISAs are routinely performed in both university and medical settings and, as such, are considered the gold standard of diagnostics. The specific antigens targeted were Phl p 2 and Phl p 5, two major timothy grass allergens [60–63]. Grass pollen allergens are among the most potent elicitors of type I allergy, affecting more than 20% of the population of industrialized countries [64]. Phl p 2 and Phl p 5 were chosen because they are recognized by nearly all grass pollen allergic people [65]. For these experiments, monoclonal antibodies for both Phl p 2 and Phl p 5 were generated. The construction, expression, and purification of these specific IgG 1 antibodies are described [62,63]. The purified recombinant (r) Phl p 1, Phl p 2, Phl p 5, and Bet v 1 were purchased. The ELISA experiments performed verified both the activity and the negligible (undetectable) cross-reactivity of the antibodies to these allergens.

A similar set of the experiments were performed using the microcavity sensor. One focused on verifying the activity of each allergen to its specific antibody; the second focused on demonstrating detection when the allergens were mixed. Pure solutions of the individual allergens were prepared at 3×10^{-16} M (300 aM); a 300 aM/300 aM mixed solution was also prepared. The solutions were injected at controlled flow rates into the volume around the sensor using a syringe pump. Experiments were repeated with many microtoroid devices ($N = 6$) and yielded highly reproducible resonant wavelength red-shifts.

As can be seen in Fig. 13-5a, when the microtoroid sensor targeted Phl p 2, only Phl p 2 bound to the surface and produced a resonant-frequency shift; Phl p 5 was rejected from the surface to the extent that no binding events were observed, even at the single-molecule level. Similarly, when the microtoroid was targeted to Phl p 5, only Phl p 5 bound and Phl p 2 was rejected from the surface, even at the single-molecule level (Fig. 13-5b). These results demonstrate that the microtoroid sensor produces the same specificity as observed in the ELISA measurements, and does so without requiring a second antibody or an alternative optical marker. These experiments were performed at exceptionally low concentrations, enabling resolution of individual binding events.

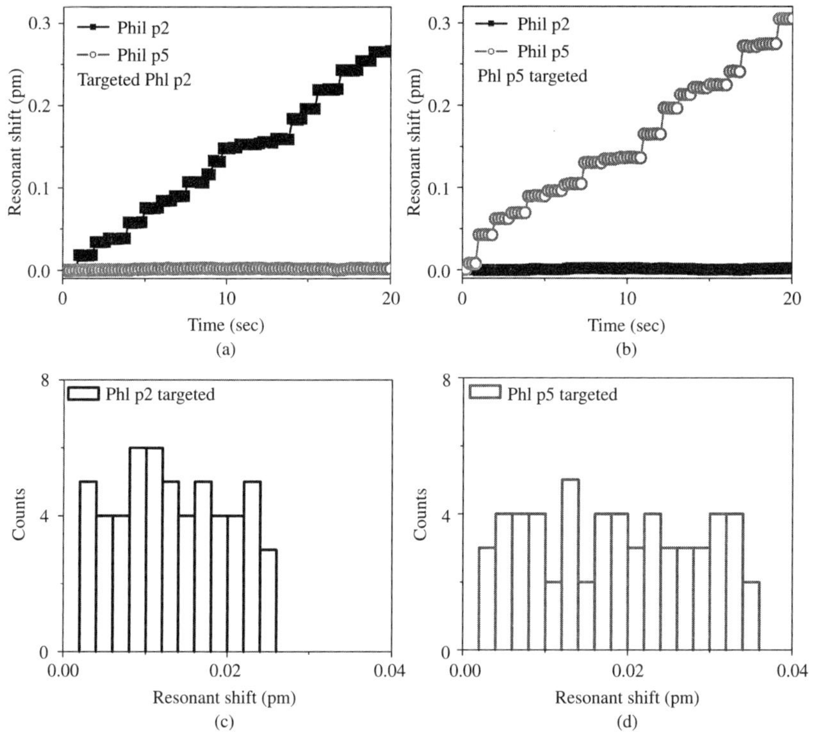

FIGURE 13-5 Verification of activity of the Phl p 2–specific and Phl p 5–specific IgG using the microtoroid sensor. Pure solutions of Phl p 2 and Phl p 5 were flowed over microtoroid sensitized using (a) Phl p 2–specific IgG or (b) Phl p 5–specific IgG. As the antigen bound to the surface of the microtoroid, the resonant wavelength red-shifted. Only the correct antigen bound, indicating no cross-reactivity between the incorrect antigen and that the correct pair had activity. (c) Histogram of the resonant wavelength shift versus time data for Phl p 2. The maximum shift induced was 0.0245 pm. (d) Histogram of the resonant wavelength shift versus time data for Phl p 5. The maximum shift induced was 0.0348 pm. Histogram bin size in both histograms was 0.002 pm. Shifts below 0.002 pm was considered noise and not included. (*A. M. Armani, "Biophotonics: resonant cavity-based biosensors", Optomechatronic Technologies, SPIE Proceedings, vol. 7266, Paper 7266A-113, 2008.*)

Figure 13-5c and 13-5d show histograms of the individual binding events compiled from the data in Fig. 13-5a and 13-5b. The maximum resonant wavelength shift, $\delta\lambda$ for Phl p 2 (0.0245 pm) differs from that observed for Phl p 5 (0.0348 pm) because Phl p 2 and Phl p 5 have different absorption cross sections (σPhl p 2 = 1.8×10^{-16} cm^2, σPhl p 5 = 2.55×10^{-16} cm^2). A highly simplified expression for the maximum resonant wavelength shift is $\delta\lambda = C\sigma$, where C is an empirically-determined constant that varies with toroid dimensions and analyte [67]. The maximum wavelength shift is induced when the molecule binds at the region of highest intensity; however, there is a distribution in resonant wavelengths shifts (as seen in Fig. 13-5c and 13-5d); the whispering gallery mode produces distribution of optical intensities about the midplane that is approximately gaussian; molecules binding to different locations on the resonator surface are, thus, probed with a range of optical intensities [67].

Figure 13-6a and 13-6b show the resonant wavelength shifts when microtoroids targeting either Phl p 2 or Phl p 5 were exposed to solutions containing both Phl p 2 and Phl p 5. In these experiments, the sensor's ability to accurately discriminate between the different allergens relied on the affinity of the antibody. As can be seen in Fig. 13-6c and 13-6d, the maximum signals for each of the targeted antigens

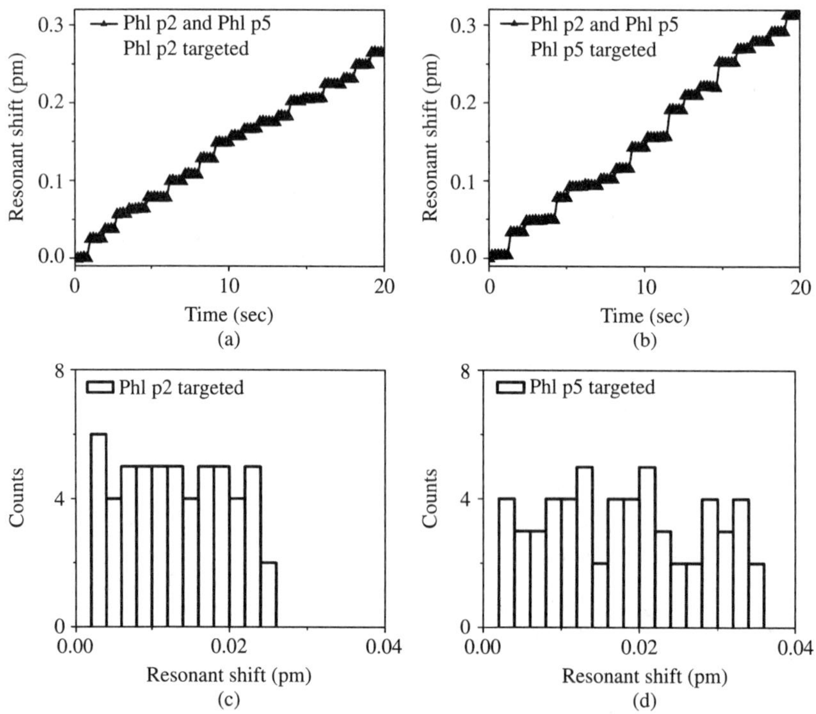

Figure 13-6 Verification of activity of the Phl p 2–specific and Phl p 5–specific IgG using the microtoroid sensor when the allergens are mixed. Mixed solutions of Phl p 2 and Phl p 5 were flowed over microtoroid sensitized using (a) Phl p 2–specific IgG or (b) Phl p 5–specific IgG. (c) Histogram of the resonant wavelength shift versus time data for Phl p 2. The maximum shift induced was 0.0246 pm. (d) Histogram of the resonant wavelength shift versus time data for Phl p 5. The maximum shift induced was 0.0347 pm. Only the correct antigen bound, as verified by comparing these maximum wavelength shifts with those in Fig. 13-5, indicating no cross-reactivity. Histogram bin size in both histograms was 0.002 pm. Shifts below 0.002 pm was considered noise and not included. (*A. M. Armani, "Biophotonics: resonant cavity-based biosensors", Optomechatronic Technologies, SPIE Proceedings, vol. 7266, Paper 7266A-113, 2008.*)

agrees well with those observed when exposed to the appropriate single antigen, indicating that there was high specificity even in the presence of a competing allergen.

13-3 Summary and Future Outlook

The optical microcavity technique is one of only a few methods capable of performing single-molecule measurements [68,69]. Within this small group of detection technologies, it is the only technique capable of performing label-free, single-molecule measurements at room temperature; thus enabling measurements of biological specimens, such as the example of allergen-detection shown here, in real time. Such systems may also prove useful for performing atmospheric measurements as well.

One of the benefits of using planar optical resonators is the numerous avenues available for integration. For example, an area of future study will be the incorporation of microfluidic control. Microfluidics will enable directed delivery of small volumes of liquid (reagents, molecules) to the sensor surface, improving sensing efficiency [70–74].

References

1. D. W. Vernooy, V. S. Ilchenko, H. Mabuchi, E. W. Streed, and H. J. Kimble, "High-*Q* measurements of fused-silica microspheres in the near infrared," *Optics Letters*, 23, 247–249, 1998.
2. M. L. Gorodetsky, A. A. Savchenkov, and V. S. Ilchenko, "Ultimate *Q* of optical microsphere resonators," *Optics Letters*, 21, 453–455, 1996.
3. C. P. Michael, K. Srinivasan, T. J. Johnson, O. Painter, K. H. Lee, K. Hennessy, H. Kim, and E. Hu, "Wavelength- and material-dependent absorption in GaAs and AlGaAs microcavities," *Applied Physics Letters*, 90, 2007.
4. P. E. Barclay, K. Srinivasan, O. Painter, B. Lev, and H. Mabuchi, "Integration of fiber-coupled high-*Q* SiNx microdisks with atom chips," *Applied Physics Letters*, 89, 2006.
5. M. Borselli, T. J. Johnson, and O. Painter, "Beyond the Rayleigh scattering limit in high-*Q* silicon microdisks: theory and experiment," *Optics Express*, 13, 1515–1530, 2005.
6. K. Srinivasan, M. Borselli, and O. Painter, "Cavity *Q*, mode volume, and lasing threshold in small diameter AlGaAs microdisks with embedded quantum dots," *Optics Express*, 14, 1094–1105, 2006.
7. D. K. Armani, T. J. Kippenberg, S. M. Spillane, and K. J. Vahala, "Ultrahigh-*Q* toroid microcavity on a chip," *Nature*, 421, 925–928, 2003.
8. Q. F. Xu and M. Lipson, "All-optical logic based on silicon microring resonators," *Optics Express*, 15, 924–929, 2007.
9. P. Rabiei, W. H. Steier, C. Zhang, and L. R. Dalton, "Polymer microring filters and modulators," *Journal of Lightwave Technology*, 20, 1968–1975, 2002.
10. C. Y. Chao and L. J. Guo, "Polymer microring resonators fabricated by nanoimprint technique," *Journal of Vacuum Science Technology B*, 20, 2862–2866, 2002.
11. B. Little, S. Chu, H. Haus, J. Foresi, and J. Laine, "Microring resonator channel dropping filter," *J. of Lightwave Technol*, 15, 998–1005, 1997.
12. T. Carmon, T. J. Kippenberg, L. Yang, H. Rokhsari, S. Spillane, and K. J. Vahala, "Feedback control of ultrahigh-*Q* microcavities: application to micro-Raman lasers and microparametric oscillators," *Optics Express*, 13, 3558–3566, 2005.

13. H. Rokhsari, T. J. Kippenberg, T. Carmon, and K. J. Vahala, "Radiation-pressure-driven micro-mechanical oscillator," *Optics Express*, 13, 5293–5301, 2005.
14. C. Hood, T. Lynn, A. Doherty, A. Parkins, and H. Kimble, "The atom-cavity microscope: Single atoms bound in orbit by single photons," *Science*, 287, 1447–1453, 2000.
15. A. Yariv, *Optical Electronics in Modern Communications*, 5th ed., Oxford University Press, New York 1997.
16. O. Painter, R. Lee, A. Scherer, A. Yariv, J. O'Brien, P. Dapkus, and I. Kim, "Two-dimensional photonic bandgap defect mode laser," *Science*, 284, 1819–1821, 1999.
17. J. Scheuer and A. Yariv, "Two-dimensional optical ring resonators based on radial Bragg resonance," *Optics Letters*, 28, 1528–1530, 2003.
18. Y. Xu, W. Liang, A. Yariv, J. Fleming, and S. Lin, "High quality-factor Bragg onion resonators with omnidirectional reflector cladding," *Optics Letters*, 28, 2144–2146, 2003.
19. A. L. Martin, D. K. Armani, L. Yang, and K. J. Vahala, "Replica-molded high-Q polymer microresonators," *Optics Letters*, 29, 533–535, 2004.
20. J. K. S. Poon, L. Zhu, G. A. DeRose, and A. Yariv, "Transmission and group delay of microring coupled-resonator optical waveguides," *Optics Letters*, 31, 456–458, 2006.
21. C. Y. Chao and L. J. Guo, "Biochemical sensors based on polymer microrings with sharp asymmetrical resonance, *Applied Physics Letters*, 83, 1527–1529, 2003.
22. G. M. Hale and M. R. Querry, "Optical-constants of water in 200-nm to 200-μm wavelength region," *Applied Optics*, 12, 555–563, 1973.
23. P. Rabiei and W. H. Steier, "Tunable polymer double microring filters," *IEEE Photonics Technology Letters*, 15, 1255–1257, 2003.
24. T. J. Kippenberg, S. M. Spillane, D. K. Armani, and K. J. Vahala, "Ultralow-threshold microcavity Raman laser on a microelectronic chip," *Optics Letters*, 29, 1224–1226, 2004.
25. L. Yang, D. K. Armani, and K. J. Vahala, "Fiber-coupled erbium microlasers on a chip," *Applied Physics Letters*, 83, 825–826, 2003.
26. S. M. Spillane, T. J. Kippenberg, and K. J. Vahala, "Ultralow-threshold Raman laser using a spherical dielectric microcavity," *Nature*, 415, 621–623, 2002.
27. H. Takeuchi, K. Natsume, S. Suzuki, and H. Sakata, "Microcavity distributed-feedback laser using dye-doped polymeric thin films," *Electronics Letters*, 43, 30–32, 2007.
28. T. J. Kippenberg, H. Rokhsari, T. Carmon, A. Scherer, and K. J. Vahala, "Analysis of radiation-pressure induced mechanical oscillation of an optical microcavity," *Physical Review Letters*, 95, 2005.
29. H. Rokhsari, S. M. Spillane, and K. J. Vahala, "Loss characterization in microcavities using the thermal bistability effect," *Applied Physics Letters*, 85, 3029–3031, 2004.
30. H. Rokhsari and K. J. Vahala, "Observation of Kerr nonlinearity in microcavities at room temperature," *Optics Letters*, 30, 427–429, 2005.
31. T. Aoki, B. Dayan, E. Wilcut, W. P. Bowen, A. S. Parkins, T. J. Kippenberg, K. J. Vahala, and H. J. Kimble, "Observation of strong coupling between one atom and a monolithic microresonator," *Nature*, 443, 671–674, 2006.
32. S. M. Spillane, T. J. Kippenberg, O. J. Painter, and K. J. Vahala, "Ideality in a fiber-taper-coupled microresonator system for application to cavity quantum electrodynamics," *Physical Review Letters*, 91, 2003.
33. K. Srinivasan, P. E. Barclay, and O. Painter, "Photonic crystal microcavities for chip-based cavity QED," *Physica Status Solidi B-Basic Solid State Physics*, 242, 1187–1191, 2005.
34. C. Monat, P. Domachuk, and B.J. Eggleton, "Integrated optofluidics: A new river of light," *Nature Photonics*, 1, 106–114, 2006.
35. E. Krioukov, D. J. W. Klunder, A. Driessen, J. Dreve, and C. Otto, "Sensor based on an integrated optical microcavity," *Optics Letters*, 27, 512–514, 2002.
36. A. M. Armani, R. P. Kulkarni, S. E. Fraser, R. C. Flagan, and K. J. Vahala, "Label-free, single-molecule detection with optical microcavities," *Science*, 317, 783, 2007. Published online July 5, 2007 [DOI: 10.1126/*science*.1145002].

37. K. J. Vahala, "Optical microcavities," *Nature*, 424, 839–846, 2003.
38. K. J. Vahala, "Optical microcavities (Advanced Series in Applied Physics)," *World Scientific Publishing Company*, 502 pages, 2005.
39. M. C. M. Lee and M. C. Wu, "MEMS-actuated microdisk resonators with variable power coupling ratios," *IEEE Photonics Technology Letters*, 17, 1034–1036, 2005.
40. S. Y. Cho and N. M. Jokerst, "A polymer microdisk photonic sensor integrated onto silicon," *IEEE Photonics Technology Letters*, 18, 2096–2098, 2006.
41. P. Dumais, C. L. Callender, J. P. Noad, and C. J. Ledderhof, "Silica-on-silicon optical sensor based on integrated waveguides and microchannels," *IEEE Photonics Technology Letters*, 17, 441–443, 2005.
42. F. Yi, X. F. Jin, S. L. Zheng, and X. M. Zhang, "Investigation on spectra of prism-coupled microdisk resonator," *Microwave and Optical Technology Letters*, 48, 1265–1269, 2006.
43. B. E. Little, J. P. Laine, and H. A. Haus, "Analytic theory of coupling from tapered fibers and half-blocks into microsphere resonators," *Journal of Lightwave Technology*, 17, 704–715, 1999.
44. V. S. Ilchenko, X. S. Yao, and L. Maleki, "Pigtailing the high-Q microsphere cavity: a simple fiber coupler for optical whispering-gallery modes," *Optics Letters*, 24, 723–725, 1999.
45. M. Cai, O. Painter, and K. J. Vahala, "Observation of critical coupling in a fiber taper to a silica-microsphere whispering-gallery mode system," *Physical Review Letters*, 85, 74–77, 2000.
46. M. Cai and K. Vahala, "Highly efficient hybrid fiber taper coupled microsphere laser," *Optics Letters*, 26, 884–886, 2001.
47. S. G. Leon-Saval, T. A. Birks, W. J. Wadsworth, P. S. J. Russell, and M. W. Mason, "Supercontinuum generation in submicron fibre waveguides, *Optics Express*, 12, 2864–2869, 2004.
48. A. M. Armani, D. K. Armani, B. Min, K. J. Vahala, and S. M. Spillane, "Ultrahigh-Q microcavity operation in H_2O and D_2O," *Applied Physics Letters*, 87, 151118, 2005.
49. A. M. Armani and K. J. Vahala, "Heavy water detection using ultrahigh-Q microcavities," *Optics Letters*, 31, 1896–1898, 2006.
50. F. Vollmer, S. Arnold, D. Braun, I. Teraoka, and A. Libchaber, "Multiplexed DNA quantification by spectroscopic shift of two microsphere cavities," *Biophysical Journal*, 85, 1974–1979, 2003.
51. S. Blair and Y. Chen, "Resonant-enhanced evanescent-wave fluorescence biosensing with cylindrical optical cavities," *Applied Optics*, 40, 570–582, 2001.
52. S. Arnold, M. Khoshsima, I. Teraoka, S. Holler, and F. Vollmer, "Shift of whispering gallery modes in microspheres by protein adsorption," *Optics Letters*, 28, 272–274, 2003.
53. A. Ksendzov and Y. Lin, "Integrated optics ring-resonator sensors for protein detection," *Optics Letters*, 30, 3344–3346, 2005.
54. R. W. Boyd and J. E. Heebner, "Sensitive disk resonator photonic biosensor," *Applied Optics*, 40, 5742–5747, 2001.
55. L. A. Klumb, V. Chu, and P. S. Stayton, "Energetic roles of hydrogen bonds at the ureido oxygen binding pocket in the streptavidin-biotin complex," *Biochemistry*, 37, 7657–7663, 1998.
56. R. A. Vijayendran and D. E. Leckband, "A quantitative assessment of heterogeneity for surface-immobilized proteins," *Analytical Chemistry*, 73, 471–480, 2001.
57. J. Yakovleva, R. Davidsson, A. Lobanova, M. Bengtsson, S. Eremin, T. Laurell, and J. Emneus, "Microfluidic enzyme immunoassay using silicon microchip with immobilized antibodies and chemiluminescence detection," *Analytical Chemistry*, 74, 2994–3004, 2002.
58. P. H. Maddox and D. Jenkins, "3-Aminopropyltriethoxysilane (Apes–a new advance in section adhesion," *Journal of Clinical Pathology*, 40, 1256–1257, 1987.
59. A. M. Armani, S. Flicker, M. Glovsky, R. Valenta, and R. C. Flagan, "Detection of attomolar concentation of allergens," *in review at Journal of Allergy and Clinical Immunology*, 2008.

60. S. Vrtala, W. R. Sperr, I. Reimitzer, R. Vanree, S. Laffer, W. D. Muller, P. Valent, K. Lechner, H. Rumpold, D. Kraft, O. Scheiner, and R. Valenta, "cDNA cloning of a major allergen from timothy grass (phleum-pratense) pollen–characterization of the recombinant phl p-v allergen," *Journal of Immunology*, 151, 4773–4781, 1993.
61. C. Dolecek, S. Vrtala, S. Laffer, P. Steinberger, D. Kraft, O. Scheiner, and R. Valenta, "Molecular characterization of phl p-ii, a major timothy grass (phleum-pratense) pollen allergen," *Febs. Letters*, 335, 299–304, 1993.
62. S. Flicker, S. Vrtala, P. Steinberger, L. Vangelista, A. Bufe, A. Petersen, M. Ghannadan, et al., "A human monoclonal IgE antibody defines a highly allergenic fragment of the major timothy grass pollen allergen, Phl p 5: Molecular, immunological, and structural characterization of the epitope-containing domain," *Journal of Immunology*, 165, 3849–3859, 2000.
63. S. Flicker, P. Steinberger, L. Norderhaug, W. R. Sperr, Y. Majlesi, P. Valent, D. Kraft, and R. Valenta, "Conversion of grass pollen allergen-specific human IgE into a protective IgG(1) antibody," *European Journal of Immunology*, 32, 2156–2162, 2002.
64. P. G. Holt and W. R. Thomas, "Sensitization to airborne environmental allergens: unresolved issues," *Nature Immunology*, 6, 957–960, 2005.
65. V. Niederberger, S. Laffer, R. Froschl, D. Kraft, H. Rumpold, S. Kapiotis, R. Valenta, and S. Spitzauer, "IgE antibodies to recombinant pollen allergens (Phl p 1, Phl p 2, Phl p 5, and Bet v 2) account for a high percentage of grass pollen-specific IgE," *Journal of Allergy and Clinical Immunology*, 101, 258–264, 1998.
66. A. M. Armani, "Biophotonics: resonant cavity-based biosensors," Optomechatronic Technologies, SPIE Proceedings, vol. 7266, Paper 7266A–113, 2008.
67. A. M. Armani, R. P. Kulkarni, S. E. Fraser, R. C. Flagan, and K. J. Vahala, "Label-free, single-molecule detection with optical microcavities," *Science Express*, 2007.
68. T. P. Burg, M. Godin, S. M. Knudsen, W. Shen, G. Carlson, J. S. Foster, K. Babcock, and S. R. Manalis, "Weighing of biomolecules, single cells and single nanoparticles in fluid," *Nature*, 446, 1066–1069, 2007.
69. F. Patolsky, G. F. Zheng, and C. M. Lieber, "Fabrication of silicon nanowire devices for ultrasensitive, label-free, real-time detection of biological and chemical species," *Nature Protocols*, 1, 1711–1724, 2006.
70. R. E. Oosterbroek, D. C. Hermes, M. Kakuta, F. Benito-Lopez, J. G. E. Gardeniers, W. Verboom, D. N. Reinhoudt, and A. van den Berg, "Fabrication and mechanical testing of glass chips for high-pressure synthetic or analytical chemistry," *Microsystem Technologies-Micro-and Nanosystems-Information Storage and Processing Systems*, 12, 450–454, 2006.
71. M. B. Kerby, R. S. Legge, and A. Tripathi, "Measurements of kinetic parameters in a microfluidic reactor," *Analytical Chemistry*, 78, 8273–8280, 2006.
72. B. H. Jo, L. M. Van Lerberghe, K. M. Motsegood, and D. J. Beebe, "Three-dimensional micro-channel fabrication in polydimethylsiloxane (PDMS) elastomer," *Journal of Microelectromechanical Systems*, 9, 76–81, 2000.
73. D. C. Duffy, J. C. McDonald, O. J. A Schueller, and G. M. Whitesides, "Rapid prototyping of microfluidic systems in poly(dimethylsiloxane)," *Analytical Chemistry*, 70, 4974–4984, 1998.
74. T. Thorsen, S. J. Maerkl, and S. R. Quake, "Microfluidic large-scale integration," *Science*, 298, 580–584, 2002.

CHAPTER 14

Optofluidic Plasmonic Devices

Boris Slutsky, Lin Pang, Joanna Ptasinski, and Yeshaiahu Fainman

Department of Electrical Engineering, University of California at San Diego

Optofluidic plasmonics, consisting of integrated microfluidics with optics and plasmonics, is an emerging research direction that enables advancement of fundamentals in surface sciences of plasmonic fields with unique implications on numerous potential applications in chemistry, biochemistry, biology, medicine, and engineering. Plasmonics possesses unique physical properties that enable localization of optical fields beyond the diffraction limit. These highly confined/nanoscale optical modes will enhance light/matter interactions in systems with free electrons in micro/nanoscale geometric structures. New applications and devices that are expected to directly benefit from these light confined modes include biochemical sensors (SERS, SECARS), optical nonlinearities (SHG, etc.), near-field probes and data storage, nanoscale lasers, left-handed materials and "perfect" lens, enhanced light extraction/detection, detectors and thermo/photovoltaics, subdiffraction-limit lithography, modulators, spectral filters, interconnects, and the like.

Many metals in the optical frequency regime behave as electron plasmas, which below the plasma resonance frequency are characterized by a negative real part of permittivity. This property is equivalent to having a positive quantum mechanical potential as opposed to negative potential corresponding to dielectric materials [1]. Metal-dielectric fluid interfaces can thus support surface plasmon polaritons (SPPs), which are electromagnetic modes interacting with free electron oscillations, and can be thought of as extending evanescent fields from both sides of the interface [2]. For properly chosen parameters, the effective index of the SPP modes can be considerably higher than the index of the surrounding dielectric media and therefore localize the optical fields in a nanoscale volume near metals [3,4]. In the next section, a brief introduction to the basics of the SPP fields and an overview of commonly used excitation techniques with special emphasis on the use of two-dimensional (2D) nanohole arrays is given. Section 14-2 describes the basic technologies involved in fabrication of integrated optofluidic chips consisting of microfabrication of plasmonic structures (e.g., 2D arrays of nanoholes perforating a thin metal film on a solid substrate) and their integration with microfluidic devices encompassing multiple layers of channels and valves used for fluid delivery and control. This optofluidic chips integration technology is also briefly described in Sec. 14-2. To realize the promise of SPP technologies, a comprehensive arsenal of devices for launching, detecting, guiding, imaging, focusing, and otherwise transforming SPP waves must be readily available. However, a challenge remains to excite and control propagating plasmonic fields in a systematic fashion, similar to optical fields in free space and in dielectric waveguides. Coupling of optical fields to excite the surface waves, modal structure of these waves and their ultrafast electrodynamics is advanced in Sec. 14-3. Depending on the geometry and the composition, it is possible, in principle, to achieve from 1D up to 4D confinement (spatial and temporal). The integrated optofluidic plasmonic chips are useful for advanced studies of various SPP modal structures, enabling demonstration of a bandgap and the existence of bright/dark states for degenerate modes (Sec. 14-3). Optofluidic plasmonic chips are further used for implementation of an optofluidic plasmonic sensor with angular and wavelength (Sec. 14-4) interrogation, demonstrating in situ, real time, label-free detection of protein-protein interaction. These experiments reveal the dynamics of protein-protein interactions, essential not only for advancement of biological research, but also for reduction of false alarms in biochemical sensing. Optofluidic plasmonic sensors can be combined into a high density 2D array to provide a very large throughput (e.g., about a million independent measurements) with high sensitivity and resolution, operating with small volumes, in real time and without any labels. These functionalities are useful for various applications including drug discovery and proteomics, in vitro diagnostics, food and drug industry, environmental and process monitoring, as well as military/homeland security applications. Summary and discussions of future optofluidic plasmonic sensors research directions in Sec. 14-5 conclude this chapter.

14-1 Basic Properties of Surface Plasmon Polaritons

Surface plasmon polariton (SPP) waves (see, e.g., Barnes et al. [3,5]) are longitudinal electron density waves propagating along a metal-dielectric interface (Fig. 14-1). Physically, the waves arise from the interplay of the mechanical inertia of the quasi–free electrons in the

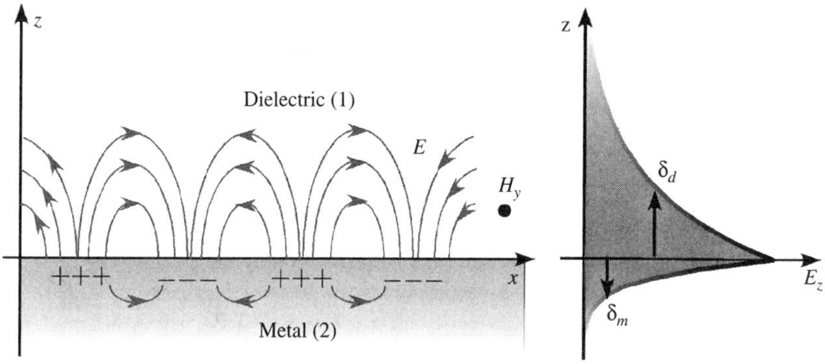

FIGURE 14-1 Schematic representation of a longitudinal electron density wave at a metal-dielectric interface. Electric field lines connect positive and negative electron density antinodes of the wave. Magnetic field is parallel to the interface. The plot on the right illustrates evanescent decay of the fields away from the interface. Typically, the fields extend a few hundred nanometers into the dielectric and a few tens of nanometers into the metal.

metal and their electrostatic repulsion, similarly to density waves in a gas of electrically charged particles (hence the term "plasmons"). Mathematically, the SPP are a solution of Maxwell's field equations at a planar interface when the relative permittivity on one side is large and negative, as is the case in many metals at optical frequencies.

14-1-1 SPP Dispersion Relation at a Metal-Dielectric Interface

The TM-like (i.e., with magnetic field parallel to the interface) SPP solutions are formally found as follows. Let x be the direction of propagation of the surface wave, and let $\mathbf{H}_j = \hat{\mathbf{y}} H_y \exp[i(k_x x + k_{jz} z - \omega t)]$, $j = 1, 2$, represent the magnetic fields in the dielectric ($j = 1$) and the metal ($j = 2$); by adopting this notation, the continuity of H_y across the interface is assured. From the Maxwell's $\partial \mathbf{D}_j/\partial t = \varepsilon_j \varepsilon_0(-i\omega)\mathbf{E}_j = \nabla \times \mathbf{H}_j$, $j = 1, 2$, and the boundary condition $E_{1x} = E_{2x}$ one finds $k_{1z}/\varepsilon_1 = k_{2z}/\varepsilon_2$. This must be solved jointly with the wavevector length constraint $k_x^2 + k_{jz}^2 = \varepsilon_j(\omega/c)^2$ (where c denotes the vacuum speed of light). When $\varepsilon_2 < 0$ and $|\varepsilon_2| >> \varepsilon_1$, the result

$$k_x = \frac{\omega}{c}\left(\frac{\varepsilon_1\varepsilon_2}{\varepsilon_1+\varepsilon_2}\right)^{1/2}, \quad k_{1z} = \frac{\omega}{c}\left(\frac{\varepsilon_1^2}{\varepsilon_1+\varepsilon_2}\right)^{1/2}, \quad k_{2z} = \frac{\omega}{c}\left(\frac{\varepsilon_2^2}{\varepsilon_1+\varepsilon_2}\right)^{1/2} \tag{14-1}$$

yields a propagating surface wave with substantially real k_x and evanescent fall-off, away from the interface. By following the same steps with a TE-like (i.e., having the electric rather than magnetic field parallel to the interface) solution, it can be shown that no such solution exists: A TE-like SPP would have required a medium with a negative magnetic permeability μ.

Field penetration depths into the dielectric and the metal are obtained as $\delta_j = \frac{1}{2}\text{Im}\left[k_{jz}^{-1}\right]$ with, respectively, $j = 1$ and $j = 2$:

$$\delta_1 = \frac{\lambda_0}{4\pi}\text{Im}\left[\left(\frac{\varepsilon_1+\varepsilon_2}{\varepsilon_1^2}\right)^{1/2}\right], \quad \delta_2 = \frac{\lambda_0}{4\pi}\text{Im}\left[\left(\frac{\varepsilon_1+\varepsilon_2}{\varepsilon_2^2}\right)^{1/2}\right]$$

where $\lambda_0 = 2\pi c/\omega$ is the free-space wavelength. When the permittivity ε_2 has an imaginary part representing Joule power dissipation, the SPP incurs propagation loss, with characteristic SPP propagation length

$$L_{\text{SPP}} = \frac{1}{2}\text{Im}\left[k_x^{-1}\right] = \frac{\lambda_0}{4\pi}\text{Im}\left[\left(\frac{\varepsilon_1+\varepsilon_2}{\varepsilon_1\varepsilon_2}\right)\right]$$

The typical length scales for water-gold interface and $\lambda_0 = 1550 \cdot \text{nm}$ are $\delta_1 \sim 700 \cdot \text{nm}$, $\delta_2 \sim 15 \cdot \text{nm}$, $L_{\text{SPP}} \sim 90 \cdot \text{nm}$.

The first of Eq. (14-1) can be rewritten as the SPP dispersion relation

$$k_{\text{SPP}}(\omega) = \frac{\omega}{c}\left(\frac{\varepsilon_1 \cdot \varepsilon_2(\omega)}{\varepsilon_1+\varepsilon_2(\omega)}\right)^{1/2} = \frac{\omega n_1}{c}\left(\frac{\varepsilon_2(\omega)}{n_1^2+\varepsilon_2(\omega)}\right)^{1/2}, \quad n_1 \triangleq \sqrt{\varepsilon_1} \tag{14-2}$$

where k_{SPP} is the SPP wavevector in the plane of the interface, and n_1 is the refractive index of the dielectric. In most applications, the behavior of $k_{\text{SPP}}(\omega)$ is governed by the dispersion $\varepsilon_2(\omega)$ of the metal material, shown explicitly in Eq. (14-2). The dispersion of the dielectric is usually negligible.

14-1-2 Optical Excitation of SPP

The conditions $\text{Re}[\varepsilon_2] < 0$, $|\text{Re}[\varepsilon_2]| >> |\text{Im}[\varepsilon_2]|$, $|\text{Re}[\varepsilon_2]| >> \varepsilon_1$ under which SPP devices normally operate, imply $\text{Re}[k_{\text{SPP}}] > \omega n_1/c$ in Eq. (14-2). On the other hand, for the in-plane wavevector projection $k_\parallel = (\omega n_1/c) \cdot \sin\theta$ of a light beam incident at angle θ from the dielectric side of the interface we have $k_\parallel < \omega n_1/c$. Therefore, additional momentum $\Delta k = k_{\text{SPP}} - k_\parallel$ is necessary in order to achieve phase matching and excite SPP with a light beam (Fig. 14-2a). This difficulty is commonly overcome with evanescent excitation (Fig. 14-2b) or excitation via a surface grating (Fig. 14-2c). The technique illustrated in Fig. 14-2b utilizes a prism made of a dense dielectric material having a refractive index $n_3 > n_1$. Owing to $n_3 > n_1$, the in-plane wavevector projection $k_\parallel = (\omega n_3/c) \cdot \sin\theta$ within the prism can match the SPP wavevector k_{SPP} at the interface of n_1. Excitation occurs via the evanescent fields reaching from the prism to the SPP interface either across a narrow air gap ("Otto configuration") or across a thin metal film ("Ketschmann configuration") [2].

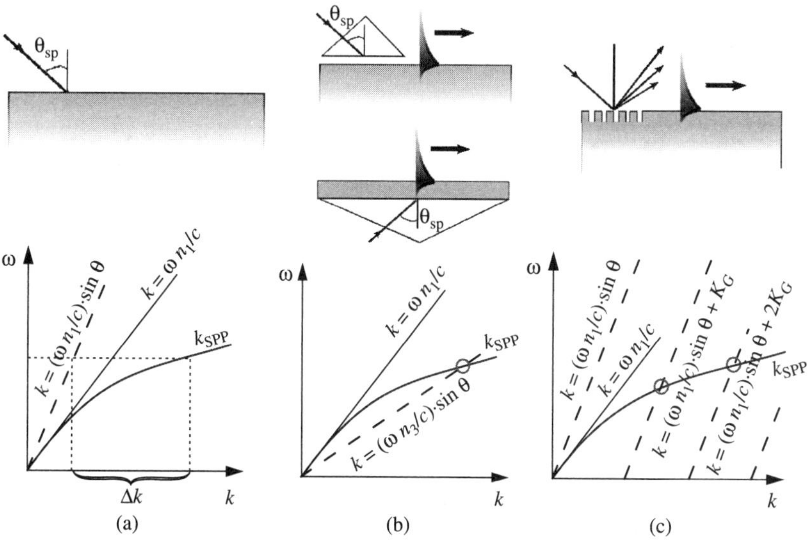

Figure 14-2 (a) Momentum mismatch between SPP and a plane wave incident at angle θ from the dielectric side. SPP excitation does not occur. (b) SPP excitation in Otto (upper diagram) and Kretschmann-Raether (lower diagram) configurations. Owing to the high refractive index $n_3 > n_1$ of the prism material, the in-plane projection $k_{||}$ of the beam wavevector can match k_{SPP} at the interface of n_1. SPP is excited by the evanescent fields reaching across the air gap (Otto configuration) or the thin metal film (Kretschmann configuration). (c) SPP excitation via a surface grating. The grating vector K_G adds to $k_{||}$ and helps match momentum at multiple frequency points (one for each grating order).

The second method, presented in Fig. 14-2c, requires that a surface grating with an appropriately chosen grating period Λ_G be constructed at the coupling point. When light is incident on a surface grating, the wavevectors $k = k_{\|} + nK_G$, $n = \pm 1, \pm 2, \ldots, K_G = 2\pi/\Lambda_G$, become accessible, some or all of which may be used to match the SPP wavevector. The approach is easily generalized to two-dimensional rectangular gratings. In that case, the matching condition takes vector form,

$$\mathbf{k}_{SPP} = \mathbf{k}_{||} + p\frac{2\pi}{\Lambda_x}\hat{\mathbf{x}} + q\frac{2\pi}{\Lambda_y}\hat{\mathbf{y}}, \quad p, q = 0, \pm 1, \pm 2, \ldots \tag{14-3}$$

where Λ_x, Λ_y are the grating periods in x and y directions, $\hat{\mathbf{x}}$, $\hat{\mathbf{y}}$ are the corresponding unit vectors, and the integers p, q identify the grating order that is mediating the coupling between the SPP and the incident light. As in Fig. 14-2b, the light beam can be brought in from below as well as from above provided the metal film is sufficiently thin for the fields to reach the upper interface. Unlike prisms, grating couplers can accommodate excitation at normal or near-normal incidence; this is an important advantage for applications, including multichannel biological sensors discussed later in this chapter, that require the excitation of SPP over a large surface area and the imaging of this area into a CCD camera.

It is important to note that both the prism-based and the grating-based coupling schemes are sharply resonant: For any given incident angle θ, phase matching is achieved and SPP is successfully excited at only one optical frequency $\omega = \omega_r$ (or, in the case of a grating, a discrete set of frequencies corresponding to individual grating orders). This is seen most clearly in Fig. 14-2b and 14-2c, where the excitation points are marked with circles. The resonance frequency ω_r is a function of material parameters, and in particular of the dielectric refractive index n_1, which enters the dispersion relation $k_{SPP}(\omega)$ [Eq. (14-2)]. Small refractive index changes δn_1 of the dielectric can therefore be deduced from changes $\delta\omega_r$ of the SPP resonance frequency ω_r. Furthermore, owing to the confinement of the SPP fields at the interface, only the dielectric properties in the immediate vicinity of the metal surface contribute to $\delta\omega_r$. This circumstance enables a new class of sensing instruments that can monitor progress of surface chemical reactions separately from reactions taking place in the volume.

An optofluidic plasmonic biosensor constructed according to these ideas is depicted schematically in Fig. 14-3. In the figure, the SPP waves propagate on the upper surface of a thin metal film, punctured with a regular pattern of nanoholes acting as a two-dimensional grating. Reagents are delivered to the metal surface through a microfluidic channel. A tunable laser is used to probe the reaction chamber from the back side of the film. When the incident optical frequency is such that the matching condition [Eq. (14-3)] is satisfied, some of the beam energy is coupled into the SPP, and some of the SPP energy subsequently reradiates off the film, leading to an increase in transmission from which the resonance frequency ω_r can be identified. The metal surface is initially activated by adsorbing a known protein (Bovine Serum Albumin, or BSA). Following the activation, a solution containing monoclonal anti-BSA is launched into the microfluidic channel. As the anti-BSA molecules bind to the BSA already resting on the metal surface, the effective refractive index above the metal is modified, and with it the resonance frequency ω_r. The progress of the binding can be monitored in real time by repeatedly scanning the tunable laser source across the relevant frequency band. To the extent that the binding is selective, the instrument can also be used to determine whether a target protein (in this case, anti-BSA) is present in the solution. Any material other than the target would be rejected by the preactivated surface and therefore would not modify the SPP resonance.

The remainder of this chapter focuses in detail on integrated plasmonic-microfluidic cells such as shown in Fig. 14-3. After discussing their fabrication, we report experiments designed to elucidate the behavior of SPP over patterned metal films, present

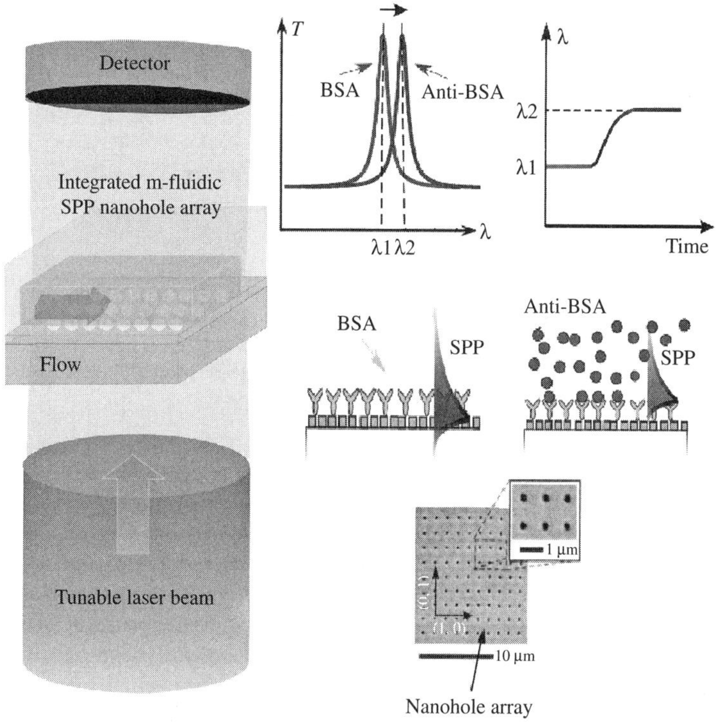

FIGURE 14-3 Sketch and principle of operation of a SPP biosensor utilizing a regularly patterned thin metal film as the SPP-coupling element. (*Reprinted with permission from L. Pang, G. M. Hwang, B. Slutsky, and Y. Fainman, "Spectral sensitivity of two-dimensional nanohole array surface plasmon polariton resonance sensor," Appl. Phys. Lett., 91,12,123112, (2007). Copyright 2007, American Institute of Physics.*)

demonstrations of complete biosensor systems, and examine their performance metrics.

We conclude this section by pointing out the polarization selectivity of SPP coupling. Because the SPP electric field has a vertical and a longitudinal component, but no in-plane component normal to the direction of propagation (Fig. 14-1), an *s*-polarized incident beam and an SPP traveling parallel to the plane of incidence have zero field overlap (Fig. 14-4a). Consequently, no SPP can be launched in this direction even if the momentum-matching condition [Eq. (14-3)] holds. An SPP aimed out of the incidence plane [provided that it also satisfies Eq. (14-3)] is excited most efficiently with the *s*-polarized beam. By contrast, *p*-polarized excitation favors SPP directed along the plane of incidence. The antenna pattern of *p*-polarized excitation has no full nulls because some overlap between the beam and the SPP always exists through the vertical field components.

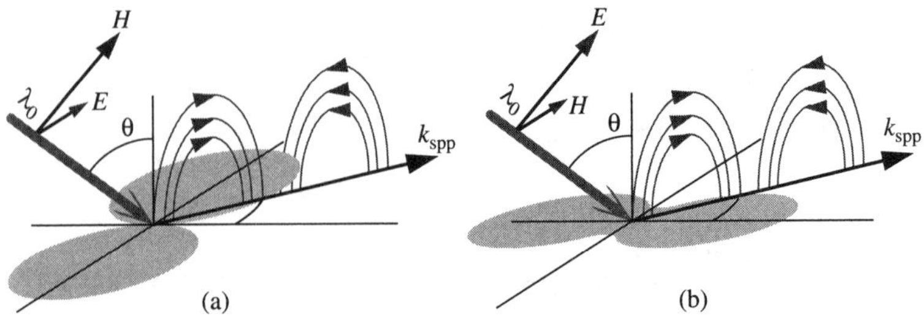

FIGURE 14-4 SPP excitation by (a) *s*-polarized and (b) *p*-polarized light. The shadow drawn in the *xy* plane is an approximate representation of the antenna pattern. The antenna pattern of an *s*-polarized beam has a null in the direction of the plane of incidence because an SPP propagating in this direction has zero field overlap with the beam.

14-2 Fabrication of Optofluidic Plasmonic Chips

An integrated plasmonic-microfluidic cell sketched in Fig. 14-3 is fabricated in several steps. First, a thin metal film is deposited on a solid substrate. A regular pattern of nanoholes is then lithographically defined and etched into the film. Separately, a system of microfluidic channels for mixing and delivery of reagents is constructed by pouring an elastomer over a mold and curing it to a rubber-like flexible form. To complete the device, the elastomer is bonded over the metal film. We take up these fabrication steps in turn.

14-2-1 Deposition of the Metal Film

The process starts with the deposition of a thin metal film on an optically transparent solid substrate. Both semiconductor and insulator substrates are acceptable; nominally undoped Silicon (100), undoped GaAs (100), and optical-quality quartz substrates have been used in our work. All substrates are cleaned prior to loading in a thermal evaporator for the deposition of a gold or aluminum film of desired thickness (generally 50–200·nm).

To achieve good gold-adhesion, a thin (several nanometers) layer of nickel, chromium, or titanium is deposited between gold and the substrate. In some cases we also prepared gold films by sputtering, which is more efficient for large area samples.

14-2-2 Lithographic Definition of the Nanohole Pattern

The two-dimensional pattern of nanoholes in the metal film is defined by either electron beam or holographic photolithography. Holographic lithography can quickly and inexpensively create periodic patterns over large sample areas; the e-beam method is slower, but offers greater flexibility in nanohole positioning.

When patterning with e-beam, we follow the standard process, consisting of the following steps: (1) an e-beam resist is spin-coated on the substrate; (2) the sample is prebaked to remove solvent from the resist; (3) the pattern is exposed on the resist by a scanning electron beam; (4) the resist is chemically developed; and (5) postbake is optionally used to enhance etch resistance of the developed mask. The e-beam resist used in our work is 950K PMMA (950 A4 from MicroChem), which we spin at a typical speed of 4000·rpm for 40·s to obtain a 200-nm-thick layer. The temperature of both prebake and postbake is 170°C and the duration is 90·min. The e-beam writer is a converted scanning electron microscope (SEM) with nanometer pattern generation system (NPGS) from JC Nabity Lithography Systems. This writer offers a resolution of approximately 100·nm over a maximum field of view (and therefore exposure area) of approximately 200 µm × 200·µm. To pattern larger fields at the same high resolution, stitching and/or multiple patterning must be employed. This makes fabrication costly and requires long writing times.

Holographic lithography utilizes interference of two ultraviolet (UV) laser beams to create a fringe pattern, and hence an exposure pattern, in the form of a linear grating. The pitch of the grating depends on the angle between the beams, and can be as small as a fraction of a micron. In this way, large areas can be patterned quickly and cost-effectively. The method is limited to specific applications, however, because only certain periodic patterns can be constructed. A number of authors are using holographic techniques, but most resists are thin and soft, and therefore limit their applicability to use as a mask for transfer of the pattern to metallic films [6].

We have developed a custom process for holographic patterning of two-dimensional nanohole arrays by two successive two-beam interference exposures [7,8]. The optical setup is pictured in Fig. 14-5. An Ar^+ ion laser operating at the wavelength $\lambda = 364$·nm is used as the UV source. Its output is expanded and collimated, and then divided into two beams with a nonpolarizing UV beam splitter. Two mirrors direct the beams at equal incident angles θ toward the sample, where they recombine and form linear interference fringes with fringe-to-fringe distance $d = (\lambda/2)/\sin\theta$. After the sample is exposed once, rotated 90° in its own plane, and exposed again, the cumulative exposure profile has an egg carton shape seen in Fig. 14-5b. In a negative photoresist, the high exposure points produce a rectangular pattern of circular holes when developed. The diameter of the holes, particularly in nonlinear resists such as SU-8, predictably depends on the exposure dose, while the hole-to-hole distance is determined by the fringe pattern and can be controlled through the incident angle θ. Various other mesh patterns can also be obtained in this manner by combining multiple exposures and different sample rotations.

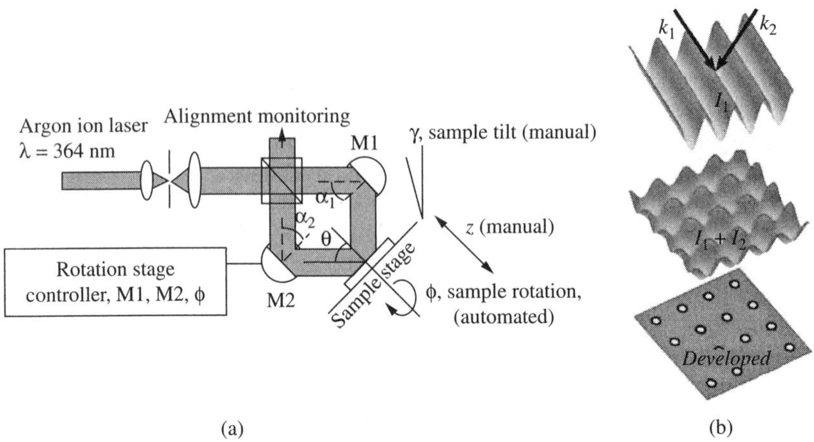

FIGURE 14-5 Multiple-exposure dual-beam holographic lithography: Optical setup (a) and basic operating principle (b). A linear fringe pattern is created by interfering two beams derived from the same 364-nm laser source. The interference angle θ, which determines the period of the pattern, is set by two computer controlled, high-precision rotation stages. The third rotation stage sets the sample azimuthal angle φ. Two successive exposures at $\phi = 0°$ and at $\phi = 90°$ imprint a rectangular mesh onto the sample, which, when developed, produces a nanohole array. Due to the nonlinear response of the photoresist, greater exposures result in larger holes.

To create a uniform hole pattern over a large area, a UV beam of uniform intensity is required. This requirement can be met, for example, by using the central section of a larger Gaussian beam, or using scanning method. Current fabrication approach of metal-dielectric periodical nanostructures can be further extended to larger areas, more complex shapes, and higher uniformity across the array by automation and phase-stabilization of optical interferometric lithography systems.

14-2-3 Etching

The nanohole pattern can be transferred onto the metal film by either wet or dry etching. Wet etching can be employed with thin (less than 100-nm) metal films and moderate size holes. For thicker films and/or narrower holes, dry etching, such as Reactive Ion Etching/Inductive Coupled Plasma (RIE/ICP) with suitable etching gases, is preferred due to its anisotropic nature.

The dry-etching process is simplified if the developed photoresist pattern can be used directly as the etching mask, avoiding the liftoff to a harder mask. In the experiments reported here, we used the chemically amplified negative SU-8 photoresist, which is relatively resistant to ion bombardment. The etching rate ratio of the resist to that of the underlying material determines the required resist thickness. This ratio varies with the material, the etching gas, chamber pressure,

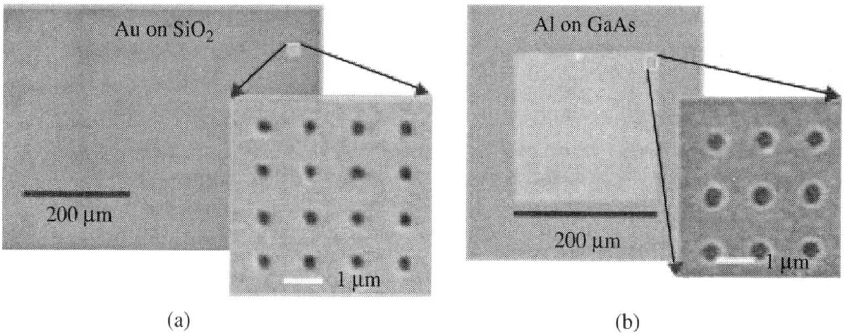

FIGURE 14-6 Optical images of nanohole arrays fabricated by holographic lithography (a) and direct e-beam writing (b). Insets show representative SEM images of typical arrays.

plasma density, RF power, feature size, and other process parameters. As an example, the figure is 8:1 for SU-8 over gallium arsenide (GaAs), meaning that only 100-nm of SU-8 is needed to etch 800-nm of GaAs [8]. Gold is relatively resistant to dry etching due to the nonvolatile reactants. However, etching rate ratios close to 8:1 can also be achieved for gold films by adjusting the chamber pressure and the proportion of chlorine and boron trichloride gases. Typical images of nanohole arrays fabricated by e-beam and holographic lithography methods are reproduced in Fig. 14-6.

14-2-4 Fabrication of Microfluidic Channels

The microfluidic part of the device is fabricated in polydimethylsiloxane (PDMS) with the rapid prototyping technique originally developed by Whitesides' group [9,10], as illustrated in Fig. 14-7. The initial hard molds are constructed on a flat wafer from two separate photoresists. First, a 75-μm-thick layer of the negative SU-8 resist (MicroChem SU8-2050) is spin-coated onto the wafer, patterned through a photomask, and developed. The second resist (Clariant AZ100XT) is spin-coated in a 100-μm layer over the first resist, patterned through a second photomask aligned with the first, and also

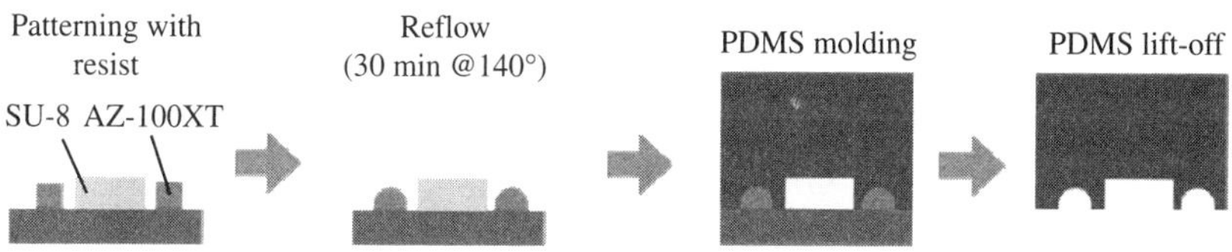

FIGURE 14-7 Fabrication of fluidic channels in PDMS. (*A. Groisman, S. Zamek, K. Campbell, L. Pang, U. Levy, and Y. Fainman, "Optofluidic 1x4 switch," Opt. Express 16, 13499-13508, 2008.*)

developed. Regular near-UV contact photolithography with 8000-dpi-resolution masks is used in both cases. The sample is then baked for 30-min on a 140°C hot plate to reflow the AZ100XT and round its edges; the bake has no effect on the SU-8. In this manner, two types of features are created in the same hard mold, one of a square and the other of a rounded cross section, and, if desired, differing in height.

Once the hard mold has been fabricated, the resin and catalyst parts of the silicon elastomer (Dow Corning Sylgard 184) mixed in 5:1 proportion are poured into it to a depth of afew millimeters. The elastomer is cured to a flexible rubber-like sheet by a 25-min, 80°C oven bake, and separated from the mold. When the sheet is placed on a flat surface, the relief grooves impressed into it by the mold become enclosed channels. To access the channels, holes are punched from the top through the PDMS. The fluidic network can have any desired in-plane layout and interconnections as determined by the mold and the original photomasks.

To assemble the plasmonic microfluidic chip, the PDMS sheet is treated with oxygen plasma and pressed to the substrate where the patterned metal film had been prepared. Because the PDMS does not bond well to the gold film, the film is fabricated only under the measurement area of the chip, with the rest of the substrate stripped bare. This ensures good mechanical connection to PDMS that can withstand fluidic pressures without leakage. A completed device can be seen in Fig. 14-8. In our experiments, fluidic inlets and outlets were connected via 1-mm ID flexible tygon tubing to 60-cm^3 plastic syringes (not shown). The syringes were open to atmosphere and mounted vertically at different heights to control the pressures and flow rates.

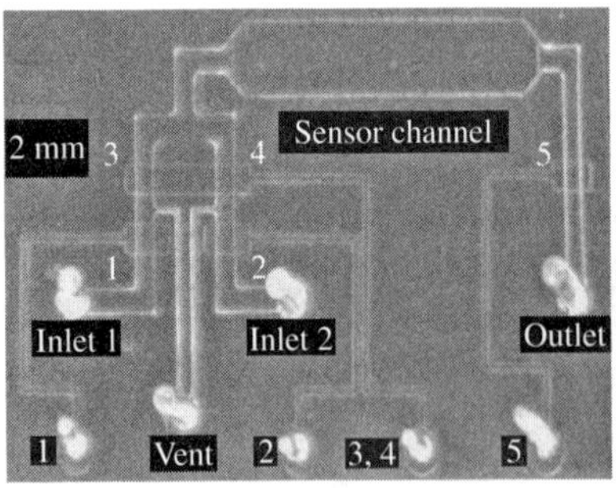

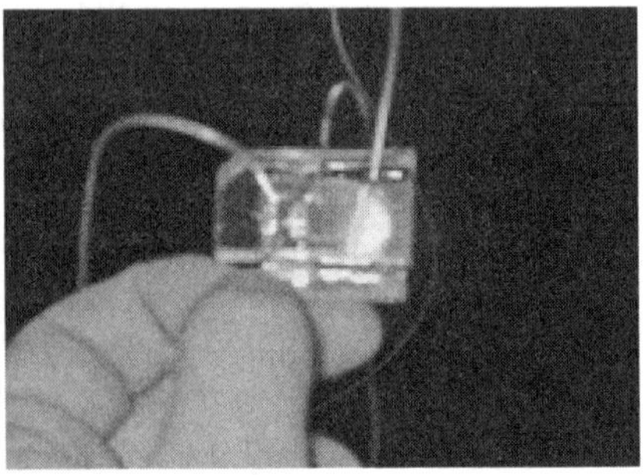

FIGURE 14-8 Integrated plasmonic-microfluidic chip. (*Reprinted with permission from K. Campbell, A. Groisman, U. Levy, L. Pang, S. Mookherjea, D. Psaltis, and Y. Fainman, "A microfluidic 2×2 optical switch," Appl. Phys. Lett., 85, 25, 6119, Dec 2004. Copyright 2004, American Institute of Physics.*) The sensor channel (reaction chamber) is 10-mm long, 2-mm wide, and 75-μm deep. A photograph of the completed device with external fluidic connections is shown on the right. (*G. M. Hwang, L. Pang, E. H. Mullen, and Y. Fainman, "Plasmonic Sensing of Biological Analytes through Nanoholes," IEEE Sensors J., 8, 2074-2079, 2008.*)

14-3 Experimental Observation of SPP Coupling, Propagation and Focusing, and SPP Mode Splitting

SPP waves at a metal-dielectric interface have recently become subject of renewed attention. The interest was sparked by the discovery of resonantly enhanced light transmission through films with regular patterns of subwavelength holes, or single holes surrounded by surface corrugations [14,15]. Light transmission through such structures was found to be several times greater than what might be expected based on the aperture size. In some cases at least, SPP excitation is believed to be responsible for this phenomenon. When conditions are favorable for the excitation of SPP at the interface, part of the incident light energy is coupled to SPP, and part of the SPP energy subsequently reradiates on either side of the film, interfering with directly transmitted light [16]. If the directly transmitted component is small relative to the reradiated component, the net transmission is enhanced; if the two are comparable, interference results in a characteristic Fano-type spectral feature with a minimum and a maximum on either side of the SPP excitation frequency.

There have been a number of studies that investigated and explained the effects of the various geometric parameters on the shape of the resonant transmission (e.g., hole size, metal film thickness, and optical properties of the metal). We note that the critical factor (assuming a relatively "thick" film) is the hole diameter, which increases the scattering rate and hence broadens the resonance linewidth [17].

14-3-1 Observation of SPP Coupling

We studied excitation and propagation of SPP over nanohole arrays using samples of gold, silver, and aluminum films prepared on glass, gallium arsenide (GaAs), and silicon (Si) substrates, with film thickness ranging from 10· to 300·nm and nanohole diameters from 50· to 350·nm. Figure 14-9 shows a typical transmittance map, obtained by illuminating the film with a collimated and polarized broadband source (tungsten halogen lamp) at various angles of incidence, and capturing the transmitted light into a monochromator for spectral analysis [18,19]. Figure 14-9 combines data from two samples with hole-to-hole distances $a = 1.4$·μm and $a = 1.6$·μm, because the accessible wavelength band was insufficient to cover the entire interval $0.7 < a/\lambda < 1.4$ using a single sample. The features in Fig. 14-9 are consistent with the momentum-matching curves computed via Eq. (14-3).

Resonant transmittivity of a nanohole array can be investigated in more detail with the setup in Fig. 14-10a [18]. Here, a beam from a tunable laser source (1520-1570·nm) collimated to ~10-mm diameter is incident on the nanohole array at a small angle θ_{yz} in the yz plane. A microfluidic channel is fabricated over the nanohole array as

FIGURE 14-9 Transmittance through a 2D array of 350-nm holes in a 100-nm-thick aluminum film on GaAs substrate. To cover the domain of normalized frequencies a/λ shown, data from two samples with hole periods $a = 1.4\cdot\mu m$ and $a = 1.6\cdot\mu m$ are combined; the horizontal white line at $a/\lambda \sim 0.95$ indicates the boundary between the two data sets. Dashed white lines represent the momentum-matching condition [Eq. (14-3)]. The two pink ovals mark the angular and spectral range of short pulses in a separate experiment described later in this section (Fig. 14-11). (*Reprinted with permission from R. Rokitski, K. A. Tetz, and Y. Fainman, "Propagation of femtosecond surface plasmon polariton pulses on the surface of a nanostructured metallic film: space-time complex amplitude characterization," Phys. Rev. Lett., 95, 17, 177401, 2005. Copyright 2005 by the American Physical Society.*) (See also color insert.)

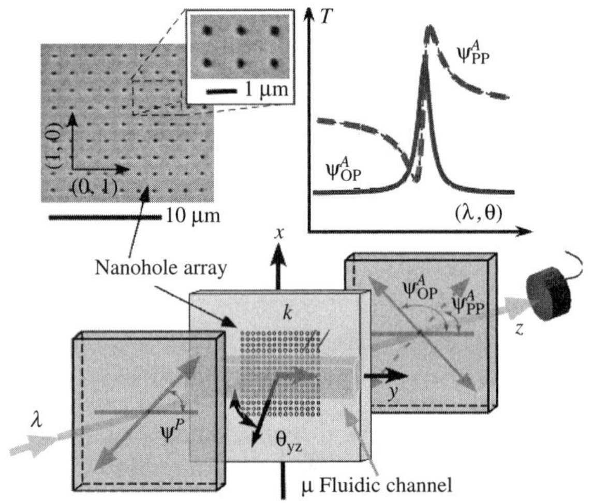

FIGURE 14-10 Polarization-resolved measurement of transmittance lineshape of a nanohole array. (a) Experimental setup; also shown is an SEM image of a representative sample. (b) Measured spectral (top) and angular (bottom) transmittance of the nanohole array when the polarizer and analyzer are aligned ("PP"), and orthogonal ("OP"). In the PP condition, a Fano lineshape is produced by the interference of light directly transmitted through the nanoholes and light reradiated via the SPP. In OP, the directly transmitted component is suppressed, resulting in a Lorentzian lineshape. Each curve has been normalized to its maximum to clearly illustrate the respective lineshape functions, except in the inset, where the same data is plotted in logarithmic scale to show the ~15–20 dB background level reduction for the Lorentzian versus Fano-type resonances. (*K. A. Tetz, L. Pang, and Y. Fainman, "High-resolution surface plasmon resonance sensor based on linewidth-optimized nanohole array transmittance," Opt. Lett., 31, 10, 1528–1530, 2006.*)

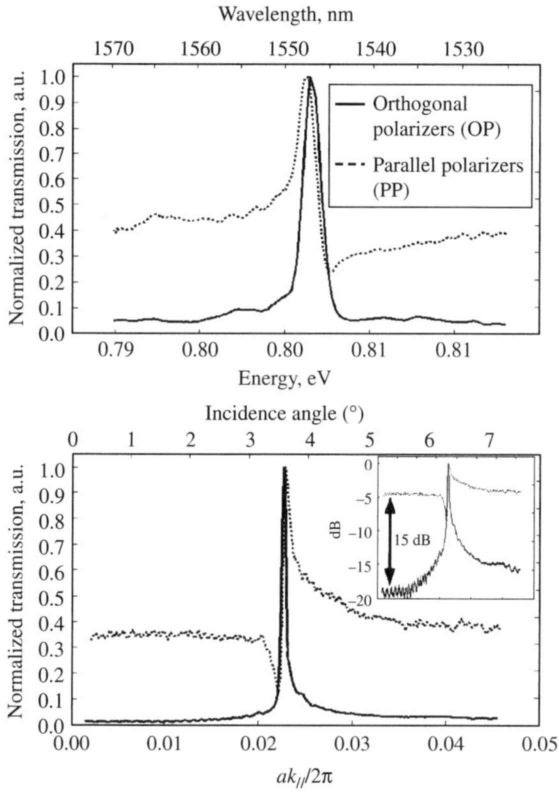

Figure 14.10 (*Continued*)

described in Sec. 14-2. By filling the channel with fluids of different refractive indices, the various SPP resonances can be brought into the wavelength range of the tunable laser, avoiding the need for multiple samples with different nanohole periods.

For the experiment in Fig. 14-10, the nanohole array spacing $a = 1.4 \cdot \mu m$ is chosen to satisfy momentum-matching Eq. (14-3) for the SPP propagating in the $-y$ direction, which couples predominantly to the p-polarized component of the incident beam (Fig. 14-4). Light reradiated by the SPP toward the detector is therefore p-polarized, while the directly transmitted background maintains the polarization of the input, which is set at 45° with respect to the incidence plane yz. When the analyzer is aligned with the input polarization ("PP"), the interference of (one half of) the reradiated light with the background produces a Fano-like lineshape. However, when the analyzer is orthogonal to the input ("OP"), the background is suppressed and a Lorentzian lineshape of the resonant component is observed (Fig. 14-10b). This effect has been explored previously with one-dimensional gratings [20] and elliptical nanohole arrays

[21], and utilized in imaging SPPs excited on two-dimensional SPP grating couplers [19,22].

The background extinction would ideally be limited by that of the polarizers (typically 60 dB), but in practice we measure ~15–20 dB which we attribute to depolarization due to surface roughness in the etched holes. Under wavelength interrogation (the upper plot in Fig. 14-10b), the background level does not drop to the same deep minimum levels within the tuning range of our laser. The measured full-width-half-maxima (FWHM) for wavelength interrogation are 1.28·meV (2.47·nm) and 2.86·meV (5.53·nm) in the OP and PP condition, respectively, and the PP transmission peak is red-shifted from that of OP by 0.40·meV (0.77·nm). Similarly, the measured FWHM for angular interrogation (the lower plot in Fig. 14-10b) are $0.0012 \cdot ak_{//}/2\pi$ (0.092°) and $0.011 \cdot ak_{//}/2\pi$ (0.87°) for OP and PP, respectively, and the corresponding red-shift is 0.0005 (0.04°).

14-3-2 Time-Resolved Imaging of SPP Propagation

Figure 14-11 illustrates an experimental design with which temporal dynamics of SPP propagation can be explored [19]. In this case, light radiated off the metal film is not collected into a single detector but instead imaged onto a CCD array, revealing the geometric paths of SPP in the sample plane. Furthermore, the SPP is excited with a short

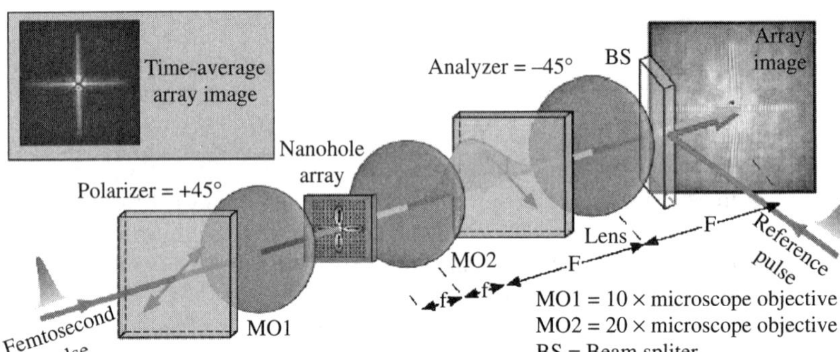

FIGURE 14-11 Schematic diagram of SPP imaging setup. A ~200·fs pulse excites SPP at the center of the nanohole array. As the SPP propagate across the sample, they reradiate part of their energy; by imaging this radiation into a CCD camera, the SPP paths can be visualized as seen in the inset. If the image is mixed with a delayed reference pulse, an interference pattern is observed, such as one shown in false color in the image plane. By varying the delay and noting the location of interference fringes, the progress of the SPP across the sample plane over time can be mapped out. The orthogonally oriented polarizer-analyzer pair suppresses directly transmitted light as explained in the text. (*Reprinted with permission from R. Rokitski, K. A. Tetz, and Y. Fainman, "Propagation of femtosecond surface plasmon polariton pulses on the surface of a nanostructured metallic film: space-time complex amplitude characterization," Phys. Rev. Lett., 95, 17, 177401,2005. Copyright 2005 by the American Physical Society.*) (See also color insert.)

pulse, and the transmitted signal is mixed with a delayed reference pulse (derived from the same laser source) to produce interference. The interference fringes form only where the signal and reference pulses temporally overlap. By varying the delay and noting fringe location, time progress of the SPP along the plane can be mapped out. As in Fig. 14-10, an orthogonally oriented polarizer-analyzer pair in the beam path suppresses the background and isolates light reradiated by the propagating SPP.

Figure 14-12a through 14-12d present snapshots of SPP spatial intensity distribution, as reconstructed from the experiment in Fig. 14-11. The plots in Fig. 14-12e are temporal cross-correlations of the reradiated pulse and the reference pulse taken at different points along the SPP path; the cross-correlations peak at the moment in time when the SPP passes the corresponding point along the path. Taken together, the cross-correlation peaks yield in-plane group velocity of the SPP (Fig. 14-12f).

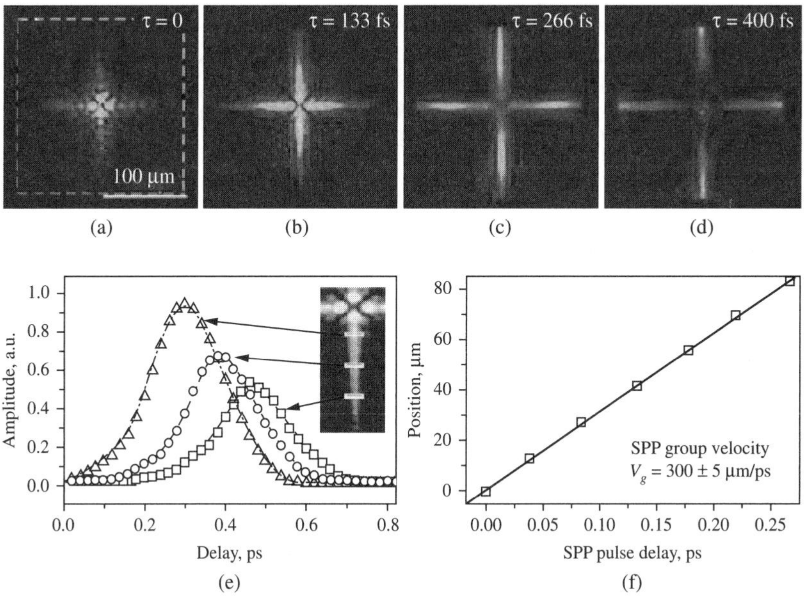

FIGURE 14-12 Temporal evolution of scattered femtosecond SPP pulse. (a through d) Spatial amplitude of the femtosecond SPP pulse (shown in false color) scattered from nanohole array with $a = 1.6$ μm hole period, measured for several time delays between SPP and reference pulses. (e) Normalized temporal cross-correlation of scattered SPP and reference pulses, measured at three spatial locations along the SPP propagation path. (f) Delay of the SPP pulse peak measured at different positions along the propagation path gives SPP pulse group velocity ~300 ± 5 μm/ps. Dashed line in (a) outlines dimensions of the nanohole array. (*Reprinted with permission from R. Rokitski, K. A. Tetz, and Y. Fainman, "Propagation of femtosecond surface plasmon polariton pulses on the surface of a nanostructured metallic film: space-time complex amplitude characterization," Phys. Rev. Lett., 95, 17, 177401,2005. Copyright 2005 by the American Physical Society.*) (See also color insert.)

14-3-3 SPP Focusing

Figure 14-13 shows amplitude and phase snapshots in several experiments at the moment in time when the leading edge of the SPP pulse has reached the outer boundary of the 200·μm × 200·μm nanohole array, while the trailing edge has not yet left its center. The nanohole array period *a* is 1.6·μm in Fig. 14-13a and 14-13b

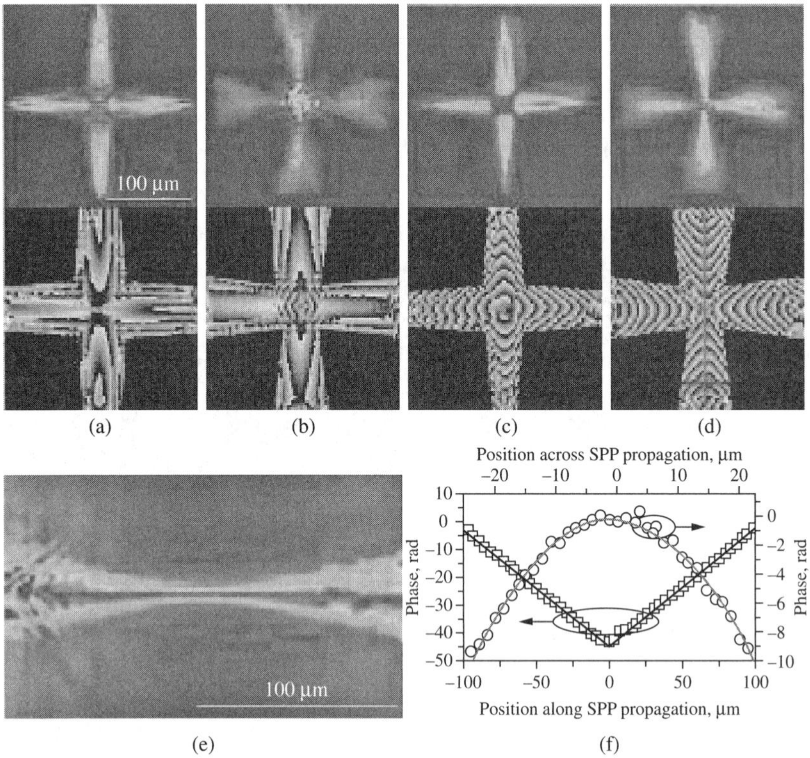

FIGURE 14-13 Snapshots of scattered SPP field over nanohole arrays with hole-to-hole distance $a = 1.6$ μm (a,b) and $a = 1.4$ μm (c,d), excited by spatially converging (a,c) and diverging (b,d) Gaussian pulses. In-plane convergence and divergence of SPP is clearly seen. (e) In-plane SPP focal waist formation, measured in nanohole array with 1.6 μm period. Amplitude is renormalized in the SPP propagation direction (left to right) to compensate for attenuation. (f) Unwrapped longitudinal (open squares) and transverse (open circles) cross sections of the spatial phase distribution from (d) with linear and quadratic least squares–fitting plots. Linear phase corresponds to 6.0° phase-matching angle for (−1,0) and (0,−1) SPP modes, while the quadratic spatial phase of the SPP field is acquired from the quadratic spatial phase of the incident optical field. (*Reprinted with permission from R. Rokitski, K. A. Tetz, and Y. Fainman, "Propagation of femtosecond surface plasmon polariton pulses on the surface of a nanostructured metallic film: space-time complex amplitude characterization," Phys. Rev. Lett., 95, 17, 177401,2005. Copyright 2005 by the American Physical Society.*) (See also color insert.)

and 1.4-μm in Fig. 14-13c and 14-13d, and the excitation pulse is converging in Fig. 14-13a and 14-13c and diverging in Fig. 14-13b and 14-13d. The incident spherical phase is imparted to the SPP in each case, so that the SPP maintains the convergence/divergence of the excitation. The image in Fig. 14-13e is taken under the same conditions as Fig. 14-13a but with the SPP excited further to the left on the nanohole array in order to observe SPP propagation over a longer distance. Convergence toward a waist and subsequent divergence are clearly seen.

It must be noted that the phase profiles captured with the apparatus of Fig. 14-11 and reproduced in Fig. 14-13 are those of the reradiated field and not of the SPP wave itself. Plotted phase fronts therefore correspond to the in-plane wavevector $\mathbf{k}_{\parallel} = \mathbf{k}_{SPP} - K_G$, where $\mathbf{K}_G = p(2\pi/a)\hat{\mathbf{x}} + q(2\pi/a)\hat{\mathbf{y}}$ is the grating vector of order (p,q) responsible for matching the SPP momentum to the free space beam [Eq. (14-3)]. Of course, $\mathbf{k}_{\parallel}$ is quite different from $\mathbf{k}_{SPP}$, and in Fig. 14-13c and 14-13d even differs from it in sign; this accounts for the counterintuitive shape of the phase fronts, which point toward the beam waist rather than away from it. The linear phase component along the SPP optical axis (line in Fig. 14-13f is consistent with expectations: The implied incidence angles $\theta = \mathbf{k}_{\parallel}\lambda/(2\pi)$ estimated from Fig. 14-13a and 14-13b and Fig. 14-13c and 14-13d are, respectively, ~1.2° and ~6.4°, compared with 1.9° and ~6.0° predicted via Eqs. (14-2) and (14-3). Achieving excitation of femtosecond SPP pulses and observing them using the time-resolved spatial heterodyne imaging are important steps toward understanding the connection between spatial and temporal characteristics of the incident optical waves and of the excited and scattered SPP waves. The focusing of the femtosecond SPP pulses leads to complete localization of the electromagnetic field in space and time—which is essential for various applications in sensing, nonlinear optics, and biomedical imaging.

14-3-4 Degenerate Mode Splitting

Up till this point, we used the dispersion relation Eq. (14-2), derived for a continuous metal surface, to also describe SPP dispersion on a surface perforated with nanoholes. Because the surface area occupied by the nanoholes is small, this is generally a good approximation. The perforation does perturb the dispersion relation, however, most noticeably by lifting degeneracy of SPP modes and creating forbidden gaps.

The effect can be explained intuitively with reference to Fig. 14-14. Figure 14-14a graphically represents the momentum-matching condition Eq. (14-3) for the excitation of SPP modes via (0,1) and (0,−1) grating orders at an air-metal interface. Because the plane of incidence in Fig. 14-14a is the xz plane, the condition Eq. (14-3) is satisfied

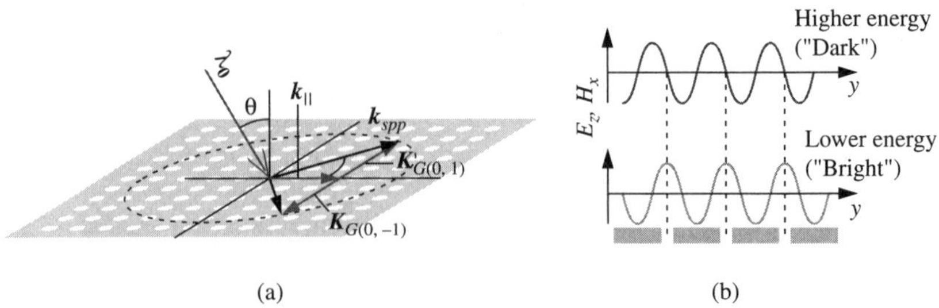

FIGURE 14-14 (a) Schematic diagram of the momentum-matching condition Eq. (14-3) for the SPP modes excited via (0,1) and (0,−1) grating orders. On a continuous metal film, these SPP modes (and all SPP modes with equal $|k_{SPP}|$), and all their linear combinations, would have the same energy. (b) In the presence of nanoholes, the linear combination that has both field antinodes over the metal has higher energy than the combination that has only one antinode over the metal.

simultaneously for these two grating orders, and two SPP modes are simultaneously launched,

$$\mathbf{k}_{SPP}^{(0,1)} = k_{||}\hat{\mathbf{x}} + K_G\hat{\mathbf{y}}, \quad \mathbf{k}_{SPP}^{(0,-1)} = k_{||}\hat{\mathbf{x}} - K_G\hat{\mathbf{y}},$$

$$k_{||} \triangleq \frac{2\pi}{\lambda_0}\sin\theta, \quad K_G \triangleq \frac{2\pi}{\Lambda_y}, \quad \left|\mathbf{k}_{SPP}^{(0,1)}\right| = \left|\mathbf{k}_{SPP}^{(0,-1)}\right| = \frac{2\pi}{\lambda_0}\left(\frac{\varepsilon_2}{1+\varepsilon_2}\right)^{1/2}$$

where λ_0 is the vacuum wavelength of the excitation, θ is the angle of incidence, and ε_2 is the permittivity of the metal material [Eqs. (14-2) and 14-3)]. On a continuous film, the SPP modes $\mathbf{k}_{SPP}^{(0,1)}$, $\mathbf{k}_{SPP}^{(0,-1)}$ would be degenerate: these modes would have equal energy, and all linear combinations of them would also have the same energy. Over the nanohole array, however, the two linear combinations

$$e^{i(k_{||}x+K_Gy)} + e^{i(k_{||}x-K_Gy)} \sim e^{ik_{||}x}\cos K_Gy, \quad e^{i(k_{||}x+K_Gy)} - e^{i(k_{||}x-K_Gy)} \sim e^{ik_{||}x}\sin K_Gy$$

have slightly different energies (Fig. 14-14b); the degeneracy is lifted. It can be shown that the sum and difference of $\mathbf{k}_{SPP}^{(0,1)}$, $\mathbf{k}_{SPP}^{(0,-1)}$ modes are in fact the new eigenmodes of the perturbed system. The eigenmode whose field antinodes are aligned with the nanoholes has lower energy; it also scatters more light into free space and is therefore called *bright*. The other, higher-energy, less scattering eigenmode is called *dark*. The terms *bright* and *dark* are due to Ropers et al. [23], who first observed the SPP mode splitting over a one-dimensional slit grating at normal incidence.

We studied SPP mode splitting over rectangular nanohole arrays, where degeneracy occurs at off-normal (as in Fig. 14-14a) as well as normal incidence. Figure 14-15a shows the experimental design [24]. The test sample is a perforated gold film on glass substrate, with a nanohole spacing of 1.0 μm. The sample is equipped with a microfluidic channel for placing the various SPP resonances at the desired spectral locations by injecting fluids of different permittivity. The SPP are excited with a collimated beam from a 1520- to 1570-nm, 1-pm wavelength resolution tunable laser source. Transmitted light is collected into a single detector (the CCD camera shown in Fig. 14-15a serves only as an alignment aid). The cross-oriented polarizer-analyzer pair is used to suppress directly

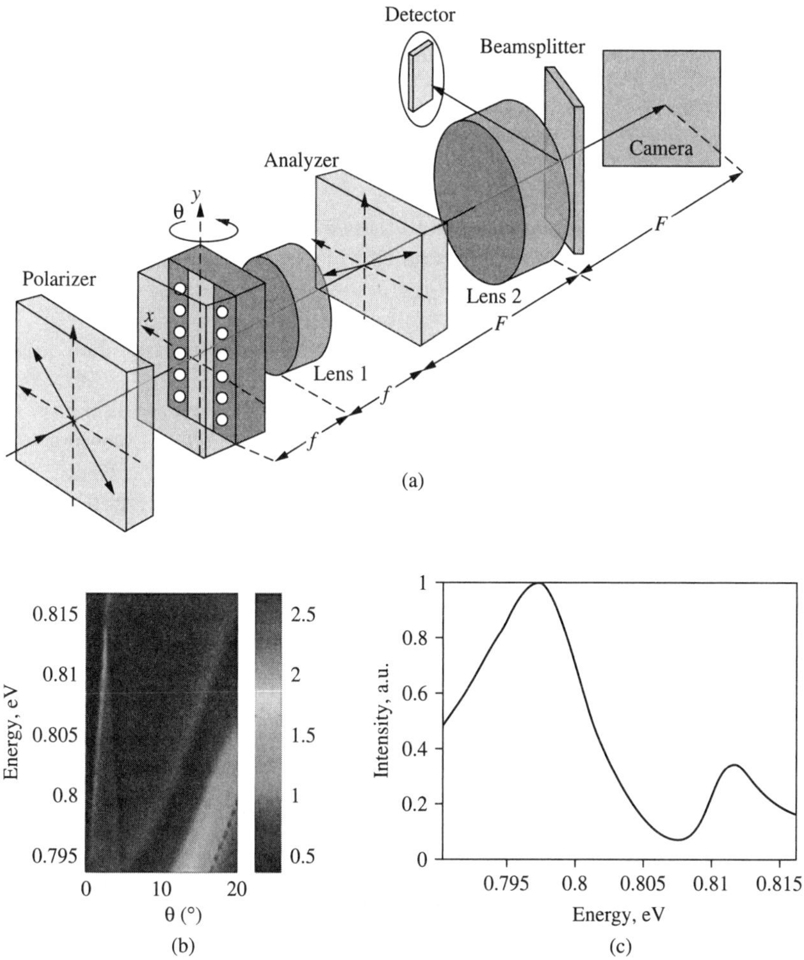

FIGURE 14-15 Observation of SPP mode splitting: (a) Experimental setup; (b) angle- and frequency-resolved transmittance data; (c) a slice through (b) at $\theta = 18°$, showing the low-energy *bright* and high-energy *dark* modes. *Reprinted with permission from L. Pang, K. A. Tetz, and Y. Fainman, "Observation of the splitting of degenerate surface plasmon polariton modes in a two-dimensional metallic nanohole array," Appl.Phys. Lett., 90, 111103, 2007. Copyright 2007, American Institute of Physics.*) (See also color insert.)

transmitted light in the same way as in Fig. 14-11. The sample is mounted on a 0.001° angular resolution rotation stage, so that both the wavelength (by tuning the laser) and the incidence angle (by rotating the stage) transmittance characteristics of the sample can be explored.

The color map in Fig. 14-15b reports measured transmittance as a function of the angle of incidence θ and the energy $E = \hbar \cdot 2\pi c/\lambda_0$. The same data is given in Fig. 14-15c as a one-dimensional plot for a fixed $\theta = 18°$. The lower-energy *bright* and the higher-energy *dark* modes, separated by ~14·meV, are clearly seen. The fluidic channel in this experiment was filled with oil, the dielectric constant of which was found by best-fit of the transmittance in Fig. 14-15b to be $\varepsilon_d = 2.57$.

14-4 Resonant SPP Sensors

Resonant SPP phenomena on metal-dielectric interfaces have been utilized for real-time quantitative analysis of chemical and biological interactions. Various surface plasmon resonance (SPR) sensors have been demonstrated. The most common configurations employ the Kretschmann geometry or a shallow grating coupler discussed in Sec. 14-1-2 and monitor the shift of the resonance wavelength with the incidence angle held constant, or the resonance angle with the wavelength held constant, or simply the change in reflected power at a constant wavelength and incidence angle [25–27]. More recent approaches have included phase-sensitive variations, demonstrated in both interferometric [28] as well as ellipsometric configurations [29]. These methods have been used for detection of surface perturbation when a liquid or a gaseous species is flown along the surface, or of specific binding events when a biomolecular recognition element is attached to the surface and an analyte solution is flown by it.

For instruments intended to simultaneously monitor multiple chemical reactions, an important drawback of the Kretschmann geometry is the limited numerical aperture (NA) afforded through a prism face, and hence limited spatial resolution and the number of resolvable spots in the measurement plane. Furthermore, the typical incidence angles necessary to achieve SPR resonance are relatively large, requiring a large depth of focus of any imaging system used to simultaneously measure large arrays of assays. Massive parallelism, and hence high throughput, is of primary importance in many potential SPR sensor applications but they are severely limited by most of the current design configurations [30–33].

The difficulties associated with narrow NA and oblique incidence are removed if a surface grating or a nanohole array is used to couple SPP to the incident light as discussed in Sec. 14-1-2. Thin metal films perforated with nanohole arrays exhibit resonant transmission [14], which, while not without some controversy [34,35], is generally attributed to the excitation of SPP waves. Nanohole-based devices can operate at substantially normal incidence, allowing larger area to be imaged and smaller surface spots to be resolved by the imager. This leads, in turn, to small interrogation volumes, high packing density, minimal analyte volumes, and large number of parallel channels. These advantages may make such devices preferable in a number of applications although the ultimate measurement resolution may not be as high as with prism-based sensor devices due to the fact that the SPR linewidth is affected by both radiative and material damping, and hence is always broader. Several authors have suggested and demonstrated the use of subwavelength hole arrays for sensing applications [36–38], and there are many numerical and experimental studies on their spectral and polarization properties. Here we demonstrate an integrated optofluidic chip SPR sensor based on a metal film, perforated by a nanohole array. Specifically, we investigate sensitivity and resolution of these chips using angular and wavelength interrogations.

The basic setup for our SPP biosensing experiments is given in Fig. 14-16 [39]. The reaction chamber is defined by a PDMS microfluidic channel routed over a perforated gold film. These structures are fabricated by holographic lithography, dry etching of nanoholes into the gold film, and plasma bonding of PDMS as described in detail in Sec. 14-2. For biosensing experiments, we use ~200-nm-thick gold films on glass substrates, perforated with ~200-nm diameter holes. The period a of the nanohole array is chosen close to the excitation wavelength λ in the fluid, so that the SPP resonance occurs at near normal incidence. The array has an overall usable area on the order of ~10 mm × 10 mm; the reaction chamber, molded in PDMS and bonded over the nanohole array, measures 10·mm × 2·mm and is 100·μm deep. The SPP are excited by a beam from a tunable laser source (1520–1570·nm) collimated to ~10-mm diameter. A ~200 μm × 200 μm area at the center of the reaction chamber is imaged onto a CCD camera for alignment (not shown in Fig. 14-16), and also onto an InGaAs photodetector for transmission measurements. Angular interrogation is achieved using a mechanical rotation stage rotating the sample in the yz plane.

14-4-1 Angular Interrogation Sensing Experiments

The sensing system in Fig. 14-16 can be characterized by injecting an index-calibrated solution through the microfluidic channel to create a controlled gold-fluid interface. Figure 14-17 presents data from a series of experiments in which Na_2CrO_4/H_2O solutions of varying concentration were used for calibration. Due to the strong absorption of water in the 1.55-μm wavelength range, the resonance under angular interrogation broadens from $0.0012 \cdot ak_{//}/2\pi$ (0.092) reported for the OP condition in Fig. 14-10b to $0.0064 \cdot ak_{//}/2\pi$ (0.52°). Under wavelength interrogation (not shown), the resonance

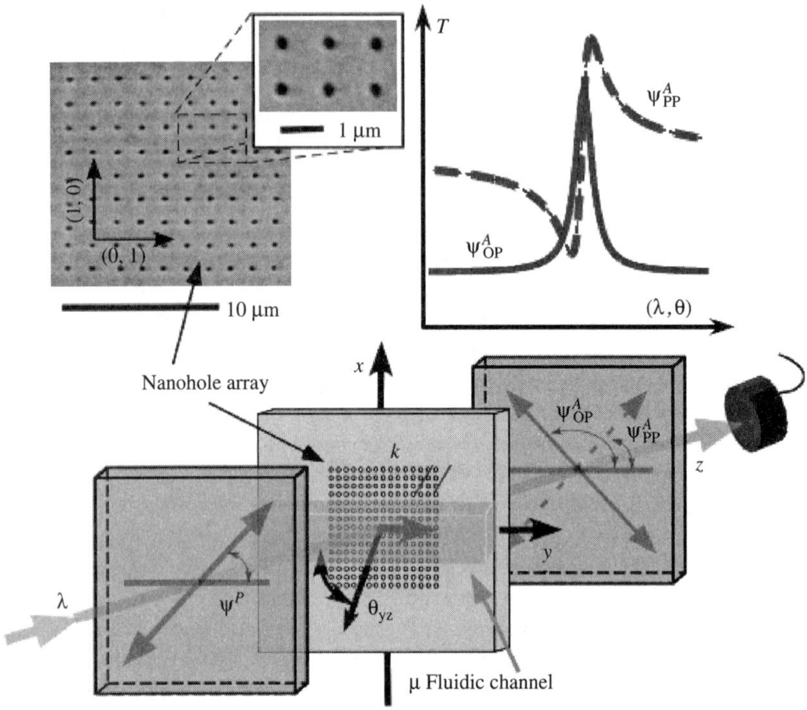

FIGURE 14-16 Conceptual diagram of the 2D nanohole array–based SPR sensor. The setup is similar to Fig. 14-10a. In most of our biosensing experiments, the polarizer and analyzer are orthogonally oriented ("OP"), producing Lorentzian spectral and angular lineshapes. A microfludic channel is used to transports the analyte fluid to the surface of the sensing area. Also shown is an SEM image of a representative sample. (*K. A. Tetz, L. Pang, and Y. Fainman, "High-resolution surface plasmon resonance sensor based on linewidth-optimized nanohole array transmittance," Opt. Lett., 31, 10, 1528–1530, 2006.*)

similarly broadens from 1.28·meV (2.47·nm) to 4.32·meV (8.31·nm). At shorter wavelengths the damping due to water would be reduced—however, the metal losses would be greater. Also, when a surface reaction is of interest, shorter wavelengths yield a greater overlap of the SPP field with the reaction, as the reach of the mode into the dielectric is reduced [37].

The data plotted in Fig. 14-17 reflect the displacement of the Lorentzian peak in the OP condition. We note that one may well monitor other features of the transmittance curve, for example, the point of highest slope of the Fano curve in the PP condition, but by usual convention we monitor the resonance maxima. Error bars in the horizontal direction are from uncertainty in calibration solution index as well as variations in temperature. Peak positions are determined by both the method of moments (centroid position) and by fitting

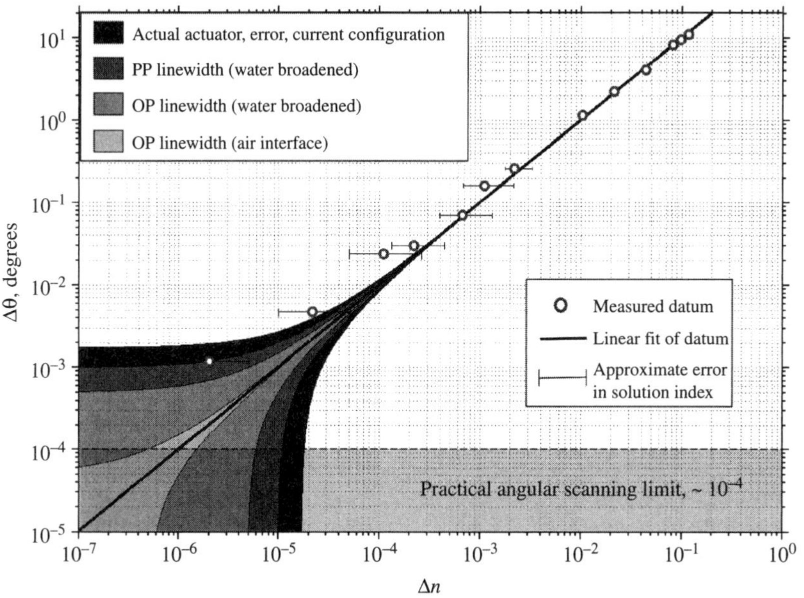

FIGURE 14-17 Resonance peak-position-shift versus refractive index change (i.e., salt concentration in water) in the fluidic overlayer. The line is a linear fit to the data. Shaded regions represent uncertainty of the curve fitting in the presence of noise for the OP and PP conditions for both air- and water-broadened linewidths as well as estimated theoretical resolution limits. (*K. A. Tetz, L. Pang, and Y. Fainman, "High-resolution surface plasmon resonance sensor based on linewidth-optimized nanohole array transmittance," Opt. Lett., 31, 10, 1528–1530, 2006.*)

Lorentzian functions, and the error bounds for these methods in the presence of our noise are shown as the various shaded regions. This procedure corresponds to estimated sensing limits of 5×10^{-6} refractive index units (RIU) for OP and 1×10^{-5}·RIU for PP. The darkest region corresponds to the observed mechanical error of 1.7×10^{-3} degree (standard deviation) due to lack of full optimization in the feedback controls, which limited our direct measurement limit to ~1.5×10^{-5} RIU. We estimate the limits for a nonabsorbing overlayer (with a gaseous species analyte, for example) with OP and an optimized rotation stage (mechanical limits of ~10^{-4} in angle [26]) to be on the order 1×10^{-6} which is shown with the lightest shading.

While peak position is typically determined more precisely, it is useful to introduce the metric $\chi_{\lambda,\theta} \equiv S_{\lambda,\theta}/\Gamma_{\lambda,\theta}$, which is a measure of the resolving power that facilitates comparisons of different sensors and interrogation methods [40]. Here S is the sensitivity (i.e., derivative of resonance position with respect to index of refraction) and Γ is the FWHM and the subscript λ or θ refers, respectively, to wavelength or angular interrogation. We experimentally determine S_λ ~ 1022 ± 8 nm/RIU and S_θ ~ 78.4 ± 0.6 degree/RIU that yield values of χ_θ ~ 850

RIU^{-1} and $\chi_\lambda \sim 410$ RIU^{-1} with an air overlayer while these values are reduced to $\chi_\theta \sim 150$ RIU^{-1} and $\chi_\lambda \sim 120$ RIU^{-1} with water-broadened transmission. Additional details on wavelength interrogation will be discussed further in the next section.

14-4-2 SPR Sensor with Wavelength Interrogation

The experimental system in Fig. 14-16 can also be used for wavelength interrogation measurements, by scanning the wavelength of the tunable laser source and monitoring the spectral location λ_r of the resonance. The sensitivity S_λ (nm/RIU) is in this case defined as the derivative of λ_r with respect to the refractive index of the dielectric we aim to determine. Unlike the resolution R, which is the minimum detectable refractive index change and which is a performance characteristic of the overall system, the sensitivity is largely determined by the physics of the measurement process, the interrogation configuration, and the metal material [40].

For sensors based on the SPR of individual metal nanoparticles ("localized SPR"), extensive electromagnetic simulations and measurements in visible spectral range indicate sensitivity values of 200–300 nm/RIU [42–44]. Experiments with Kretschmann configuration systems reveal sensitivities $S_\lambda \sim 15000$·nm/RIU near 850 nm excitation wavelength [26,45]. For SPR sensors based on 2D nanohole arrays, S_λ depends on the periodicity of the array and the grating order involved in SPP excitation. To obtain this relationship analytically, we establish a Cartesian coordinate system aligned with the lattice vectors of the nanohole array, and express the parallel projection of the incident wavevector as

$$\mathbf{k}_{\parallel} = \frac{2\pi c}{\lambda_0}(\hat{\mathbf{x}} \cdot \sin\theta\cos\phi + \hat{\mathbf{y}} \cdot \sin\theta\sin\phi) \tag{14-4}$$

where λ_0 is the vacuum wavelength and θ, ϕ are the polar and azimuthal angles of incidence. Inserting Eq. (14-4) and the SPP dispersion relation Eq. (14-2) into the momentum-matching condition, Eq. (14-3) leads to an equation that implicitly defines the SPP resonance frequency ω_r as a function of incident angles θ, ϕ, the grating order indices p, q, and the refractive index n of the dielectric:

$$k_{\text{spp}}^2 = \left(\frac{\omega_r}{c}\right)^2 \frac{n^2\varepsilon_m(\omega_r)}{n^2 + \varepsilon_m(\omega_r)} = \left[\frac{\omega_r}{c}\sin\theta\cos\phi + p\frac{2\pi}{d}\right]^2 + \left[\frac{\omega_r}{c}\sin\theta\sin\phi + q\frac{2\pi}{d}\right]^2 \tag{14-5}$$

where $d=\Lambda_x=\Lambda_y$ is the period of the nanohole array, and ε_m is the permittivity of the metal [denoted by ε_2 in Eq. (14-2)]. Under wavelength interrogation, the angles θ, ϕ, and the indices p, q are presumed fixed, and differentiation of Eq. (14-5) yields the relationship between increments dn of the refractive index n and $d\omega_r$ of the resonance frequency ω_r,

$$\frac{1}{\omega_r}\frac{d\omega_r}{dn} = \frac{\frac{n\varepsilon_m^2}{(n^2+\varepsilon_m)^2}}{-\left(\frac{n^2\varepsilon_m}{n^2+\varepsilon_m}+\frac{n^4}{2(n^2+\varepsilon_m)^2}\omega_r\frac{d\varepsilon_m}{d\omega}\Big|_{\omega_r}\right)+\left(\sin\theta\cos\phi+p\frac{2\pi}{d}\frac{c}{\omega_r}\right)\sin\theta\cos\phi+\left(\sin\theta\sin\phi+q\frac{2\pi}{d}\frac{c}{\omega_r}\right)\sin\theta\sin\phi}$$

that can be equivalently expressed in terms of the resonance wavelength $\lambda_r \triangleq 2\pi c/\omega_r$, as the sensitivity

$$S_\lambda \triangleq \frac{d\lambda_r}{dn} = -\lambda_r \frac{\frac{n\varepsilon_m^2}{(n^2+\varepsilon_m)^2}}{-\left(\frac{n^2\varepsilon_m}{n^2+\varepsilon_m}-\frac{n^4}{2(n^2+\varepsilon_m)^2}\lambda_r\frac{d\varepsilon_m}{d\lambda}\Big|_{\lambda_r}\right)+\left(\sin\theta\cos\phi+p\frac{\lambda_r}{d}\right)\sin\theta\cos\phi+\left(\sin\theta\sin\phi+q\frac{\lambda_r}{d}\right)\sin\theta\sin\phi} \quad (14\text{-}6)$$

Equation (14-6) explicitly includes the material-dispersion term $d\varepsilon_m/d\lambda$ of the metal; as in Eq. (14-2), the material dispersion of the dielectric has been neglected.

Figure 14-18a shows the sensitivity S_λ computed via Eq. (14-6) at a gold-fluid ($n = 1.32$) interface for various grating orders (p, q). The sensitivity is shown as a function of the polar incidence angle θ, with the azimuthal angle $\phi = 0$. Although q does not explicitly enter Eq. (14-6) when $\phi = 0$, different q imply different values of ω_r, θ, p, and/or n necessary to satisfy the momentum constraint Eq. (14-5), and consequently the resulting sensitivity curves also differ. The sensitivity in $(p, 0)$ type configurations is largely independent of the angle θ, whereas in $(0, q)$ configurations it increases with θ. The sensitivity also slightly increases with θ when p, q are both nonzero. The sign of the order indices p, q has no effect on S_λ.

For comparison, Fig. 14-18a also shows the sensitivity of SPR involving a 1D linear grating, obtained numerically with the Rigorous Coupled-Wave Analysis (RCWA) method [46]. In the simulations, the metal and dielectric materials and the periodicity of the linear grating were set to match those used in evaluating Eq. (14-6). The sensitivities S_λ obtained for 1D grating orders $p = \pm 1$ and $p = 2$ agree closely with those for $(\pm 1, 0)$, $(\pm 2, 0)$ in the 2D case. Additionally, S_λ for both 1D gratings and $(p, 0)$ orders of nanohole arrays does not strongly depend on the excitation wavelength; this is in contrast to SPR devices utilizing reflected rather than transmitted light [40]. Finally, it merits repeating that the sensitivity S_λ reflects only the

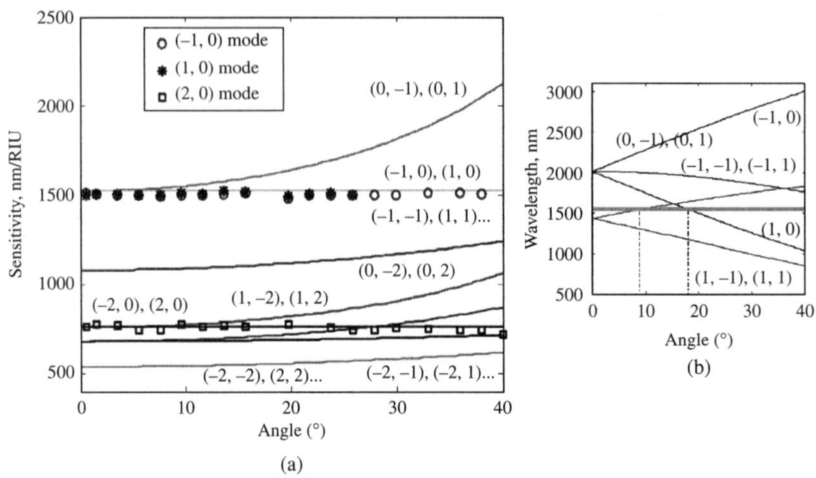

FIGURE 14-18 (a) Spectral sensitivity Sλ as a function of the polar angle of incidence θ (azimuthal angle φ = 0) for a 2D nanohole array SPR sensor utilizing different grating orders (*p,q*) to excite SPP. Solid lines show sensitivities computed with Eq. (14-6). The array period *d* = 1.53 mm, the metal is gold, and the dielectric refractive index *n* = 1.32. Circles mark values obtained by numerical simulation of a 1D metallic grating with the same periodicity *d* and the same metal and dielectric parameters. For the simulation, the duty ratio of the grating is 0.2, and the sensitivity is extracted by comparing resonances with *n* = 1.32 and *n* = 1.36. (b) Momentum-matching condition Eq. (14-3) at a gold interface with the dielectric *n* = 1.32. The grey area shows the 1520- to 1570-nm wavelength range experimentally accessible with the current setup. Dash-dot lines indicate momentum matching points corresponding to (−1,−1) and (1,0) grating orders with *n* = 1.32 and λ_r =1533 nm. (*Reprinted with permission from L. Pang, G. M. Hwang, B. Slutsky, and Y. Fainman, "Spectral sensitivity of two-dimensional nanohole array surface plasmon polariton resonance sensor," Appl. Phys. Lett., 91, 12, 123112, 2007. Copyright 2007, American Institute of Physics.*)

displacement of the resonance due to changes in refractive index *n*; it does not reflect the width and contrast of the resonance, both of which ultimately affect the signal-to-noise ratio and the resolution limit of the instrument [39].

Experimental determination of the sensitivity S_λ of the device depicted in Fig. 14-16 is illustrated in Fig. 14-19. Figure 14-19a is a real-time record of calibration sequences in two device configurations, one aimed at exciting SPP via the (1,0) grating order, the other via the (−1,−1) order. The angles of incidence were fixed in the first case at θ ~9°, and in the second case at θ ~18°. These values are in good agreement with the 8.3° and 18.5° calculated from the momentum-matching condition Eq. (14-3) (dash-dot lines in Fig. 14-18b); slight differences may be attributed to errors in the estimation of permittivities of gold and water. During calibration, water solutions of ethylene glycol with volume concentrations of 0, 1.96, 3.85, 5.60, 7.41,

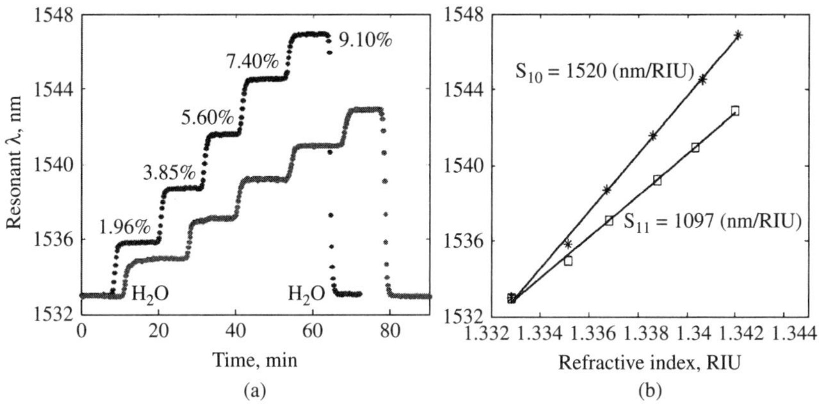

FIGURE 14-19 (a) Time-evolution of the resonance wavelength during sensitivity calibration of the SPP sensor. During calibration, progressively more concentrated water solutions of ethylene glycol were flown through the fluidic channel fabricated over the gold film perforated with a 300-nm hole size, 1.53-μm period nanohole pattern. The grey and black curves correspond, respectively, to SPP excitation via the (1,0) grating order (angle of incidence ~9°), and (−1,−1) grating order (~18°). (b) Resonance wavelength plotted against the refractive index n of the solution in the fluidic channel. The solid lines are linear fits for the (1,0) and (−1,−1) configurations. (*Reprinted with permission from L. Pang, G. M. Hwang, B. Slutsky, and Y. Fainman, "Spectral sensitivity of two-dimensional nanohole array surface plasmon polariton resonance sensor," Appl. Phys. Lett., 91, 12, 123112, 2007. Copyright 2007, American Institute of Physics.*)

and 9.10% were flown, in succession, through the fluidic channel and over the measurement area of the device. The flow rate of the solution through the channel was 260·μL/min. The location of the SPP resonance was monitored by continuously sweeping tunable laser source across the 1520- to 1570-nm band and capturing the spectral transmittance characteristic of the device. Because the concentration increments are unequal (1.96, 1.89, 1.75, 1.81, and 1.68%), the resonance wavelength increments seen in Fig. 14-19a are also unequal.

The relationship between the resonance wavelength and the refractive index of the solution, captured in Fig. 14-19a, is shown explicitly in Fig. 14-19b. The solid lines are linear fits to the data, which yield sensitivities $S_\lambda = 1520$·nm/RIU in the (1,0) condition and $S_\lambda = 1097$·nm/RIU in the (−1,1) condition. These measured values compare well with the 1526·nm/RIU and 1095·nm/RIU computed from Eq. (14-5) with $\lambda_r = 1533$·nm.

Finally, Fig. 14-20 shows an experiment in which the refractive index of reacting biological material, rather than a passive solution, is monitored in real time [47]. The sequence in Fig. 14-20b begins with the cleaning of the reaction chamber and the activation of it by flowing a solution of bovine serum albumin (BSA) through the chamber

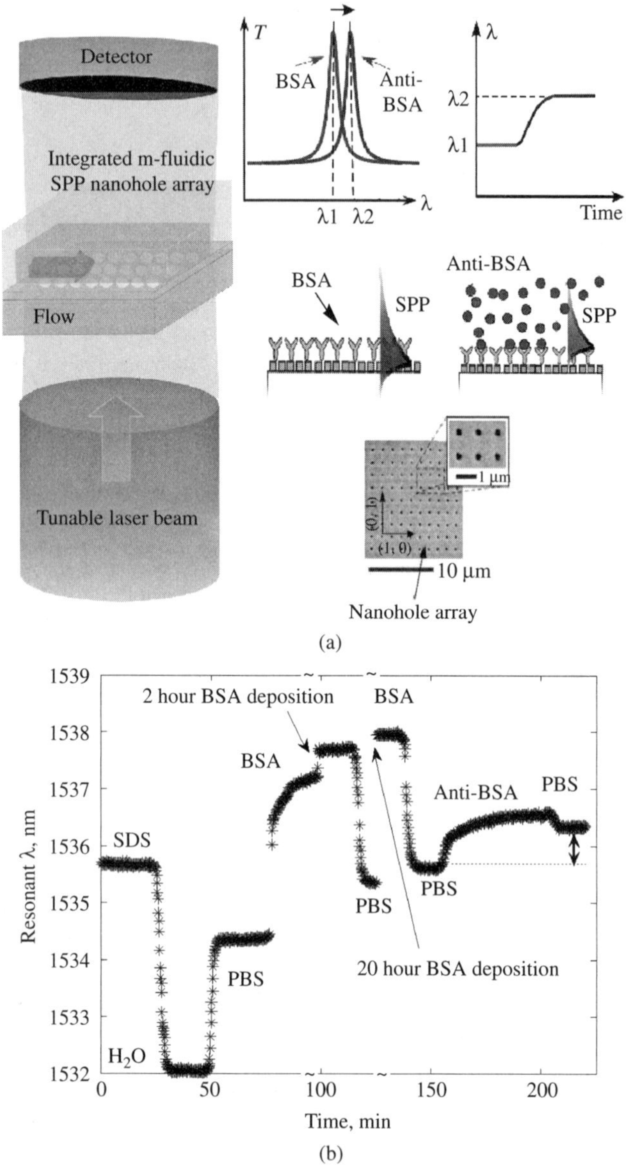

Figure 14-20 (a) The schematic illustration of label-free, real-time nanohole array SPR biosensor. The SPP resonant wavelength is extracted and plotted to show real-time measurement. A nanohole array is also shown. (b) SPP resonant wavelength evolves with time to monitor the process of a series of biological reactions from surface cleaning, protein attachment, and specific binding. (*Reprinted with permission from L. Pang, G. M. Hwang, B. Slutsky, and Y. Fainman, "Spectral sensitivity of two-dimensional nanohole array surface plasmon polariton resonance sensor," Appl. Phys. Lett., 91, 12, 123112, 2007. Copyright 2007, American Institute of Physics.*)

over a period of time sufficient for the BSA to adsorb at the gold surface. It can be noted in Fig. 14-20b that the SPP resonance does not return to its original position after this step, hence some amount of BSA has indeed been adsorbed. Next, a solution of anti-BSA is flown over the activated surface. As anti-BSA from the solution chemically binds to the BSA already resident at the metal surface, the refractive index in the vicinity of the surface is affected, and the SPP resonance frequency is also affected. The progress of this reaction is seen in real time in Fig. 14-20b. The resonance remains shifted by approximately 0.7·nm after the anti-BSA solution has been removed, due to the 190·nM of anti-BSA trapped at the BSA-activated surface. Given the 0.1-nm wavelength resolution limit of the device, as little as 26·nM of anti-BSA can be detected.

It is worth noting that the time scale in Fig. 14-19a is much shorter than in Fig. 14-20b. In the former case, equilibrium is reached as soon as the new fluid replaces the old in the reaction chamber. In the BSA/anti-BSA experiment, on the other hand, the protein-binding reaction continues until all adsorbed BSA reaction sites are filled with anti-BSA molecules.

In conclusion, we present an analytical expression of wavelength sensitivity obtained from the SPP dispersion relation for 2D nanohole array SPR sensor. The sensitivity of nanohole array SPR sensor depends on the periodicity of the array and the order of the SPP modes. The analytical expression is confirmed by numeric results using electromagnetic simulation and also validated by the experiments. Real-time monitoring of protein-protein specific bonding between BSA and monoclonal anti-BSA is performed to demonstrate the integrated optofluidic nanohole array SPR biosensor. The detection resolution of the system can be increased by employing self-assembled linker layer on the gold surface and improving detection limit of the optical and electrical detection system. The clarified analysis and the demonstration of the sensitivity for SPP fields in 2D nanohole array not only elucidate the mechanism of the nanohole array SPR sensor, but also would facilitate the improvement of the sensitivity of SPR sensors [49].

Multichannel versions of the device in Fig. 14-16 can be realized by subdividing the active area into reaction cells and utilizing a CCD camera to separately monitor the optical transmission of each cell. In such arrangements, the SPP propagation length may have to be artificially limited to reduce cross-talk between the cells and to pack more cells in a unit area. A design tradeoff thus exists between higher sensitivity and smaller interrogation volumes; the choice will depend on the particular application. There are a number of interesting variations to explore in the future, including design of the periodic structure [36] such that the SPR can be tuned to a molecular resonance of interest. In addition, one can break the in-plane symmetry and use, for example, elliptical [41] or chiral-shaped holes to produce polarization dependence even at normal incidence. These results will help in designing future grating-coupled surface plasmon resonance sensors, both in the transmission (a nanohole) and the traditional (reflection surface grating relief) geometries.

14-5 Summary and Discussion

Optofluidic plasmonics, consisting of integrated microfluidics with optics and plasmonics, is introduced. We made an introduction and overview of demonstrating experimentally coupling to SPP modes in a cubic array of holes and the direct observation of radiation leakage from such arrays. A wide variety of propagating waves, with different frequencies and in-plane wave vectors can be excited and observed for various sample geometries and under variable excitation conditions. Such techniques may prove useful for investigating the properties of SPP waves for a variety of applications and in interfacing with various nanoplasmonic devices. We also gave a review on excitation and observation of femtosecond surface plasmon polariton wavepackets using time-resolved spatial heterodyne imaging approach. It is an important step toward understanding relationship between spatial and temporal characteristics of the incident optical waves and the excited and scattered femtosecond SPP fields. Demonstrated in-plane focusing of femtosecond SPP pulses leads to complete 3D and temporal localization of electromagnetic field, which will find applications in nonlinear surface studies, sensing, surface plasmon polariton waveguiding, and information processing. Due to optical field correlation nature of our measurements, only spatial amplitude and phase information of the SPP field can be measured precisely. We envision, however, two-photon absorption realization of our measurements with the possibility of characterizing temporal amplitude and phase of the femtosecond surface plasmon polariton pulses.

We have also investigated the mode interference among the SPP modes in a 2D metallic nanohole array integrated with microfluidic channel for delivery and precise control of the index of refraction of overlaying layer using spectroscopy with a polarizer-analyzer pair, high-resolution wavelength and angle scan. We observed the strong coupling among SPP modes at the normal excitation, and more importantly, the splitting of the two degenerate (0, ±1) modes, leading to the formation of the symmetric and antisymmetric modes with an energy separation of ~14 meV. The collinear propagating directions of uncoupled SPP (0, ±1) modes contribute to the pronounced splitting in the dispersion relation.

A high resolution SPR sensor based on transmission through nanohole arrays has also been described and evaluated. The transmission lineshape function was shown to vary with the input and output polarization states—being minimal when these two states are orthogonal. In these structures (and gratings in general), the propagation length may be reduced to specification and can therefore increase the relative system resolution (limit the crosstalk between channels). This leads to a design tradeoff: the sensitivity may be sacrificed for smaller interrogation volumes depending on the particular application.

We also derive an analytical expression of wavelength sensitivity obtained from the SPP dispersion relation for 2D nanohole array SPR sensor. The sensitivity of nanohole array SPR sensor depends on the periodicity of the array and the order of the SPP modes. The analytical expression is confirmed by numeric results using electromagnetic simulation and also validated by the experiments. Real-time monitoring of protein-protein specific bonding between BSA and monoclonal anti-BSA is performed to demonstrate the integrated optofluidic nanohole array SPR biosensor. The detection resolution of the system can be increased by employing self-assembled linker layer on the gold surface and improving detection limit of the optical and electrical detection system. The clarified analysis and the demonstration of the sensitivity for SPP fields in 2D nanohole array not only elucidate the mechanism of the nanohole array SPR sensor, but also would facilitate the improvement of the sensitivity of SPR sensors.

The optofluidic plasmonic systems used for implementation of a biochemical sensor with angular and wavelength interrogation, demonstrate in situ, real-time, label-free detection of protein-protein interaction. These experiments reveal the dynamics of protein-protein interactions, essential not only for advancement of biological research, but also for reduction of false alarms in biochemical sensing. Optofluidic plasmonic sensors can be combined into a high-density 2D arrays to provide a very large throughput (e.g., about a million independent measurements) with high sensitivity and resolution, operating with small volumes, in real time and without any labels. These functionalities are useful for various applications including drug discovery and proteomics, in vitro diagnostics, food and drug industry, environmental and process monitoring, as well as military/homeland security applications.

References

1. I. R. Hooper, T. W. Preist, and J. R. Sambles, *Phys. Rev. Lett.*, **97**, 053902 (2006).
2. H. Raether, *Surface Plasmons on Smooth and Rough Surfaces and on Gratings*, Springer, Berlin (1998).
3. W. L. Barnes, A. Dereux, and T. W. Ebbesen, *Nature*, **424**, 824, 2003.
4. E. Ozbay, *Science*, **311**, 189 (2006).
5. W. L. Barnes, "Surface plasmon polariton length scales: a route to sub-wavelength optics," *J. Opt. A*, **8**, S87–S93 (2006).
6. H. C. Guo, D. Nau, A. Radke, X. P. Zhang, J. Stodolka, X. L. Yang, S. G. Tikhodeev, N. A. Gippius, and H. Giessen, "Large-area metallic photonic crystal fabrication with interference lithography and dry etching," *Applied Physics B-Lasers and Optics*, **81**(2–3), 271–275 (2005).

7. L. Pang, W. Nakagawa, and Y. Fainman, "Fabrication of two-dimensional photonic crystals with controlled defects by use of multiple exposures and direct write," *Applied Optics*, **42**, 5450–5456 (2003).
8. L. Pang, M. Nezhad, U. Levy, and Y. Fainman, "Form-birefringence structure fabrication in GaAs by use of SU-8 as a dry-etching mask," *Applied Optics*, **44**, 2377–2381 (2005).
9. Y. N. Xia, E. Kim, X. M. Zhao, J. A. Rogers, M. Prentiss, and G. M. Whitesides, "Complex optical surfaces formed by replica molding against elastomeric masters," *Science*, **273**, 347–349 (1996).
10. Y. N. Xia, and G. M. Whitesides, "Soft lithography," *Annual Review of Materials Science*, **28**, 153–184 (1998).
11. A. Groisman, S. Zamek, K. Campbell, L. Pang, U. Levy, and Y. Fainman, "Optofluidic 1 × 4 switch," *Opt. Express*, **16**, 13499–13508 (Sep 2008).
12. K. Campbell, A. Groisman, U. Levy, L. Pang, S. Mookherjea, D. Psaltis, and Y. Fainman, "A microfluidic 2×2 *optical switch*," *Appl. Phys. Lett.*, **85**, 25, 6119 (Dec 2004).
13. G. M. Hwang, L. Pang, E. H. Mullen, and Y. Fainman,"Plasmonic Sensing of Biological Analytes through Nanoholes," *IEEE Sensors* J., **8**, 2074–2079 (2008).
14. T. W. Ebbesen, H. J. Lezec, H. F. Ghaemi, T. Thio, and P. A. Wolff, "Extraordinary optical transmission through sub-wavelength hole arrays," *Nature*, **391**, 667–669 (1998).
15. C. Genet and T. W. Ebbesen, "Light in tiny holes," *Nature*, **445** (2007).
16. W. L. Barnes, W. A. Murray, J. Dintinger, E. Devaux, and T. W. Ebbesen, "Surface plasmon polaritons and their role in the enhanced transmission of light through periodic arrays of subwavelength holes in a metal film," *Phys. Rev. Lett.*, **92**, 10 (2004).
17. R. Muller, V. Malyarchuk, and C. Lienau, "Three-dimensional theory on light-induced near-field dynamics in a metal film with a periodic array of nanoholes," *Phys. Rev. B*, **68**, 20 (2003).
18. K. A. Tetz, R. Rokitski, M. Nezhad, and Y. Fainman, "Excitation and direct imaging of surface plasmon polariton modes in a two-dimensional grating," *Appl. Phys. Lett.*, **86**, 11, 111110 (2005).
19. R. Rokitski, K. A. Tetz, and Y. Fainman, "Propagation of femtosecond surface plasmon polariton pulses on the surface of a nanostructured metallic film: space-time complex amplitude characterization," *Phys. Rev. Lett.*, **95**, 17, 177401 (2005).
20. S. J. Elston, G. P. Bryanbrown, and J. R. Sambles, "Polarization conversion from diffraction gratings," *Phys. Rev. B*, **44**, 12, 6393–6400 (1991).
21. J. Elliott, Smolyaninov, II, N. I. Zheludev, and A. V. Zayats, "Polarization control of optical transmission of a periodic array of elliptical nanoholes in a metal film," *Opt. Lett.*, **29**, 12, 1414–1416 (2004).
22. E. Altewischer, M. P. van Exter, and J. P. Woerdman, "Polarization analysis of propagating surface plasmons in a subwavelength hole array," *J. Opt. Soc. Am. B-Opt. Phys.*, **20**, 9, 1927–1931 (2003).
23. C. Ropers, D. Park, G. Stibenz, G. Steinmeyer, J. Kim, D. Kim, and C. Lienau, "Femtosecond light transmission and subradiant damping in plasmonic crystals," *Phys. Rev. Lett.*, **94**, 113901, 4 (2005).
24. L. Pang, K. A. Tetz, and Y. Fainman, "Observation of the splitting of degenerate surface plasmon polariton modes in a two-dimensional metallic nanohole array," *Appl.Phys. Lett.*, **90**, 111103 (2007).
25. C. Nylander, B. Liedberg and T. Lind, "Gas detection by means of surface plasmon resonances," *Sens. Actuators*, **3**, 79 (1982).
26. J. Homola, S. S. Yee, and G. Gauglitz, "Surface plasmon resonance sensors: review," *Sens. Actuator B-Chem.*, **54**, 1–2, 3–15 (1999).
27. J. Homola, "Present and future of surface plasmon resonance biosensors," *Anal. Bioanal. Chem.*, **377**, 3, 528–539 (2003).
28. S. Y. Wu, H. P. Ho, W. C. Law, C. L. Lin, and S. K. Kong, "Highly sensitive differential phase-sensitive surface plasmon resonance biosensor based on the Mach-Zehnder configuration," *Opt. Lett.*, **29**, 20, 2378–2380 (2004).

29. I. R. Hopper and J. R. Sambles, "Differential ellipsometric surface plasmon resonance sensors with liquid crystal polarization modulators," *Appl. Phys. Lett.*, **85**, 15, 3017–3019 (2004).
30. N. Bassil, E. Maillart, M. Canva, Y. Levy, M. C. Millot, S. Pissard, W. Narwa, and M. Goossens, "One hundred spots parallel monitoring of DNA interactions by SPR imaging of polymer-functionalized surfaces applied to the detection of cystic fibrosis mutations," *Sens. Actuator B-Chem.*, **94**, 3, 313–323 (2003).
31. J. Dostalek, J. Homola, and M. Miler, "Rich information format surface plasmon resonance biosensor based on array of diffraction gratings," *Sens. Actuator B-Chem.*, **107**, 1, 154–161 (2005).
32. Y. D. Su, S. J. Chen, and T. L. Yeh, "Common-path phase-shift interferometry surface plasmon resonance imaging system," *Opt. Lett.*, **30**, 12, 1488–1490 (2005).
33. R. Rella, J. Spadavecchia, M. G. Manera, P. Siciliano, A. Santino, and G. Mita, "Liquid phase SPR imaging experiments for biosensors applications," *Biosens. Bioelectron.*, **20**, 6, 1140–1148 (2004).
34. H. J. Lezec and T. Thio, "Diffracted evanescent wave model for enhanced and suppressed optical transmission through subwavelength hole arrays," *Opt. Express*, **12**, 16, 3629–3651 (2004).
35. Q. Cao and P. Lalanne, "Negative role of surface plasmons in the transmission of metallic gratings with very narrow slits," *Phys. Rev. Lett.*, **88**, 5 (2002).
36. S. M. Williams, K. R. Rodriguez, S. Teeters-Kennedy, S. Shah, T. M. Rogers, A. D. Stafford, and J. V. Coe, "Scaffolding for nanotechnology: extraordinary infrared transmission of metal microarrays for stacked sensors and surface spectroscopy," *Nanotechnology*, **15**, 10, S495–S503 (2004).
37. A. G. Brolo, R. Gordon, B. Leathem, and K. L. Kavanagh, "Surface plasmon sensor based on the enhanced light transmission through arrays of nanoholes in gold films," *Langmuir*, **20**, 12, 4813–4815 (2004).
38. T. Rindzevicius, Y. Alaverdyan, A. Dahlin, F. Hook, D. S. Sutherland, and M. Kall, "Plasmonic sensing characteristics of single nanometric holes," *Nano Lett.*, **5**, 11, 2335–2339 (2005).
39. K. A. Tetz, L. Pang, and Y. Fainman, "High-resolution surface plasmon resonance sensor based on linewidth-optimized nanohole array transmittance," *Opt. Lett.*, 31, 10, 1528–1530 (2006).
40. J. Homola, I. Koudela, and S. S. Yee, "Surface plasmon resonance sensors based on diffraction gratings and prism couplers: sensitivity comparison," *Sens. Actuator B-Chem.*, **54**, 1–2, 16–24 (1999).
41. J. Elliott, I. I. Smolyaninov, N. I. Zheludev, and A. V. Zayats, "Polarization control of optical transmission of a periodic array of elliptical nanoholes in a metal film," *Opt. Lett.*, 29, 12, 1414–1416 (2004).
42. P. Hanarp, M. Käll, and D. S. Sutherland, Optical Properties of Short Range Ordered Arrays of Nanometer Gold Disks Prepared by Colloidal Lithography *J. Phys. Chem. B*, **107**, 5768, (2003).
43. A. J. Haes, S. Zou, G. C. Schatz, and R. P. Van Duyne, Nanoscale Optical Biosensor: Short Range Distance Dependence of the Localized Surface Plasmon Resonance of Noble Metal Nanoparticles *J. Phys. Chem. B*, **108**, 109 (2004).
44. M. M. Miller, A. A. Lazarides, Sensitivity of Metal Nanoparticle Surface Plasmon Resonance to the Dielectric Environment *J. Phys. Chem. B*, **109**, 21556 (2005).
45. http://www.biacore.com/lifesciences/index.html.
46. I. Richter, P. -C. Sun, F. Xu, and Y. Fainman, Design considerations of form birefringent microstructures *Appl. Opt.*, 34, 2421 (1995).
47. L. Pang, G. M. Hwang, B. Slutsky, and Y. Fainman, "Spectral sensitivity of two-dimensional nanohole array surface plasmon polariton resonance sensor," *Appl. Phys. Lett.*, **91**, 12, 123112 (2007).
48. L. S. Jung, Charles T. Campbell,T. M. Chinowsky, M. N. Mar, and S. S. Yee, Quantitative Interpretation of the Response of Surface Plasmon Resonance Sensors to Adsorbed Films *Langmuir*, **14**, 5636 (1998).
49. A. Lesuffleur, H. Im, N. C. Lindquist, and S-H. Oh, Periodic nanohole arrays with shape-enhanced plasmon resonance as real-time biosensors *Appl. Phys. Lett.*, **90**, 243110 (2007).

CHAPTER 15

Optical Manipulation and Applications in Optofluidics

Kishan Dholakia and Tomáš Čižmár

SUPA, School of Physics and Astronomy, University of St Andrews, North Haugh, Fife, Scotland

15-1 Introduction to Optical Manipulation

The application of optical forces in the microscopic and nanoscopic world is an enabling technique in the natural sciences. Though such forces are small they may trap, guide, and in general manipulate samples; they suffice to realize noninvasive mechanical control over atomic, biological, and colloidal systems. The techniques of such "optical manipulation" are compatible with modern microscopy and enhance the reconfigurability of the trap while the accuracy achieved in a calibrated optical trap presents itself as a very precise and quantitative force probe. Typically ultraprecise motional and force measurements for molecular motors or cell mechanotransduction studies are achievable. The applications are not restricted to biology. Optical traps have provided seminal studies in colloidal and optical physics including the phase dynamics of thermodynamic systems, Brownian diffusion, aspects of microfluidics, and fundamental issues related to optical angular momentum. There is little doubt regarding the seminal advances we have seen across the natural sciences in the last few decades that are based upon the light-matter interaction. From a wider viewpoint, optical force and momentum-exchange with atomic ensembles has paved the way for the very powerful methods of laser assisted cooling [1–3] and the achievement of ultracold quantum gases and onset of Bose-Einstein condensation [4,5]. However we will concentrate here on aspects of optical manipulation that are pertinent to the emerging area of optofluidics.

The origin of optical forces lies in the fact that light may be considered as quanta of energy that possesses momentum. Overall, the exchange of momentum with matter may result in a force and thus physical motion: in this manner, light may move, hold or more generally manipulate material objects and, importantly for our purposes, objects the size of a single cell or smaller. As many fields in photonics, this has been enabled and propelled forward by the advent of the laser and its inherent properties. Importantly, the applied force can be readily calibrated lending itself to measurements in the piconewton to femtonewton region complementing atomic force microscopy. Importantly, laser light is naturally reconfigurable and may be sculpted or adapted in variety of ways that is a key issue for the topic of microfluidic and notably optofluidic applications.

The de Broglie relation shows us that the momentum of light is very small and thus exchange of momentum with matter naturally results in a very small force, typically of the order of piconewtons. The interaction between light and the particle (microscopic or smaller) produces a change of photon momentum upon the particle at a rate that would lead to small forces that are sufficient enough to move or hold a microparticle. At such a size level, this concept can be utilized particularly by the biological sciences in pursuit of studying several macromolecular and cellular processes in a quantifiable manner. The field of fundamental physics too has benefited in numerous ways using optical traps: seminal studies in the last 15 years include a deeper understanding of the optical angular momentum of light, and exploration of colloidal hydrodynamics or microfluidics. Such experimental studies of light-matter interaction have, in turn, advanced physicists toward a more complete appreciation of the theoretical basis for optical forces.

We have to go back four decades to explore the first optical trapping experiments and over 20 years since the inception of the popular "optical tweezers" [6]. Arthur Ashkin, the key pioneer of this field, in his first study, dispersed microparticles in water within a chamber which were then exposed to a single horizontally propagating visible laser beam [7]. The microparticles aligned themselves along the propagation axis and were guided along the beam axis: this was the first observation of "optical guiding." Introducing a second beam (of equivalent optical power) at counter-propagating geometry halted the motion of a sphere along the beam axis while retaining its position within the bright region of the two beams: the first optical trap was formed. This counter-propagating beam optical trapping geometry [7] has been realized with fiber [8] and is returning to prominence as we shall discuss later in the form of the optical stretcher [9] and for manoeuvring large cells [10]. Such traps may also be of importance for longitudinal optical binding [11,12]. We note that such fiber traps are particularly amenable to studies in microfluidic chambers and the concept of integrating fluidic chambers with optical addressing [13]. Sixteen years after the realization of counter-propagating trap, Ashkin and colleagues realized the single-beam gradient trap (popularly known as *optical tweezers*) [6] that is the most widespread and popular method for applying optical forces for moving microscopic particles. This trap has now been well recognized as having the largest impact to date within the field of optical manipulation.

This chapter is not intended to give a comprehensive overview of this field. Rather, it is directed to give the reader an insight into the basic aspects of optical trapping and manipulation with the emphasis toward emergent applications and some recent and relevant experiments in microfluidics and optofluidics. We also emphasize the new technologies that offer true reconfigurability of light using sculpted or shaped light fields. For the reader interested in the broader remit of this field we note that this chapter is complemented by other reviews of this subject area [14–17]. Prior to the discussion of optical trapping within microfluidics and optofluidics, we shall begin with a theoretical perspective upon optical trapping looking at how we may describe the optical forces and a very brief consideration of some of the experimental issues for the implementation of the widely used "optical tweezers." We observe that this field has made major impacts within single-molecule studies, namely, the study of molecular motors and other biological macromolecules as well as cellular material. Optical tweezers have produced some seminal studies within single-molecule biophysics and allowed an insight into this field in a manner hitherto unforeseen and other reviews cover this topic very well [18]. Many of the applications within biology and chemistry are very active current areas of research which are continually evolving, so the aim will be to give the reader a grounding in the various techniques to facilitate the reasoning behind the use of optical trapping and manipulation in optofluidics. The chapter is structured as follows: firstly, we give an overview of how to understand theoretically the optical forces exerted upon a particle, paying attention to the various particle size scales with respect to the trapping wavelength. We then progress to more advanced trapping schemes that use multiple trapping geometries. Finally, we look at the use of optical trapping in microfluidic environments and illustrate this with some examples of recent experimental work pertinent to the optofluidics community.

15-2 Theoretical Considerations

A theoretical insight into the optical forces may be understood in a number of ways. In the simplest form the optical forces can be demonstrated on Rayleigh particles. Here we assume that the optical field is uniform over the volume of the object, so this description is valid for very small particles (usually with diameter up to $\lambda_{vac}/20$, where λ_{vac} is the vacuum wavelength). In this case, the object can be described as a scattering dipole and the total time-averaged optical force can be expressed as [19]

$$< F_\iota > \equiv F_\iota = \frac{1}{2}\varepsilon_0 \varepsilon_m \Re \left\{ \sum_{\gamma=1}^{3} \alpha E_\gamma \nabla_\iota E_\gamma^* \right\} \tag{15-1}$$

where $\nabla_\iota \equiv (\partial / \partial r_\iota)$, E_γ is the γ component of the electric field, ε_0 is the permittivity of vacuum, $\varepsilon_m \equiv n_m^2$ is the relative permittivity of the surrounding medium, $\Re$ the real value of the subsequent expression in brackets, $*$ denotes the complex conjugated value, α is a complex valued polarizability of the object given by [20]

$$\alpha = \frac{\alpha_{ll}}{1 - \dfrac{2}{3}\dfrac{ik^3 a_{ll}}{4\pi}} \equiv \alpha' + i\alpha'' \tag{15-2}$$

α' and α'' are the real and the imaginary parts of α, and α_{ll} is done by Lorentz-Lorenz relation:

$$\alpha_{ll} = 4\pi a^3 \frac{m^2 - 1}{m^2 + 2} \tag{15-3}$$

where a is the radius of the spherical object and $m \equiv n_p / n_m$ is the ratio of the refractive indexes of the particle and the surrounding medium. Traditionally it is established, that the part of force related to the real part of α is called the gradient force because it results from the gradient of optical intensity. For high-index particles ($m > 1$), it causes particle attraction to the intensive places of the ($\alpha' > 0$), whereas for low-index particles ($m < 1$) this force brings repelling of particles from high intensity ($\alpha' < 0$). The imaginary part of α causes the scattering force—the force of radiation pressure in the direction of the wave propagation. As demonstrated in Fig. 15-1 one can achieve a stable confinement when producing optical field with a high three-dimensional intensity gradient. The classic geometry of optical tweezers uses a tightly focused Gaussian beam produced by a high numerical aperture (NA) microscope objective. The gradient force

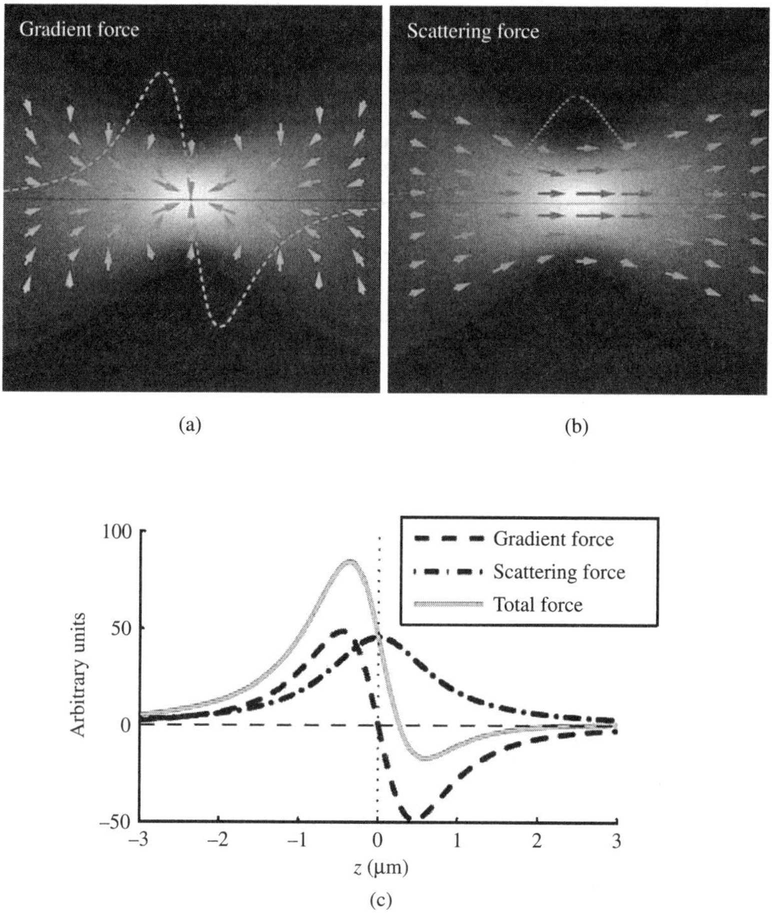

FIGURE 15-1 An example of optical forces acting on a Rayleigh particle in a focused Gaussian beam. The particle is attracted to the focus under the influence of (a) the gradient force. (b) The scattering force pushes the particle in the direction of the beam propagation so the stable position appears behind the focus where the scattering and gradient forces are balanced and the total force is equal to zero. Plot (c) shows the axial components of the scattering, gradient, and total force.

then attracts the particle to the beam focus and the scattering force pushes the object downstream, so the stable position appears on the axis behind the focus, where the scattering and gradient forces are balanced.

The scattering force is directly proportional to the trapping laser intensity and the gradient (or dipole) force upon the object is due to the inhomogeneous field gradient created by the tightly focused light beam [16,21]. In general, we see that the gradient force is proportional to the polarizability, and when considering a dielectric particle it scales with its volume. This is a very important point and means it is quite difficult in practice to trap very small dielectric objects (e.g., diameter of 50 nm). These considerations also explain why gold nanoparticles (with their very large polarizability) may be readily trapped at sizes of 100 nm and below [22], though absorption, denoted by their complex refractive index, is a key consideration for such metal nanoobjects [23].

Optical trapping, however, is not restricted to the area of object sizes in the Rayleigh regime and one can readily hold dielectric objects from ~0.5 to 5 μm in diameter. For large particles within the Mie regime, where the microparticle radius is much larger than the trapping wavelength, the use of geometrical ray optics may be used to picture the forces involved. Ashkin employs ray tracing and the well-known Fresnel equations at the sphere-medium boundary [24] to determine the optical forces. Figure 15-2 elucidates this approach in a

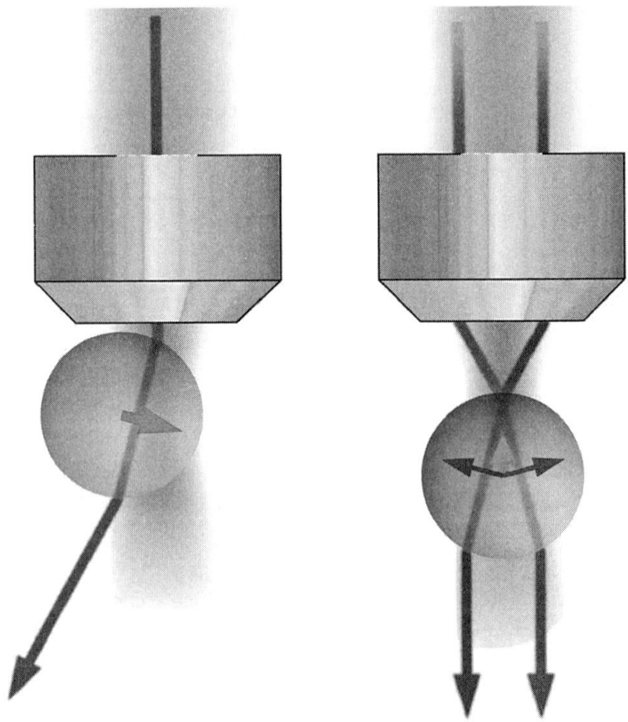

(a) (b)

Figure 15-2 Optical trapping in Mie regime. (a) Lateral confinement: the off-axis particle refracts the beam thus providing a change in the original momentum of the beam. As a reaction the beam exerts a force on a particle in the opposite direction attracting the particle to the axis. (b) Longitudinal confinement: the particle behind the focus acts like a collimating lens. Since the axial momentum of the collimated beam is larger when compared to the original diverging beam, the reaction of light acting on particle attracts the particle toward the focus. More illustrative demonstrations of these principles may be found at R. DiLeonardo, http://glass.phys.uniroma1.it/dileonardo/Applet.php?applet=TrapForcesApplet.

basic form where no reflections on the liquid-particle boundary are shown for the sake of simplicity within the figure.

Other studies looked at different aspects of the problem. Barton and coworkers [25,26] derived a fifth order corrections to the focused gaussian beam such as to compute the forces using a Maxwell stress tensor approach. Rohrbach and Stelzer [27] extended the Rayleigh theory to make it valid for large particles by inclusion of second-order scattering terms. The incident field is expanded in terms of constituent plane waves allowing apodization and aberration transformations (due to the high numerical aperture microscope objective) to be incorporated in the theoretical model to yield the resultant optical forces upon the dipole, in this instance without resorting to use of the Maxwell stress tensor method. The vast majority of optical tweezers and trapping experiments are performed where the particle size is comparable to the wavelength of the trapping laser beam. In this region, the key studies by Rohrbach reported good quantitative agreement between the theory calculations and experimental measurements pertaining to the strength of the optical tweezers. His theoretical approach for trapping forces computed the Lorentz force density. He found that the optimal trapping performance is reached when the wavelength of light (within the viscous medium) is comparable to the diameter of the particle, $d \approx \lambda_{\text{vac}}/n_m$ [28]. Overall, it is important to note that the detailed numerical and theoretical modelling of optical forces is an ongoing topic of research.

15-3 Experimental Considerations for Single-Beam Optical Tweezers

The single-beam optical tweezer is the simplest system to consider and indeed the most popular experimentally. We give an overview of considerations for such a trap and a more detailed explanation for construction and assembly of such an optical tweezer may be found elsewhere [29]. For such optical tweezers we usually need the use of a high numerical-aperture (NA $\geq$ 1) microscope objective (in upright or inverted geometries) to get the lateral and axial gradients required for three-dimensional trapping. Usually this means the use of a microscope body to add rigidity to a trapping system, though a system may be assembled from off-the-shelf optomechanical components. The input beam is usually expanded to slightly overfill the back aperture of the microscope objective thus ensuring a very tight beam focus. Conjugate beam-steering systems are also employed to ensure that the input beam does not deviate or "walk off" the back aperture of the objective, provided that when multiple traps or steering of a single trap is performed no beam clipping occurs. The choice of laser is crucial and usually monochromatic continuous wave lasers operating at near-infrared wavelengths are used (e.g., 750–1100 nm), which lie in the so-called therapeutic window. This minimizes the laser damage to biological samples [30]. Pointing stability and amplitude noise are crucial too, especially while performing single-molecule studies. It is worth remarking (though not described in detail here) that a particle within a single optical tweezers is an elegant overdamped simple harmonic oscillator that may be calibrated for force measurements: the system acts like a microscopic version of a spring and thus obeys Hooke's law [29]. The trap stiffness is related to the laser beam's quality and power. Incredibly, forces as small as a few 10s of femtonewtons [31] may be recorded as well as displacements of only a few angstroms [32] making this an exceptional technique for single-molecule biophysics and the study of molecular motors. Powers of a few milliwatt upward are desired to confine objects in three dimensions, though, of course, when one uses multiple traps, and considering the efficiency of various optical components in the optical train, one might desire laser powers of several watts. Samples are usually composed of two thin (100 μm) cover slips between which ~10 μL of the colloidal or biological specimen is dispersed. Index-matching fluid is used with the high-numerical-aperture objectives (typically oil immersion) to avoid any refractive index mismatch and reduce aberrations, particularly spherical aberration [29].

15-4 The Counter-Propagating Beam Trap

While the single-beam trap has a multitude of uses, the original trapping geometry already described used a counter-propagating beam trap [7]. This was later modified to fiber geometry in 1993 [8]. This has a large potential in the domain of optofluidics due to its inherent compatibility with microfluidics and the fact that it lends itself directly to integration. It also removes the need for high numerical-aperture optics for implementation (though, of course, these may be desired for particle observation). Various recent experiments have shown how such integrated fibers can be used along with fluidics for new geometries and applications. A good example has been the development of the optical stretcher [9,33]. Here a cell is held in the counter-propagating beam trap and analysed within a microfluidic chamber. One might intuitively think that increasing the power in the fiber trap will actually compress the held object if deformable but, in fact, the reverse is true. To appreciate this point one must consider the momentum of light in the medium and within the cell: whenever light passes from a medium of lower to higher refractive index, one finds there is a force exerted away from the high index particle [34]. This allows one to stretch or deform cells in a fiber trap and this deformability can be used to characterize cells and distinguish neoplastic cells from their healthy counterparts. The differences in response here lie in the fact that the actin cytoskeleton differs in these cell types leading to an intrinsic manner by which to diagnose in situ abnormal cell types. The counter-propagating beam trap has other notable attributes that include the ability to hold large objects: recent work has shown it can move and hold objects up to ~100 μm in diameter [10]. Due to its large capture range and the fact that the light is distributed over the whole cell, this makes it more amenable to holding and manoeuvring large objects. In turn this can lead to combining such traps with spectroscopy for optofluidic applications such as *in situ* Raman analysis [10]. Other work has used novel photonic crystal fiber for creating counter-propagating beam traps, allowing one to deliver multiple wavelengths and indeed white light supercontinuum sources to trap and move particles in an "interference-free" environment [35] and perhaps in the future perform spectroscopy in tandem with the trapping. Other very recent studies have explored the details of fiber optics/microfludic integration, exploring a number of several flow-dependent particle-trapping mechanisms by controlled rotational and lateral displacements of the trapping fibers [36]. Geometries of parallel and offset fibers (orthogonally oriented to the fluid flow direction) showed a cyclic cross-stream particle motion. Fibers angled upstream, again with flow present, exhibit a circulatory trajectory for the particles. Asymmetric angled fibers resulted in continuous particle circulation in these studies. A significant step forward for the integration agenda showed how one could marry the lasers directly with the fluidics for particle trapping and detection which were seen by Cran-McGreehin and colleagues [37,38]. Both interrogation and manipulation are made more amenable through such monolithic integration. The laser diodes created a monolithic counter-propagating beam trap that can hold and manoeuvre the particles. These lasers were coupled directly into the microfluidic channel, allowing dispersed particles to pass through the laser's output beams. Isolating the electrical p-n junctions from the fluid is the key challenge, achieved by careful use of a photodefinable SU8-2000 polymer [37,38]. The interaction between particles in the channel and the lasers, operated in either forward or reverse bias, allows particle detection and ultimately counting as they transit this section of the chip. These very small devices require no external optical components and intrinsically have perfect alignment. Trap operation does not require an experience with optical systems adding a further advantage. This opens up the possibility of truly automated optical manipulation and even particle/cell counting.

Notably, each integrated device may contain multiple traps (see Fig. 15-3) and has a footprint of only a few millimeters in a given direction. This paves the way for fully transportable optical traps and sensors. From an optofluidics viewpoint we note here that every part is defined lithographically onto a single piece of semiconductor material, obviating the need for alignment, and removing coupling losses. The laser beams enter directly into the test chamber where they

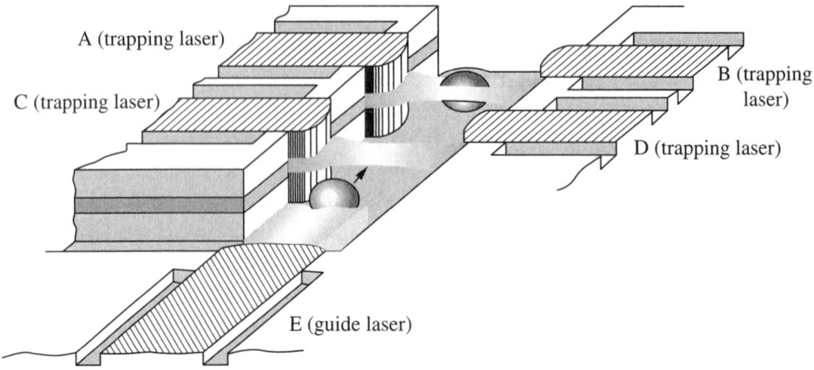

FIGURE 15-3 A microchip comprising several integrated trapping lasers and a microfluidic channel. Particles can be guided along the channel and confined at several locations by pairs of counter-propagating beams. (*Source: S. J. Cran-McGreehin, K. Dholakia, and T. F. Krauss, "Monolithic integration of microfluidic channels and semiconductor lasers," Optics Express, 14, 2006, 7723–7729.*)

interact with the specimen. This contrasts with conventional systems that use discrete, macroscopic optical components to couple laser beams from external sources into the sample chamber. A more broad general review exploring issues of integration for microanalysis in an optofluidics environment has been recently published [39]. Now we have described the basic forms of trapping, optical tweezers, and the counter-propagating beam trap, and we progress to the topic of trap multiplexing and, in general, the use of advanced light fields in optical manipulation.

15-5 Advanced Light Fields

Optical tweezers is very powerful in its own right: we need to establish the motivation to move away from such a single optical trap and use advanced light patterns, either as sculpted traps on their own or replicated in multiple trap arrays. While optical tweezers have engendered excellent studies across the sciences, notable limitations persist with the method. For example, a single trap limits the manoeuvring of single molecules and in several instances (e.g., studying the actin-myosin system), a dual beam (two-trap) system is desirable. The standard optical tweezing geometry uses a microscope objective lens, employing a zeroth-order gaussian laser beam. This delivers a single tweezers with an approximately ellipsoidal volume at beam focus. In turn this restricts the ratio of trap stiffness in lateral and axial directions. Such a setup offers little flexibility for tailoring an arbitrary-designed optical potential energy landscape in the sample chamber. A gaussian beam has a given Rayleigh range and divergence given by the beam parameters that may be restrictive given the use of high numerical aperture optics, for example, optical guiding of particles separated over large distances (millimeters) would be problematic. How might we rotate or spin trapped objects? It would be advantageous to apply known torques to trapped objects within optical tweezers so that a rotational torque [40,41] can be applied upon biomolecules. A rotating particle may be applied for microrheology applications where the terminal angular velocity attained can be used to measure the rotational stokes drag and ultimately the local viscosity or to measure a range of viscoelastic behaviour in different media. Further, rotating single particles or groups of particles can induce pumping action within the microfluidic chamber [42] which has potential for controlling flow rates or mixing of small amounts of fluids relevant for optofluidics. By increasing the plurality of the optical tweezers—creating multiple trap arrays—researchers may start to explore multiplexed trapping experiments or indeed probe and use larger patterned arrays of cells or colloidal microparticles. It is important to stress that a large array of optical tweezers is appropriately considered as an "optical potential energy landscape" which can exert forces over its used area. Such engineered landscapes enable the generation of two- and three-dimensional quasi-crystal structures of colloids within which one may tune the interparticle interaction, for example, by surface chemistry or use of appropriate solvents. Such an ensemble enables important research to be undertaken in material science and thermodynamics [43,44]. We group our discussion here under the heading of *advanced light fields*. We firstly mention the different ways of multiplexing a single-beam optical tweezer into many traps. After this we look at two key nonzero-order light fields of use for this community, namely the "nondiffracting" or propagation invariant Bessel beams [45] and Laguerre-Gaussian (LG) beams [46].

15-5-1 Multiple Trapping Techniques

With the increasing complexity of optical trapping applications, new requirements are placed upon optical tweezers. In particular, the ability of multiple trapping and precise delivery of confined objects is required. One topical area in the last decade has thus been to move toward a plurality of traps for a number of applications. As we shall see this has a major impact upon applications within optofluidics and in general microfluidics environments. It is important to stress that a plurality of traps offers a lot more than just a simple multiplexing of studies. The interconnectivity and availability of multiple traps allows a variety of interesting studies in colloidal physics as well as biophysics. The optical trap multiplicity was enabled by dividing the laser beam into several beams by a beam-splitter, gratings, or hologram, or by fast switching of a single-beam trap between several positions of focus (time-sharing). Let us briefly focus here on two recent and very popular techniques using acousto-optical deflectors (AOD), and spatial light modulators (SLM), these latter devices acting as arbitrary phase or amplitude holograms.

AODs can provide fast steering of a laser beam that leads to the formation of a number of time-shared optical traps with switching rates typically of 10 to 100 of kHz (see Fig. 15-4). Positioning of the traps can be easily controlled by altering the RF signal applied to a piezoelectric transducer. This transducer produces an acoustic signal inside a glass material where based on acousto-optic effect the beam is declined with an angle dictated by the applied frequency. This way one might generate up to several hundreds of optical traps that might be arbitrarily spread and rapidly and precisely positioned along a single-axial plane (this method provides only two-dimensional control of optical traps unless combined with another technique) [47].

The second and, perhaps, the most powerful method of multiple trapping and generation of advanced light fields is holographic optical trapping (HOT) [48]. HOTs, in contrast to acousto-optically time-shared traps, produce all of the traps simultaneously as the beam is multiplexed directly between them. Furthermore, the trapping is not restricted to a single plane only but using an appropriate algorithm one can generate structures in three dimensions [49–53]. This method can be extended by the use of spatial light modulators (see Fig. 15-5) where these features may be implemented in a reconfigurable way [54]. Besides multiplexing and positioning of optical traps this technique offers generation and control of special optical fields like Laguerre-Gausian or "nondiffracting" Bessel light modes that will be introduced later in this section. The diffractive optic element (DOE) can also be encoded to compensate any inherent aberration present [55]. Other recent papers describe the development of a HOT system in detail [56,57].

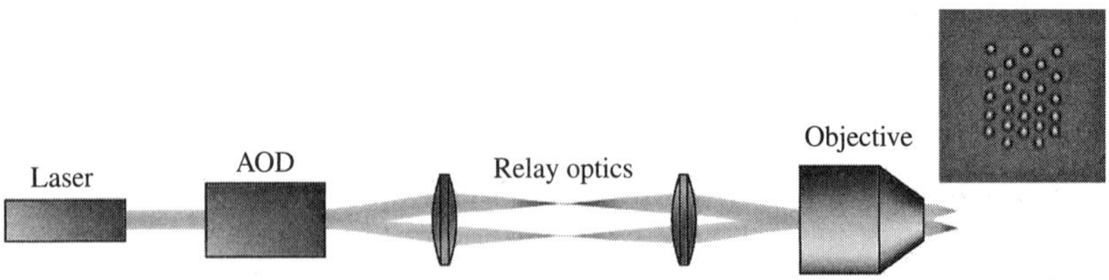

FIGURE 15-4 Generation of multiple optical traps using an acousto-optical deflector (AOD). Steering of the beam is provided by the acousto-optic effect in the glass material where acoustic signal is generated by piezoelectric transducer. The steering angle is controlled by RF harmonic signal applied on the transducer. Using an AOD one is able to send the light only to one trapping position at a time; however, the ability of very fast switching between the trapping positions enables one to create up to several hundred stable time-shared optical traps in one axial plane.

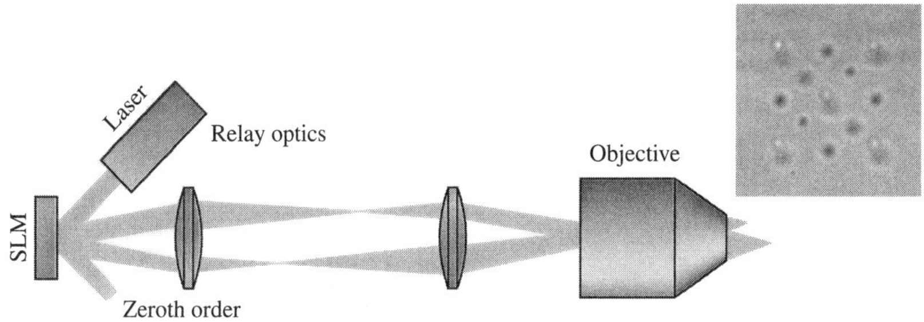

FIGURE 15-5 Holographic optical tweezers. Spatial light modulators are devices producing arbitrary holographic phase or amplitude modulation of the illuminating beam that can be used for beam steering, focusing, or multiplexing. These devices can be efficiently used to produce several arbitrarily placed optical traps at the same time. Inset picture was taken from Glasgow University Optical Trapping web page: http://www.physics.gla.ac.uk/Optics/projects/tweezers/movies/Diamond% 20Morph%2036s.mpg. [*Source: G. Sinclair, P. Jordan, J. Courtial, M. Padgett, J. Cooper, and Z. Laczik, "Assembly of 3-dimensional structures using programmable holographic optical tweezers," Optics Express, 12(22), pp. 5475–5480 (2004).*]

The drawback of this method lies in the large consumption of computational time required for producing the hologram encoding. The way around this might be placing the hologram readout off-axis in Fresnel regime rather than in the typically used on-axis Fourier regime [58] (as presented in Fig. 15-5), where positioning of the traps can be done real-time by moving a hologram window at the SLM display. Another way how to deal with this problem is the use of the generalized phase contrast (GPC). This technique is an alternative to the HOT method that does not use the SLM as a hologram device but rather more directly as a phase element. This encoded phase-modulation is then converted to the amplitude-modulation on a phase filter with a very high efficiency [59,60].

Dual beam traps may use acousto-optic devices or simple beam splitting. The generation of two steerable traps [61] has enabled a number of novel achievements in single-molecule biology [32], hydrodynamics between two trapped spheres [62], and in fusion studies in chemistry within a microfludic environment [63]. Spatial light modulators combined with AOD systems have organized three-dimensional heterotypic networks of living cells in hydrogel [64]. SLMs with advanced camera technology have explored hydrodynamic coupling between a number of colloidal particles [65]. The combination of microfluidic and multiple trapping was used for direct monitoring of a cell response to environmental changes like an increase of osmolarity [66]. The used microfluidic system allowed two different media to be merged in a Y-shaped channel. Microscale channel dimensions ensured purely laminar flow and, as a result, an environmental gradient was created between the two media. Groups of cells confined in a system of optical traps were repeatedly exposed to these two different media that resulted in changes of the cells' volume. Other applications in microfluidics, this time using AODs, were presented by Terray et al. [67], where a number of confined particles were used to create and control micropumps and valves in the laminar flow in microchannel, each the size of human blood cell.

Advanced light fields are not solely restricted to creating 2D or 3D arrays of traps. Importantly a variety of light fields in the optofluidic context enable studies such as mixing, pumping of fluids, droplet manipulation and mixing, optical sorting, and long-distance guiding. These beams may include a component of sculpting the output wavefront and include Bessel light beams [68] and Laguerre-Gaussian light fields [46] with embedded vortices or phase singularities. We will look at both briefly again with the emphasis on their potential for optofluidic applications.

15-5-2 Bessel Light Modes

Bessel beams have the unusual property of propagation invariance over a limited region, enabling one to generate optical features with immunity to diffractive spreading. Such beams enable the creation of long-distance guiding of particles [69], conveyor belts [70], and sorting of microscopic objects [71]. Originally they were proposed by Durnin in 1987 [72] and the first experimental verification was shown in the same year [45]. As solutions of the Helmholtz equation, they are of the form of a Bessel function and higher-order versions have phase singularities at a beam center. Such beams may be created with an annulus placed in the back focal plane of a lens (though not efficient) or by use of a conical glass element known as an axicon or by way of a spatial light modulator. The Bessel beam does offer extended "nondiffracting" optical features but the price is a distribution of the optical intensity across the whole profile which is a series of concentric rings. A more extended beam demands a larger number of such rings with the power distributed almost equally amongst all of these rings. However, for optical manipulation in a microfluidic environment such beams may offer extended transport of particles [69] and simultaneous trapping in multiple microfluidic sample chambers [73]. This latter experiment makes use of the self-healing of the beam that arises from the conical wave-vectors that constitute the Bessel beam profile. For optofluidic applications such light modes can transport particulate matter over long distances and also initiate passive optical-sorting due to the selective response of both dielectric and biological samples to the periodicity of the Bessel profile (the optical potential energy landscape). In such studies Paterson et al. [71] made use of the outer rings of the Bessel mode to engineer the motion of red versus white blood cells. The erythrocytes moved to beam center whereas the lymphocytes migration was halted in one of the outer rings where they aligned. Optically engineered Bessel modes can be used for cell sorting and photoporation within a biophotonics workstation [74]. As we move toward more optofluidic geometries we note that several groups are exploring on-chip Bessel mode generation with microaxicons [75] to utilize the properties of Bessel light modes directly next to a microfluidic chamber.

15-5-3 Laguerre-Gaussian Light Modes

We now turn our attention to the Laguerre-Gaussian laser modes [46]. A given mode (denoted LG^l_p) is described with the two integer indices l and p. The azimuthal index l is the most important for our purposes and refers to the number of 2π phase cycles around the circumference of the mode and $(p + 1)$ indicates the number of radial nodes in the mode profile. LG modes with $l \neq 0$ and $p = 0$ (a single annulus in form) have garnered a large interest owing to their azimuthal phase term $(-il\phi)$ and that gives rise to a well defined orbital angular momentum (OAM) of $l\hbar$ per photon, which is distinct from and may be larger in magnitude than any angular momentum associated with the spin-angular momentum of the field $\pm\hbar$. The physical interpretation of the orbital angular momentum is due to the inclined optical wavefront [46,76] and resulting azimuthal component of the Poynting vector. A general description of the electric field of a LG mode $E(LG^l_p)$ of indices l and p may be written as [77]:

$$E(\mathrm{LG}^{l}{}_{p}) \propto \exp\left[\frac{-ikr^2z}{2(z_r{}^2+z^2)}\right].\exp\left[\frac{-r^2}{\omega^2}\right].\exp\left[-i(2p+l+1)\arctan\left(\frac{z}{z_r}\right)\right].\exp[-il\phi]$$

$$\times(-1)^p.\left(\frac{r\sqrt{2}}{\omega}\right)^l.L^l_p\left(\frac{2r^2}{\omega^2}\right) \tag{15-4}$$

where z denotes the distance from the beam waist, z_r is the Rayleigh range, k is the wave number, ω is the radius at which the Gaussian term $e^{(-r^2/\omega^2)}$ falls to $1/e$ of its on-axis value, r is the radius, ϕ is the azimuthal angle and L^l_p is the generalized Laguerre polynomial. The term $(2p+l+1)\arctan(z/z_r)$ is the Guoy phase of the LG mode that varies with the mode indices.

How are Laguerre-Gaussian modes created in a practical situation? The most practical and versatile method is the generation of LG modes directly from a fundamental TEM_{00} Gaussian beam, external to the laser cavity. Two popular methods that satisfy this requirement, each using diffractive optical elements, have been established. These are the use of a spiral phase element or a computer-generated hologram, respectively. When considering a spiral phase elements, a high refractive-index substrate is shaped into the spiral phase ramp [78,79] that directly imparts the helical phase structure upon the input Gaussian beam. With recent microfabrication techniques, the spiral phase element has been miniaturized [80,81], making it compatible for microfluidic applications. The form of the output mode here is in fact best described as a superposition of LG modes [78]. The computer-generated holographic generation method involves mathematically encoding the spiral phase with a given input field at an angle onto a computer-generated pattern, and indeed this computer-generated hologram may be sent directly to a spatial light modulator or be written into a fused silica substrate. The origin of orbital angular momentum of Laguerre-Gaussian light fields can be appreciated by careful consideration of the helical wavefronts of an LG beam. The inclined helical wavefront leads one to consider the energy flow in such fields: the Poynting vector moves in a corkscrew like manner [46]. This angular momentum is therefore linked with the azimuthal component of the Poynting vector. A trapped particle placed in such a field (e.g., LG_0^1) would rotate continuously around the beam's circumference.

In an optical trapping geometry, such orbital angular momentum may be transferred by a number of means with scattering and absorption, the most typical mechanisms. In 1995, He and colleagues [82] set absorptive copper oxide particles into rotation using LG modes: in a broader context, this was one of the first ever implementations of holographic optical trapping. The experiment trapped particles in two dimensions and the authors were able to rule out rotation due to any asymmetric scattering that might have been present. In a three-dimensional trap, Simpson and coworkers [83] rotated absorptive objects using linear and circularly polarized Laguerre-Gaussian light modes. By observing the rates of rotation due to each form of angular momentum, they experimentally decoupled the spin angular momentum of light from the orbital angular momentum of light. Friese et al. achieved very similar results using higher-order ($l = 3$) LG modes [84]. Optically absorptive particles showed clearly the physics of these light modes but naturally would not be ideal for biological applications nor for applications within microfluidics.

Subsequently, it was realised that one could actually transfer orbital angular momentum onto nonabsorbing dielectric particles simply by the trapped particles scattering light off the inclined wavefronts of the LG beams [85,86]. Particles situated off-axis within the circumference of the LG beams were seen to respond to the spin or orbital component of the light field in a different manner. In turn this gives insight into the intrinsic and extrinsic nature of spin and orbital angular momentum and the detailed studies of the angular momentum density [85,86]. Laguerre-Gaussian beams have had a significant influence in the advancement of optical trapping.

Their phase structure can initiate rotation or actuation in a microfluidic environment. In the context of microfluidics, droplet, and optofluidics their annular intensity profile is also of equal importance and it is this we now discuss.

The manipulation of droplets is a challenging, but important, area for optofluidics and LG beam trapping is a powerful method for this end, as the droplet's refractive index may be lower than their surrounding medium [87,88]. Ashkin first observed that low refractive-index particles are repelled from the high-intensity region of light while high refractive-index particles are drawn into the trap [7]. Further experiments with a high-order mode laser beam (TEM^*_{01}) levitated a low-index particle against gravity [89,90]. Fast scanning mirrors allowed Sasaki and his colleagues [91,92] to show that they could cage and propel both reflective metallic particles or low index microdroplets. Using tightly focused LG beams one may thus manipulate low refractive-index microparticles, where specifically high azimuthal order single-ringed LG beams ($l > 1$) may confine a low-index particle within its smooth annular intensity profile (see Fig. 15-6) [93]. To this end, in recent work, Lee et al. [94] explored the optical field generated from gradual lateral displacement of a phase element from the center of an incident Gaussian beam. This manipulated the position of the dark vortex core creating an off-axis optical vortex. They showed that a low-index microparticle can be trapped in such a vortex manipulated around the beam's central axis without moving the entire beam. Figure 15-7 shows relative position of the hologram and the illuminating beam as well as the resulting optical structures. In subsequent studies, Lorenz et al. [95] adapted this technique with the use of two such off-axis LG beams to controllably fuse two aqueous droplets.

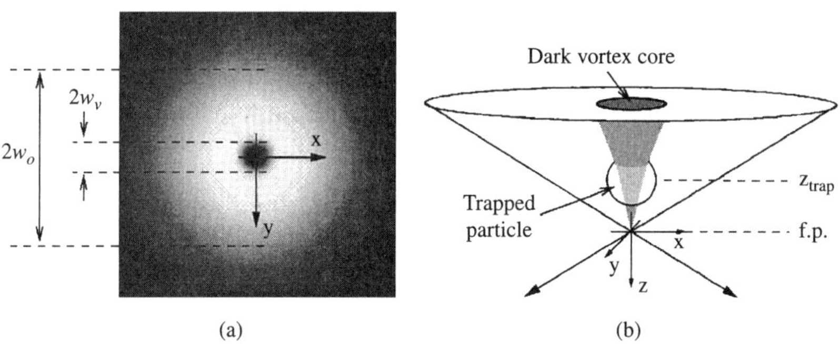

Figure 15-6 Trapping geometry for low-index particles such as water droplets in a higher-index medium (oil). Particles are expelled by optical forces from the high-intensity ring that leads to a stable localization on the axis in front of the focal plane. (*Source: K. T. Gahagan and G. A. Swartzlander, "Trapping of low-index microparticles in an optical vortex," Journal of Optical Society of America B, 15, 1998, 524–534.*)

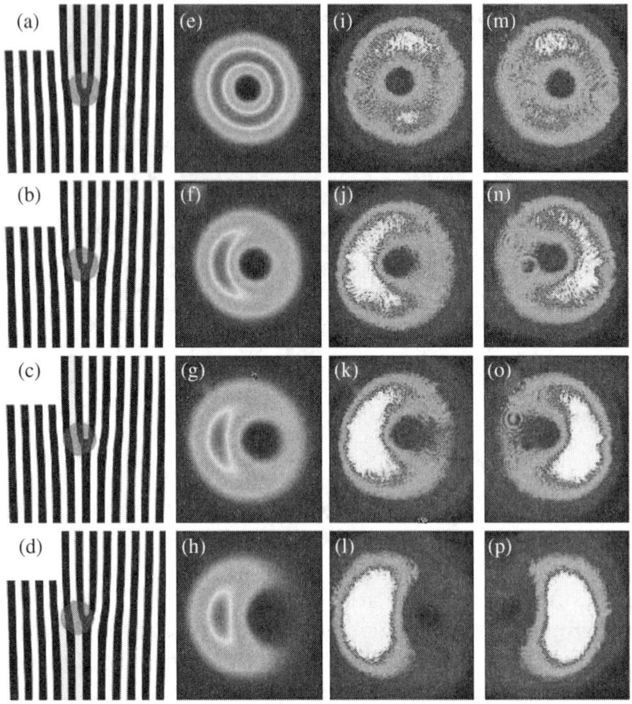

FIGURE 15-7 Optical vortices with an off-axis singularity suitable for fusion of liquid droplets. [*Source: R. M. Lorenz, J. S. Edgar, G. D. M. Jeffries, Y. Zhao, D. McGloin, and D. T. Chiu, "Vortex-trap-induced fusion of femtoliter-volume aqueous droplets," Analytical Chemistry, 79(1), 2007, 224–228.*]

15-6 Optical Manipulation for Optofludics

In this section, we look at some of the relevant optical geometries and experiments within optical manipulation for the area of optofluidics building upon our discussion of the single- and dual-beam traps as well as the array of advanced light fields we have mentioned. In a microfluidic chamber the fluid flow is typically laminar. The flow velocities are typically 10 to 100 μm/s and the chamber dimensions usually 10 to 100 μm. This is the low Reynolds number regime where one cannot rely on processes such as turbulence for mixing or sorting. Optical manipulation may assist in such an environment for developing techniques for microrheology, where one studies the deformation of viscoelastic materials or fluid flow in response to applied force. Separately, researchers are interested in actuating microcomponents and controlled passive or active (labelled) sorting schemes. In this section, we give examples where the linear and angular momentum of lights have played a key role in actuating components in an optofluidics environment. We explore examples where one may make rheology measurements in microfluidic chambers as well as the use of optical manipulation for creating microsensors.

15-6-1 Optical Actuation, Microrheology, and Optically Trapped Sensors

Optical trapping is ideal for actuation. Recent exciting developments include the optofluidic microscope (OFM), which is essentially on-chip imaging. In this study, a closely spaced 2D grid of nanoapertures (each of typical diameter ~100 nm) provides a patterned illumination of the sample. A near-infrared optical tweezers was used to hold and translate the chosen sample and thus "actuate" it over the nanoaperture grid and image the sample [97].

Optical control of microparticles can be used to create a further range of novel optofluidic devices. One interesting example is the optofluidics beam-manipulator from the work of Domachuk et al. [13]. In this study an optically trapped microsphere is placed in front of the exit port of a positioned fiber. The sphere acts as an optically movable lens for beam manipulation. By steering the microsphere, the output beam can be deflected in a range of directions. Domachuk et al. used the method to create an all-optical switch by steering the microsphere between two mutually facing fiber waveguides that are separated by a small microfluidic flow channel (see Fig. 15-8). The transmission of light from one waveguide to the other is enhanced when the trapped microsphere is well centered and functions as a focussing lens.

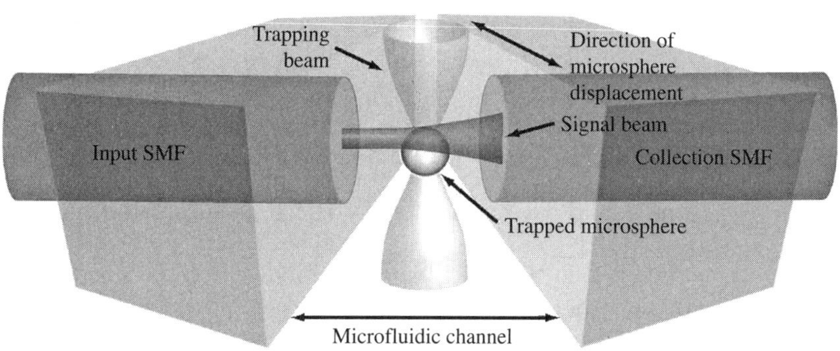

FIGURE 15-8 The all-optical switch. Optically confined particle can be positioned between two ports of a waveguide thus changing the coupling efficiency of the signal beam at the output port of the waveguide. (*Source: P. Domachuk, M. Cronin-Golomb, B. Eggleton, S. Mutzenich, G. Rosengarten, and A. Mitchell, "Application of optical trapping to beam manipulation in optofluidics," Optics Express, 13, 2005, 7265–7275.*)

We now turn to rheology in the context of optofluidics. Studying the motion of tracer particles in complex fluids leads to viscosity measurements [98]. Optical traps are ideal candidates to make such local viscosity measurements. In the context of optical manipulation, one may exploit the trapping beam itself as a confocal probe and may study the response of the microparticles to periodic motion of the tweezers to yield information about the medium viscosity, particle properties, and trap stiffness. The authors controllably forced the trapped sphere back and forth, the resulting particle motion was seen to be periodic, with a frequency equal to the forcing frequency but critically exhibiting a measurable phase lag due to the hydrodynamic drag that, in turn, is related to the local viscosity. This method is a form of scanning photonic force microscopy for applications in which a high spatial and temporal resolution of the medium viscosity is desired [99]. Rotating trapped objects with optical tweezers is alternative for microrheology as already mentioned [100,101]. In such experiments, the rotational Stokes drag reaches equilibrium with the rotating birefringent object yielding a local measurement of viscosity. Recently, Brau et al. [102] summarized the wider range of microrheology applications open to the techniques of optical manipulation.

Besides microrheology, rotation is key for actuating structures and driving pumps or even valves: photopolymerized structures [103] may be set into rotation by asymmetric scattering. Colloidal microparticles may act as pumps too as already described [67]. In this study, we see simultaneous trapping and rotating microspheres held by multiple time-shared optical tweezers—the AOD [67]. Optical angular momentum can initiate rotation of trapped objects. In the case of spin angular momentum a trapped birefringent particle may start rotating due to the transfer of spin angular momentum for the light field to the particle, which in this instance acts as a microscopic waveplate forming a microscopic version of Beth's famous experiment [104]. Friese et al. [42,105] rotated calcite particles with a circularly polarized trapping beam. Two birefringent microspheres may be set into rotation in opposite directions to one another, creating an optical pump though the flow rates and speed of particle motion are slow [106,107]. When considering actuation for optofluidics or other applications, one would like not to rely upon such intrinsic birefringence. To address this, Neale et al. [108] engineered birefringence into SU-8 polymer [109]. This concept allowed rotation of cog-like structures with a circularly polarized light tweezing beam. Optical torques and rotation may occur with orbital angular momentum as well. The helically phased LG beams may generate optically driven pumps. Rows of alternating single-ringed high-order LG beams generated with spatial light modulators allowed K. Ladavac and D. Grier to spin large numbers of trapped microspheres around the LG circumference [110] and create a pumping action. Jesacher et al. [111] also saw large rates of rotation for particles confined in holographic optical vortex traps. Furthermore, they demonstrated interactive particle-flow steering with arrays of optical vortex pumps. Exerting a small but significant torque upon a biological specimen often requires that the beams are tailored to the shape of the biological particles. Orientation of particles with optical traps is also a desirable quantity in this respect and another way of actuating samples in an optofluidic environment. With higher-order Hermite-Gaussian laser modes and the interference of LG modes with plane waves, asymmetric and moving light fields can rotate cells and chromosomes [112,113]. In the studies with LG beams, chromosomes were trapped in the asymmetric light field and set into rotation by controlled adjustment of the relative optical path length in the LG beam interferometer. Actuation need not be limited to rotation. Various micro-electromechanical systems may use cantilevers for both sensing and optical switching. Recent studies have used optical tweezers to actuate a tapered optical fiber used as a cantilever. This is then driven as a micromechanical oscillator. The authors used a fiber optic confocal detection system to record both the position and oscillation characteristics of the cantilever using the backscattered component of the trapping beam [114].

One important area where optical manipulation can assist in optofluidics is in the development of sensors. Such sensors may take various forms and indeed the trapped particle itself may act as a sensor for parameters such as viscosity or temperature. Local viscosity measurements as already described are important and there are a number of ways optically trapped particles may be used for this purpose. Particle position may be oscillated to determine the local viscosity [115] or a popular method is the use of the rotation of trapped particles to make this measurement. Here we spin a trapped birefringent object (using, for example, spin angular momentum) and the particle reaches a terminal angular velocity dictated by the rotational Stokes drag.

The motion of an optically trapped microsphere in an oscillating laser trap may measure velocity fields in fluid flow with a resolution at the micron-size scale. The authors obtained a two-dimensional map of the flow past a microscopic wedge. Importantly, no fluid-dependent calibration is required since the velocity is measured simultaneously with the trap-relaxation time. The technique is also independent of the trap stiffness and the size of the microparticle [116]. Leach et al. used a different method employing holographic optical tweezers and a "trap and release" scheme to record local fluid velocities [106] around a rotating object in a microfluidic environment.

Trapped microparticles that are modified with fluorescent dyes may act as sensors within a microfluidic channel. In this manner we can measure parameters in a microfluidic chip such as pH and temperature. In recent work, Kluake et al. [117] functionalized amine-modified polystyrene spheres with the pH-sensitive fluorochrome

SNARF-1. The spheres were subsequently trapped at various positions close to a pair of integrated planar gold microelectrodes. Suitable applied electrochemical potentials created changes in the local pH. The fluorescence signal from these spheres-functionalized beads indicated the pH changes in the channel. The bead size dictated spatial resolution of the probe system [117].

15-6-2 Microfluidic Sorting

In an optofluidic application one might wish to separate or sort particles or cells. Microfluidic cell sorting is an important area and may be performed in a number of schemes. A fluorescence-activated cell sorter may indeed be miniaturized [118] and we may use an optical trap to remove particles at a Y-junction after they pass through a detection region. Integrated fiber-based devices can also lead to such sorting [119] or other forms of cell sorting [120]. An emergent theme in the last 5 years has been to look at a form of sorting that is independent of markers or any attached tags to chosen particles. Such passive sorting is a newly emergent theme and makes use of the variation in response of a given sample object to an imposed optical potential energy landscape. A wide variety of methods have now appeared in the literature including multiple trap scheme, Bessel light modes, or even interfering, propagating, or evanescent fields [71,121–127]. This method relies on a difference in response of the object to the pattern that is a result of a variation, for example, in polarizability between particles. Ultimately, in this way one can sort based upon size, shape, and refractive index with good selectivity. In Fig. 15-9, we see the geometry for sorting in an optical lattice where this varying affinity of latex and silica particles to the field structure dictates if the particle propagates in the original direction or if it follows the lattice

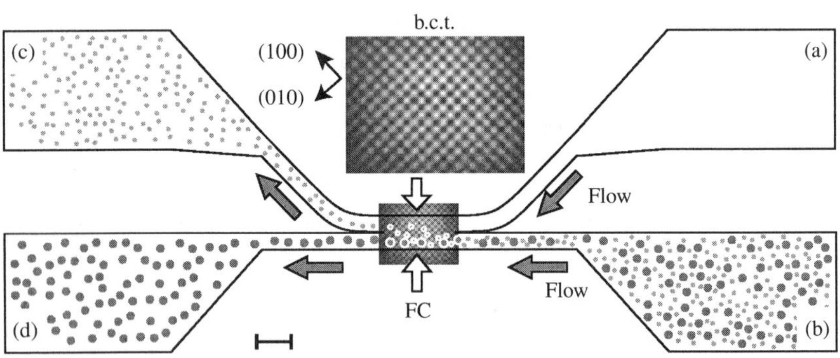

FIGURE 15-9 Optical sorting in optical lattices. This device efficiently uses the varying affinity of different micro-objects to the periodic structure of optical lattice. For specific combinations of particle size and refractive index, the particles are deviated from their original path while others are not influenced. (*Reprinted with permission from Macmillan publishers Ltd M. P. MacDonald, G. C. Spalding and K. Dholakia, "Microfluidic sorting in an optical lattice," Nature, 426, 2003, 421–424.*)

structure and diverts to a separate channel. A detailed discussion of this topic is given in Ref. 128.

15-6-3 Optical Trapping in Near-Field Waveguides

The application of a variety of optical waveguides for signal transmission is an established area. These waveguides have refractive index differences and analysis of such structures leads to an understanding of the specific light modes such structures can support. If coupled in well to such a waveguide, the light is restricted to the higher refractive-index region, though the light may leak into the rarefied medium as what is commonly termed an evanescent wave. More broadly this fits within the area of near-field optical trapping and manipulation which has emerged as a powerful method in the last 5 years [129,130] where numerous experiments have been performed using total internal reflection objectives or the well-known Kretschmann geometry for trapping and sorting [130–132]. From the perspective of microfluidic systems and optofluidics, such waveguides may provide interesting new integrated geometries for transport, confinement, and sorting of microparticles with the ability to use modern micro- and nanofabrication procedures to tune the interaction as well as develop potential observations into real devices. We review some experiments for optical waveguides used with particular emphasis on studies pertinent to a microfluidic environment. The transport and trapping of Mie particles (polystyrene spheres up to 5 μm in diameter) along channelled waveguides was seen in 1996 (see Fig. 15-10) [133]. In this study, the optical gradient force localizes

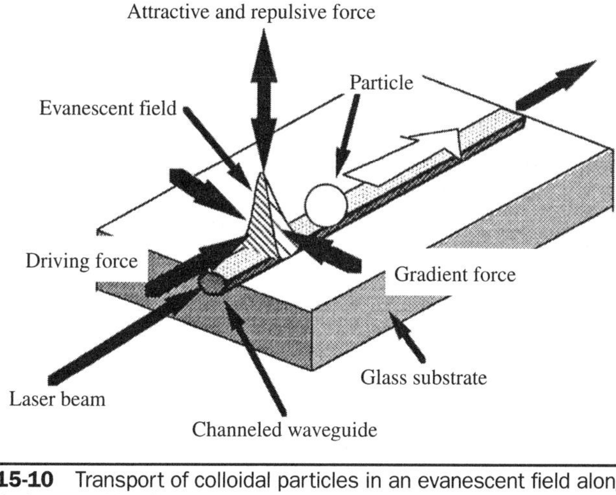

Figure 15-10 Transport of colloidal particles in an evanescent field along a channeled waveguide. The gradient force keeps the particle near the waveguide while the scattering force provides the transport along the channel. (*Source: S. Kawata and T. Tani, "Optically driven Mie particles in an evanescent field along a channeled waveguide," Optics Letters, 21, 1996, 1768–1770.*)

the particles to the waveguide region laterally whereas the evanescent field couples to and propels the particle along the guide. The near-infrared (1047 nm) laser had a power in excess of 2 W. Particle velocities up to 14 μm/s were observed. Submicron metallic microparticles (~500 nm diameter) were also transported along the channel guide. Gold possesses a high polarizability, so one can readily trap small gold nanoparticles [134]. One may also exploit the plasmon resonance for enhanced trapping. In the similar context of waveguide manipulation, Hole et al. studied the behaviour of submicron (250 nm) gold nanoparticles on a caesium ion-exchange waveguide [135]. Absorption is a consideration since gold has a complex refractive index and heating may result. The sample chamber was formed in a moulded polydimethylsiloxane (PDMS) elastomer that was placed upon the surface of the waveguides. A laser operating at 1066 nm was used to transport the particles. This laser was butt-coupled into the end of a waveguide using a single-mode fiber. In the study of Grujic and Helleso, the formation and propulsion of chains of dielectric microparticles upon a caesium ion-exchange waveguide was investigated [135,136]. Long one-dimensional chains of particles in this waveguide geometry were seen where hydrodynamics and coupling between the microspheres was deemed to play an important role in this behaviour and optical binding was a consideration [11,137,138].

Gaugiran et al. explored the polarization and particle-size dependence of radiation forces acting on gold nanoparticles that were guided on the surface of silver ion and silicon nitride waveguides [138,139]. Their particular interest was in identifying the conditions under which the force normal to the surface becomes repulsive (theoretically predicted by Ariaz-Gonzalez et al. [139,140]. Experimentally, Gaugiran et al. discovered a wide variation between the guiding velocity of 600-nm gold particles for the two orthogonal TM and TE polarizations. They found that the guiding velocity for the TM polarization case was significantly larger than for TE. Numerical calculations supported their observations and they attributed their observations to the presence of a repulsive force and an attractive force for the TE and TM cases, respectively.

In waveguide geometry, near-field sorting can also be initiated. For this purpose, Grujic et al. used a Y-branched optical waveguide for the separation of microparticles [141] (see Fig. 15-11). The experiments employed polystyrene microparticles. These were optically transported along the waveguide's evanescent field. This field was controllable and could be directed down either output branch of the Y-shaped chamber. The relative position of the fiber to the waveguide input face dictates the power distribution between the two output branches. This is a form of "active" sorting, contrasting with the passive sorting schemes using optical potential energy landscapes we described earlier in this chapter. Microspheres can be efficiently and reliably sorted with very high probability of success

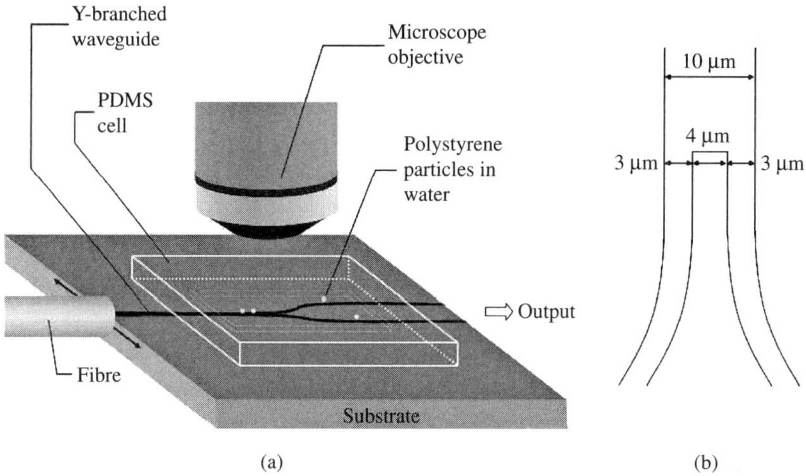

FIGURE 15-11 Y-branched waveguide for transport of micro-objects. Additional functionality such as sorting and routing maybe achieved through use of optical circuits. By controlling the coupling of the laser into the waveguide, particles may be switched between the upper and lower branch of the y-junction. (*Source: K. Grujic, O. Hellesø, J. Hole, and J. Wilkinson, "Sorting of polystyrene microspheres using a Y-branched optical waveguide," Optics Express, 13, 2005, 1–7.*)

under appropriate conditions. To realise such sorting in a biological context, the biological sample should exhibit a sufficiently high refractive-index mismatch relative to the buffer medium, otherwise a sufficient gradient force would not be exerted upon the particles. If sorting of biological macromolecules was required, these could be made to adhere to suitably functionalized latex spheres for subsequent selection in the Y-sorter.

15-7 Conclusion

Optical manipulation of biological and colloidal particles has sustained a very high and widening profile since 1970. The field has truly delivered in a variety of areas including single-molecule biophysics, colloidal dynamics, microrheology, and optical angular momentum. In the context of the newly emerging area of optofluidics, this chapter has described some of the new directions optical manipulation offers in terms of integration of components within microfluidics, optical actuation, new schemes for precise and highly localized measurements of physical parameters, and new near-field geometries for particle control and manipulation. This exciting marriage of concepts and fields offers exciting new possibilities in the emergent area of optofluidics.

15-8 Acknowledgments

We thank the UK Engineering and Physical Sciences Research Council for funding. KD is a Royal Society-Wolfson Merit Award Holder.

References

1. C. N. Cohen-Tannoudji, "Manipulating atoms with photons," *Reviews of Modern Physics*, **70**, 1998, 707–719.
2. S. Chu, "The manipulation of neutral particles," *Reviews of Modern Physics*, **70**, 1998, 685–706.
3. W. D. Phillips, "Laser cooling and trapping of neutral atoms," *Reviews of Modern Physics*, **70**, 1998, 721–741.
4. K. B. Davis et al., "Bose-Einstein condensation in a gas of sodium atoms," *Physical Review Letters*, **75**, 1995, 3969–3973.
5. M. H. Anderson et al., "Observation of Bose-Einstein condensation in a dilute atomic vapour," *Science*, **269**, 1995, 198–201.
6. A. Ashkin et al., "Observation of a single-beam gradient force optical trap for dielectric particles." *Optics Letters*, **11**, 1986, 288–290.
7. A. Ashkin, "Acceleration and trapping of particles by radiation pressure," *Physical Review Letters*, **24**, 1970, 156–159.
8. A. Constable et al., "Demonstration of a fiberoptic light-force trap," *Optics Letters*, **18**, 1993, 1867–1869.
9. J. Guck et al., "The optical stretcher: A novel laser tool to micromanipulate cells," *Biophysical Journal*, **81**, 2001, 767–784.
10. P. R. T. Jess et al., "Dual beam fiber trap for Raman microspectroscopy of single cells," *Optics Express*, **14**, 2006, 5779–5791.
11. S. A. Tatarkova, A. E. Carruthers, and K. Dholakia, "One-dimensional optically bound arrays of microscopic particles," *Physical Review Letters*, **89**, 2002, 283901.
12. W. Singer et al., "Self-organized array of regularly spaced microbeads in a fiber-optical trap," *Journal of the Optical Society of America B-Optical Physics*, **20**, 2003, 1568–1574.
13. P. Domachuk, M. Cronin-Golomb, B. Eggleton, S. Mutzenich, G. Rosengarten, and A. Mitchell "Application of optical trapping to beam manipulation in optofluidics," *Optics Express*, **13**, 2005, 7265–7275.
14. J. E. Molloy and M. J. Padgett, "Lights, action: optical tweezers," *Contemporary Physics*, **43**, 2002, 241–258.
15. K. Dholakia, P. Reece, and M. Gu, "Optical micromanipulation," *Chemical Society Reviews*, **37**, 2008, 42–55.
16. K. C. Neuman and S. M. Block, "Optical trapping," *Review of Scientific Instruments*, **75**, 2004, 2787–2809.
17. D. G. Grier, "A revolution in optical manipulation," *Nature*, **424**, 2003, 810–816.
18. S. M. Block et al., "Molecule by molecule, the physics and chemistry of life: SMB 2007," *Nat Chem Biol.*, **3**, 2007, 193–197.
19. P. C. Chaumet and M. Nieto-Vesperinas, "Time-averaged total force on a dipolar sphere in an electromagnetic field," *Optics Letters*, **25**, 2000, 1065–1067.
20. B. T. Draine, "The discrete-dipole approximation and its application to interstellar graphite grains," *Astrophysical Journal*, **333**, 1988, 848–872.
21. K. Svoboda and S. M. Block, "Optical trapping of metallic rayleigh particles," *Optics Letters*, **19**, 1994, 930–932.
22. P. M. Hansen et al., "Expanding the optical trapping range of gold nanoparticles," *Nano Letters*, **5**, 2005, 1937–1942.
23. Y. Seol, A. E. Carpenter, and T. T. Perkins, "Gold nanoparticles: enhanced optical trapping and sensitivity coupled with significant heating," *Optics Letters*, **31**, 2006, 2429–2431.

24. A. Ashkin, "Forces of a single-beam gradient laser trap on a dielectric sphere in the ray optics regime," *Biophysical Journal*, **61**, 1992, 569–582.
25. J. P. Barton and D. R. Alexander, "Fifth-order corrected electromagnetic field components for a fundamental Gaussian beam," *Journal of Applied Physics*, **66**, 1989, 2800–2802.
26. J. P. Barton, D. R. Alexander, and S. A. Schaub, "Theoretical determination of net-radiation force and torque for a spherical-particle illuminated by a focused laser-beam," *Journal of Applied Physics*, **66**, 1989, 4594–4602.
27. A. Rohrbach and E. H. K. Stelzer, "Trapping forces, force constants, and potential depths for dielectric spheres in the presence of spherical aberrations," *Applied Optics*, **41**, 2002, 2494–2507.
28. A. Rohrbach, "Stiffness of optical traps: Quantitative agreement between experiment and electromagnetic theory," *Physical Review Letters*, **95**, 2005, 168102.
29. W. M. Lee et al., "Construction and calibration of an optical trap on a fluorescence optical microscope," *Nature Protocols*, **2**, 2007, 3226–3238.
30. K. C. Neuman et al., "Characterization of photodamage induced by optical traps," *Biophysical Journal*, **76**, 1999, A96–A96.
31. A. Rohrbach, "Switching and measuring a force of 25 femtoNewtons with an optical trap," *Optics Express*, **13**, 2005, 9695–9701.
32. E. A. Abbondanzieri et al., "Direct observation of base-pair stepping by RNA polymerase," *Nature*, **438**, 2005, 460–465.
33. J. Guck et al., "Stretching biological cells with light," *Journal of Physics-Condensed Matter*, **14**, 2002, 4843–4856.
34. J. Guck et al., "Optical deformability of soft biological dielectrics," *Physical Review Letters*, **84**, 2000, 5451–5454.
35. D. M. Gherardi et al., "A dual beam photonic crystal fiber trap for microscopic particles," *Applied Physics Letters*, **93**, 2008, 041110.
36. J. T. Blakely, R. Gordon, and D. Sinton, "Flow-dependent optofluidic particle trapping and circulation," *Lab On A Chip*, **8**, 2008, 1350–1356.
37. S. Cran-McGreehin, T. F. Krauss, and K. Dholakia, "Integrated monolithic optical manipulation," *Lab On A Chip*, **6**, 2006, 1122–1124.
38. S. J. Cran-McGreehin, K. Dholakia, and T. F. Krauss, "Monolithic integration of microfluidic channels and semiconductor lasers," *Optics Express*, **14**, 2006 7723–7729.
39. H. C. Hunt and J. S. Wilkinson, "Optofluidic integration for microanalysis," *Microfluidics and Nanofluidics*, **4**, 2008, 53–79.
40. A. La Porta and M. D. Wang, "Optical torque wrench: angular trapping, rotation, and torque detection of quartz microparticles," *Physical Review Letters*, **92**, 2004, 190801.
41. G. Volpe and D. Petrov, "Brownian motion in a nonhomogeneous force field and photonic force microscope," *Physical Review E (Statistical, Nonlinear, and Soft Matter Physics)*, **76**, 2007, 061118–10.
42. M. E. J. Friese et al., "Optically driven micromachine elements," *Applied Physics Letters*, **78**, 2001, 547–549.
43. C. Bechinger, M. Brunner, and P. Leiderer, "Phase behavior of two-dimensional colloidal systems in the presence of periodic light fields," *Physical Review Letters*, **86**, 2001, 930–933.
44. P. T. Korda, G. C. Spalding, and D. G. Grier, "Evolution of a colloidal critical state in an optical pinning potential landscape," *Physical Review B*, **66**, 2002, 024504.
45. J. Durnin, J. J. Miceli, and J. H. Eberly, "Diffraction-free beams," *Physical Review Letters*, **58**, 1987, 1499–1501.
46. L. Allen et al., "Orbital angular-momentum of light and the transformation of Laguerre-Gaussian laser modes," *Physical Review A*, **45**, 1992, 8185–8189.
47. K. Svoboda and S. M. Block, "Biological applications of optical forces," *Ann. Rev. Biophys. Biomol. Struct.*, **23**, 1994, 247–285.
48. G. C. Spalding, J. Courtial, and R. D. Leonardo, *Holographic Optical Trapping*, in *Structured Light and its Applications: An Introduction to Phase-Structured Beams and Nanoscale Optical Forces*, D. L. Andrews, ed. 2008, Elsevier Press.

49. J. Liesener et al., "Multifunctional optical tweezers using computer-generated holograms," *Optics Communications*, **185**, 2000, 77–82.
50. E. R. Dufresne and D. G. Grier, "Optical tweezer arrays and optical substrates created with diffractive optics," *Review of Scientific Instruments*, **69**, 1998, 1974–1977.
51. H. Melville et al., "Optical trapping of three-dimensional structures using dynamic holograms," *Optics Express*, **11**, 2003, 3562–3567.
52. J. Leach et al., "3D manipulation of particles into crystal structures using holographic optical tweezers," *Optics Express*, **12**, 2004, 220–226.
53. J. E. Curtis, B. A. Koss, and D. G. Grier, "Dynamic holographic optical tweezers," *Optics Communications*, **207**, 2002, 169–175.
54. G. Sinclair, P. Jordan, J. Courtial, M. Padgett, J. Cooper, and Z. Laczik, *"Assembly of 3-dimensional structures using programmable holographic optical tweezers,"* *Optics Express*, **12**(22), pp. 5475–5480 (2004).
55. K. D. Wulff et al., "Aberration correction in holographic optical tweezers," *Optics Express*, **14**, 2006, 4169–4174.
56. E. Martin-Badosa et al., "Design strategies for optimizing holographic optical tweezers set-ups," *Journal of Optics A: Pure and Applied Optics*, **9**, 2007, S267–S277.
57. M. Polin et al., "Optimized holographic optical traps," *Optics Express*, **13**, 2005, 5831–5845.
58. A. Jesacher et al., "Diffractive optical tweezers in the Fresnel regime," *Optics Express*, **12**, 2004, 2243–2250.
59. R. L. Eriksen, P. C. Mogensen, and J. Gluckstad, "Multiple-beam optical tweezers generated by the generalized phase-contrast method," *Optics Letters*, **27**, 2002, 267–269.
60. P. J. Rodrigo et al., "Interactive light-driven and parallel manipulation of inhomogeneous particles, *Optics Express*, **10**, 2002, 1550–1556.
61. E. Fallman and O. Axner, "Design for fully steerable dual-trap optical tweezers," *Applied Optics*, **36**, 1997, 2107–2113.
62. J. C. Meiners and S. R. Quake, "Direct measurement of hydrodynamic cross correlations between two particles in an external potential," *Physical Review Letters*, **82**, 1999, 2211–2214.
63. S. Kulin et al., "Optical manipulation and fusion of liposomes as microreactors," *Langmuir*, **19**, 2003, 8206–8210.
64. G. M. Akselrod et al., "Laser-guided assembly of heterotypic three-dimensional living cell microarrays," *Biophysical Journal*, **91**, 2006, 3465–3473.
65. R. Di Leonardo et al., "Eigenmodes of a hydrodynamically coupled micron-size multiple-particle ring," *Physical Review E.*, **76**, 2007.
66. E. Eriksson et al., "Holographic optical tweezers combined with a microfluidic device for exposing cells to fast environmental changes—art. no. 65920P," in *Bioengineered and Bioinspired Systems III.*, 2007, Spie-Int Soc Optical Engineering: Bellingham. pp. P5920–P5920.
67. A. Terray, J. Oakey, and D. W. M. Marr, "Microfluidic control using colloidal devices," *Science*, **296**, 2002, 1841–1844.
68. D. McGloin and K. Dholakia, "Bessel beams: diffraction in a new light," *Contemporary Physics*, **46**, 2005, 15–28.
69. J. Arlt et al., "Optical micromanipulation using a Bessel light beam," *Optics Communications*, **197**, 2001, 239–245.
70. T. Čižmár et al., "Optical conveyor belt for delivery of submicron objects," *Applied Physics Letters*, **86**, 2005, 174101.
71. L. Paterson et al., "Light-induced cell separation in a tailored optical landscape," *Applied Physics Letters*, **87**, 2005, 123901.
72. J. Durnin, "Exact-solutions for nondiffracting beams.1. The scalar theory," *Journal of The Optical Society of America A-Optics Image Science and Vision*, **4**, 1987, 651–654.
73. V. Garcés-Chávez et al., "Simultaneous micromanipulation in multiple planes using a self-reconstructing light beam," *Nature*, **419**, 2002, 145–147.

74. T. Čižmár et al., "Generation of multiple Bessel beams for a biophotonics workstation," *Optics Express*, **16**, 2008, 14024–14035.
75. G. Milne, G. D. M. Jeffries, and D. T. Chiu, "Tunable generation of Bessel beams with a fluidic axicon," *Applied Physics Letters*, **92**, 2008, 261101.
76. L. Allen, M. J. Padgett, and M. Babiker, "The orbital angular momentum of light," in *Progress in Optics*, Vol XXXIX, 1999. 291–372.
77. M. A. Clifford et al., "High-order Laguerre-Gaussian laser modes for studies of cold atoms," *Optics Communications*, **156**, 1998, 300–306.
78. M. W. Beijersbergen et al., "Helical-wave-front laser-beams produced with a spiral phaseplate," *Optics Communications*, **112**, 1994, 321–327.
79. S. S. R. Oemrawsingh et al., "Production and characterization of spiral phase plates for optical wavelengths," *Appl. Opt.*, **43**, 2004, 688–694.
80. W. C. Cheong et al., "Direct electron-beam writing of continuous spiral phase plates in negative resist with high power efficiency for optical manipulation," *Applied Physics Letters*, **85**, 2004, 5784–5786.
81. W. M. Lee, X. C. Yuan, and W. C. Cheong, "Optical vortex beam shaping by use of highly efficient irregular spiral phase plates for optical micromanipulation," *Optics Letters*. **29**, 2004, 1796–1798.
82. H. He et al., "Direct observation of transfer of angular-momentum to absorptive particles from a laser-beam with a phase singularity," *Physical Review Letters*, **75**, 1995, 826–829.
83. N. B. Simpson et al., "Mechanical equivalence of spin and orbital angular momentum of light: An optical spanner," *Optics Letters*, **22**, 1997, 52–54.
84. M. E. J. Friese et al., "Optical angular-momentum transfer to trapped absorbing particles," *Physical Review A.*, **54**, 1996, 1593–1596.
85. A. T. O'Neil et al., "Intrinsic and extrinsic nature of the orbital angular momentum of a light beam," *Physical Review Letters*, **88**, 2002, 053601.
86. V. Garcés-Chávez et al., "Observation of the transfer of the local angular momentum density of a multiringed light beam to an optically trapped particle," *Physical Review Letters*, **91**, 2003, 093602.
87. K. T. Gahagan and G. A. Swartzlander, "Optical vortex trapping of particles," *Optics Letters*, **21**, 1996, 827–829.
88. P. A. Prentice et al., "Manipulation and filtration of low index particles with holographic Laguerre-Gaussian optical trap arrays," *Optics Express*, **12**, 2004, 593–600.
89. A. Ashkin and J. M. Dziedzic, "Stability of optical levitation by radiation pressure," *Applied Physics Letters*, **24**, 1974, 586–588.
90. A. Ashkin, "Optical trapping and manipulation of neutral particles using lasers," *Proceedings of the National Academy of Sciences of the United States of America*, **94**, 1997, 4853–4860.
91. K. Sasaki et al., "Pattern-formation and flow-control of fine particles by laser-scanning micromanipulation," *Optics Letters*, **16**, 1991, 1463–1465.
92. K. Sasaki et al., "Optical trapping of a metal-particle and a water droplet by a scanning laser-beam," *Applied Physics Letters*, **60**, 1992, 807–809.
93. K. T. Gahagan and G. A. Swartzlander, "Simultaneous trapping of low-index and high-index microparticles observed with an optical-vortex trap," *Journal of the Optical Society of America B-Optical Physics*, **16**, 1999, 533–537.
94. W. M. Lee et al., "Optical steering of high and low index microparticles by manipulating an off-axis optical vortex," *Journal of Optics a-Pure and Applied Optics*, **7**, 2005, 1–6.
95. R. M. Lorenz, J. S. Edgar, G. D. M. Jeffries, Y. Zhao, D. McGloin, and D. T. Chiu, "Vortex-trap-induced fusion of femtoliter-volume aqueous droplets," *Analytical Chemistry*, **79**(1), 2007, 224–228.
96. K. T. Gahagan and G. A. Swartzlander, "*Trapping of low-index microparticles in an optical vortex*," *J. Opt. Soc. Am.* B 15, 1998, 524–534.
97. X. Heng et al., "An optical tweezer actuated, nanoaperture-grid based optofluidic microscope implementation method," *Optics Express*, **15**, 2007, 16367–16375.

98. T. G. Mason and D. A. Weitz, "Optical measurements of frequency-dependent linear viscoelastic moduli of complex fluids," *Physical Review Letters*, **74**, 1995, 1250–1253.
99. B. A. Nemet and M. Cronin-Golomb, "Measuring microscopic viscosity with optical tweezers as a confocal probe," *Applied Optics*, **42**, 2003, 1820–1832.
100. A. I. Bishop et al., "Optical microrheology using rotating laser-trapped particles," *Physical Review Letters*, **92**, 2004, 198104.
101. A. LaPorta and M. D. Wang, "Angular trapping of micro-particles: Rotating and applying torque to biological molecules with optical tweezers," *Biophysical Journal*, **86**, 2004, 599A–599A.
102. R. R. Brau et al., "Passive and active microrheology with optical tweezers," *Journal of Optics A: Pure and Applied Optics*, **9**, 2007, S103–S112.
103. L. Kelemen, S. Valkai, and P. Ormos, "Parallel photopolymerisation with complex light patterns generated by diffractive optical elements," *Optics Express*, **15**, 2007, 14488–14497.
104. R. A. Beth, "Mechanical detection and measurement of the angular momentum of light" *Physical Review*, **50**, 1936, 115.
105. M. E. J. Friese et al., "Optical alignment and spinning of laser-trapped microscopic particles," *Nature*, **394**, 1998, 348–350.
106. R. Di Leonardo et al., "Multipoint holographic optical velocimetry in microfluidic systems," *Physical Review Letters*, **96**, 2006, 134502.
107. J. Leach et al., "An optically driven pump for microfluidics," *Lab on a Chip*, **6**, 2006, 735–739.
108. S. L. Neale et al., "All-optical control of microfluidic components using form birefringence," *Nature Materials*, **4**, 2005, 530–533.
109. A. I. Bishop et al., "Optical application and measurement of torque on microparticles of isotropic nonabsorbing material," *Physical Review A.*, **68**, 2003, 033802.
110. K. Ladavac and D. G. Grier, "Microoptomechanical pumps assembled and driven by holographic optical vortex arrays," *Optics Express*, **12**, (2004) 1144–1149.
111. A. Jesacher et al., "Holographic optical tweezers for object manipulations at an air-liquid surface," *Optics Express*, **14**, 2006, 6342–6352.
112. S. Sato, M. Ishigure, and H. Inaba, "Optical trapping and rotational manipulation of microscopic particles and biological cells using higher-order mode nd-yag laser-beams," *Electronics Letters*, **27**, 1991, 1831–1832.
113. L. Paterson et al., "Controlled rotation of optically trapped microscopic particles," *Science*, **292**, 2001, 912914.
114. P. Domachuk et al., "Actuation of cantilevers by optical trapping," *Applied Physics Letters*, **89**, 2006, 071106.
115. B. A. Nemet, Y. Shabtai, and M. Cronin-Golomb, "Imaging microscopic viscosity with confocal scanning optical tweezers," *Optics Letters*, **27**, 2002, 264–266.
116. B. A. Nemet and M. Cronin-Golomb, "Microscopic flow measurements with optically trapped microprobes" *Optics Letters*, **27**, 2002, 1357–1359.
117. N. Klauke et al., "Characterisation of spatial and temporal changes in pH gradients in microfluidic channels using optically trapped fluorescent sensors," *Lab on a Chip*, **6**, 2006, 788–793.
118. M. M. Wang et al., "Microfluidic sorting of mammalian cells by optical force switching," *Nature Biotechnology*, **23**, 2005, 83–87.
119. H. I. Kirei et al., "An all optical microfluidic sorter," *Acta Biologica Hungarica*, **58**, 2007, 139–148.
120. R. W. Applegate et al., "Optical trapping, manipulation, and sorting of cells and colloids in microfluidic systems with diode laser bars," *Optics Express*, **12**, 2004, 4390–4398.
121. M. P. MacDonald, G. C. Spalding, and K. Dholakia, "Microfluidic sorting in an optical lattice," *Nature*, **426**, 2003, 421–424.
122. P. T. Korda, M. B. Taylor, and D. G. Grier, "Kinetically locked-in colloidal transport in an array of optical tweezers," *Physical Review Letters*, **89**, 2002, 128301.

123. K. Ladavac, K. Kasza, and D. G. Grier, "Sorting mesoscopic objects with periodic potential landscapes: Optical fractionation," *Physical Review E.*, **70**, 2004, 010901.
124. T. Čižmár et al., "Optical sorting and detection of submicrometer objects in a motional standing wave," *Physical Review B.* **74**, 2006, 035105.
125. I. Ricardez-Vargas et al., "Modulated optical sieve for sorting of polydisperse microparticles," *Applied Physics Letter*, **88**, 2006, 121116.
126. P. Jakl et al., "Static optical sorting in a laser interference field," *Applied Physics Letters*. **92**, (2008) 161110.
127. G. Milne et al., "Fractionation of polydisperse colloid with acousto-optically generated potential energy landscapes," *Optics Letters*, **32**, 2007, 1144–1146.
128. K. Dholakia et al., "Cellular and colloidal separation using optical forces," in *Laser Manipulation of Cells and Tissues*, 2007, Elsevier Academic Press Inc: San Diego, pp. 467–495.
129. S. Kawata and T. Sugiura, "Movement of micrometer-sized particles in the evanescent field of a laser-beam," *Optics Letters*, **17**, 1992, 772–774.
130. M. Gu et al., "Laser trapping and manipulation under focused evanescent wave illumination," *Applied Physics Letters*. **84**, 2004, 4236–4238.
131. V. Garcès-Chávez, K. Dholakia, and G. C. Spalding, "Extended-area optically induced organization of microparticles on a surface," *Applied Physics Letters*, **86**, 2005, 031106.
132. T. Čižmár et al., "Optical sorting and detection of submicrometer objects in a motional standing wave," *Physical Review B.*, **74**, 2006, 035105.
133. S. Kawata and T. Tani, "Optically driven Mie particles in an evanescent field along a channeled waveguide," *Optics Letters*, **21**, 1996, 1768–1770.
134. M. Dienerowitz, M. Mazilu, and K. Dholakia, "Optical manipulation of nanoparticles: a review," *Journal of Nanophotonics*, **2**, 2008, 021875.
135. J. P. Hole et al., "Velocity distribution of gold nanoparticles trapped on an optical waveguide," *Optics Express*, **13**, 2005, 3896–3901.
136. K. Grujic and O. G. Helleso, "Dielectric microsphere manipulation and chain assembly by counter-propagating waves in a channel waveguide," *Optics Express*, **15**, 2007, 6470–6477.
137. M. M. Burns, J. M. Fournier, and J. A. Golovchenko, "Optical Binding," *Physical Review Letters*, **63**, 1989, 1233–1236.
138. C. D. Mellor and C. D. Bain, "Array formation in evanescent waves," *Chemphyschem*. **7**, 2006, 329–332.
139. S. Gaugiran et al., "Polarization and particle size dependence of radiative forces on small metallic particles in evanescent optical fields: evidence for either repulsive or attractive gradient forces," *Optics Express*, **15**, 2007, 8146–8156.
140. J. R. Arias-Gonzalez and M. Nieto-Vesperinas, "Radiation pressure over dielectric and metallic nanocylinders on surfaces: polarization dependence and plasmon resonance conditions," *Optics Letters*, **27**, 2002, 2149–2151.
141. K. Grujic, O. Hellesø, J. Hole, and J. Wilkinson, "Sorting of polystyrene microspheres using a Y-branched optical waveguide," *Optics Express*, **13**, 2005, 1–7.

CHAPTER 16

Optofluidic Chemical Analysis and Synthesis

Dominik G. Rabus
Baskin School of Engineering, University of California, Santa Cruz

Optofluidic chemical analysis and synthesis is both a new and known field depending on one's view of the subject. New in the sense that microfluidics and optics have merged to enable the realization of miniaturized devices demonstrating well-known procedures like flow injection analysis or fluorescent spectroscopy. The field of optofluidic chemical analysis and synthesis is thriving due to the availability of appropriate manufacturing processes and the availability of matching light sources which can be integrated and combined with microfluidics. This is the essential difference to conventional lab-on-a-chip devices, which are around already for quite some time. Commercially available lab-on-a-chip devices require an additional apparatus for performing the necessary functions and analysis. The aim of optofluidic chemical analysis and synthesis devices is to perform as much as possible on chip. Of course fluid control and handling systems are required, but the main detection, sensing, or synthesis mechanism is on chip. In the future, even fluidic handling and control systems will be available on chip as integration of these devices progresses. This chapter serves as an introduction into this young optofluidics field and presents procedures and devices. The chapter is organized as follows: after an introduction into the subject, details on flow injection analysis systems and fluorescence-based methods are given. The section is complemented by a selection of devices, which highlight recent advances in the field. The section is concluded by a short summary.

16-1 Optofluidic Chemical Analysis and Synthesis

The merger of optics and fluidics in optofluidic devices provides the foundation for the interaction of matter and light as it is well known in several applications like spectroscopy. Spectroscopy is used for chemical analysis for quality purposes, in medicine, water treatment, and food, just to name a few. Chemical analysis can be performed using several methods. The wet-chemical-analysis-based methods and the instrument-based methods are the most common ones. Optofluidic methods belong to the instrument-based methods. The questions that need to be solved in chemical analysis are the quantitative, the qualitative, and the structural analysis of a substance, which means: what is in the media to be detected, how much is in it, and what is the chemical structure of the substance to be detected.

Quantitative chemical analysis is performed after a certain procedure. To start with, a sample has to be taken either fluidic in nature or solid. If the sample is solid, it needs to be brought into solution state. In the following step, the unwanted ingredients need to be separated, and in order to detect the correct substance, an ideal environment has to be made, for example, temperature adjustment or the addition of a reagent. In order to verify if a substance to be detected is present, it is essential to know if the method used is specific or nonspecific. In the nonspecific case, it has to be made certain that the substance to be detected can be isolated from those giving a similar result. Otherwise unwanted side effects can disturb the analysis.

Qualitative analysis is done using specific chemical analysis tools. For example, let's consider two solutions white in color—one passes the conductivity test, the other does not, and both solutions are neutral regarding the litmus test. The two solutions are salt and sugar.

Structural analysis is done using several measurement methods like density, refractive index, conductivity, and the like, and due to the variety of methods and the availability of detailed literature, only a brief introduction is given here.

Optofluidic analysis methods can be distinguished as direct and indirect methods, where specific physical parameters are detected or the substances to be detected need to be made visible through other optically detectable substances.

The interaction of matter and light in the form of electromagnetic waves was discovered by Max Planck and is given by

$$\Delta E = h\upsilon$$

where E = energy
h = Planck's constant
υ = frequency

Energy is absorbed/taken up or emitted in discrete portions. This is known as the birth of quantum physics. In optofluidic analysis systems, absorption and emission play a vital role. The emitted or absorbed wavelengths are originated in discrete energy states. Different wavelength ranges like x-rays, infrared, or microwaves have different effects on matter and can be used to analyze different substances. The absorption of light is advantageously used in optofluidic photometers to detect a variety of substances. This absorption is wavelength dependent, material specific, and subject to the concentration. A so-called spectrophotometer provides more details on the wavelength-dependent absorption over a broader spectrum. The extinction of light traveling through a media is given by the Beer–Lambert law:

$$T = \log \frac{I_0}{I} = \varepsilon(\lambda) c d$$

where T = transmissivity
I = intensity entering the media
I_0 = intensity exiting the media
ε = extinction coefficient
c = concentration
d = distance that the light travels in the media

If the absorption wavelength is known, then ε is constant. The concentration c is determined by using a constant distance d, which the light has to travel in the media. This principle is used in so-called photometers, where due to the increase of the color the concentration of the sample to be analyzed increases proportionally.

A standard setup mainly consists of two lightpaths where two samples are analyzed. The concentration of one sample is known whereas that of the other is determined. In this way only a relationship between the known and the unknown concentration needs to be calculated, which is directly related to the absorption of the used wavelength.

The wavelength needs to be chosen adequately in order to obtain accurate concentration values. This means that the wavelength should ideally match the absorption wavelength without any bandwidth and additional wavelengths, which is practically not possible. Therefore, monochromators or band-pass filters are used to select the correct wavelength. Recent advancements in LED and laser technology have enabled the fabrication of wavelengths across the visible and infrared spectrums, which in turn enable the implementation of substance-specific photometers and, speaking in terms of optofluidics, flow-through photometers. One example of a part of an integrated photometer is demonstrated by Mogensen and coworkers [1] where a polymer SU-8 waveguide is crossed by a fluidic channel (Fig. 16-1).

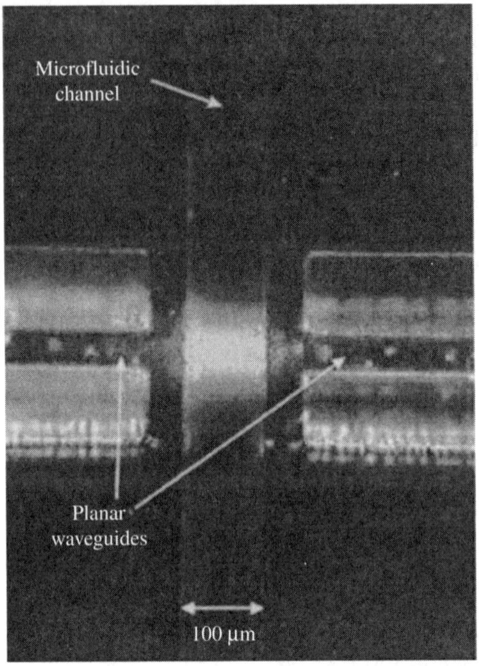

FIGURE 16-1 Light launched across a 100-μm-wide microfluidic channel. (*K. B. Mogensen, J. El-Ali, A. Wolff, and J. P. Kutter, "Integration of polymer waveguides for optical detection in microfabricated chemical analysis systems," Appl Opt, 2003, 42, 19, 4072–08.*)

The principle of photometry is used in chemical analysis to make substances of interest visible by adding "color" which binds specific to the substance of interest. There are numerous examples of detection mechanisms using photometers in literature.

In order to analyze substances in a flow-through manner, the so-called flow injection analysis, which will be explained in the following section, has been developed and is a known and accepted procedure for performing chemical analysis in combination with a photometric readout.

16-1-1 Flow Injection Analysis

The basic flow injection analysis system consists of a flow path where a pump is being used to transport the substance or the media to be analyzed, a reagent injection section, a mixing section, and finally the detection mechanism, usually a flow-through photometer. This system has several advantages. It is fast, flexible, and can be automatable. The essential part is that using micro- and optofluidics, it is possible to miniaturize these kinds of systems. The concept of flow injection analysis systems depends on a combination of three factors:

(1) reproducible sample injection volumes required in the case of automated analysis; (2) an appropriate sample-measurement device, controllable sample dispersion; and (3) reproducible timing of the injected sample through the flow system, which requires adequate valves with short opening and closing times. The detection mechanism, which is considered in this case to be optical, has a distinct profile (the y axis is usually the intensity and the x axis is the detection time) that is characterized by the peak height, the peak width, the area under the peak, and finally the time the sample passes through the detector. This obtained profile is an essential part of flow injection analysis systems and is referred to as dispersion. It is important that the dispersion of a fluid zone is reproducibly introduced into a nonsegmented flow stream (carrier) during transport of the zone to the detector. This concentration profile depends on convection and diffusion, whereas the effects of convection dominate. Both effects lead to the dilution of the sample. In order to quantify the dispersion, the so-called dispersion coefficient D is introduced, which is given by

$$D = \frac{C_0}{C}$$

where C_0 is the original concentration of the reagent before dispersion and C is the concentration of this specific fluid, which passes the detector unit and the data is obtained.

Parameters that are important for the layout of a flow injection analysis system are the entire length of the fluidic channel, the surrounding temperature, the volume of the required reagent and sample, the fluidic layout of the mixing zone, the concentration of the reagent used, the length of the detection zone, and the position.

Complex flow injection analysis systems with several samples and reagents can be constructed using microfabrication and microfluidic technology and are used for different analysis methods, which can be found in literature. The advantage of flow injection analysis systems is the possibility to integrate several process steps into a single system that enables highly automated devices and procedures. Miniaturized systems that can be densely integrated enable small sample and reagent volumes, thus producing smaller amounts of waste, and possess a faster reaction and detection time. The drawback in miniaturization is the fact that valves and pumps need to be developed, which can be integrated into the system. Small reagent and sample volumes lead to low detection signals which require more sophisticated readout electronics and detection methods, often impossible to integrate. Despite these drawbacks, microfluidic flow injection analysis has a great potential for the future as advancing optofluidic technology overcomes these hurdles.

Flow injection analysis systems are well known and have numerous applications. Several publications and books exist, and the intention of this section is to provide a brief introduction.

16-1-2 Fluorescence-Based Analysis

Fluorescent-based detection systems are known in literature and have numerous applications in medical, biological, and chemical detection systems. This section provides a basic understanding of the principal detection methods and serves as an introduction into this vast area.

In absorption and emission processes, the absorbed and the emitted wavelengths are usually the same. In fluorescent absorption and emission processes, this is not the case, which is why it is extremely interesting for detection applications. Several examples of fluorescent substances are quinine, fluorescein, and rhodamine.

A basic setup consists of an emitter (LED, laser, white light source), a filter for selecting the appropriate wavelength, which is absorbed by the fluorescent substance, a sample container, a filter (e.g., a bandpass filter) for filtering out the emitter wavelength, and finally the detector. The advantage of fluorescent-based systems is the fact that the wavelength of the emitter is different from the emitted fluorescent wavelength. The sample itself provides the light to be detected. This method is highly sensitive and enables single-molecule spectroscopy. In order to enable fluorescent detection, appropriate reagents showing a fluorescent behavior are required. Several important factors need to be taken into consideration. The absorption and emitting wavelengths need to be as far from each other in terms of nanometers in order to separate the two wavelengths from each other with the help of optical filters. The photon energy of the detection signal is required to be sufficient enough to be able to be detected. Two effects that work against these requirements are quenching of the fluorescent reagent and photobleaching. Current research is focused on the elimination of these effects in order to be able to design fluorescent markers with high energy transfer, which enables a high-optical-output signal. Fluorescent markers exist in a large variety and can explicitly be engineered to bind to the required target substance, which can be chemical or biological in nature.

The fact that fluorescent markers can be engineered to be biospecific is used advantageously in flow cytometry, which is a technology that is used to measure characteristics of biological particles. An example of a flow cytometer is shown in Fig. 16-2. Single particles are sent past a light source. The fluorescent scattered light is collected by a photodetector and analyzed. Flow cytometry is used not only for analyzing single molecules, particles, and the like, but also for sorting of these particles. Cell sorting is an example of such a flow cytometry system. An introduction into flow cytometry is given in the review paper by Jaroszeski and Radcfiff [2]. A further source of information can be found online (see Refs. 3 and 4).

Fluorescent-based analysis systems including the required fluorescent markers are commonly used in laboratories. Focus of technology is to automate these processes in order to reduce the time of the

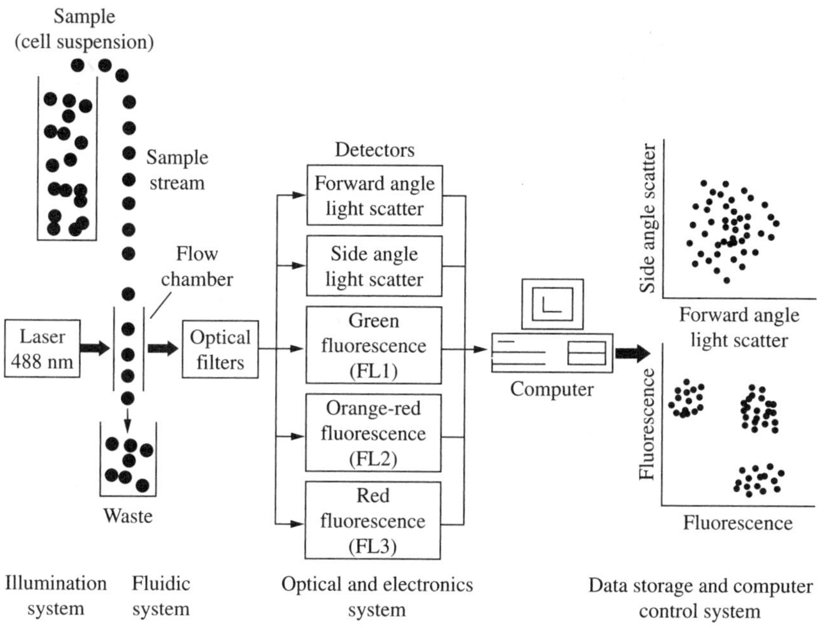

Figure 16-2 A schematic of the primary components that comprise a flow cytometer. Dark arrows indicate the flow of particles and information. A fluidic system transports particles or cells from a prepared suspension past a focused laser beam that is generated by an illumination system. Particle-interrogation takes place, one cell at a time, in a flow chamber. The resulting scattered light and fluorescence are gathered by an optical and electronics system that translates the light signals into information, which is saved by the data storage and computer control system. After data from a sample have been stored, retrospective graphical data analysis can be performed with the aid of software. (*With kind permission from Springer Science+Business Media: Mol Biotechnol, "Fundamentals of flow cytometry," 11(1), 1999, 37–53, M. J. Jaroszeski and G. Radcfiff, Figure 1.*)

measurement, the amount of substances which includes markers and the analyte, and what is from the technical point of view regarded as inevitable is the reduction of apparatus necessary to perform this kind of detection. Optofluidic devices offer a large potential in solving these issues. In order to perform analysis and synthesis, a great deal of device variety is required including the necessary peripheral devices like valves and pumps.

16-1-3 Devices

The driving forces behind the automation, miniaturization, and detection can be found in the life sciences where several analysis experiments up to more than hundreds have to be carried out in parallel or sequentially. The same effort has to be made for synthesizing chemicals or medicines, for example. Both analysis and synthesis rely on each other due to the fact that one is only as good as one can analyze

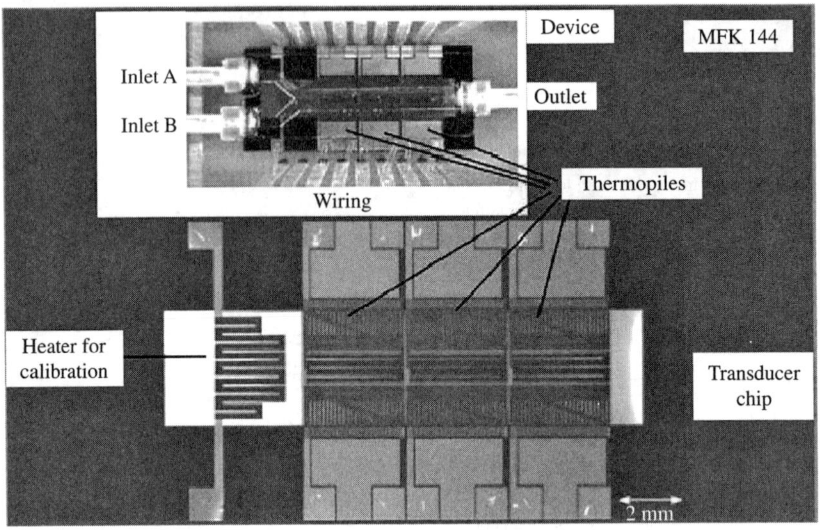

FIGURE 16-3 Example of a microflow calorimeter: MFK 144 consisting of cover chip with Y-shaped microchannel and a Si-chip containing three blocks with a total of 144 micropatterned thin-film thermocouples (Sb/BiSb) and thin-film heaters for calibration. (*With kind permission from Springer Science+Business Media: Appl Microbiol Biotechnol, "Chip devices for miniaturized biotechnology," 69, 2005, 113–125, J. M. Koehler and T. Henkel, Figure 2.)*

the targets one has synthesized. A review on the evolution of analysis in life science research and molecular medicine is given by Regnier in Ref. 5, where the current state and possible future of separation methods in the rapidly developing field of bioscience is presented. In order to be able to perform parallel or sequential analysis and synthesis in micro- and optofluidic devices, necessary platforms need to be available, which provide the basic tools. Fluid-control systems are the backbone of every micro- and optofluidic chip. In Ref. 6 Koehler et al. present a review on-chip devices for miniaturized biotechnology. An example of an integrated device is shown in Fig. 16-3, which demonstrates the high integration level of state-of-the-art devices.

Polymer and silicon-based materials are used primarily in realizing microfluidic devices due to the availability of standardized fabrication processes. A design for three-dimensional microfluidic structures consisting of three stacked glass wafers is presented in Ref. 7. A fabricated device is shown in Fig. 16-4.

The aim of microfluidic integrated devices is to provide several functions on chip. Historically, chips have been designed to provide one of a kind of solution. It would be advantageous, if compared to the digital world of transistors and log gates, to be able to configure one's "circuit" oneself. Fair addresses this challenge in the paper "Digital microfluidics: is a true lab-on-a-chip possible" in Ref. 8, where the suitability of electrowetting-on-dielectric (EWOD)

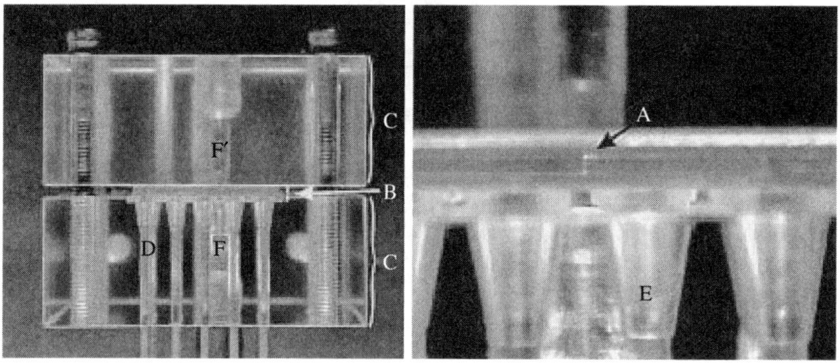

FIGURE 16-4 Photograph of the microfluidic chip and chip holder. (Right) Closeup view of the microfluidic chip, microfluidic connections, and optical connections. (A) Optical cuvette drilled by SAE, (B) microfluidic chip made of three Pyrex layers, (C) PMMA plates, (D) PEEK tubing 0.020 in i.d.×1/16 in o.d., (E) flangeless ferrule 1/16 in, (F and F′) collection and illumination optical fibers, respectively. (*With kind permission from Springer Science+Business Media: Fresenius J Anal Chem, "Multilayer microfluidic glass chips for microanalytical applications," 371, 2001, 261–269, A. Daridon, V. Fascio, J. Lichtenberg, R. Wütrich, H. Langen, E. Verpoorte, ·and N. F. de Rooij, Figure 2.*)

microfluidics for true lab-on-a-chip applications is discussed. Another paper addressing these issues is presented in Ref. 9, where the aim was to develop microsystems immediately usable by biologists for complex protocol integrations. All fluid operations are performed on nanoliter droplets independently handled solely by EWOD actuation.

Microfluidic chips are the basic building blocks of future optofluidic devices. The optical part relies on mature microfluidics technology. A review of optical techniques implemented into microfluidic devices is presented in Ref. 10. The study addresses the progress made toward the miniaturization of optical functions in the lab-on-a-chip devices, where significant advances are being made in both miniaturization of the instrumentation and on the chips themselves. Future effort has to be made in fully integrating light sources and detectors on chip. Recent developments in detection in microfluidic chips are addressed by Mogensen et al. in Ref. 11. Detection methods are essential for the identification and quantification not only of chemical species that are being analyzed but also for all analytes optically detectable. Detection methods are as good as the obtained signals from the analyte. An increase in the signal-to-noise ratio can be achieved by guiding light directly by the integrated waveguide on microfluidic chips. In this way optical losses are minimized and the detector can directly be coupled to the fluidic channel. Device design and implementation of optofluidic waveguides are presented by Schmidt and Hawkins in Refs. 12 and 13. The focus is on liquid-core optical waveguides for creating fully planar optofluidic lab-on-a-chip.

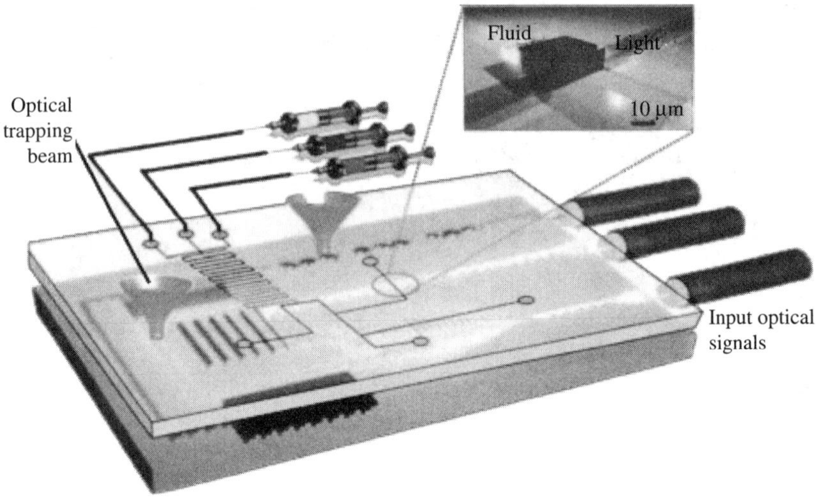

FIGURE 16-5 Illustration of a multifunctional optofluidic chip. The chip includes (i) a photonic chip (bottom part) that is coupled to optical fibers, (ii) a microfluidic circuit (on top of it) that is connected to fluid sources with different refractive indices, and (iii) optical-trapping beams to manipulate micromechanical components. The schematic on the top right-hand side shows a close-up of the chip, where the optical signal intersects with a fluidic channel at the micrometer scale. (*With kind permission from Springer Science+Business Media: Microfluid Nanofluid, "Optofluidics: a novel generation of reconfigurable and adaptive compact architectures," 4, 2008, 81–95, C. Monat, P. Domachuk, C. Grillet, M. Collins, B. J. Eggleton, M. Cronin-Golomb, S. Mutzenich, T. Mahmud, G. Rosengarten, and A. Mitchell, Figure 1.*)

Optofluidics not only enables new analysis methods but also addresses the need for reconfigurable and adaptive compact architectures [14]. The illustration of a multifunctional optofluidic chip is shown in Fig. 16-5.

The tuning and reconfiguration of optofluidic devices is one step toward flexible and application-oriented device design, which enables the use of one system for several different analysis and synthesis variations. Enhancement of optofluidic analysis and synthesis is achieved by the implementation of colloids and their assemblies. Details on the use of colloids in optofluidic applications are given in Refs. 15 and 16.

16-2 Summary

Optofluidic chemical analysis and synthesis has been briefly addressed in this section to provide a starting point for further reading. The focus in current state-of-the-art devices is application driven and focused on the detection and analysis more than on the synthesis. Optofluidic detection has vast applications especially for biosensing [17] and for analyzing chemical substances [18]. Different technologies, approaches, and functionalities like detection, configuration, reconfiguration, and integration of functions have been described and future research activities will have to focus on the integration of these functions in order to fully address the idea of analysis and synthesis for various applications.

References

1. K. B. Mogensen, J. El-Ali, A. Wolff, and J. P. Kutter, "Integration of polymer waveguides for optical detection in microfabricated chemical analysis systems," *Appl Opt*, 2003, 42, 19, 4072–08.
2. M. J. Jaroszeski and G. Radcfiff, "Fundamentals of flow cytometry," *Mol Biotechnol*, 1999, 11(1), 37–53.
3. Introduction to flow cytometry, http://probes.invitrogen.com/resources/education/tutorials/4Intro_Flow/player.html.
4. Flow cytometry principles, http://biology.berkeley.edu/crl/flow_cytometry_basic.html.
5. F. Regnier, "The evolution of analysis in life science research and molecular medicine: the potential role of separations," *Chromatographia Supplement I*, 1999, 49, S56–S64.
6. J. M. Koehler and T. Henkel, "Chip devices for miniaturized biotechnology," *Appl Microbiol Biotechnol.*, 2005, 69, 113–125.
7. A. Daridon, V. Fascio, J. Lichtenberg, R. Wütrich, H. Langen, E. Verpoorte, and N. F. de Rooij, "Multilayer microfluidic glass chips for microanalytical applications," *Fresenius J Anal Chem*, 2001, 371, 261–269.
8. R. B. Fair, "Digital microfluidics: is a true lab-on-a-chip possible," *Microfluid Nanofluid*, 2007, 3, 245–281.
9. Y. Fouillet, D. Jary, C. Chabrol, P. Claustre, and C. Peponnet, "Digital microfluidic design and optimization of classic and new fluidic functions for lab on a chip systems," *Microfluid Nanofluid*, 2008, 4, 159–165.
10. H. C. Hunt and J. S. Wilkinson, "Optofluidic integration for microanalysis," *Microfluid Nanofluid*, 2008, 4, 53–79.
11. Klaus B. Mogensen, Henning Klank, and Jörg P. Kutter, "Recent developments in detection for microfluidic systems," *Electrophoresis*, 2004, 25, 3498–3512.
12. Holger Schmidt and Aaron R. Hawkins, "Optofluidic waveguides: I. Concepts and implementations," *Microfluid Nanofluid*, 2007.
13. Aaron R. Hawkins and Holger Schmidt, "Optofluidic waveguides: II. Fabrication and structures," *Microfluid Nanofluid*, 2007.
14. C. Monat, P. Domachuk, C. Grillet, M. Collins, B. J. Eggleton, M. Cronin-Golomb, S. Mutzenich, T. Mahmud, G. Rosengarten, and A. Mitchell, "Optofluidics: a novel generation of reconfigurable and adaptive compact architectures," *Microfluid Nanofluid*, 2008, 4, 81–95.
15. S.-H. Kim, S.-J. Jeon, and S.-M. Yang, "Optofluidic encapsulation of crystalline colloidal arrays into spherical membrane," *J Am Chem Soc*, 2008, 130, 6040–6046.
16. S.-K. Lee, S.-H. Kim, J.-H. Kang, S.-G. Park, W.-J. Jung, S.-H. Kim, G.-R. Yi, S. and M. Yang, "Optofluidics technology based on colloids and their assemblies," *Microfluid Nanofluid*, 2008, 4, 129–144.
17. D. Erickson, S. Mandal, A. H. J. Yang, and B. Cordovez, "Nanobiosensors: optofluidic, electrical and mechanical approaches to biomolecular detection at the nanoscale," *Microfluid Nanofluid*, 2008, 4, 33–52.
18. H. Y. Tan, N.-T. Nguyen, W.-K. Loke, and Y. T. Tan, "Microfluidic chip with optical sensor for rapid detection of nerve agent Sarin in water samples," *Proc of SPIE*, 2006, 6416, 64160M.

CHAPTER 17

Optofluidic Maskless Lithography and Guided Self-Assembly

Wook Park, Su Eun Chung, Seung Ah Lee, and Sunghoon Kwon

Department of Electrical Engineering, Seoul National University, Seoul, Republic of Korea

In this chapter, concepts in optofluidics are applied to an advanced manufacturing technology based on self-assembled microparts. The "optical" aspect of optofluidics will be described in the context of photolithography, and the "fluidic" aspect will be discussed in the context of self-assembly. First, optofluidic maskless lithography will be introduced as a dynamic fabrication method to generate microparticles in microfluidic channels. Next, the history and application of optofluidic lithography will be presented. Finally, optofluidic-guided self-assembly using railed microfluidics will be introduced as a new method to assemble microparticles into complex systems.

17-1 Optofluidic Maskless Lithography

Microparticles have been widely used in various applications, that is, microbeads for biological molecular handling, microcapsules for drug delivery, microabrasives for chemical mechanical polishing, and so on. In this light, the control of particle size, shape, and composition is of particular importance. Recent developments in various optofluidic devices added great flexibility regarding the synthesis of various microparticles. In this section, droplet-based fabrication of microparticles will be discussed as a method to fabricate spherical particles. Next, pattern-based particle fabrication using a lithographic approach will be introduced for the fabrication of particles with arbitrary shapes. Finally, optofluidic maskless lithography (OFML) will be discussed as a dynamic fabrication method that provides advanced controllability in microparticle generation.

17-1-1 Droplet-Based Fabrication of Microparticles

Conventional fabrication of spherical microparticles is based on batch-fabrication using emulsion of two immiscible fluids. Although this emulsion-based method can generate spherical particles in large volumes, the fabricated particles are dispersed in size. To address this issue, the microfluidic concept has been recently applied to generate monodispersed microparticles. Compared with the emulsion method, particle generation based on microdevices can mean better controllability and uniformity with regard to particle size and material composition.

Two typical methods based on T-junction and flow-focusing have been rapidly developed over the past decade [1–5]. The concepts of the two methods are shown in Fig. 17-1a. In the T-junction method, the channel of the dispersed phase perpendicularly intersects a continuous phase channel [6–10]. The continuous phase stream breaks the dispersed stream into a droplet at the neck of the dispersed phase via shear-force. In the flow-focusing method, the dispersed and continuous phases narrow into a focused region in the microfluidic device [11–17]. This singular point in the focused region ensures that the break-off of droplets from the fluid stream occurs consistently at that point. Soft lithography [18], or the insertion of capillary sheaths into microchannels, is used as a fabrication method for flow-focusing channels. Figure 17-1b shows microdroplet encapsulation via the flow-focusing method. After generating microdroplets in a microdevice using these methods, monodisperse spherical liquid droplets are solidified by polymerizing a liquid monomer.

An example of microparticle fabrication based on the T-junction method is shown in Fig. 17-1c. After the generation of droplets in the T-junction, the shape of the particle can be modified by changing channel dimensions. For instance, the plugs are formed by solidifying the droplets in the narrow channel whereas the disks are fabricated by illuminating UV to the liquid droplet in the wide reservoir [19]. Similarly, alteration of channel dimensions in flow-focusing method enables the capture of microspheres, rods, disks, ellipsoids, and fibers [20–22]. Various microcapsule morphologies such as core-shell droplets [15] or

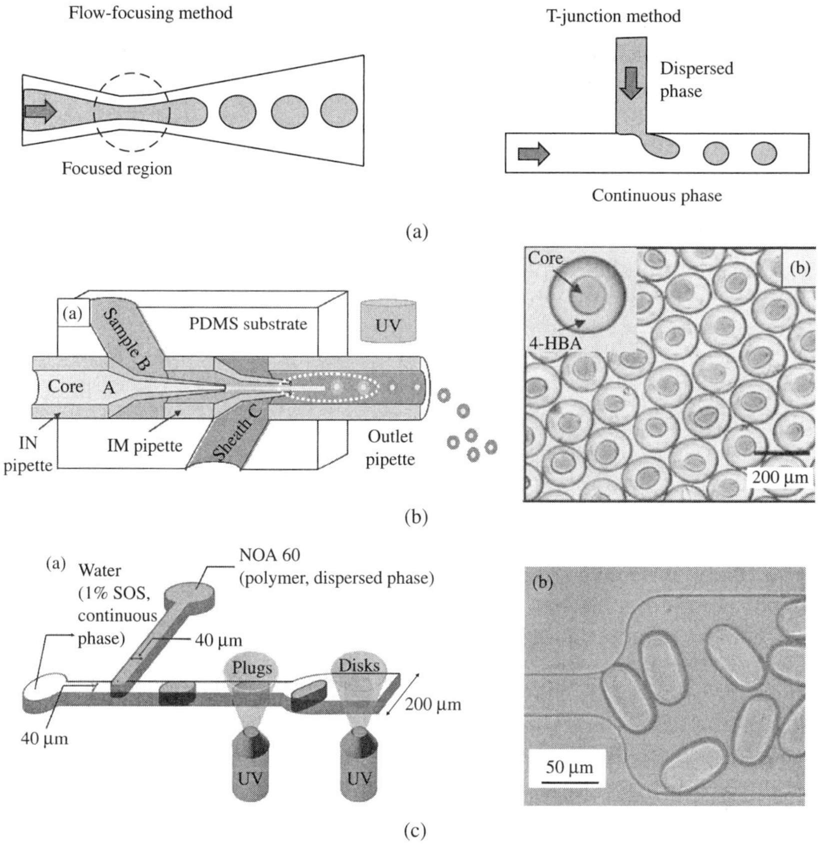

Figure 17-1 Droplet-based fabrication of microparticles. (a) Flow-focusing microchannel (left). T-junction microchannel (right). (b) An example of flow-focusing method. Schematic illustration of both the microcapsule-fabrication apparatus (MCFA) and the production microcapsules via *'on-the-fly'* photopolymerization (left). Polymerized microcapsules and zoomed micrograph of a single microcapsule (right). (c) An example of T-junction method. Schematic of channel with plug and disk creation zones marked (left). Polymerized plugs (right). (*Reprinted with permission from D. Dendukuri, K. Tsoi, T. A. Hatton, and P. S. Doyle, "Controlled synthesis of nonspherical microparticles using microfluidics," Langmuir, 21, 2005, 2113–2116. Copyright 2005 American Chemical Society; H.-J. Oh, S.-H. Kim, J.-Y. Baek, G.-H. Seong, and S.-H. Lee, "Hydrodynamic micro-encapsulation of aqueous fluids and cells via 'on the fly' photopolymerization," Journal of Micromechanics and Microengineering 16(2), 2006, 285–291. Institute of Physics and IOP Publishing.*)

spherical colloidal clusters [23–26] were demonstrated based on multi-compositional droplets with sacrificial components. However, these droplet-based fabrication methods are all limited to the generation of rounded particles. For generation of microparticles with complex shapes, optofluidic fabrication methods can be employed.

17-1-2 Patterned Microparticle Generation

Applying photolithography in microfluidic particle generation means great flexibility in particle shape control. Photolithography used in microfabrication transfers a geometric pattern in a photomask onto a photoresist-coated substrate. Using photoresist-filled microfluidic channel as a substrate for photolithography, various polymeric structures can be fabricated inside a microfluidic channel by in situ photopolymerization [27] (Fig. 17-2a). Using this technique, various microfluidic active components that are anchored in the channel can be formed. Figure 17-2a shows a hydrogel microvalve that swells and shrinks in response to the solution's pH changes.

Microscope projection photolithography is a simple photolithography technique in which a transparent photomask is inserted in front of a microscope objective lens to pattern a photoresist-coated substrate. This method results in a rapid prototyping solution for generating small numbers of prototype test structures quickly and inexpensively. Doyle et al. employed microscope projection photolithography to photopolymerize oligomeric diacrylate monomers flowing in a polydimethylsiloxane (PDMS) microfluidic device, a process known as continuous-flow lithography [28]. As shown in Fig. 17-2b, free-floating microparti-cles of triangular, cuboidal, cylindrical, and other irregular shapes can be fabricated in a microfluidic channel.

In the vicinity of the PDMS channel walls, polymerization is inhibited, meaning that nonpolymerized liquid is left as a lubrication layer. This nonpolymerized region near the channel wall is called the oxygen inhibition layer [28]. Since PDMS is highly oxygen-permeable, the concentration of oxygen is high near the PDMS surface. Since oxygen takes up the initiator radicals for photopolymerization, the polymerization is locally inhibited in the channel surface, enabling fabrication of the free-floating particles [29].

Continuous-flow lithography provides high-throughput production of microparticles with various shapes due to its continuous synthesizing process. Continuous-flow lithography has another advantage of enabling fabrication of multifunctional particles that are synthesized by adding new functional groups one by one. When a multilaminar stream flows through a microfluidic device at a low Reynolds number, multicomposite microparticles such as Janus particles can be fabricated across the interface of fluids with one-step lithography. Figure 17-2c shows an application example of such Janus particles with distinct regions for analyte-encoding, target-capture, and control-reference for DNA hybridization assay [30]. A composite number in this multifunctional microparticle depends on the number of streams flowing through the microfluidic channel.

To increase the throughput of the particle production, one needs to increase flow-speed in the microfluidic channel and decrease the exposure time. At high flow-rates, the particle fabricated by

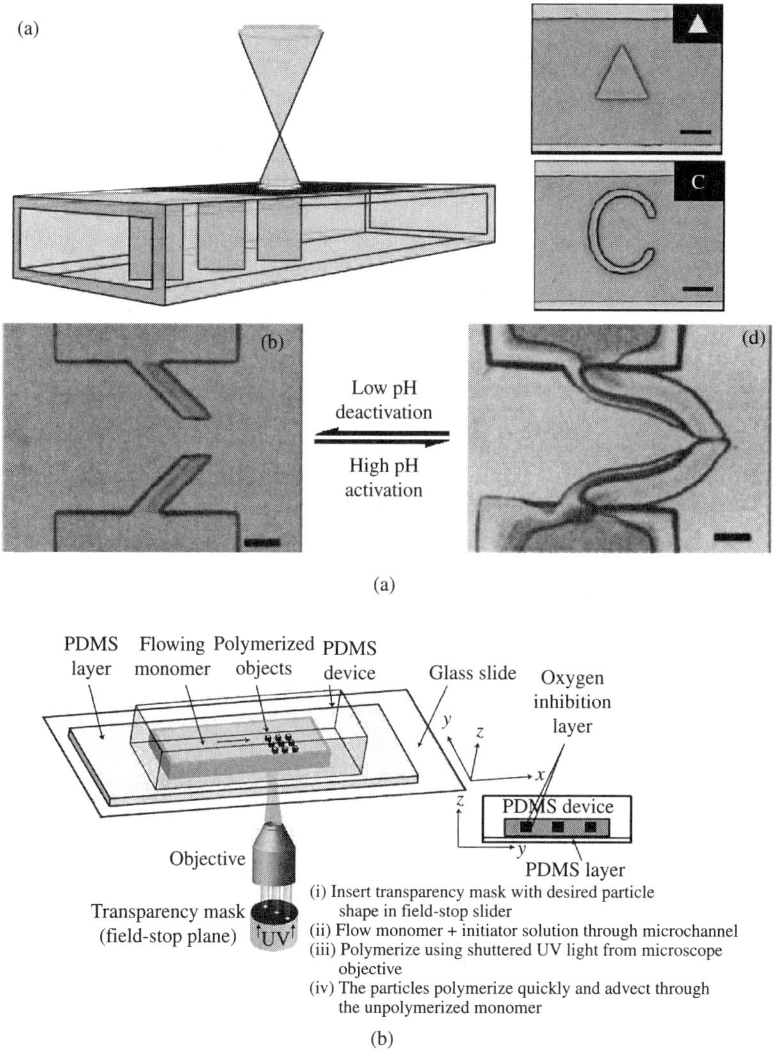

FIGURE 17-2 Photolithographical approaches for generation of microstructures in a microfluidic channel. (a) Fabrication of fixed hydrogel structures via patterned photopolymerization. (Top) Hydrogel structure fabricated inside a microfluidic device can be used for flow-regulating pH-sensitive valves. (Bottom) (b) Continuous flow lithography method for generation of free-floating microparticles. (Top) Microparticles with various shapes can be generated from the corresponding transparency mask patterns. (Bottom left) Multifunctional shape-encoded particles have been demonstrated for the application of free-floating microparticles in bioanalysis. (Bottom right). (*Reprinted by permission from Macmillan Publishers Ltd: Nature [27], copyright (2000); Yu, Q., Bauer, J. M., Moore, J. S. and Beebe. D. J., Responsive biomimetic hydrogel valve for microfluidics. Applied Physics Letters, 78(17), 2001, 2589–2591. American Institute of Physics; reprinted by permission from Macmillan Publishers Ltd: Nature Materials, P. S. Doyle, D. Dendukuri, D. C. Pregibon, J. Collins, and T. A. Hatton, "Continuous-flow lithography for high-throughput microparticle synthesis," Nature Materials, 5, 2006, 365–369, copyright (2006); from D. C. Pregibon, M. Toner, and P. S. Doyle, "Multifunctional encoded particles for high-throughput biomolecule analysis," Science, 315, 2007, 1393–1396. Reprinted with permission from AAAS.*)

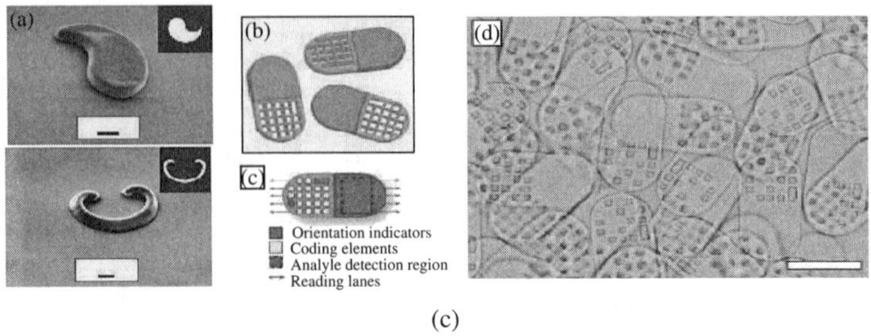

FIGURE 17-2 (*Continued*)

continuous-flow lithography is smeared and blurred since the exposure is done on moving fluid. Doyle et al. introduced an advanced method known as stop-flow lithography to overcome this problem. In stop-flow lithography, the oligomer flow is briefly halted by an external solenoid valve during the exposure [31]. This stop-flow technique allows for increased throughput without sacrificing patterning resolution. It is applied for generating cell-laden hydrogels because diluted hydrogels required for cell culture need more exposure time [32].

17-1-3 Optofluidic Maskless Lithography (OFML)

Background: Maskless Lithography

Photolithographic fabrication techniques incorporate transparent photomasks composed of a transparent plate and metal or black film with a defined pattern. During exposure, a light beam is reshaped by the patterns on the photomask and generates corresponding structures on the resin. The use of fixed photomasks is advantageous for mass production of devices with the same geometry. However, for applications that require a large number of photomasks or frequent mask exchanges, fixed photomasks are expensive and bulky. Dynamic masks using electronic spatial light modulator are introduced as an alternative solution, mainly for the inexpensive generation of a large number of masks. Bertsch et al. demonstrated one of the first dynamic mask generators using a liquid crystal display (LCD) device [33]. Later, digital micromirror devices (DMD) replaced the LCDs, with higher resolution, high contrast, and low energy-loss during UV exposure. A DMD is a spatial light modulator invented by Dr. Larry Hornbeck and Dr. William E. Nelson of Texas Instruments (Fig. 17-3a). It is developed mainly for digital light processing (DLP) projectors, one of the leading technologies in rear-projection display. A DMD chip is composed of a two-dimensional rectangular array of microscopic mirrors, each representing a pixel of displayed image [34]. Each mirror is supported by a torsion hinge, which allows ±10 ~12° rotation for generating on/off states. An on-state mirror on a DMD chip reflects the light to the projection

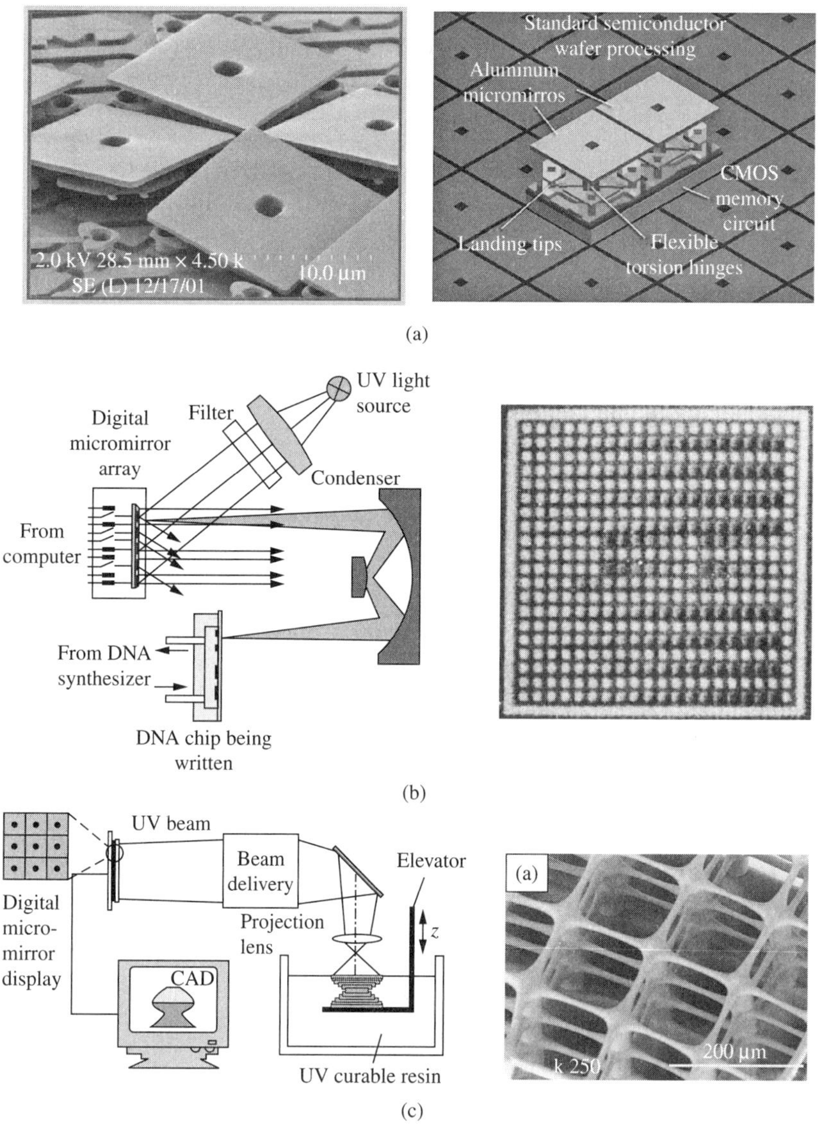

FIGURE 17-3 (a) Texas Instruments DMDTM chips. SEM image of micromirror array (left) and detailed structure of a micromirror (right). (b and c) Lithography systems utilizing DMD as dynamic mask generator. (b) The maskless array synthesizer (MAS) for DNA microarray (right). (c) The projection micro-stereolithography (PSL) apparatus and fabricated 3D microstructure (right). (*Texas Instruments DLP Discovery TM, http://www.ti.com; reprinted by permission from Macmillan Publishers Ltd: Nature Biotechnology S. Singh-Gasson, R. D. Green, Y. Yue, C. Nelson, F. Blattner, M. R. Sussman, and F. Cerrina, "Maskless fabrication of light-directed oligonucleotide microarrays using a digital micromirror array," Nature Biotechnology, 17, 1999, 974–978., copyright (1999); reprinted from Sensors and Actuators A: Physical, 121(1), C. Sun, N. Fang, D. M. Wu, and X. Zhang, "Projection micro-stereolithography using digital micromirror dynamic mask," 113–120, Copyright (2005), with permission from Elsevier.*)

screen and the pixel appears bright. In the off-state, light is deflected onto a heat sink, and the pixel appears dark on the projection screen. Gray scale is also produced using pulse-width modulation by toggling the mirror at a high frequency.

As the DMD chips became commercially available, maskless lithography systems utilizing a DMD chip as a dynamic mask generator were developed mainly to reduce the time and the cost of making photomasks. Singh-Gasson et al. demonstrated this concept in DNA microarray, where light-directed chemistry requires a large number of photomasks to generate an array of different DNA sequences on a substrate [35] (Fig. 17-3b). Use of DMD greatly reduces the time and cost for DNA chip fabrication, and this maskless array synthesizer (MAS) technology is now one of the core technologies of NimbleGen Inc. [36].

In addition, maskless lithography systems are advantageous in layer-by-layer micro-stereolithography, replacing the need of fixed masks corresponding to the cross section of each unit layer. Sun et al. developed a projection micro-stereolithography system using DMD and demonstrated various 3D complex microstructures as shown in Fig. 17-3c [37]. The DMD-based dynamic mask exhibits many advantages over previous LCD-based dynamic masks: higher efficiency, UV compatibility, faster switching time, higher resolution, and contrast. The micro-stereolithography technique incorporating DMD is applied for the fabrication of complex 3D scaffolds in tissue engineering [38]. Due to its flexibility, DMD-based maskless lithography systems are currently being developed for photolithography [39] and maskless gray-scale lithography [40].

Concept of OFML

OFML is a technique that uniquely combines maskless and continuous-flow lithography techniques in microfluidic channels. It provides real-time control of the in situ polymerization process to dynamically synthesize extruded polymeric microstructures with various two-dimensional shapes. In continuous-flow lithography, photomasks in the optical-projection lithography system are not dynamically changeable in real time; therefore, it is difficult to control the exposure pattern and timing of the microstructure fabrication process in a unified manner. For continuous high-throughput fabrication of a large number of distinctive microparticles, maskless lithography techniques with programmable exposure patterns can significantly improve the performance and flexibility of the fluidic lithography systems. Figure 17-4a describes the schematic diagram of the experimental setup, which combines a high-speed maskless lithography system for dynamic UV photopatterning, a microfluidic channel for control of photocurable acrylate oligomer stream, and a microscopic imaging system for inspection and monitoring.

The top view of the polymerized microstructures is defined by the exposure pattern of the UV light as schematically depicted in

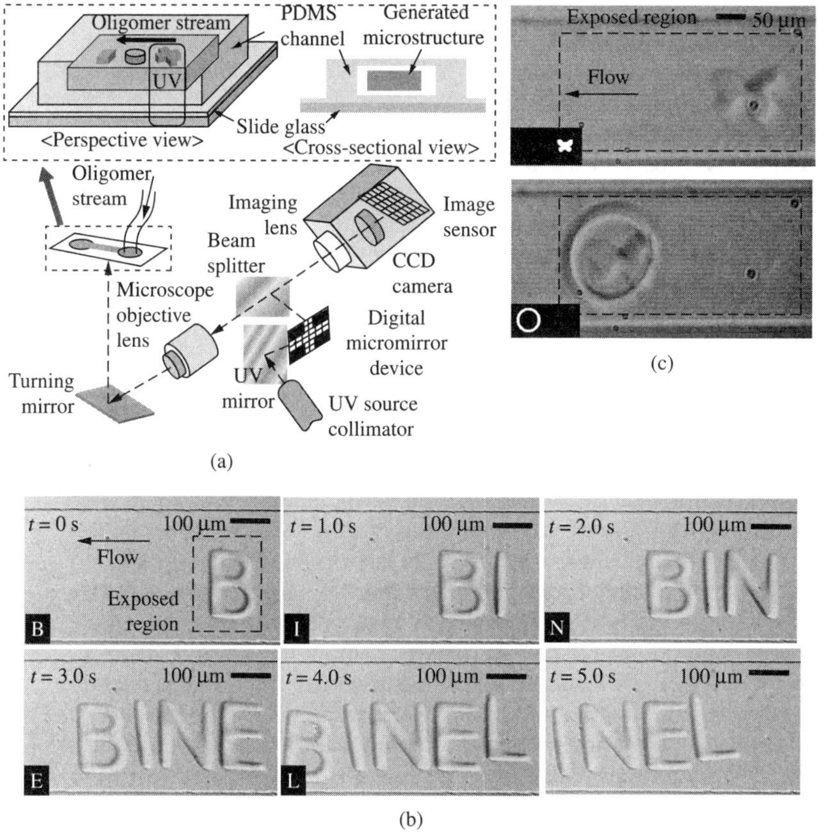

FIGURE 17-4 Concept of optofluidic maskless lithography system. (a) Schematic of optofluidic maskless lithography system. (b) Dynamic illumincation with a programmable SLM. Different UV patterns are exposed onto a flowing photopolymer inside a microfluidic channel to generate various microstructures at every second. (c) Dynamic targeting of a free-floating microstructure and multiple UV exposures. (*Reprinted with permission from S. E. Chung, W. Park, H. Park, K. Yu, N. Park, and S. Kwon, "Optofluidic maskless lithography system for real-time synthesis of photopolymerized microstructures in microfluidic channels," Applied Physics Letters, 91, 2007, 041–106. Copyright 2007, American Institute of Physics.*)

Fig. 17-4a, while their thickness is determined by the height of the channel and the lubricating inhibition layer. The microfluidic channels are fabricated using a standard soft lithography technique with PDMS material [41]. All four inner faces of the microfluidic channels are made of PDMS (cross-sectional view in the inset of Fig. 17-4a) to generate free-floating particles.

Characteristics of OFML

The OFML technique can flexibly control the shapes and sizes of the extruded polymeric microstructures as well as their relative positions and orders. For example, different polymeric character patterns were generated at 1-s intervals, as shown in Fig. 17-4b. The dynamic SLM controls the spatial pattern of UV light exposure onto the microfluidic channel with high timing accuracy.

In Fig. 17-4b, a small portion of the SLM was used to make relatively simple structures with a single UV exposure. However, it is also possible to improve the fabrication resolution and to synthesize more complicated structures with multiple-exposure techniques [42]. For example, it is possible for an adaptive multiple-exposure lithography scheme to monitor and correct the fabricated structures iteratively in real time within the field of view of the proposed lithography system. Figure 17-4c demonstrates a dynamically targeted free-floating butterfly-shaped microstructure, the subsequent augmentation of the targeted structure with a second UV exposure by a computer vision system, and the formation of an additional ring-shaped structure around the original microstructure with the second UV exposure (Fig. 17-4c). The ring ("butterfly trap") and butterfly structures are connected with each other, and float with the uncured prepolymer stream. This also demonstrates accurate temporal and spatial control of microparticle generation.

In the traditional projection-based maskless lithography schemes with a limited number of SLM pixels, there is a trade-off between the spatial resolution and the photopatternable area. The OFML technique overcomes this limitation by continuously translating the fabricated structures with fluidic forces, and thus enables the fabrication of very long structures while maintaining the patterning accuracy and flexibility of maskless lithography.

Figure 17-5a(i) shows that long polymeric microwire structures, whose lengths are much longer than the exposure area, can be synthesized by exposing a time-varying UV pattern in a continuously flowing photocurable resin. Large polymer structures with fine resolutions can be fabricated in a cost-effective manner without expensive equipment, such as a stepper [43]. Figure 17-5a(ii) shows a curved polymeric wire fabricated by moving the circular exposure pattern up and down. The shape and width of the microwires can also be easily controlled by changing the center position and size of the exposure pattern as a function of time, respectively. It is also possible to use multiple exposure spots oscillating up and down to form interwoven microwire structures as shown in Fig. 17-5a(iii). By changing relative offsets between two exposure spots, the relative position can be controlled between two twisted polymeric microwires.

Taking advantage of microfluidic control and dynamic exposure, material composition of microstructures can be controlled using OFML. As the example in Fig. 17-5b, multicomposite microwires can be fabricated by moving the exposing spot between two laminar streams of different photocurable polymers. Another example of material composition control is demonstrated in Fig. 17-5c, which shows the embedding of inorganic materials in organic packaging

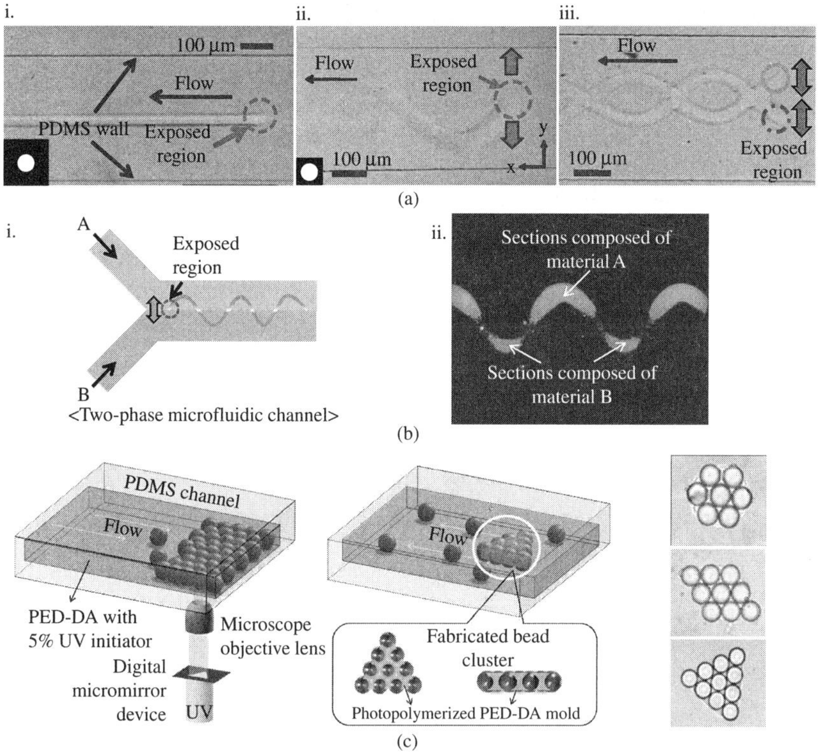

FIGURE 17-5 Characteristics of optofluidic maskless lithography system. (a) Fabricated high-aspect ratio structures larger than the exposed region, straight lines (left) curved polymer wire (center) overlapped two curved polymer wires (right). (b) Multiple material compositions. Schematic diagram of fabrication method of compositional wire in two-phase microfluidics (left). Image of multi-compositional wire (right). (c) Microbead cluster fabrication via OFML by selectively polymerizing a close-packed monolayer of microbeads.

using OFML. Inorganic materials such as microchips or beads introduced to a microfluidic channel can be captured by OFML, allowing for organic packaging to be formed around the material. Doyle et al. also demonstrated that inorganic magnetic beads embedded in hydrogel are able to capture cells in the microfluidic channel [44]. Magnetic microbeads are first assembled via magnetic force and next polymeric packaging is cured upon the magnetic particles. A microbead monolayer assembled in UV-curable polymer resin can be selectively cured by OFML, generating bead clusters with polymeric adhesive that holds the beads together. Also, using OFML, mask patterns can be dynamically generated through image processing techniques. With the feedback from the charge-coupled device (CCD) camera, inorganic materials flowing through the microfluidic channel can be dynamically captured (Fig. 17-5c).

OFML as a Platform for Fluidic Self-Assembly

Due to its ability to fabricate microparticles with great temporal and spatial control in a fluidic environment, the OFML system is an attractive platform to study fluidic self-assembly and to manufacture scalable systems based on self-assembly. For instance, a large number of free-floating particles generated by OFML can be handled by the flow and subsequently assembled into a system of particles in the same microfluidic channel. By separating the fabrication area and the assembly area in the channel, different material systems and different manufacturing processes can be integrated into an assembled system (Fig. 17-6a). In fabrication of the building block for fluidic self-assembly,

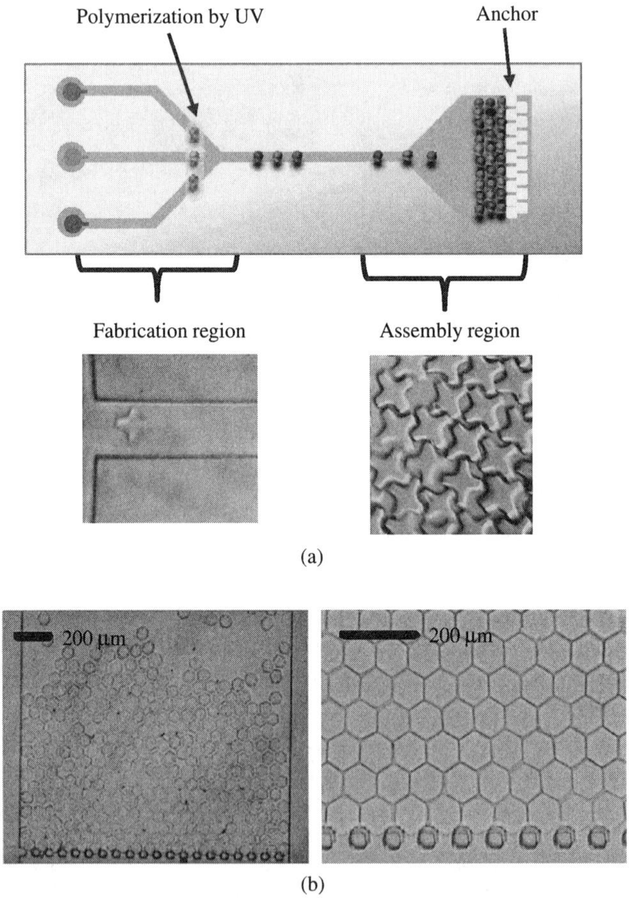

Figure 17-6 OFML system as a platform for self-assembled scalable systems. (a) A microfluidic channel can have a fabrication region and an assembly region, so that the particles continuously generated at a fixed area are assembled simultaneously by the flow. (b) Fluidic self-assembly of hexagonal microstructures in a microfluidic channel. (See also color insert.)

OFML provides high flexibility in generating free-floating microparticles with various shapes such as triangles, squares, and hexagons for a high-density monolayer of microparticles by fluidic self-assembly. A potential application of this self-assembled scalable system is particle-based biosensor systems, where the detection time and yield can be greatly improved by arranging the particles into a high-density array. However, although highly parallel, particle-manipulation by laminar flow lacks precise control over the order of assembly and particle selectivity, thereby sacrificing flexibility for high-yield. The ability to control a single particle in a scalable system requires an additional approach, which will be introduced in the next section as the rail-guided fluidic self-assembly technique.

In this section, various microparticle fabrications in a flow and optofluidic maskless lithography have been discussed. The unique combination of high-speed maskless lithography and microfluidics allows us to control the timing and location of the photopolymerization process. This technique would be a versatile platform to investigate the fluidic self-assembly process. Once microstructures with various shapes and compositions are fabricated inside the microfluidic channels, they can be fluidically self-assembled to form large-scale systems. Rail-guided fluidic self-assembly based on OFML will be discussed in the next section.

17-2 Optofluidic-Guided Self-Assembly: Railed Microfluidics

17-2-1 Self-Assembly

Self-assembly is a promising pathway for parallel fabrication of devices made up of many small components. It takes place at all scales, from nanoscale elements such as atoms of materials to the arrangement of galaxies in the universe. This self-assembly process can be categorized into two types according to the energy dissipation of components: static self-assembly and dynamic self-assembly [45]. Static self-assembly is the most common assembly type subject to significant research, and systems comprised of static self-assembly are at equilibrium state and do not dissipate energy once they are formed. However, systems made of dynamic self-assembly dissipate energy, and mostly are related to biological or environmental topics. This chapter will mainly discuss static self-assembly.

In all assembly processes, components need to first be transported to specified assembly sites in a substrate (or to other components), and then be assembled together by proper driving forces. Many pioneering self-assembly works using various driving forces such as gravity [46–49], surface energy [50–54], electrostatic force [54–57], electromagnetic force [58,59], fluidic force [60–64], or capillary force [65,66] have been reported for a variety of applications, such as heterogeneous assembly of electronic parts [67], LEDs on silicon wafers [68], RFID chips on antennas [69], lead bump assembly [70], bead packing [71,72], and so on. These driving forces are not the only forces to self-assemble various components. Several driving forces can be combined together to assemble many components. For instance, fluidic self-assembly is mainly dominated by fluidic force, but gravity also affects the assembly process.

For larger components, gravity is a main driving force by conveying a component from an upper position to a lower position. If a component has a shape that matches the hole(s) of the substrate, then gravitational force will force the component down the corresponding hole of the object.

Two driving forces of self-assembly, surface force and capillary force, are closely related to each other. The intermolecular attraction between the liquid and solid materials is called capillary force. In detail, liquid tends to be drawn in a narrow tube due to the capillary force. The cohesion force, or the attraction force between liquid and solid materials, produces surface tension, which allows for objects denser than the liquid to be supported on the surface of liquid, much like a water strider. The self-assembly process using surface energy or capillary force is related to the surface characteristics of the components to be assembled. For example, hydrophobic surfaces tend to be attracted to other hydrophobic surfaces over large distances by minimizing interfacial free energy, as shown in Fig. 17-7a. Furthermore, millimeter and mesoscale components are mostly affected by surface force as well as gravity [73–76].

Capillary force can be combined with physical templates to assemble microcomponents or to form assembled structures [75–78]. For example, electromagnetic force induced a solder to be molten and capillary action between movable micromachined metal structures and molten alloy forced the structure to be lifted up, as shown in Fig. 17-7c.

Many self-assembly processes can be executed under fluidic environments. Most examples of these are executed within a liquid, making the process easier. In Fig. 17-7e, the fluid is the main driving force of self-assembly. The substrate is patterned chemically, and liquid containing nanowires flow on the templates. Due to the patterned chemical template, nanowires can be aligned and assembled on the substrate.

Fluidic self-assembly is a technique using fluidic force as the driving force for assembly of components. Components float around the assembly sites and some of them will meet with the holes on the template due to gravity force, as described in Fig. 17-7f. The process can be easily applied to a heterogeneous assembly by using various shapes of holes and matching microcomponents. This self-assembly technique is often massively parallel and therefore faster and cheaper than serial pick-and-place robotic assembly. However, assembly yield is not as high as conventional robotic assembly owing to the probabilistic nature of self-assembly. In order to have a

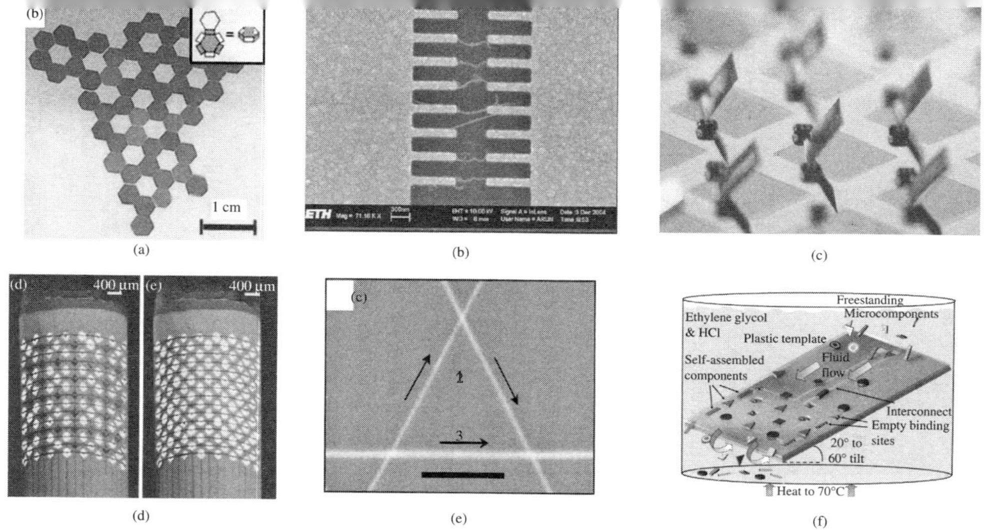

FIGURE 17-7 Self-assembly. (a) Mesoscale components self-assembly using surface force and capillary force. (b) Carbon nanotubes self-assembly by applying electrostatic force. (c) Lifting-up movable metal structure using electromagnetic force and capillary force. (d) Gravity-controlled self-assembly on the flexible substrates. (e) Layer-by-layer nanowire self-assembly using fluidic force. (f) Heterogeneous self-assembly by fluidic self-assembly method. (*From N. Bowden, A. Terfort, J. Carbeck, and G. M. Whitesides, "Self-assembly of mesoscale objects into ordered two-dimensional arrays," Science, 276, 1997, 233–235. Reprinted with permission from AAAS; A. Subramanian, B. J. Nelson, D. Lixin, and D. Bell. "Dielectrophoretic nanoassembly of individual carbon nanotubes onto nanoelectrodes," The 6th IEEE International Symposium on Assembly and Task Planning: From Nano to Macro Assembly and Manufacturing, Montreal, Canada. copyright 2005 IEEE; Y. Hsueh-An, L. Chiung-Wen, and F. Weileun, "Wafer level self-assembly of microstructures using the global magnetic lifting and localized induction welding," Solid-State Sensors, Actuators and Microsystems, copyright 2005 IEEE; from H. O. Jacobs, A. R. Tao, A. Schwartz, D. H. Gracias, and G. M. Whitesides, "Fabrication of a cylindrical display by patterned assembly," Science, 296(5566), 2002, 323–325. Reprinted with permission from AAAS; from Y. Huang, X. F. Duan, Q. Q. Wei, and C. M. Lieber, "Directed assembly of one-dimensional nanostructures into functional networks," Science, 291, 2001, 630–633. Reprinted with permission from AAAS; S. A. Stauth and B. A. Parviz, "Self-assembled single-crystal silicon circuits on plastic," Proceedings of the National Academy of Sciences, 103, 2006, 13922–13927. Copyright 2006 National Academy of Sciences, U.S.A.*)

higher probability of matching and assembling, it is necessary to increase the chance of matching components by overloading the number of parts floating around the assembly sites as demonstrated in fluidic self-assembly of RFID [47] or assembly of DNA origami [79]. Mass production of microcomponents is required in most fluidic self-assembly processes. However, if microstructures can be accurately guided in a fluidic environment, an efficient fluidic self-assembly process combining the advantages of both high-yield robotic assembly and high-throughput fluidic self-assembly would be achieved.

17-2-2 Rail-Guided Fluidic Self-Assembly

Motivation for Guided Self-Assembly

The most common way to put multicomponents together in one place is serial pick-and-place robotic assembly. It has long been a fundamental manufacturing process to build complex systems out of various mechanical and electrical components. In this assembly process, a robotic arm picks up each part individually and places it on a designated position on a substrate. Thus, robotic assembly is a very deterministic process for components larger than several hundred micrometers, allowing for high assembly yield with great flexibility in component choice.

For component sizes smaller than 200 μm, however, robotic assembly is extremely slow and expensive due to the difficult control requirements needed to position parts with high accuracy. Additionally, strong parasitic stiction forces make pick-and-place assembly even more difficult at the microlevel [67], because at such scales surface force wins over gravity. When the robotic arm picks up the component to convey it to the other position, the component will normally adhere to the robotic arm due to the stiction between the arm and the particle. Therefore, when a large number of microstructures need to be assembled, the conventional robotic assembly process becomes too costly for most applications.

Concept of Railed Microfluidics

Railed microfluidics can be thought of as a microscale version of a monorail where a train follows the monorail due to the matching shapes of the train body and the rail [80]. This shape-matching concept was implemented in microfluidic channels by uniquely combining a grooved microfluidic channel [81] with previously demonstrated OFML [82] developed from continuous-flow lithography [28]. The basis of railed microfluidics is cross-sectional shape matching between a microfluidic channel and the microstructures flowing through the channel.

Instead of using a conventional microfluidic channel with a flat-top channel surface, a groove ("rail") was formed on the top surface of the channel using two-step mold fabrication by twice repeating photolithography in the mold preparation phase and standard soft lithography. The groove functions as a guide rail track inside the microfluidic channel. After the channel was filled with UV-curable oligomer solution, a polymeric microstructure with a fin ("microtrain") was created using in situ photopolymerization via OFML. The top-view shape of the polymeric particle is dynamically controlled by patterns on a digital micromirror device [82]. The fin structure is shown in the cross-sectional diagram in Fig. 17-8a(ii). The fin is an exact fit with the rail since it is "molded" by the rail itself during the polymerization process. The fin is a little bit smaller than the rail because polymerization is inhibited owing to the high oxygen

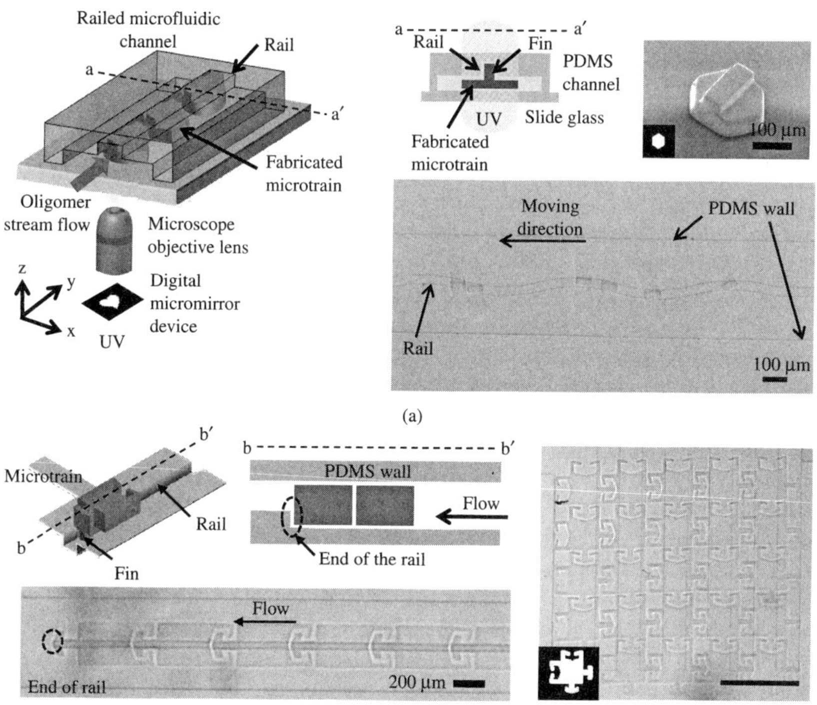

FIGURE 17-8 (a) Concept of railed mcirofluidics. Schematic diagram of railed microfluidics (left). Cross section of the PDMS channel and a finned microtrain cut at a-a' (top center). 3D SEM image of microtrain (top right). (b) Assembly mechanism using railed microfluidics. The end of the rail as an assembly site (top left). One-dimensional chain of guided self-assembly (bottom left). Two-dimensional self-assembly forming 5 × 5 matrix (right). (*Reprinted by permission from Macmillan Publishers Ltd: Nature Materials S. E. Chung, W. Park, S. Shin, S. A. Lee, and S. Kwon, "Guided and fluidic-self-assemlby of microstructures using railed microfluidic channels," Nature Materials, 7, 2008, 581–587, copyright 2008.*)

concentration near the PDMS surface. The finned microtrain is a polymeric microstructure that normally should follow the flow field in the microfluidic channel. However, the matching of the fin and the rail enables the finned microtrains to deviate from the flow field and instead follow the rail.

This guided movement is shown in Fig. 17-8a(IV). Although the flow in the channel is a left-to-right linear movement, the microtrains move sinusoidally, according to the sinusoidally designed rail. In short, the movement of the finned microtrain is controlled by the rail rather than the flow field of the channel. Therefore, by designing the rail, one can control the movement of the particle inside a microfluidic channel.

Concept of Rail-Guided Assembly

The railed microfluidic process is an innovative technique to guide the movement of in situ photopolymerized microstructures. Rail-based guiding was applied to self-assembly in order to overcome many limitations of conventional free-flow self-assembly. In general robotic assembly, the parts are first moved to the assembly site. Similarly, parts were sent to the assembly site at the end of the rail by microtrain (Fig. 17-8b). At the end of the rail, the microtrains are blocked and are unable to move forward due to the rail's structural geometry (Fig. 17-8b). Therefore, the end of the rail works as a barrier to block the movement of the microtrains and to initiate the assembly process.

To illustrate the assembly process, a one-dimensional self-assembly of multiple polymeric microlatches on a single rail was executed. Microlatches are similar to latches people use in daily life. Multiple microlatches fabricated on the rail are simultaneously pushed by flow and assembled at the end of the rail (Fig. 17-8b).

The assembly process can easily be expanded to two-dimensional assembly. As shown in Fig. 17-8c, two-dimensional self-assembly of microlatches is executed using horizontal and vertical rails concurrently. This process is scalable to large area assembly. A relatively simple assembled structure composed of a single type of microstructure has been presented. Even for this type of simple assembly with repeating motifs, a conventional fluidic self-assembly process would only be possible after a large quantity of extra parts were wasted. Even so, conventional fluidic self-assembly would produce only a limited yield. In comparison, railed microfluidics demonstrates a high-yield assembly without wasting even a single part, owing to the rail guiding of the microstructures.

Rail-Guided Complex Self-Assembly

Previously, all fluidic self-assembly was a thermodynamically driven process that was probabilistic in nature. Rail-guided self-assembly is fundamentally different since this assembly process achieves zero error and is completely deterministic. Thus, it enables to efficiently assemble the very complex structures. Such assembly would not be possible with conventional fluidic self-assembly.

The real benefit of railed microfluidics over the conventional fluidic self-assembly lies in its capability to assemble complex systems made up of a large number of different parts. By increasing the number of rails and kinds of microstructures involved in the assembly process, very complex microsystems are easily assembled. Figure 17-9a through 17-9c shows the Greek temple self-assembly as an example. Each microstructure forming a complex self-assembled system is

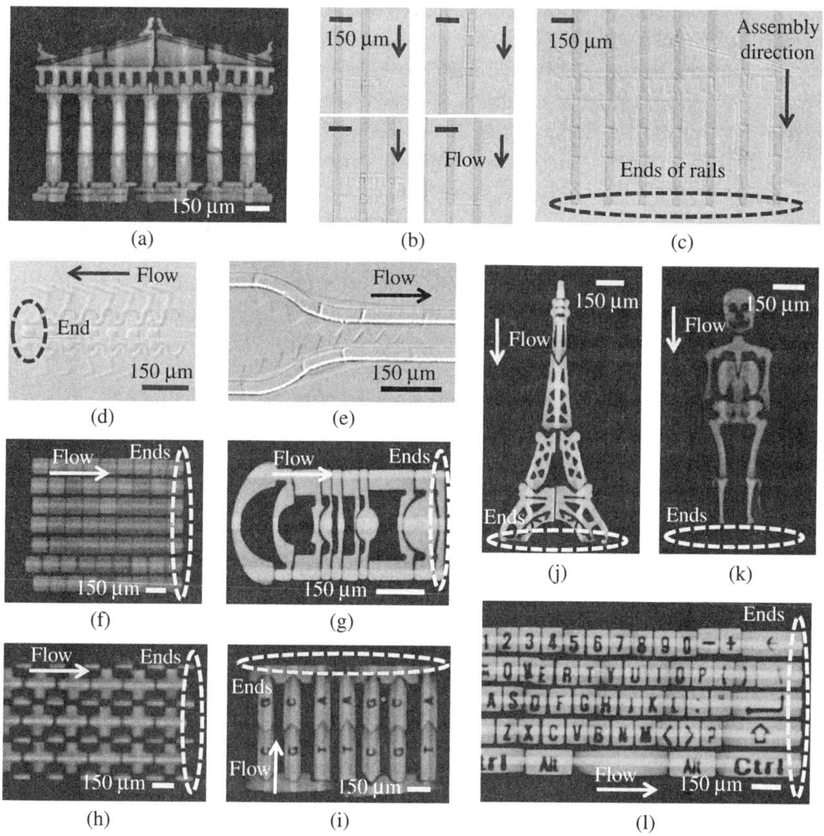

Figure 17-9 Guided complex self-assembly using railed microfluidics. (a) Greek temple assembly. (b) Components fabrication examples. (c) Assembled structures in DIC image. (d) Microcentipede assembly. (e) Microzipper assembly. (f) Heart-shaped mosaic assembly. (g) Fish-eye lens assembly. (h) Alternative particle packing. (i) DNA assembly. (j) Eiffel tower assembly. (k) Skeleton assembly. (l) Microkeyboard assembly. (*Reprinted by permission from Macmillan Publishers Ltd: Nature Materials S. E. Chung, W. Park, S. Shin, S. A. Lee, and S. Kwon, "Guided and fluidic-self-assemlby of microstructures using railed microfluidic channels," Nature Materials, 7, 2008, 581–587, copyright 2008.*)

fabricated independently on a corresponding rail. Next, all the microstructures are assembled together at the end of rails by applying fluidic force. The structures shown in Fig. 17-9a through 17-9l are examples of complex structures assembled using railed microfluidic channels. Fully assembled complex structures can be easily actuated by controlling flow field as shown in the dancing skeleton assembly (Fig. 17-9k) or the crawling microcentipede assembly (Fig. 17-9d). After assembly, such structures can also be fixed together via ultraviolet exposure without a mask.

In complex self-assembly, using railed microfluidics, no more than the exact number of constituent microstructures is needed to form the complex system, unlike in the conventional fluidic self-assembly technique. For instance, in the assembly of a microcomputer keyboard composed of 68 keypads (Fig 17-9l), exactly 68 keypads are created and perfectly assembled. This is an apt illustration of the unique benefits of railed microfluidic assembly.

In addition to the end of the rail, any barrier can also work as an assembly site. As shown in the assembly of a microzipper (Fig. 17-9e), two curved rails could initiate the assembly process. If the length of the microstructures is longer than the bending radius of the bending rail, the bend works as a topological barrier, much like the end of the rail. The upper and lower zippers are synched, engaged, and assembled on the corresponding upper and lower rails.

Rail-Guided Heterogeneous Assembly

The true advantage of the guided self-assembly is in its capacity to assemble parts made out of different materials, or heterogeneous assembly. In conventional lithography, patterning three different materials in a single substrate requires three separate photolithography steps, three separate alignment steps, and three separate material patterning steps. In contrast, a much simpler process has been created to self-assemble microstructures made up of many different materials via "cross-solution movement." This technique greatly simplifies heterogeneous patterning by eliminating multiple alignments and material depositions.

Heterogeneous assembly of microlatches composed of three different materials is achieved on the basis of this cross-solution movement scheme, as shown in Fig. 17-10a. In addition, taking advantage of the laminar flow in microfluidics, simple fabrication of a two-dimensional heterogeneous pattern with a single lithographic exposure is demonstrated in Fig. 17-10a. As shown in this figure, five vertical channels are intersected with a horizontal channel, with multiple assembly rails in the center. First, five oligomer streams composed of two different fluorescently labeled oligomer solutions flow from top to bottom to form multiple vertical laminar streams over the assembly rails. Next, microstructures are fabricated on the assembly rail by a single UV exposure. Since the exposure area covers multiple

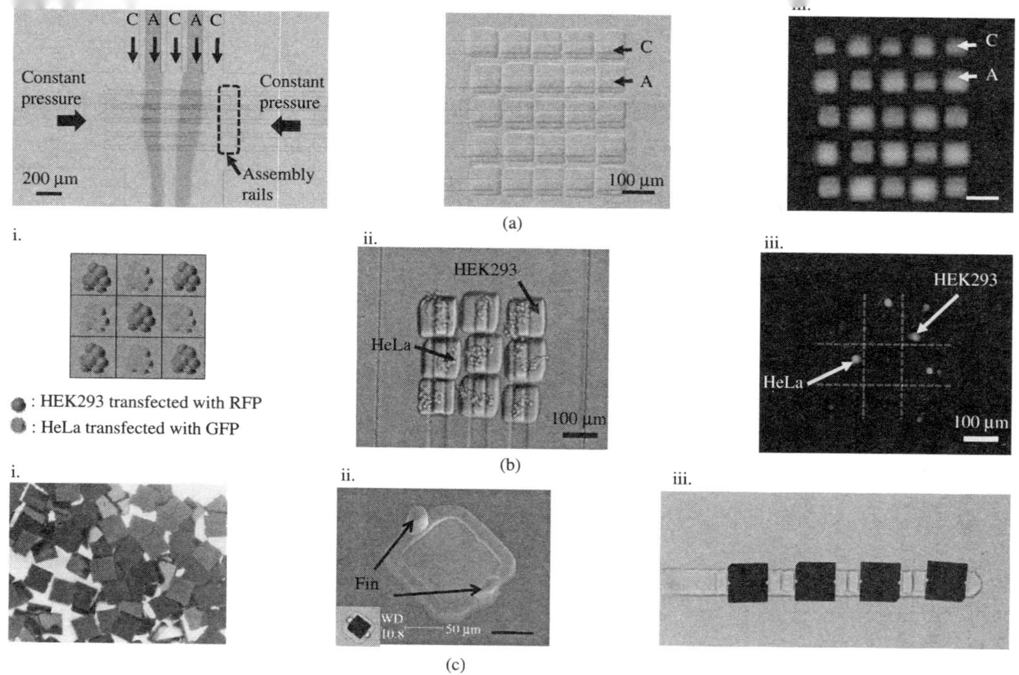

Figure 17-10 Applications of railed mcirofluidics. (a) Heterogeneous self-assembly of microstructures. Two-dimensional heterogeneous assembly process (left). 5 × 5 matrix assembly of squared microstructures (center). Corresponding fluorescent microscopic image displaying a checkerboard pattern of two different colors (right). (b) Heterogeneous cell patterning. 3 × 3 microscale hydrogel matrix (left), 3 × 3 patterns with two different living cells (center). Corresponding fluorescent image (right). (c) Microchips polymer packaging. Externally fabricated silicon microchips (left). SEM image of a silicon microchip in the polymer package (center). Packaged silicon chips are fluidically placed with desired uniform spacing (right). (*Reprinted by permission from Macmillan Publishers Ltd: Nature Materials S. E. Chung, W. Park, S. Shin, S. A. Lee, and S. Kwon, "Guided and fluidic-self-assemlby of microstructures using railed microfluidic channels," Nature Materials, 7, 2008, 581–587, copyright 2008.*) (See also color insert.)

laminar streams, microstructures made of different materials are formed with a single lithographic exposure as shown in Fig. 17-10a(ii). The fluorescent image shown in Fig. 17-10a(iii) displays a checkerboard pattern of two different colors. The different colors in these experiments may conceivably represent a broad range of materials, such as organic or inorganic beads, living cells, nanoparticles, magnetic particles, and so forth.

Application Examples of Rail-Guided Assembly

As macroscale trains can carry various cargos, the rail-based transportation and assembly is not limited to photocurable material but applicable to a wide variety of material systems. For example, railed microfluidics can easily be applied to assemble living cells for biomedical tissue engineering. Using railed microfluidics, many different types of cells with an exact specified configuration can be assembled. The patterning of many different cells in a hydrogel substrate is an important task in tissue engineering and in cell-based biochips [83–86]. It is difficult to form microscale heterogeneous assembly of different cells because one has to first form hydrogel pieces containing different cell types and then manually assemble them [87]. Even the most advanced hydrogel-based cell micropatterning is done by serially repeating multiple photolithographic processes with many exposure and alignment steps [88]. Application of rail-based heterogeneous assembly can greatly simplify this process. In Fig. 17-10b, formed is a 3×3 microscale hydrogel matrix with two different living cells, HeLa transfected with green fluorescent protein (GFP) and HEK293 transfected with red fluorescent protein (RFP). Rather than using manual pick-and-place or serial photolithography, the same method in Fig. 17-10b(ii) was applied to fabricate a microscale matrix of two different living cells in PEG-DA solution. Figure 17-10b and 17-10c shows the assembled hydrogel matrix. Note that this heterogeneous matrix is formed in single-step lithography, precluding the need for multiple alignment steps. As shown in the complex system assembly, complex assembly of different cells with various shapes would be easily achieved.

Railed microfluidics also has applicability in industrial processes such as integrated chip packaging. In the integrated circuit industry, the cost-effective packaging of small chips is of utmost importance. The chip size of radio frequency identification (RFID) and light-emitting devices (LEDs) currently in commercial production are smaller than 200 μm. Conventional serial pick-and-place in this small-scale suffers from high costs, and fluidic self-assembly is demonstrated to be a great alternative [47]. Accurate fluidic manipulation of such small chips would be a great improvement over the currently used fluidic self-assembly process. Rail-guided assembly technique was applied to locate and assemble externally fabricated silicon chips.

First, silicon microchips are fabricated externally, as shown in Fig. 17-10c(i). These chips can be thought of as a conceptual substitute for CMOS devices or LEDs. When a chip reaches the rail, a rectangular mask pattern slightly larger than the silicon chip is exposed. With a single exposure of the simple rectangular pattern, polymeric packaging and three-dimensional fins are fabricated around the chip, as shown in Fig. 17-10c(ii). Therefore, silicon chips in the polymer package can be guided along the rail and assembled at the end of rail, as shown in Fig. 17-10c(iii).

In this chapter, various types of fluidic self-assembly and railed microfluidics as a method for guiding and assembling microstructures inside a microfluidic channel were demonstrated. While all self-assemblies were thermodynamically driven processes that were probabilistic in nature, rail-based assembly is deterministic enough to achieve near-zero assembly error. Since no extra parts are required during assembly, this enables us to demonstrate the efficient assembly of complex microsystems. In addition, assembling heterogeneous microsystems made out of different materials using cross-solution movement and rail-based microfluidic design was also demonstrated. Immediate application areas by using railed microfluidics for the patterning of different living cells in MEMS-integrated cellular micromanipulations such as large or small molecule drug-screening process for tissue engineering as well as manipulating externally fabricated silicon devices for microchip packaging are identified. Due to its simplicity and flexibility, railed microfluidics will not only impact current self-assembly but also encourage innovation in a wide range of application areas in the future.

References

1. L. F. Cheow, L. Yobas, and D. L. Kwong, "Digital microfluidics: Droplet-based logic gates," *Applied Physics Letters*, **90,** 2007, 054107-3.
2. G. F. Christopher and S. L. Anna, "Microfluidic methods for generating continuous droplet streams," *Journal of Physics D-Applied Physics*, **40,** 2007, R319–R336.
3. A. D. Griffiths and D. S. Tawfik, "Miniaturising the laboratory in emulsion droplets," *Trends in Biotechnology*, **24,** 2006, 395–402.
4. B. T. Kelly, J. C. Baret, V. Taly, and A. D. Griffiths, "Miniaturizing chemistry and biology in microdroplets," *Chemical Communications*, **18,** 2007, 1773–1788.
5. T. Thorsen, R. W. Roberts, F. H. Arnold, and S. R. Quake, "Dynamic pattern formation in a vesicle-generating microfluidic device," *Physical Review Letters*, **86,** 2001, 4163–4166.
6. T. Nisisako, T. Torii, and T. Higuchi, "Droplet formation in a microchannel network," *Lab on a Chip*, **2,** 2002, 24–26.
7. J. D. Tice, H. Song, A. D. Lyon, and R. F. Ismagilov, "Formation of droplets and mixing in multiphase microfluidics at low values of the Reynolds and the capillary numbers," *Langmuir*, **19,** 2003, 9127–9133.
8. T. Nisisako, T. Torii, and T. Higuchi, "Novel microreactors for functional polymer beads," *Chemical Engineering Journal*, **101,** 2004, 23–29.

9. B. Zheng, J. D. Tice, and R. F. Ismagilov, "Formation of droplets in microfluidic channels alternating composition and applications to indexing of concentrations in droplet-based assays," *Analytical Chemistry*, **76,** 2004, 4977–4982.
10. S. Okushima, T. Nisisako, T. Torii, and T. Higuchi, "Controlled production of monodisperse double emulsions by two-step droplet breakup in microfluidic devices," *Langmuir*, **20,** 2004, 9905–9908.
11. S. L. Anna, N. Bontoux, and H. A. Stone, "Formation of dispersions using flow focusing in microchannels," *Applied Physics Letters*, **82,** 2003, 364–366.
12. P. Garstecki, I. Gitlin, W. Diluzio, G. M. Whitesides, E. Kumacheva, and H. A. Stone, "Formation of monodisperse bubbles in a microfluidic flow-focusing device," *Applied Physics Letters*, **85,** 2004, 2649–2651.
13. W. J. Jeong, J. Y. Kim, J. Choo, E. K. Lee, C. S. Han, D. J. Beebe, G. H. Seong, and S. H. Lee, "Continuous fabrication of biocatalyst immobilized microparticles using photopolymerization and immiscible liquids in microfluidic systems," *Langmuir*, **21,** 2005, 3738–3741.
14. P. C. Lewis, R. R. Graham, Z. H. Nie, S. Q. Xu, M. Seo, and E. Kumacheva, "Continuous synthesis of copolymer particles in microfluidic reactors," *Macromolecules*, **38,** 2005, 4536–4538.
15. Z. H. Nie, S. Q. Xu, M. Seo, P. C. Lewis, and E. Kumacheva, "Polymer particles with various shapes and morphologies produced in continuous microfluidic reactors," *Journal of the American Chemical Society*, **127,** 2005, 8058–8063.
16. A. S. Utada, E. Lorenceau, D. R. Link, P. D. Kaplan, H. A. Stone, and D. A. Weitz, "Monodisperse double emulsions generated from a microcapillary device," *Science*, **308,** 2005, 537–541.
17. T. Nisisako, T. Torii, T. Takahashi, and Y. Takizawa, "Synthesis of monodisperse bicolored janus particles with electrical anisotropy using a microfluidic coflow system," *Advanced Materials*, **18,** 2006, 1152–1156.
18. Y. N. Xia and G. M. Whitesides, "Soft lithography," *Angewandte Chemie-International Edition*, **37,** 1998, 551–575.
19. D. Dendukuri, K. Tsoi, T. A. Hatton, and P. S. Doyle, "Controlled synthesis of nonspherical microparticles using microfluidics," *Langmuir*, **21,** 2005, 2113–2116.
20. M. Seo, Z. H. Nie, S. Q. Xu, P. C. Lewis, and E. Kumacheva, "Microfluidics: From dynamic lattices to periodic arrays of polymer disks," *Langmuir*, **21,** 2005, 4773–4775.
21. W. Jeong, J. Kim, S. Kim, S. Lee, G. Mensing, and D. J. Beebe, "Hydrodynamic microfabrication via 'on the fly' photopolymerization of microscale fibers and tubes," *Lab on a Chip*, **4,** 2004, 576–580.
22. S. Q. Xu, Z. H. Nie, M. Seo, P. Lewis, E. Kumacheva, H. A. Stone, P. Garstecki, D. B. Weibel, I. Gitlin, and G. M. Whitesides, "Generation of monodisperse particles by using microfluidics: Control over size, shape, and composition," *Angewandte Chemie-International Edition*, **44,** 2005, 724–728.
23. O. D. Velev, A. M. Lenhoff, and E. W. Kaler, "A class of microstructured particles through colloidal crystallization," *Science*, **287,** 2000, 2240–2243.
24. Y. D. Yin and Y. N. Xia, "Self-assembly of monodispersed spherical colloids into complex aggregates with well-defined sizes, shapes, and structures," *Advanced Materials*, **13,** 2001, 267–271.
25. V. N. Manoharan, M. T. Elsesser, and D. J. Pine, "Dense packing and symmetry in small clusters of microspheres," *Science*, **301,** 2003, 483–487.
26. Y. S. Cho, G. R. Yi, J. M. Lim, S. H. Kim, V. N. Manoharan, D. J. Pine, and S. M. Yang, "Self-organization of bidisperse colloids in water droplets," *Journal of the American Chemical Society*, **127,** 2005, 15968–15975.
27. D. J. Beebe, "Functional hydrogel structures for autonomous flow control inside microfluidic channels," *Nature*, **404,** 2000, 588–590.
28. P. S. Doyle, D. Dendukuri, D. C. Pregibon, J. Collins, and T. A. Hatton, "Continuous-flow lithography for high-throughput microparticle synthesis," *Nature Materials*, **5,** 2006, 365–369.
29. P. S. Doyle, D. Dendukuri, P. Panda, R. Haghgooie, J.M. Kim, and T.A. Hatton, "Modeling of oxygen-inhibited free radical photopolymerization in a pdms microfluidic device," *Macromolecules*, **41,** 2008, 8547–8556.

30. D. C. Pregibon, M. Toner, and P. S. Doyle, "Multifunctional encoded particles for high-throughput biomolecule analysis," *Science*, **315**, 2007, 1393–1396.
31. P. S. Doyle, D. Dendukuri, S. S. Gu, D. C. Pregibon, and T. A. Hatton, "Stop-flow lithography in a microfluidic device," *Lab on a Chip*, **7**, 2007, 818–828.
32. P. Panda, S. Ali, E. Lo, B. G. Chung, T. A. Hatton, A. Khademhosseini, and P. S. Doyle, "Stop-flow lithography to generate cell-laden microgel particles," *Lab on a Chip*, 8, 2008, 1056–1061.
33. A. Bertsch, J. Y. Jézéquel, and J. C. André, "Study of the spatial resolution of a new 3D microfabrication process: The microstereophotolithography using a dynamic mask-generator technique," *Journal of Photochemistry and Photobiology A: Chemistry*, **107**, 1997, 275–281.
34. L. J. Hornbeck, "Digital light processing TM for high-brightness, high-resolution applications," *Projection Displays III*, **3013**, 1997, 27–40.
35. S. Singh-Gasson, R. D. Green, Y. Yue, C. Nelson, F. Blattner, M. R. Sussman, and F. Cerrina, "Maskless fabrication of light-directed oligonucleotide microarrays using a digital micromirror array," *Nature Biotechnology*, **17**, 1999, 974–978.
36. http://www.nimblegen.com/.
37. C. Sun, N. Fang, D. M. Wu, and X. Zhang, "Projection micro-stereolithography using digital micro-mirror dynamic mask," *Sensors and Actuators A: Physical*, **121**, 2005, 113–120.
38. Y. Lu, G. Mapili, G. Suhali, S. Chen, and K. Roy, "A digital micro-mirror device-based system for the microfabrication of complex, spatially patterned tissue engineering scaffolds," *Journal of Biomedical Materials Research Part A*, **77A**, 2006, 396–405.
39. M. Klosner and K. Jain, "Massively parallel, large-area maskless lithography," **84**, 2004, 2880.
40. K. Totsu, K. Fujishiro, S. Tanaka, and M. Esashi, "Fabrication of three-dimensional microstructure using maskless gray-scale lithography," *Sensors and Actuators A: Physical*, **130-131**, 2006, 387–392.
41. G. M. Whitesides, E. Ostuni, S. Takayama, X. Y. Jiang, and D. E. Ingber, "Soft lithography in biology and biochemistry," *Annual Review of Biomedical Engineering*, **3**, 2001, 335–373.
42. A. M. Biswas, J. Li, J. A. Hiserote, and L. S. Melvin Iii, "Extension of 193 nm dry lithography to 45-nm half-pitch node: Double exposure and double processing technique," *Proceedings of SPIE*, **6349**, 2006, 63491P.
43. C. J. Hernandez and T. G. Mason, ""Colloidal alphabet soup: Monodisperse dispersions of shape-designed LithoParticles," *Journal of Physical Chemistry C*, **111**, 2007, 4477–4480.
44. P. S. Doyle, D. C. Pregibon, and M. Toner, "Magnetically and biologically active bead-patterned hydrogels," *Langmuir*, **22**, 2006, 5122–5128.
45. G. M. Whitesides and B. Grzybowski, "Self-assembly at all scales," *Science*, **295**, 2002, 2418–2421.
46. H. J. Yeh and J. S. Smith, "Fluidic self-assembly for the integration of GaAs light-emitting diodes on Si substrates," *IEEE Photonics Technology Letters*, **6**, 1994, 706–708.
47. E. J. Snyder, J. Chideme, and G. S. W. Craig, "Fluidic self-assembly of semiconductor devices: A promising new method of mass-producing flexible circuitry," *Japanese Journal of Applied Physics*, **41**, 2002, 4366–4369.
48. J. C. Hulteen and R. P. V. Duyne, "Nanosphere lithography: A materials general fabrication process for periodic particle array surfaces," *Journal of Vacuum Science Technology*, **13**, 1995, 1553–1558.
49. A. D. Ormonde, E. C. M. Hicks, J. Castillo, and R. P. V. Duyne, "Nanosphere lithography: Fabrication of large-area Ag nanoparticle arrays by convective self-assembly and their characterization by scanning UV-visible extinction spectroscopy," *Langmuir*, **20**, 2004, 6927–6931.
50. D. Dendukuri, T. A. Hatton, and P. S. Doyle, "Synthesis and self-assembly of amphiphilic polymeric microparticles," *Langmuir*, **23**, 2007, 4669–4674.
51. N. Bowden, A. Terfort, J. Carbeck, and G. M. Whitesides, "Self-assembly of mesoscale objects into ordered two-dimensional arrays," *Science*, **276**, 1997, 233–235.

52. T. L. Breen, J. Tien, S. R. J. Oliver, T. Hadzic, and G. M. Whitesides, "Design and self-assembly of open, regular, 3D mesostructures," *Science*, **284,** 1999, 948–951.
53. D. H. Gracias, J. Tien, T. L. Breen, C. Hsu, and G. M. Whitesides, "Forming electrical networks in three dimensions by self-assembly," *Science*, **289,** 2000, 1170–1172.
54. J. J. Boote, K. Critchley, and S. D. Evans, "Surfactant mediated assembly of gold nanowires on surfaces," *Journal of Experimental Nanoscience*, **1,** 2006, 125–142.
55. K. F. Bohringer, K. Goldberg, M. Cohn, R. Howe, and A. Pisano. "Parallel micro-assembly with electrostatic force fields," *International Conference on Robotics and Automation*, 1998, Leuven, Belgium.
56. A. Subramanian, B. J. Nelson, D. Lixin, and D. Bell. "Dielectrophoretic nano-assembly of individual carbon nanotubes onto nanoelectrodes," *The 6th IEEE International Symposium on Assembly and Task Planning: From Nano to Macro Assembly and Manufacturing*, 2005, Montreal, Canada.
57. X. Duan, Y. Huang, Y. Cui, J. Wang, and C. M. Lieber, "Indium phosphide nanowires as building blocks for nanoscale electronic and optoelectronic devices," *Nature*, **409,** 2001, 66–69.
58. M. Tanase, D. M. Silevitch, A. Hultgren, L. A. Bauer, P. C. Searson, G. J. Meyer, and D. H. Reich, "Magnetic trapping and self-assembly of multicomponent nanowires," *Journal of Applied Physics*, **91,** 2002, 8549–8551.
59. Y. Hsueh-An, L. Chiung-Wen, and F. Weileun, "Wafer level self-assembly of microstructures using the global magnetic lifting and localized induction welding," *Solid-State Sensors, Actuators and Microsystems, 2005, Digest of Technical Papers, TRANSDUCERS '05. The 13th International Conference on*, 2005. Seoul, Korea.
60. Y. Huang, X. F. Duan, Q. Q. Wei, and C. M. Lieber, "Directed assembly of one-dimensional nanostructures into functional networks," *Science*, **291,** 2001, 630–633.
61. W. Salalha and E. Zussman, "Investigation of fluidic assembly of nanowires using a droplet inside microchannels," *Physics of Fluids*, **17,** 2005, 063301.
62. S. A. Stauth and B. A. Parviz, "Self-assembled single-crystal silicon circuits on plastic," *Proceedings of the National Academy of Sciences*, **103,** 2006, 13922–13927.
63. J. S. Smith, "High density, low parasitic direct integration by fluidic self-assembly (FSA)," *International Electron Devices Meeting*, 2000, San Francisco.
64. U. Srinivasan, M. A. Helmbrecht, R. S. Muller, and R. T. Howe, "MEMS: Some self-assembly required," *Optics & Photonics News*, **13,** 2002, 20–56.
65. U. Srinivasan, D. Liepmann, and R. T. Howe, "Microstructure to substrate self-assembly using capillary forces," *Journal of Microelectromechanical Systems*, **10,** 2001, 17–24.
66. K. L. Scott, T. Hirano, H. Yang, R. T. Howe, and A. M. Niknejad, "High-performance inductors using capillary-based fluidic self-assembly," *Journal of Microelectromechanical Systems*, **13,** 2004, 300–309.
67. C. J. Morris, S. A. Stauth, and B. A. Parviz, "Self-assembly for microscale and nanoscale packaging: Steps toward self-packaging," *IEEE Electron Device Letters*, **12,** 2005, 792–794.
68. D. H. Gracias, J. Tien, T. L. Breen, C. Hsu, and G. M. Whitesides, "Forming electrical networks in three dimensions by self-assembly," *Science*, **289,** 2000, 1170–1172.
69. J. S. Smith, *Fluidic self-assembly of active antenna*, A. Technology, Editor. 2003, The Regents of the University of California, US 2002/0093396 A1.
70. W. Reinert. "Industrial wafer ball attach by gravity feeding in comparison to competing technologies," *International Exhibition and Conference for Electronic Packaging and Assembly Technology*, 2007, Shanghai, China.
71. Y. Lu, Y. Yin, and Y. Xia, "A self-assembly approach to the fabrication of patterned, two-dimensional arrays of microlenses of organic polymers," *Advanced Materials*, **13,** 2001, 34–37.

72. Y. Xia, Y. Yin, Y. Lu, and J. Mclellan, "Template-assisted self-assembly of spherical colloids into complex and controllable structures," *Advanced Functional Materials*, **13,** 2003, 907–918.
73. A. Terfort, N. Bowden, and G. M. Whitesides, "Three-dimensional self-assembly of millimetre-scale components," *Nature*, **386,** 1997, 162–164.
74. H. K. Wu, N. Bowden, and G. M. Whitesides, "Selectivities among capillary bonds in mesoscale self-assembly," *Applied Physics Letters*, **75,** 1999, 3222–3224.
75. W. Zheng, P. Buhlmann, and H. O. Jacobs, "Sequential shape-and-solder-directed self-assembly of functional microsystems," *Proceedings of the National Academy of Sciences of the United States of America*, **101,** 2004, 12814–12817.
76. J. C. Love, D. B. Wolfe, H. O. Jacobs, and G. M. Whitesides, "Microscope projection photolithograpy for rapid prototyping of masters with micron-scale features for use in soft lithography," *Langmuir*, **17,** 2002, 6005–6012.
77. Y. Yin, Y. Lu, B. Gates, and Y. Xia, "Template-assisted self-assembly: A practical route to complex aggregates of monodispersed colloids with well-defined sizes, shapes, and structures," *Journal of American Chemical Society*, **123,** 2001, 8718–8729.
78. M. Boncheva and G. M. Whitesides, "Templated self-assembly: Formation of folded structures by relaxation of pre-stressed, planar tapes," *Advanced Materials*, **17,** 2005, 553–557.
79. P. W. K. Rothemund, "Folding DNA to create nanoscale shapes and patterns," *Nature*, **440,** 2006, 297–302.
80. S. E. Chung, W. Park, S. Shin, S. A. Lee, and S. Kwon, "Guided and fluidic-self-assemlby of microstructures using railed microfluidic channels," *Nature Materials*, **7,** 2008, 581–587.
81. A. D. Stroock, S. K. Dertinger, G. M. Whitesides, and A. Ajdari, "Patterning flows using grooved surfaces," *Analytical Chemistry*, **74,** 2002, 5306–5312.
82. S. E. Chung, W. Park, H. Park, K. Yu, N. Park, and S. Kwon, "Optofluidic maskless lithography system for real-time synthesis of photopolymerized microstructures in microfluidic channels," *Applied Physics Letters*, **91,** 2007, 041–106.
83. W. G. Koh, A.Revzin, and M. V. Rishko, "Poly (thylene glycol) hydrogel microstructures encapsulating living cells," *Langmuir*, **18,** 2002, 2459–2462.
84. W. G. Koh, L. J. Itle, and M. V. Pishko, "Molding of hydrogel microstructures to create multiphenotype cell microarrays," *Analytical Chemistry*, **75,** 2003, 5783–5789.
85. V. A. Liu and S. N. Bhatia, "Three-dimensional photopatterning of hydrogels containing living cells," *Biomedical Microdevices*, **4,** 2002, 257–266.
86. J. El-Ali, P. K. Sorger, and K. F. Jensen, "Cells on chips," *Nature*, **442,** 2006, 403–411.
87. J. Yeha, Y. Ling, J. M. Karp, J. Gantz, A. Chandawarkar, G. Eng, J. B. Iii, R. Langer, and A. Khademhosseini, "Micromolding of shape-controlled, harvestable cell-laden hydrogels," *Biomaterials*, **27,** 2006, 5391–5398.
88. D. R. Albrecht, V. L. Tsang, R. L. Sah, and S. N. Bhatia, "Photo-and electropatterning of hydrogel-encapsulated living cell arrays," *Lab on a Chip*, **5,** 2005, 111–118.

CHAPTER 18

Reconfigurable Photonic Crystal Circuits Using Microfluidics

Christian Karnutsch, Snjezana Tomljenovic-Hanic, Christelle Monat, and Benjamin J. Eggleton

Institute of Photonics and Optical Science (IPOS), Centre for Ultrahigh-Bandwidth Devices for Optical Systems (CUDOS), School of Physics, University of Sydney, Sydney, Australia

18-1 Introduction

18-1-1 From the Infiltration of Photonic Crystals to the Concept of Reconfigurable Circuits

Tunable Photonic Crystals and the Premises of Infiltration

Photonic crystals (PhCs) represent a class of materials that display a periodic arrangement—along one, two, or three directions—of their internal dielectric structure [1]. For an appropriate choice of the dielectric constant and the PhC period (nearly equal to the optical wavelength), the propagation of light through the PhC is strongly affected. Forbidden light propagation—photonic band gaps (PBGs)—light with dramatically reduced speed (slow light [2]) or light localization are all various phenomena that become possible within PhCs and at a very compact length scale, hence the idea that PhCs can effectively "mold the flow of light" [1].

As opposed to a homogeneous medium, the peculiar dispersion of a PhC is highly dependent on the optical frequency. The abrupt spectral variations in the associated photonic band structure imply that moderate shifts in the refractive index—obtained through an external perturbation—can substantially modify the optical properties of the PhC at a particular frequency. This offers the potential for creating flexible and dynamic optical functionalities as required for many practical applications [3]. Most of the relevant properties of PhCs (such as complete PBGs) require a high index contrast that is provided for instance by air/semiconductor (like silicon or III-V). On the one hand, the tuning range provided by the direct modulation of the semiconductor refractive index is limited ($\Delta n \sim 10^{-4}$ per kelvin [4] for thermal tuning and $\Delta n \sim 10^{-4}$ per kelvin [5] for electro-optical tuning). On the other hand, the intrinsic porous structure of PhCs leaves the space to infiltrate with another material possessing larger tuning properties.

In 1999, Busch and John suggested the idea to infiltrate a PhC structure with electro-optic liquid crystals (LCs) to expand the degree of tunability of PhCs [6]. Nematic-phase LCs form an optically birefringent material, the refractive index of which can be widely ($\Delta n \approx 0.05$) and reversibly modulated by changing the orientation of the LC molecules, commonly via an external electric field. In addition, LCs possess a strong refractive index response to temperature ($\Delta n \sim 10^{-3}$ per kelvin [7]). Since the first experimental demonstrations in three-dimensional (3D) opals [8], and two-dimensional (2D) PhCs [9], many studies have reported the thermal tuning of the PBG of LC-PhC structures in a variety of materials and geometries [7,10–12]. Exploiting the LC anisotropy in PhC structures has been suggested to modify the optical modes supported by a PhC [13,14] or to modulate the refraction of light [15]. Dynamically tuning the resonance of a LC-PhC cavity with temperature has also been successfully reported [7,16–18]. The capability to modify the resonance of a cavity after its fabrication is of particular interest either for tuning the cavity properties on demand or for cavity trimming—that is, compensating for the fabrication imperfections a posteriori. The electrical tuning of LC-PhC structures has been widely demonstrated in 3D opals [19–21] as well as in PhC cavities [16,22–25]. The associated shifts (5.5 nm [20], 6 nm [16], and 1.2 nm [25]) are generally limited by surface anchoring effects of the LC molecules at the sidewalls of the nanometer-size PhC pores, although this effect could be minimized via an appropriate electrode configuration [26,27]. The integration of electronics in microfluidic PhC structures opens the path to further functionalities, like electrically tunable filters [22,27] and lasers [23–25] with a potential interest for optical interconnects [28]. Besides thermal and electrical control, the optical tuning of PhC infiltrated with a photoresponsive LC mixture has been demonstrated [29–32]. In Ref. 33, it was used to produce an optically triggered Q-switched PhC laser, where either of the two laser modes could be selected by shining the photoadressable polymer top layer with the appropriate polarization, providing a 7-nm spectral tuning range. Note that the relatively slow response time of LC-based devices (a few milliseconds through optical [29] or thermal tuning, tens of microseconds for electrical tuning [20]) will not provide rapid switching, although this could be achieved through the direct and fast tuning of the PhC semiconductor matrix [34]. The unique and large tuning range achievable via LCs is well adapted for reconfiguration applications, where one needs to widely adjust the optical response of filter devices or to switch between different functionalities or output ports [35] at low modulation frequencies. This class of devices could provide the basis for reconfigurable network (protection and restoration) applications, while LC-PhC laser arrays have been envisaged as optical reprogrammable read-only memory elements [33].

Since the premises of infiltration, other materials than LCs have been introduced in PhCs, such as liquids [36–39], organic liquids [29,31,32], polymers [40–42], nanoparticle-based composites [43], colloidal quantum dots [44–46], and fluorescent organic dyes [41,47–52]. In particular, when active materials are combined with PhC structures, their spontaneous emission can be inhibited or amplified [46,48] while stimulated emission (lasing) can be achieved [41,44,47]. This has led for instance to a new class of widely tunable microfluidic dye lasers [49–52]. The large optical nonlinearity of PbSe QDs has been envisaged for optical switching applications [45]. Because there exists a range of liquid materials featuring a wide array of optical properties, PhC infiltration opens up many different opportunities associated to the particular characteristics of the infused material [53,54].

Selective Infiltration of Planar Photonic Crystals for Reconfigurable Photonic Circuits

The idea of PhC infiltration has been expanded through the concept of selective fluid filling. In this scheme, introducing LCs into individual air pores of a planar PhC was proposed for creating various tunable photonic elements (Y-junctions, bends, waveguide intersections, and beam splitters) integrated in a PhC circuit [55,56]. Planar PhCs (see Fig. 18-1) are a particular class of periodic structures that can confine light in three directions by combining a 2D PhC lattice (typically air holes) and a 1D step index waveguide (e.g., a thin silicon slab) [57,58]. As such, their realization is compatible with the mature microelectronic fabrication techniques while they provide a suitable platform for creating a variety of optical devices that can be readily integrated onto a single chip [59]. Planar PhC components—for example, waveguides and cavities—are realized by inducing a local (linear or point) "defect" in the periodic lattice. While these defects generally consist of air-hole removal and displacement, they can be alternatively created through

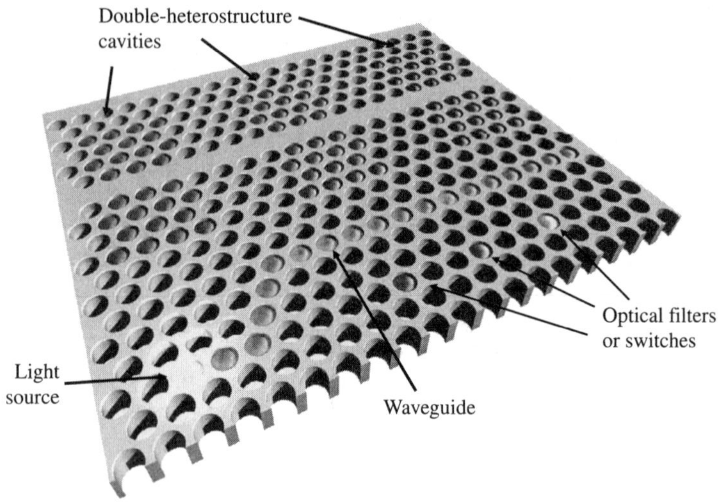

FIGURE 18-1 Schematic of a microfluidic reconfigurable photonic circuit made of a planar photonic crystal (air holes etched into a thin high-index membrane). It integrates many different components onto the same platform that are all achieved through selective infiltration of air holes along specific (point or linear) patterns and with particular liquids. The individual elements are meant to be highly tunable when specific liquid is infused (liquid crystals, photosensitive liquid, etc.) and the whole circuit is reconfigurable.

the selective infiltration of air holes with a liquid. Light beam can be effectively routed along these fluidically defined and arbitrarily written paths [56,60]. Although the resulting light confinement is shallower due to the reduction of the PhC index contrast, effective light guiding is possible even around tight bends, while the design can be made tolerant upon the missed infiltration of some intended holes [60]. Figure 18-1 shows the schematic vision of such a "microfluidically defined" photonic circuit, where the infiltrated holes would act as waveguides, microcavities, light sources, filters, or switches, depending on the infiltrated pattern and the specific properties of the infused liquid. Even more remarkably, the different components of such a photonic circuit could be either modulated based on the tunable—optical or electro-optic—properties of the fluid (see the previous section) or fully erased and then rewritten due to fluid mobility. This flexible platform would therefore provide a dynamic control over guiding the flow of light as well as full reconfigurability of the resulting photonic integrated circuit.

The actual breakthrough toward rewritable PhC circuits was achieved with the recent advances in the development of controlled microinfiltration techniques [61,62] for locally infiltrating selected air holes of a planar PhC. This was performed using an elongated micropipette (diameter < 1 μm [61]), or tapered microtip (diameter ~220 nm [62]), controlled through micropositioning and with a size being comparable to the targeted PhC air holes. It has been shown that the infused liquid could be completely removed [61,63], providing the basis for full reconfigurability. Another "writing" method that has been demonstrated relies on the selective polymerization of UV-curable polymers infiltrated into the PhC air holes by locally illuminating the intended PhC region with a focused argon-ion laser [64]. In parallel, the progress in optofluidic integration has led to the capability of addressing selected air holes using an actual microfluidic circuit, controlled with valves and pumps, and bonded on top of the PhC structure [65,66], thus allowing sophisticated, hybrid functionality.

18-1-2 Optofluidics and Planar Photonic Crystals

Increasing the Liquid-Light Interaction and Sensing Applications

Among the attractive properties of periodic PhC structures lies the ability to tightly confine light within highly compact microcavities [67,68]. Planar PhC microcavities in particular represent a versatile platform for realizing various and small-scale optical components such as low-threshold lasers [69,70], optical switches [5,34], narrow filters [71], and slow light structures [72] that can be all integrated onto the same chip. When combined with quantum emitters, they also find applications within quantum electrodynamics and quantum information processing [73,74]. For this wide range of applications, design rules generally aim at generating high Q-factors and small modal volumes ($V < \lambda^3$) to trap light for a long time and in a tiny fraction of space [75,76].

In the context of optofluidics [77,78], the strong optical confinement within planar PhC microcavities can enhance the interaction between light and the material (gas or liquid) that is infiltrated into their air pores. Besides, they can be designed so as to increase specifically the mode field intensity that overlaps with the (infused) pores [25,33]. These properties along with the device compactness have driven an entire research field dedicated to the use of PhC microcavities for chemical and biosensing applications—without requiring radioactive or fluorescent labels [79]. The cavity resonance is highly sensitive to the properties of the surrounding environment, providing the basis for detection in most proposed sensor schemes [80]. While a large liquid-light interaction and high Q-factors are required for improving both the sensitivity and the limit of detection of the sensor, there is generally a trade-off between these two parameters [80]. Firstly, the cavity Q-factor tends to decrease after the infiltration step due to the reduction of the PhC index contrast

[16,33,36,38]. Also, the liquid may absorb light, which limits the degree to which increasing the light-liquid interaction can improve the sensor performance (through a larger sensitivity) until the Q-factor of its resonance becomes critically degraded (compromising the sensor resolution) [80].

There have been numerous studies about the use of planar PhC microcavities for sensing applications based on the quasi-linear dependence of the resonance wavelength (λ_0) on the refractive index of the surrounding environment (n_{fluid}). The reported sensitivities ($\Delta\lambda_0/\Delta n_{fluid}$) range between 200 nm/RIU [81] and 350 nm/RIU [82], while larger values (512 nm/RIU) have been predicted in sophisticated PhC microcavity designs [83]. The device compactness permits the detection of optical changes within femtolitre volumes of analytes [36,82] and potentially at a low concentration level. In that respect, the sensor resolution is crucial. Since the first demonstrations where the minimum detectable index change (Δn) was ~0.002 in passive (and moderate $Q \approx 400$) PhC microcavities [81], the resolution could be notably improved by employing PhC microlasers ($\Delta n < 0.001$ [36] and $\Delta n \approx 9{\cdot}10^{-5}$ [82]) or optimized cavity designs ($\Delta n{\sim}7{\cdot}10^{-5}$ [84]).

PhC microcavities are also useful for biomolecule detection and recognition, provided that a layer with appropriate receptors is coated onto the device air pores. Essentially, the specific capture of biomolecules at the sensor surface is detected from the induced small changes in the local refractive index. Functionalized 2D PhC microcavities ($Q \sim 7000$) have been employed to monitor the binding of a protein monolayer (0.7 nm), as well as to detect the selective attachment of targeted proteins [85]. Using a specific polymer-film coating, PhC microcavities have also been applied to the detection of micromolar ion concentrations in solution [86]. As a first proof of concept, planar PhC microcavities have shown promise for probing single particles through the detection of a single 370-nm latex sphere, the size of which is comparable to various viruses of interest [87]. The small interrogation volume makes PhC microcavities more sensitive to extensive properties like the protein mass, with a record detection of 35 attograms (ag) [84].

Besides PhC microcavities, dispersive PhC waveguides have also received significant attention in the context of miniaturized (refractometry and absorption) spectroscopy [88–94]. In these devices, increasing the light-matter interaction can be achieved through optimized PhC waveguide designs [89,95] or slow light-based PhC structures [90–93,96]. Reducing the speed of light holds the promise of increasing the effective interaction length with the analytes without compromising the device compactness. The associated studies also cover the terahertz (microwaves) range [95–97] where many biomolecules have specific fingerprints.

Device Multiplexing and Integration with Microfluidic Networks

Because of their compactness and 2D geometry, planar PhC components are conducive to dense photonic integration. In the context of sensing applications, recently there have been strong efforts to integrate many compact PhC sensors onto a single chip in order to increase the device throughput, through parallel- and multianalyte-detection schemes. This is particularly relevant for lab-on-a-chip technology, in which many analytical functions are miniaturized and integrated onto a platform for both diagnostic and biochemical detection purposes. In some of the demonstrated approaches, the readout system consists of scanning the chip under free-space illumination and detecting the reflection spectra of each of the multiple PhC sensing areas to form a 2D map of the device [98–100]. Another approach relies on a dense array of PhC microlaser-based sensors having predetermined and slightly distinct spectral signatures so that the measurement of the collective spectrum provides the response of the individual sensors simultaneously [66,82]. A first demonstration of this concept has been achieved with four nanolasers of a size that would enable the integration of 1000 devices within a 340 × 340 μm^2 area [82]. In addition, the laser configuration does not necessarily require a spectrometer as the "reading" can be performed through imaging the laser spots (number and position) through a specific band-pass filter. Another multiplexed sensor platform has been demonstrated where an array of slightly detuned PhC microcavities are side coupled along a single bus waveguide [84]. In this configuration, the individual responses of the multiple detection sites can be read at the same time by analyzing the transmission through the bus waveguide.

However, the performance of such dense photonic platforms implies the additional integration of a sophisticated microfluidic network to address individually the different optical components in a flexible manner. In the case of sensing, this is essential to precisely deliver the targeted molecules onto the distinct sensing areas and to effectively lower the sample volumes. Integrating microfluidic channels also enables kinetic measurement of chemical reactions [100]. The full integration of PhC platforms with microfluidic networks has been recently reported [38,65,66,84,99,100] using two basic approaches. The first one relies on a two-level platform of microfluidic and photonic functions that is obtained after aligning and bonding two separate microfluidic and photonic chips together [38,39,65,66,84]. In the second case, both microfluidic and photonic functionalities are fabricated during the same process providing a self-aligned integrated platform [99–101].

In brief, generic microfluidic PhC circuits offer a dual application for (a) chemical and biosensing in the broader context of the lab-on-a-chip concept, and (b) dynamic and reconfigurable complex photonic chips. Note that both applications face the same issues, which are inherent to their hybrid microfluidic-photonic nature. In particular, increasing the light-liquid interaction through PhC design optimization in the context of sensing is also relevant for producing widely tunable photonic functions. In addition, the mentioned advances on PhC sensors related to photonic dense integration and integration with microfluidic networks are also significant for reconfigurable microfluidic-photonic circuits.

Although one long-term goal of the present studies remains the realization of reconfigurable photonic circuits, as emphasized through the title of this chapter, the following sections will be mainly focused on both the theoretical and experimental works carried out on a particular class of planar PhC microcavities, namely, microfluidic double-heterostructures (DH). Besides the versatility of planar PhC microcavities in general, these particular components possess unique properties that are attractive to optofluidics. In particular, we will show that a PhC resonator can be directly created by infusing a liquid into any section of a uniform PhC waveguide [102]. This self-aligned approach, which exploits the microfluidic equivalent of the DH concept [103], relaxes the constraint on both the fabrication and infiltration accuracies while ensuring the interaction between the confined light and the infused liquid. As such, we believe that these components will play a central part in achieving the photonic circuit depicted on Fig. 18-1, and represent one of the first milestones toward the realization of reconfigurable PhC circuits using microfluidics.

18-2 Designing High-*Q* Cavities Using Air-Hole Infiltration

A PhC slab cavity is usually formed in either of the two ways: forming a point cavity or forming a DH. Double-heterostructures are composed of regions of slightly different PhCs in a single slab (see Fig. 18-2). These structures can be formed in many different ways but they all rely on an increase of the average refractive index within the central PhC_2, compared to PhC_1. This has the effect of shifting the band-structure features to lower frequencies. Therefore, the waveguide, introduced across the PhC slab, has a lower dispersion curve within PhC_2 than in the surrounding PhC_1. Both curves are within the same photonic band gap, but there is a gap between them. If the resonant frequency falls within this mode-gap the mode propagates in PhC_2 waveguide and is evanescent in PhC_1 waveguide. The part of the waveguide within PhC_2 then acts as a cavity due to the mode-gap effect [103]. The highest measured quality factors in PhC slabs were achieved using this type of cavity [104,105]. In these designs, PhC_2 is

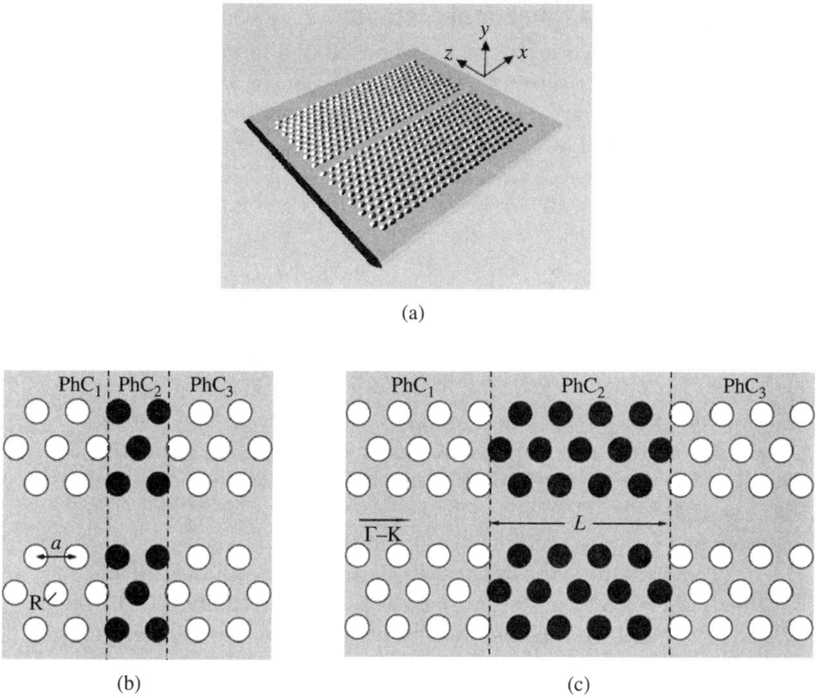

Figure 18-2 (a) Schematic of PhC slab with a W1 waveguide in the Γ-K direction and refractive index distribution in the plane of the structures considered (b) $m = 1$ and (c) $m = 4$. (*S. Tomljenovic-Hanic, C. M. de Sterke, and M. J. Steel, "Design of high-Q cavities in photonic crystal slab heterostructures by air-holes infiltration," Opt. Express 14(25), 12451–12456 (2006).*)

formed either by longitudinal [104] or lateral [105] hole displacement so that the air-filling factor decreases, thus increasing the average refractive index. However, these designs need to be finalized at the fabrication stage and they rely on extremely precise control of holes' size and position through nanolithography techniques. There are other DH designs that take advantage of the postprocessing techniques that do not require any change in the geometry of the regular structure [102,106,107].

One way to induce the refractive index change is air-hole infiltration of the central part of the homogenous structure [102]. For example, the air in the holes of PhC_2 can be replaced with material of refractive index $n > 1$. We consider materials having refractive index in the range $n = 1.1$ to 1.7, such as liquids [61,62], liquid crystal (LC) [12,25], polymer [32,40], or nanoporous silica [108]. As discussed in the Sec. 18-1, infiltration of PhC slabs with liquids, LC, and polymers have been demonstrated experimentally.

18-2-1 Model and Numerical Methods

Our numerical model is a PhC slab composed of a hexagonal array of cylindrical air holes in a silicon slab, as illustrated in Fig. 18-2a. The structure has holes of radius R, a is the lattice constant, and h is the thickness of the slab. Across the PhC slab there is a line defect, a waveguide, in the Γ-K direction. A W1 waveguide, formed by omitting one row of the airholes, is used in almost all numerical simulations presented here unless stated otherwise. In order to optimize the experiment, we briefly consider the effect of using a slightly narrow waveguide.

We start with a homogeneous slab, as illustrated in Fig. 18-1a, and design the DH by changing the holes' refractive index in the central region of the slab (indicated by the darker circles in Fig. 18-1b and 18-2c).

First we consider a silicon-based ($n = 3.4$) PhC slab that is infinite in the plane in order to obtain PBGs and associated eigenstates of a waveguide introduced in the Γ-K direction. As the second step in the design, a finite PhC slab, with $25a$ in the x direction and $25a$ in the z direction, is considered with the cavity in the center.

For both structures the hole radius is $R = 0.29a$ and the thickness of the slab is $h = 0.6a$. We start our analysis with the cavity illustrated in Fig. 18-2b. Next we consider structures that have longer cavities as illustrated in Fig. 18-2c. The PhC_2 length is denoted by L, $L = ma + 2R$, where m is an integer.

In order to design a cavity, two numerical methods are used: the 3D plane-wave expansion method for the PBG calculations and associated eigenstates of the photonic crystal waveguide, and the 3D finite-difference time-domain (FDTD) method, combined with techniques of fast harmonic analysis [109] for the quality factor calculations. This method exploits the knowledge that for a signal consisting of one or a few resonant modes, the electric field at an arbitrary point as a function of time can be represented as a sum of complex exponentials. By projecting the signal onto a Fourier basis in a narrow range around the resonant frequency, the complex frequencies can be found to very high accuracy, much greater than would be extracted from a standard Fourier transform. The error in the complex frequency is dominated entirely by the spatial grid resolution rather than the length of the simulation. The numerical parameters such as grid size, perfectly matched layer (PML) width, and height of the computational window strongly affect the convergence. In most calculations, the PML width is $2a$ and the height of the computational window is $4a$. The grid size that provides satisfactory convergence depends on the quality factor. For $Q \sim 10^5$, 28 points per period suffice, whereas 32 points per period are needed when $Q \sim 10^6$. The resonant mode's volume is

$$V = \iiint U dV / \max(U) \tag{18-1}$$

where $U = \varepsilon |E|^2 / 2$ is the electric energy density. The in-plane and out-of-plane quality factor components are obtained through post-processing that involves the use of power monitors.

18-2-2 Numerical Results

Mode-Gap

The concept of the cavity design in heterostructures relies on the mode-gap effect [103]. Therefore, we first examine if there is a sufficient mode-gap between structures having materials other than air within the holes. In Fig. 18-3a we plot the dispersion curves for the regular structure (PhC_1) and infiltrated structure (PhC_2). Both structures have two guided modes below the light line in the lowest PBG, one in the middle of the band gap and the other one in the lower part of the band gap. The lower mode is the mode of interest because it provides high-Q cavities [102]. The dispersion curves of this mode for the regular structure, PhC_1, and PhC_2 where air holes are infiltrated with material having refractive index $n = 1.5$, are plotted in Fig. 18-3a. In the same figure the lower band edge is denoted both for the regular and modified structure. In practice, the high-Q cavity modes that originate in the mode-gap can be excited by using evanescent coupling from a fiber taper directly to the cavity [110,111]. Therefore, along with the numerical results, we plot a dispersion curve of the tapered fiber used to excite the cavity modes in our experiments.

Obviously, filling the holes with a material of higher refractive index than air increases the refractive index of the structure as a whole and consequently lowers the dispersion curve. The gap between these dispersion curves, measured at the edge of the Brillouin zone, is $\Delta\tilde{\omega} = 3 \times 10^{-3}$, $\tilde{\omega} = \omega a / 2\pi c$. The size of the mode-gap is comparable with the mode-gap of the DH formed of different lattice constants PhCs [103]. However, there is another important factor for the design of high-Q cavities and that is the relative position of the mode-gap within the PBG [102]. The mode-gap should not be too close to the PhC band edge as it is the case for the W1 waveguide. We consider a waveguide with the two PhC sections to either side of the PhC waveguide shifted closer together, with a waveguide width that is 0.9 times the width of a W1 waveguide, hence it is called W0.9 waveguide. As shown in Fig. 18-3b both dispersion curves move up in the PBG as the waveguide width is reduced. This modification increases the frequency range between the fundamental waveguide mode and the low-frequency edge of the PhC band gap. The mode-gap corresponding to the infiltrated W0.9 is not too close to the PBG edge and furthermore allows for a broad selection of fluid indices to configure our devices. In our experiments it is also noticed that evanescent coupling to the structure is improved by using waveguides narrower than W1 [63].

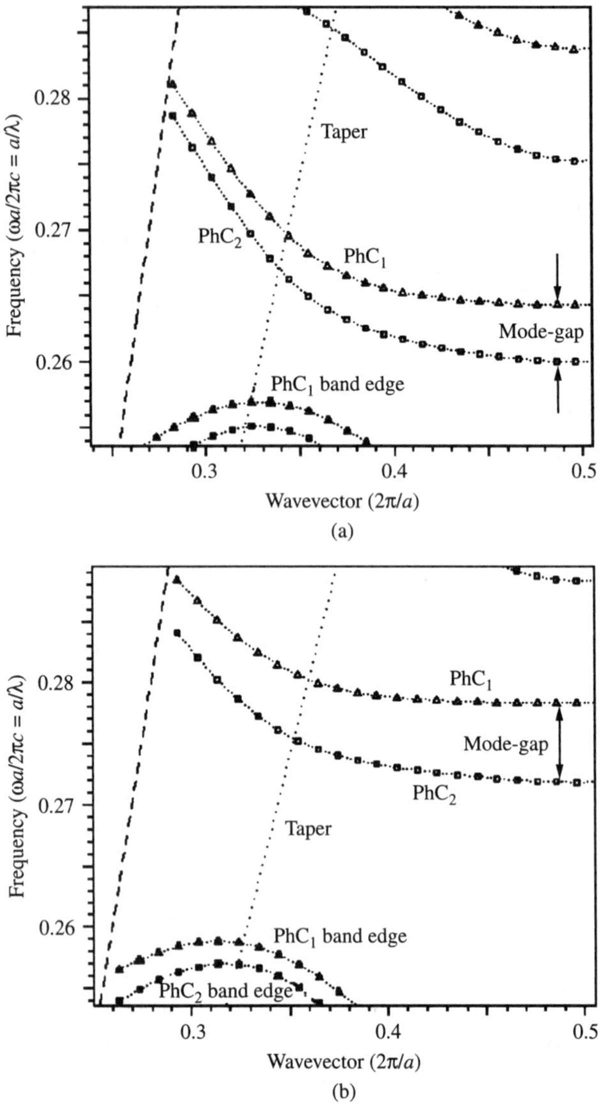

FIGURE 18-3 Dispersion curves for (a) W1 and (b) W0.9 waveguide within the region of the lowest band gap of the noninfiltrated structure PhC_1 (empty triangles) and of the PhC_2 (empty rectangles), n_{holes}=1.5; the dashed line represents the light line, and the dotted line represents the dispersion curve of the evanescent coupling taper used in the experiment with 1.3-μm diameter.

Cavity Design

Now we combine PhC_1 and PhC_2 in a single structure and evaluate the properties of the cavity, PhC_2 waveguide, using the FDTD method. First we calculate the quality factors for the structure shown in Fig. 18-2b. The refractive index of the holes in PhC_2 is varied between $n = 1.1$ and $n = 1.7$. The results for the quality factors and modal volumes of the resonant modes are plotted in Fig. 18-4a. The maximum quality factor of $Q = 2.5 \times 10^5$ appears at $n = 1.4$. As the holes' refractive

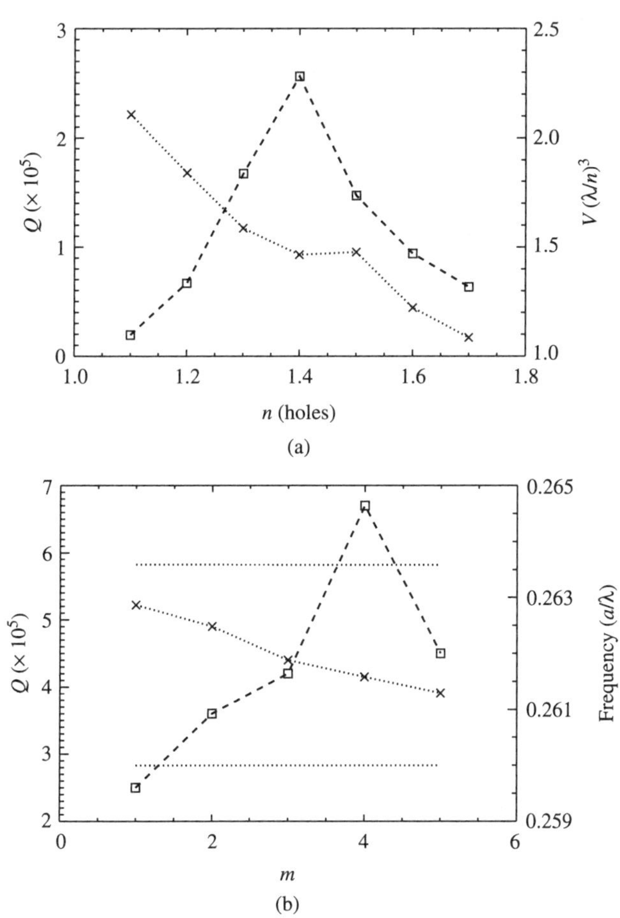

FIGURE 18-4 (a) Quality factor Q (rectangles) and modal volume V (crosses) as a function of the refractive index of the central holes for $m = 1$; (b) quality factor Q (rectangles) and resonant frequencies (crosses) as a function of the number of periods within the cavity m, for fixed $n_{holes} = 1.4$. (*S. Tomljenovic-Hanic, C. M. de Sterke, and M. J. Steel, "Design of high-Q cavities in photonic crystal slab heterostructures by air-holes infiltration," Opt. Express 14(25), 12451–12456 (2006).*)

index is increased, the average refractive index of the structure increases. This results in better out-of-plane confinement and therefore smaller out-of-plane losses, increasing Q. However, at $n = 1.4$ Q starts to decrease. This happens because the dispersion curves for higher refractive indices shift lower while the lower band edge for PhC_1 is fixed. Consequently with increasing the index, the dispersion curve of PhC_2 approaches the lower band gap edge of PhC_1. This further confirms that the relative position of the mode-gap within the PBG is an important factor when designing high-Q cavities. Even with the unfavorable refractive index, it is still possible to engineer the optimal position by adjusting the waveguide width as explained in the previous section. This additional degree of freedom allows for the use of the refractive indices that are larger than $n = 1.4$. However, the Q-factor can be significantly smaller for narrow waveguides. For example, we compare the quality factors of the W1- and W0.7-based cavities with otherwise same parameters, $n_{holes} = 1.3$ and $m = 4$. The quality factor of the W1-based cavity decreases from $Q = 7.6 \times 10^5$ to $Q = 5.2 \times 10^4$ for the W0.7-based cavity.

It is worth pointing out that there is a large range of refractive indices, $n = 1.25$ to 1.6, where the quality factors are of the order of 10^5. This coincides with the refractive indices of liquids [61,62], polymer materials [112], liquid crystals [25], and nanoporous silica [108].

The results for modal volumes of these resonances, expressed in $(\lambda/n)^3$ with $n = 3.4$, are also plotted in Fig. 18-4a. As the refractive index in the central holes increases, the modal volume decreases from $V = 2.11\ (\lambda/n)^3$ to $V = 1.17\ (\lambda/n)^3$. This is expected behavior as the resonant mode becomes better confined with the increased difference between the two PhCs. The ratio Q/V, important for many applications, still has the maximum at $n = 1.4$.

Next we investigate the effect of the cavity length on the quality factor and modal volume. Filling more holes changes the cavity length; an example is illustrated in Fig. 18-2c. We calculate quality factors for the cavities $L = ma + 2R$, where $m = 1, 2, \ldots, 5$. The results for the fixed refractive index $n = 1.4$ are shown in Fig. 18-4b. Up to $m = 4$, increasing the length increases the quality factor with the maximum exceeding $Q = 6 \times 10^5$. The drop in the quality factor in Fig. 18-4b is due to the decrease of the in-plane component that can be changed by increasing the size of the PhC slab in the waveguide direction. This effect of decreased quality factor for the longer cavities is also observed in our experiments (see Fig. 18-14). The modal volume does not change significantly with m, as the field is mainly concentrated in the central part of the cavity.

The resonant frequencies are plotted in Fig. 18-4b and the mode-gap edges are indicated by the horizontal dotted lines. As the refractive index is fixed, the mode-gap that ranges from $\tilde{\omega} = 0.2636$ to $\tilde{\omega} = 0.2607$ does not change as m changes. The resonant frequency for $m = 1$ occurs just below the upper mode-gap edge. As m increases the

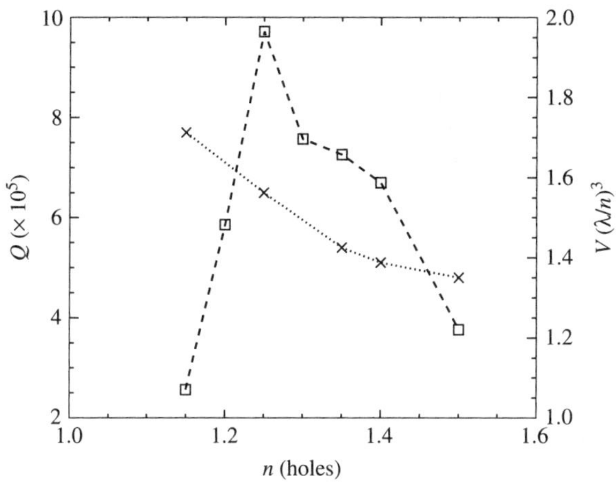

FIGURE 18-5 Quality factor Q (rectangles) and modal volume V (crosses) as a function of the refractive index of the central holes for $m = 4$. (*S. Tomljenovic-Hanic, C. M. de Sterke, and M. J. Steel, "Design of high-Q cavities in photonic crystal slab heterostructures by air-holes infiltration," Opt. Express 14(25), 12451–12456 (2006).*)

frequency crosses over the mode-gap almost linearly, passing the mid mode-gap closest to $m = 3$. Consequently, for the longer cavities there is a possibility of inducing more than one mode within the mode-gap as it is observed in our experiments (see Fig. 18-9b). Note again that there is a large range of the cavity length where the quality factor exhibits high values. This effect is also observed in our experiments (see Fig. 18-14).

Now we fix the cavity length at $m = 4$, see Fig. 18-2c, and vary the holes' refractive index in the range $n = 1.15$ to 1.5. The results are shown in Fig. 18-5. The maximum value of $Q = 9.7 \times 10^5$ is achieved at $n = 1.25$. In practice this structure can be attained by filling the holes with nanoporous silica [108]. If liquids, polymers or LCs are used; the quality factor decreases but still remains high. For example, filling the holes with a liquid (such as water) having a refractive index of $n = 1.3$ provides a high-Q cavity $Q = 7.6 \times 10^5$. The modal volume plotted in the same figure decreases as the holes' refractive index increases, as is the case for the cavity that consists of one period. The modal volume that corresponds to the maximum Q is $V = 1.56\ (\lambda/n)^3$.

We compare these results with the results presented in Fig. 18-4a. The maximum occurs at different refractive index values, for $m = 1$ at $n = 1.4$ and for $m = 4$ at $n = 1.25$. However, there is no contradiction as the resonant frequencies are very close, for $m = 1$, $\tilde{\omega} = 0.2628$ and for $m = 4$, $\tilde{\omega} = 0.2625$, in other words, in both cases close to the mid mode-gap. As expected, elongating the infiltrated region lowers the resonant frequency.

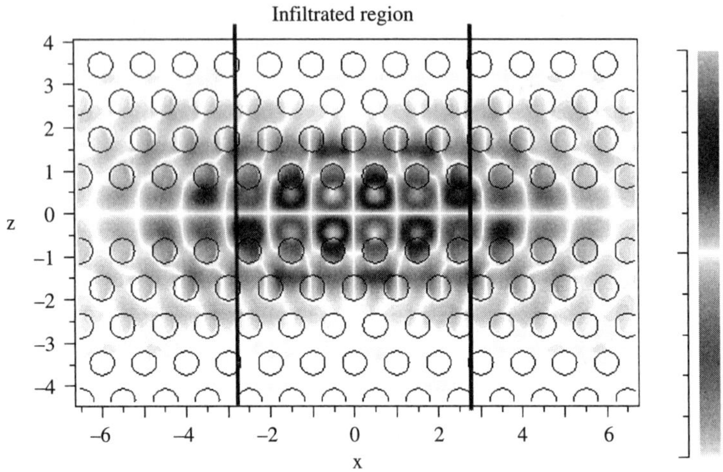

FIGURE 18-6 The calculated cross section of the major electric field component amplitude, E_x, of the resonant mode in the lower mode-gap. Circles indicate the holes; the infiltrated region is denoted by the vertical lines. (See also color insert.)

For many applications, in particular for PhC cavity sensing applications, it is important to have a large overlap of the field and the sample in addition to high-*Q*s [91]. In our case the sample resides within the holes. Therefore, in Fig. 18-6 we plot the major electric field component, E_x, at the center of the PhC slab for $m = 4$ and $n_{holes} = 1.25$. The circles in the figure represent the holes and the infiltrated region is denoted by the vertical lines. The electric field is symmetric in the *y*-direction and antisymmetric in the *x*- and *z*-directions. It is mainly concentrated in the high-index region. This is not surprising as the mode is close to the lower band gap edge, the dielectric band, where the field is mainly localized within the dielectric [1].

18-2-3 Discussion—Theory

The use of liquids, polymers, and liquid crystals for a point-cavity design decreases the quality factor because of the weaker vertical confinement that increases the out-of-plane losses [25,36]. On the other hand, a DH-type cavity formed by air-hole infiltration enables ultrahigh-*Q* heterostructures. This happens because it allows for the *mode-gap* operation that relies on the refractive index perturbation. An additional advantage of the heterostructure-cavity design over the point-cavity design is the large range of parameters that provide high-*Q* cavities. Even though the modal volume *V* is slightly higher for heterostructures than for the point cavities, the Q/V ratio remains much larger.

We designed high-Q cavities with quality factors that are comparable with those obtained for heterostructures with geometry variation [103]. The main advantage of our design is that it does not require changes in the geometry with nanometer precision [102]. The processing of air-hole infiltration can be done at any time after fabrication. If the structure is filled with LC, electro-optic, or nonlinear polymer, there is also the possibility of tuning these structures when voltage is applied. Quality factors of order $Q{\sim}10^6$ can be obtained by filling the holes in the central region of the homogenous PhC slab with nanoporous silica. The maximum values of this design achievable by using polymer materials or LCs are higher than $Q = 7 \times 10^5$.

18-3 Microfluidic PhC Components

As outlined in Sec. 18-1, microfluidic PhC devices exploit the characteristics of liquids to achieve a dynamic manipulation of their optical properties. The use of liquids allows for the optical functionalities of PhC structures to be generated, reconfigured, or tuned. In the following subsections, we will introduce a microinfiltration method (Sec. 18-3-1) that we employed to fill air pores of the PhCs. We will elucidate the evanescent coupling technique used to perform optical characterization of the fabricated microfluidic PhC components (Sec. 18-3-2), and then highlight the benefit of our approach with specific reference to microfluidic optical microcavities (Sec. 18-3-3).

18-3-1 Infiltration Method

Infiltrating PhC air pores with typical diameters of less than 300 nm is a big challenge. At these small dimensions, interface forces such as surface tension and capillary forces are the dominating factors governing the infiltration of liquids into the holes. Hence, care has to be taken in the choice of combination of substrate material and liquid and their respective wetting properties. If the liquid does not wet the surface, it will sit on top of the holes without infiltrating. If the liquid does wet the surface but has a low viscosity, it will flood the structure, and no controlled infiltration process will be achievable. It is possible to change and control the wetting properties of the surface and liquid to optimize the infiltration process, for example, by applying an oxygen or nitrogen plasma treatment to the surface or adding a surfactant to the liquid [113,114].

Our microinfiltration technique (illustrated in Fig. 18-7) uses a tapered glass microtip with an apex diameter of $\varnothing \approx 220$ nm. The microtip is fabricated by heating a borosilicate glass capillary while pulling on both ends (Micropipette Puller Sutter Instruments P-97). The heated glass softens and the cross-sectional area decreases until the two ends of the capillary separate, yielding two pointed microtips.

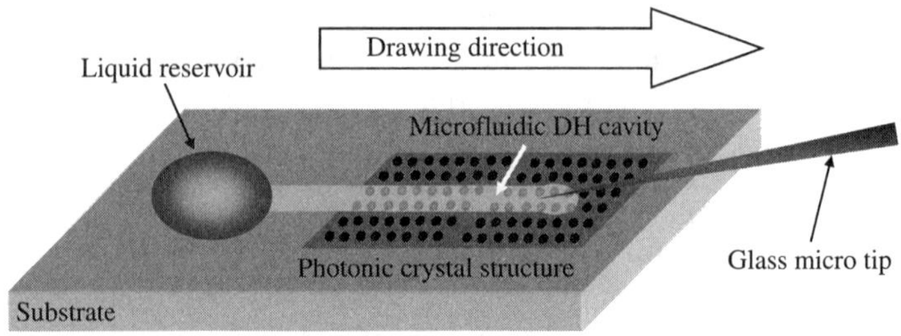

FIGURE 18-7 Schematic of our liquid infiltration process: a glass microtip is immersed into a liquid and is then drawn across a PhC structure to create a microfluidic optical component, in this example a microfluidic DH cavity. (*U. Bog, C. L. C. Smith, M. W. Lee, S. Tomljenovic-Hanic, C. Grillet, C. Monat, L. O'Faolain, et al., "High-Q microfluidic cavities in silicon-based 2D photonic crystal structures," Opt. Lett. 33(19), 2206–2208 (2008).*)

The movement of the microtip during the infiltration is controlled by a piezo-actuated translation stage (Thorlabs NanoMax-HS equipped with DRV181 actuators) with a positioning accuracy of ±20 nm. The microtip is initially inserted within a meniscus of the infiltration liquid. Both polar liquids (such as water, ethanol, and acetone) and nonpolar liquids (such as toluene, chloroform, and microscopy immersion oil) can be used to infiltrate the structure, offering a wide range of refractive indices and wetting properties.

When the microtip is withdrawn from the liquid meniscus, droplets remain attached along its length due to adhesive forces between the glass and the liquid. These droplets are then deposited on the substrate in close proximity to the PhC structure of interest; this process is monitored with a microscope (Olympus BX61). During the infiltration step, a 100× objective (0.8 NA, working distance 3.4 mm) is used, offering the best compromise between the large magnification required to resolve the PhC structure and a sufficient working distance to allow for the insertion of the microtip. Lastly, the microtip is used to draw a chosen droplet across the PhC area to create infiltrated regions where the liquid enters the holes by capillary action. We note that it is possible to fill single holes using a slightly modified technique, whereby the microtip is not drawn across the PhC but is brought in contact with the intended hole to infiltrate.

18-3-2 Evanescent Coupling

Coupling to PhC optical components—such as waveguides or microcavities—is a challenging task due to the very small mode-field dimensions of these components. To facilitate an optical characterization of the fabricated microfluidic PhC components, we employ an evanescent coupling technique using a silica nanowire [110,111,115–121].

Evanescent Coupling Setup

For the evanescent coupling, we use a silica fiber that has a tapered region where its diameter has been reduced to less than 1.5 μm. The complex fabrication process of this silica nanowire is explained in more detail in the next section. Due to the reduced dimensions of the nanowire, the electromagnetic field of the propagating mode extends significantly beyond the boundary of the wire, allowing its evanescent field to interact with the PhC structure. Coupling between the nanowire and the PhC modes can occur when phase matching is achieved [111,122]. Examples of resulting transmission spectra are displayed in the experimental section (Sec. 18-3-3) on page 25.

In our experimental setup (Fig. 18-8), light from a broadband source (Agilent EELED83437A or Fianium Femtopower1060SC450) is launched into a single-mode silica fiber connected to the nanowire. The transmission spectrum through the nanowire is recorded with an optical spectrum analyzer. The nanowire is aligned to the PhC structure using a nanopositioning setup (Luminos) and a CCD-based imaging system (Navitar).

Nanowire Fabrication

The nanowire for the evanescent coupling is fabricated using a standard single-mode silica fiber (SMF-28). The fiber is held under controlled tension in a computer-driven taper assembly while a butane flame heats the fiber locally [123]. The diameter of the fiber is then adiabatically [124] reduced from 125 to less than 1.5 μm over a length of approximately 2 mm. The finished nanowire is glued onto a microscope glass slide for mechanical support. Typical nanowires manufactured in this way present practically no insertion loss and have a transmission loss of the order of 0.1 dB (at 1550 nm). The nanowires have an induced shape to localize coupling to the micronscale liquid-filled sections of the PhC. This is achieved by bringing the wire ends together after it has been tapered, twisting one end and then stretching the nanowire ends apart. This forms a "loop" shape with a radius

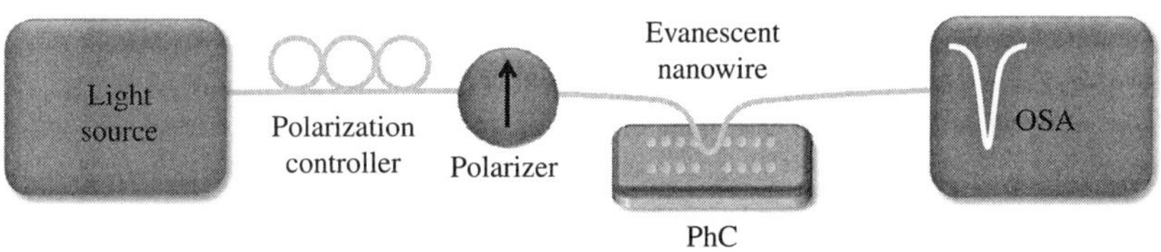

FIGURE 18-8 Schematic of the evanescent coupling setup. A polarization controller and polarizer select TE-like light from the broadband light source. The evanescent nanowire couples light to the photonic crystal (PhC) sample and its end is connected to an optical spectrum analyzer (OSA) for monitoring the transmission signal. (*C. L. C. Smith, U. Bog, S. Tomljenovic-Hanic, M. W. Lee, D. K. C. Wu, L. O'Faolain, C. Monat, C. Grillet, T. F. Krauss, C. Karnutsch, R. C. McPhedran, and B. J. Eggleton, "Reconfigurable microfluidic photonic crystal slab cavities," Opt. Express 16(20), 15887–15896 (2008).*)

of approximately 30 μm, and the looped nanowire is then annealed with a flame to fix the induced shape. After looping, the transmission loss increases to ≈1 dB (at 1550 nm), and the annealing process adds another ≈1 dB to the total transmission loss. The loop is then pried open, leaving a "u"-shaped profile. This nanowire shape has been found to present the best compromise between transmission loss, mechanical stability during measurements, and localization of the evanescent coupling [111].

18-3-3 Microfluidic Cavities

As outlined in Sec. 18-1, PhC microcavities represent a versatile platform for realizing various micron-scale optical functionalities. In this section, we show that PhC microcavities can indeed experimentally be created by infusing a liquid into a selected section of a uniform PhC waveguide (see also Sec. 18-2-1).

In order to provide a proof-of-concept demonstration, we infiltrated a chalcogenide glass-based PhC W1 waveguide with a liquid, achieving a DH microfluidic cavity with a quality factor of $Q \approx 4300$ [62]. The moderate quality factor of these cavities can be mainly attributed to high fabrication-related propagation losses in the chalcogenide PhC waveguide [125].

To improve this result, we have reverted to silicon as the PhC background material. In addition to the larger refractive index of silicon ($n_{silicon} = 3.52$, $n_{chalcogenide} = 2.68$ at 1550 nm), it also offers a highly mature PhC fabrication technology, which results in significantly reduced waveguide-propagation losses and hence the potential for higher cavity quality factors. For our experiments, we use suspended silicon membranes with a slab thickness of 220 nm ($0.537a$), into which a triangular PhC lattice with nominal period of $a = 410$ nm and hole diameter of $2R = 265$ nm ($0.646a$) has been etched. The PhC structures are typically 34 periods (14 μm) wide and 61 periods (25 μm) long. The fabrication process of these PhCs is detailed in Ref. 126. We investigated a W0.9 slab waveguide geometry, that is, a W1 waveguide—formed by omitting a single row of holes in the Γ-K direction—where the PhC structure has been shifted inward such that the waveguide width is only 90% of a regular W1 (see also Fig. 18-2). Figure 18-9 displays measured transmission spectra from infiltration experiments that we obtained via evanescent coupling. It is important to note that the nanowire was not in contact with the PhC structure [63]. First, we took a reference transmission spectrum of the waveguide before infiltration (Fig. 18-9a). We then started with infiltrating a small DH cavity (originally 2 μm long) and incrementally increased the infiltrated region on the same PhC waveguide structure in steps of ~2 μm (see Fig. 18-9b). We imaged the infiltrated cavities with a microscope objective (150×, NA 0.9, working distance 1.0 mm) using a color temperature conversion filter to improve resolution. The resulting images are displayed in the insets of Fig. 18-9.

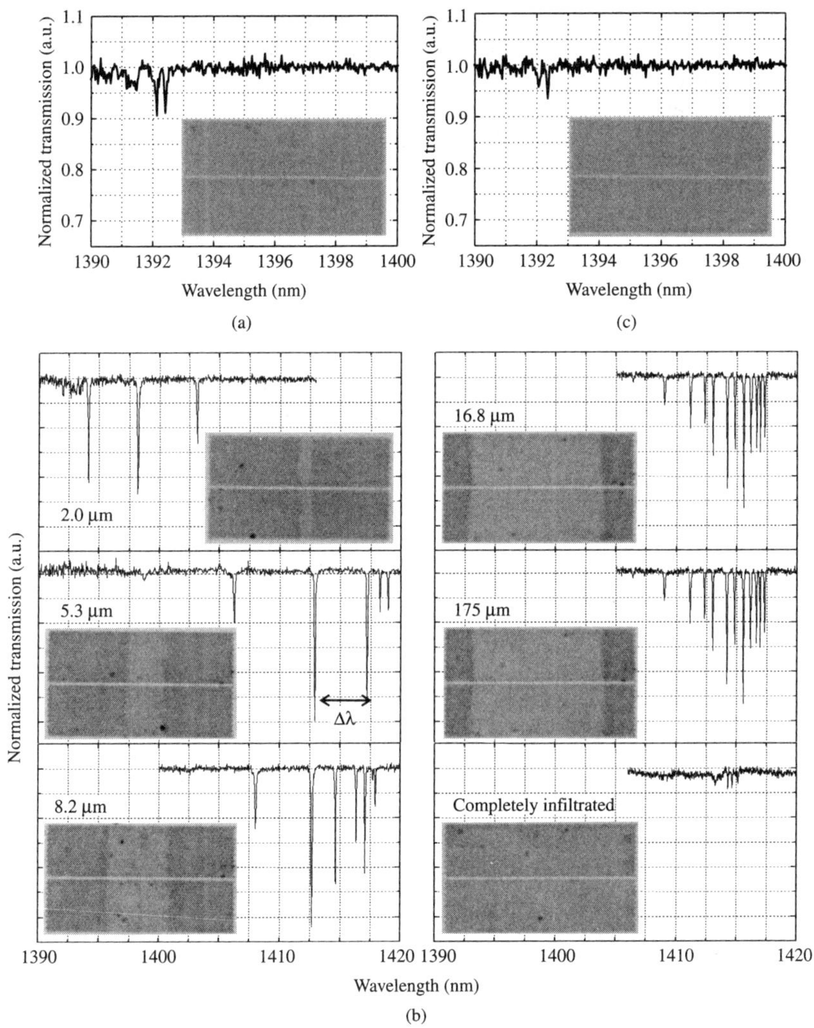

Figure 18-9 Normalized transmission through the nanowire when probing PhC waveguide structures. (a) Reference spectrum of the uninfiltrated waveguide. (b) Spectra for varying infiltrated cavity lengths. (c) Spectrum of the reconfigured (cleaned) structure. Insets are 150× microscope images of the corresponding waveguide/cavities. (*C. L. C. Smith, U. Bog, S. Tomljenovic-Hanic, M. W. Lee, D. K. C. Wu, L. O'Faolain, C. Monat, C. Grillet, T. F. Krauss, C. Karnutsch, R. C. McPhedran, and B. J. Eggleton, "Reconfigurable microfluidic photonic crystal slab cavities," Opt. Express 16(20), 15887–15896 (2008).*)

The reference spectrum of the uninfiltrated waveguide has a spectral signature at 1392 nm that is associated with coupling to the fundamental TE-like waveguide mode (Fig. 18-9a). The discrete features are attributed to Fresnel reflections at the open ends of the PhC waveguide. After the first infiltration (Fig. 18-9b), the spectral features appear at longer wavelengths. This is due to the increased effective refractive index of the guided modes caused by the presence of the fluid (refer to Fig. 18-3 in "Mode-Gap"). The observed fringe spectra are attributed to Fabry-Pérot (FP) modes sustained by the microfluidic cavity (see "Cavity Design" in section 18-2-2). As revealed by the envelope to the transmission dips of Fig. 18-9b, the coupling strength between the cavity resonances and the nanowire is at a maximum when the phase-matching is optimum [111,127]. It can be seen that—as the cavity length increases from 2 to 17.5 μm—the fringe spacing, $\Delta\lambda$ becomes smaller, which is consistent with an increased spectral density of modes for larger cavities.* Also, the fringe spacing within a particular spectrum becomes smaller at longer wavelengths for all investigated cavity lengths, which results from the dispersive nature of the PhC waveguide.

In a final infiltration step, we completely filled the PhC region with liquid. In this case, the fringes associated with the FP resonances disappear as the mode-gap effect no longer exists. Now we only couple to the fluid-filled fundamental PhC waveguide mode, which displays a spectral signature similar to the uninfiltrated case of Fig. 18-9a but shifted to longer wavelengths.

Complete reconfigurability of optofluidic circuits is highly desirable, as it enables the creation and tuning/trimming of optical functional elements from the same uniform PhC platform. One approach to achieve this is to remove the infiltrated liquid from the PhC substrate by washing it with organic solvents. We cleaned the infiltrated PhC sample by immersing it in a bath of toluene for several minutes. The transmission spectrum recorded after this cleaning step (Fig. 18-9c) shows that the spectral signature is nearly identical to the reference spectrum of the original uninfiltrated PhC structure, showing the viability of this reconfiguration approach.

We now compare the measured spectra to the calculated dispersion relation for a W0.9 infiltrated PhC waveguide using a 3D plane-wave expansion method† and the PhC parameters mentioned above. The result of this analysis represents the dispersion relation of the waveguide modes (see Fig. 18-10). The experimental group velocity is derived from the fringe spacing between the neighboring resonances measured on the spectra of Fig. 18-9 and using the equation [128]:

$$V_g = \frac{2Lc\Delta\lambda}{\lambda^2}$$

*We note that the experimentally investigated cavity lengths are larger than the ones considered in the theory section. Hence we observe several modes in our experiments.

†BandSOLVE from RSoft.

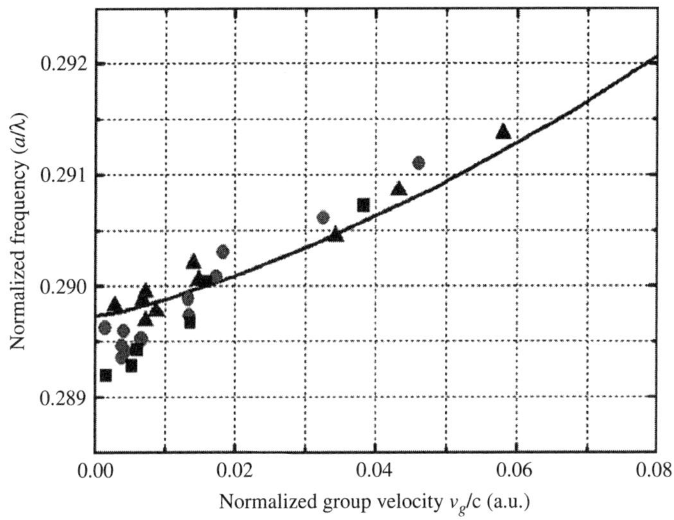

FIGURE 18-10 Comparison of measured and calculated group velocity for the fluid-filled PhCs. DH cavities with lengths of 8.2 μm (squares), 16.8 μm (circles), and 20.1 μm (triangles) are plotted along with numerical data (solid line). (*C. L. C. Smith, U. Bog, S. Tomljenovic-Hanic, M. W. Lee, D. K. C. Wu, L. O'Faolain, C. Monat, C. Grillet, T. F. Krauss, C. Karnutsch, R. C. McPhedran, and B. J. Eggleton, "Reconfigurable microfluidic photonic crystal slab cavities," Opt. Express 16(20), 15887–15896 (2008).*)

where L is the cavity length, c is the speed of light, $\Delta\lambda$ is the fringe spacing, and λ denotes the wavelength where the resonance occurs. The calculated group velocity is extracted from the gradient of the numerical dispersion relation (see, e.g., Fig. 18-3).

The experimentally measured dispersion curves derived for three different cavity lengths are reasonably superimposed, which is consistent with the FP modes all originating from the same dispersion relation, namely, the one associated with the infiltrated W0.9 fundamental waveguide mode. In addition, this experimentally retrieved dispersion is in good agreement (within the limits of fabrication tolerances) with the calculated one.

Quality Factors of Optofluidic DH Cavities

In this section we present investigations on the Q-factor associated with the resonances of optofluidic cavities. Figure 18-11 shows the trend of the intrinsic Q-factor* measured for two different cavity

*Taking the transmission T into account, the intrinsic Q-factor $Q_{\text{intrinsic}}$ can be calculated by first approximation coupled mode theory in the time domain from the measured Q-factor Q_{measured} to be $Q_{\text{intrinsic}} = Q_{\text{measured}}/\sqrt{T}$.

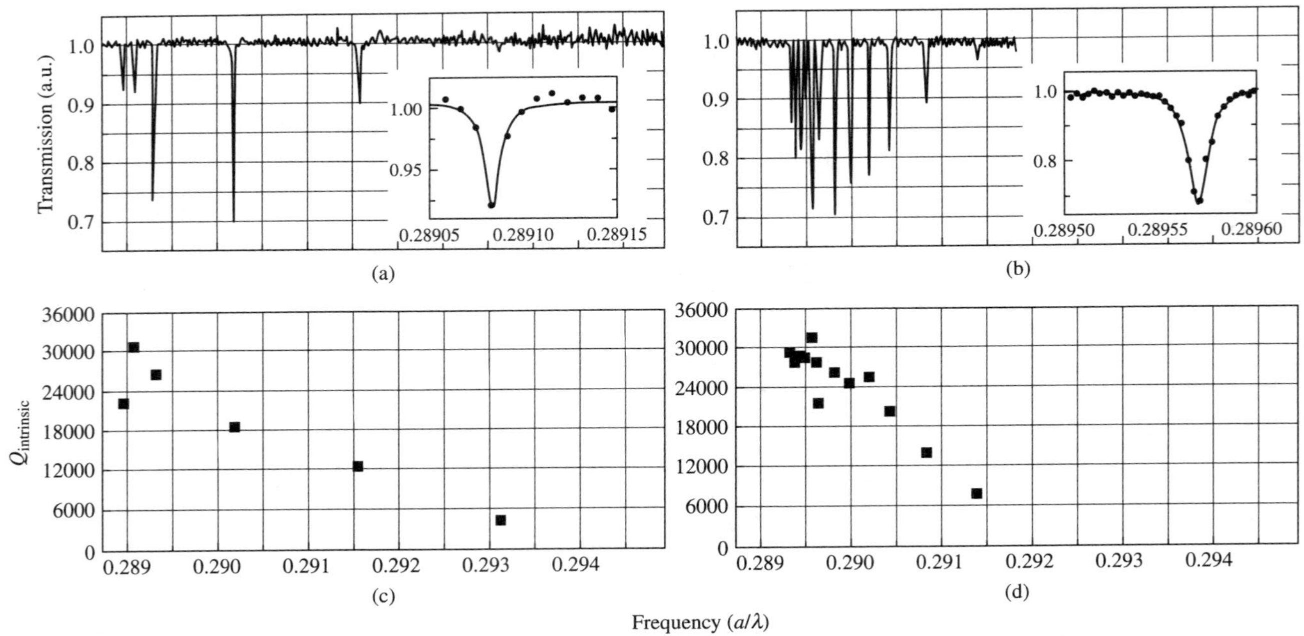

Figure 18-11 Normalized transmission spectra and intrinsic quality factor $Q_{intrinsic}$ as a function of normalized frequency. (a and b) Represent spectra from 5.3 and 16.8 μm cavity lengths, respectively. Insets are close-ups of the resonances exhibiting the highest Q-factors. (c and d) Display the Q-factors associated with the respective FP-resonances in the spectra from (a) and (b). (*C. L. C. Smith, U. Bog, S. Tomljenovic-Hanic, M. W. Lee, D. K. C. Wu, L. O'Faolain, C. Monat, C. Grillet, T. F. Krauss, C. Karnutsch, R. C. McPhedran, and B. J. Eggleton, "Reconfigurable microfluidic photonic crystal slab cavities," Opt. Express 16(20), 15887–15896 (2008).*)

lengths (5.3 and 16.8 μm) as frequency increases. For both cavities, we plot their spectral signature (Fig. 18-11a and 18-11b) and the corresponding Q-factors for each of the resonances evident in the spectra (Fig. 18-11c and 18-11d).

We observe that the Q-factors increase with decreasing frequency, and we note that this trend is representative for all investigated cavity lengths. This is expected behavior [129], because the guided modes at lower frequencies experience a higher effective refractive index and thus a better vertical confinement within the slab, reducing the out-of-plane losses. The intrinsic quality factors obtained in this set of experiments have values up to $Q_{intrinsic} = 3.5 \times 10^4$, but we note that the measurements were limited by the resolution of the optical spectrum analyzer (OSA) used in this experiment (Agilent 86140B).

High-*Q* Optofluidic DH Cavities

In order to gain an insight into the full potential of optofluidic DH cavities, we repeated our initial experiments employing a high-resolution OSA (Ando AQ6317B). We investigated cavity lengths of 3.3 μm (Fig. 18-12) and 16 μm (Fig. 18-13) as typical representatives for a short and a long cavity [130].

Figure 18-12 shows the normalized transmission spectrum associated with the 3.3-μm microfluidic DH cavity when measured with the high-resolution OSA [130]. The resonances exhibit measured Q-factors ranging from $Q_{measured} = 19{,}300$ [for resonance (1) with a transmission of

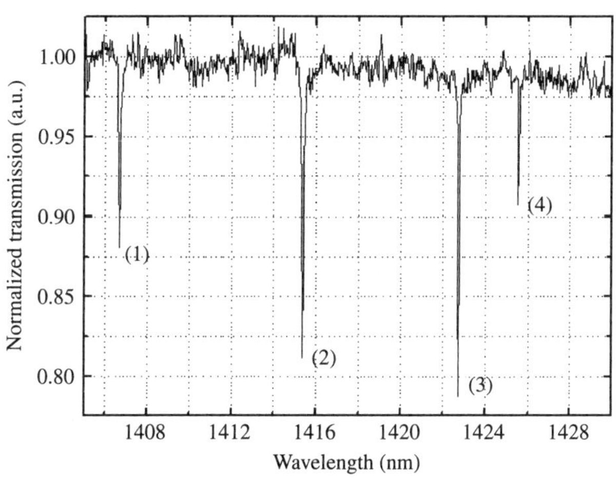

FIGURE 18-12 Normalized transmission spectrum while probing a microfluidic DH cavity of 3.3-μm length. Cavity mode (4) exhibits a measured Q-factor of $Q_{measured} = 36{,}300$. (*U. Bog, C. L. C. Smith, M. W. Lee, S. Tomljenovic-Hanic, C. Grillet, C. Monat, L. O'Faolain, et al., "High-Q microfluidic cavities in silicon-based 2D photonic crystal structures," Opt. Lett. 33(19), 2206–2208 (2008).*)

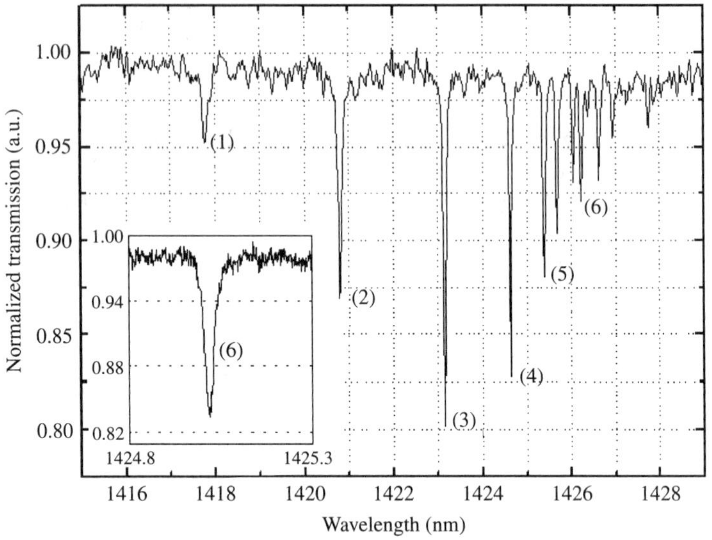

FIGURE 18-13 Normalized transmission spectrum when a 16-μm cavity is probed. The measured Q-factors for resonances (5) and (6) are $Q_{measured}$ = 45,740 and $Q_{measured}$ = 52,050, respectively. The inset shows a close-up view of resonance (6). (*U. Bog, C. L. C. Smith, M. W. Lee, S. Tomljenovic-Hanic, C. Grillet, C. Monat, L. O'Faolain, et al., "High-Q microfluidic cavities in silicon-based 2D photonic crystal structures," Opt. Lett. 33(19), 2206–2208 (2008).*)

T = 0.88] up to $Q_{measured}$ = 36,300 [resonance (4), T = 0.91]. Hence, for these two modes, intrinsic Q-factors of $Q_{intrinsic}$ = 20,810 and $Q_{intrinsic}$ = 38,050 are derived. This short 3.3-μm PhC cavity corresponds to a modal volume of only ~1.5 $(\lambda/n)^3$ [102], which highlights the potential for generating high Q-factors in very compact microfluidic devices.

Figure 18-13 shows the normalized transmission spectrum when probing the longer 16-μm cavity. The associated Q-factors are higher than for the short cavity, showing that the loss is dominated by the reflection losses at the interfaces between the infiltrated and uninfiltrated regions. For example, resonances (5) and (6) in Fig. 18-13 exhibit measured Q-factors of $Q_{measured}$ = 45,740 (T = 0.88) and $Q_{measured}$ = 52,050 (T = 0.90). The derived intrinsic Q-factors for these two resonances are $Q_{intrinsic}$ = 50,430 and $Q_{intrinsic}$ = 57,080, respectively.

In contrast with earlier demonstrations of liquid-infiltrated PhC cavities, where the Q-factor was usually degraded after the infiltration [131], the high Q-factors presented here demonstrate that the DH cavity can be applied as a highly sensitive microfluidic sensor. PhC cavity sensors typically exploit the resonance shift $\Delta\lambda$ that occurs when the refractive index of the analyte in the PhC holes changes by a value Δn. The shift of the PhC waveguide band structure induced by the fluid infiltration, as calculated by a plane-wave expansion method, allows us to estimate a potential sensitivity of $\Delta\lambda/\Delta n$ = 60 nm/RIU.

The sensitivity is limited by the relatively small overlap of the electric field with the air holes of approximately 6% (estimated from first-order approximation electromagnetic perturbation theory [132]). The overlap of the liquid with the PhC air holes could potentially be improved by optimizing the cavity geometry [133]. However, considering the full-width-half-maximum of the cavity resonance as the limit, a minimum refractive index resolution of $\delta n_{\text{analyte}} = 4.5 \cdot 10^{-4}$ could be achieved by exploiting the high-Q resonance (6) in Fig. 18-13. This number compares favorably with the values ($\delta n_{\text{analyte}} = 2 \cdot 10^{-3}$) demonstrated in previous work on passive PhC-based sensors [134].

Quality Factor as a Function of Cavity Length

Reproducibility and tolerance to misalignment and inaccuracies are important parameters for any kind of fabrication process. When infiltrating our microfluidic cavities, we frequently observe discrepancies between the targeted cavity length and the infiltrated length. We therefore investigated the dependence of the Q-factor on the cavity length (see Fig. 18-14). We analyzed the measured transmission spectra of varying cavity lengths at a fixed frequency of $\tilde{\omega} = 0.291$, which is high enough to avoid resolution-limited Q-factor values.

To obtain the data, we applied a linear fit between two neighboring spectral dips, as the resonances typically occurred to either side of

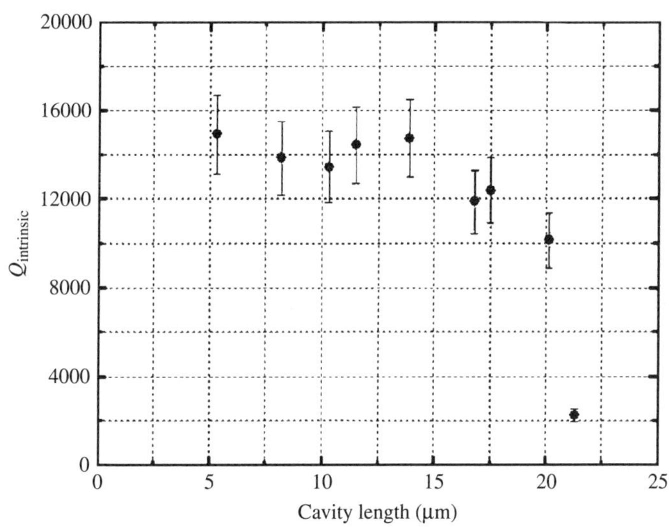

FIGURE 18-14 Evolution of the quality factor as a function of cavity length at a fixed frequency $\tilde{\omega} = 0.291$. The PhC structure has a length of 25 μm. (*C. L. C. Smith, U. Bog, S. Tomljenovic-Hanic, M. W. Lee, D. K. C. Wu, L. O'Faolain, C. Monat, C. Grillet, T. F. Krauss, C. Karnutsch, R. C. McPhedran, and B. J. Eggleton, "Reconfigurable microfluidic photonic crystal slab cavities," Opt. Express 16(20), 15887–15896 (2008).*)

the defined frequency. The resonances of the 2-μm cavity occur far away from this frequency; we therefore omitted them in this synopsis. We observe (Fig. 18-14) that the quality factor remains almost constant for cavity lengths between 5.3 and 13.8 μm. When the cavity length is increased beyond ~15 μm, the Q-factor drops as the in-plane confinement is reduced. This is due to the reduced length of the remaining uninfiltrated PhC sections that provide the mode-gap effect. Note that this is not an intrinsic limitation, as the total length of the PhC waveguide can easily be increased. The Q-factor continues to drop through to the largest investigated cavity length of 21.3 μm, before the sample is finally completely infiltrated. This suggests that 5 μm (12 periods in Γ-K direction) of uninfiltrated PhC section on each side of the microfluidic cavity is just large enough to maintain sufficient in-plane confinement. The Q-factor remains larger than 10,000 for all cavity lengths ≤ 20 μm, pointing out the large tolerance on the size of the infiltrated region for generating reasonably high-Q-factor cavities with this approach.

Single Hole Infiltration

In the previous sections we have presented microfluidic DH cavities fabricated with our current infiltration technique. The precision of this infiltration process is mainly limited by the amount of fluid residing on the microtip. A prerequisite for realizing highly complex optofluidic circuits (depicted schematically in Fig. 18-1) is the ability to precisely control the applied fluid volume. Figure 18-15 shows promising preliminary results of infiltration experiments with increased control of the fluid volume. We can clearly observe that it is indeed

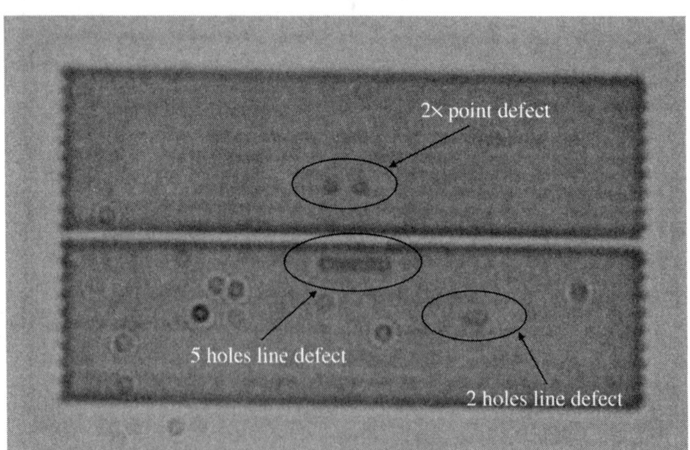

FIGURE 18-15 Microscope image of a PhC waveguide with infiltrated point defects and line defects of varying length.

possible to create single-hole point defects and line defects with varying lengths, which constitute the basic building blocks for optofluidic circuits.

18-4 Conclusion and Outlook

Merging microfluidics with photonics is a promising route to tune and reconfigure photonic circuits. Planar PhCs are well suited to this application, with the central role played by versatile microcavities in general and microfluidic double-heterostructures in particular that possess this unique advantage of confining light at the exact location where the liquid that defines the cavity is infused.

Increasing the light-liquid interaction within these microcavities is crucial for creating both dynamic and tunable functions as well as sensitive detectors. As an example, Kwon et al. have proposed a design that includes a central slot in the PhC waveguide to improve the overlap of the electric field within the infiltrated part of the PhC microcavity, thereby increasing the sensitivity of the resulting device [83].

By introducing an active material—for example, colloidal quantum dots—into the infiltration liquid, the demonstrated microcavities could be exploited to generate reconfigurable light sources. Optical properties (nonlinearity, fluorescence, etc.) associated with the particular liquid introduced during the "writing" of the double-heterostructure cavity will be potentially enhanced by the strong cavity fields.

Note that the presented microfluidic double-heterostructure component is just one basic building block that will provide the starting point to realize more complex functions, all of which could be integrated within a reconfigurable photonic circuit as depicted in Fig. 18-1. In the future, we intent to realize more complex functionalities, for instance by combining several of these cavities together to realize coupled resonator systems, which are promising for controlling the speed of light [72].

Another important expansion of the presented work will be to develop a controlled technique to "write" the envisaged reconfigurable photonic circuit and effectively benefit from its dynamic and rewritable properties that are uniquely linked to its microfluidic nature. A possible approach would be to exploit hybrid microfluidic-photonic platforms, following the preliminary demonstrations of the full integration between planar PhC components and a microfluidic network [38,39,65,66,84]. In this scenario, the valves and pumps of the sophisticated microfluidic circuitry could be remotely actuated to write a specific circuit and potentially erase and reconfigure it later on. Another possibility would be to develop an automated writing tool based on the microtip infiltration method presented above.

The desired complex photonic circuit could be designed in a CAD software environment, from where the design/drawing could then be converted into electronic signals that actuate the microtip to "write" the optofluidic circuit pixel-by-pixel.

Today's microfluidic lab-on-a-chip systems possess numerous capabilities; however, they typically require external light sources and photodetectors coupled to the chip by bulky and expensive conventional optics. The lack of an integrated light source is a significant shortcoming, as it is essential for any kind of optical measurement. Hence the integration of optical devices is believed to be the next step to further improve the functionality and portability of lab-on-a-chip systems [135]. To achieve this integration, PhC-based reconfigurable fluid-controlled optical circuits could be combined with organic semiconductor technology in lab-on-a-chip systems. Organic semiconductors have recently attracted much attention due to the possibility to engineer their molecular structure and their simplicity of deposition and processing. Optically pumped organic semiconductor lasers have been demonstrated for a wide spectral range [136–145], and they can be pumped by low-cost laser diodes and LEDs [146–150].

To advance optofluidic lab-on-a-chip technology, organic laser sources and photodetectors could be monolithically integrated on a single substrate. These innovative devices could potentially hold an array of organic semiconductor lasers, organic photodetectors, fluid-controlled optical elements and microfluidic channels. The combination of organic photonic devices with the flexibility offered by microfluidic PhC circuits will result in on-chip tunable light sources and detectors, enabling the development of novel devices that will become vital components of next-generation sensors, medical diagnostics, and biotechnological systems that are disposable, portable, and affordable.

18-5 Acknowledgments

We thank Christian Grillet, Eric Magi, Ross McPhedran, Martijn de Sterke, Michael J Steel, Cameron Smith, Uwe Bog, Michael Lee, and Darran Wu for their input to this project. We also thank Darren Freeman, Steve Madden, and Barry Luther-Davies for providing the chalcogenide PhC samples, and Liam O'Faolain and Thomas Krauss for providing the silicon PhC samples.

The support of the Australian Research Council through its Federation Fellow, Centre of Excellence and Discovery Grant programs is gratefully acknowledged. Additional acknowledgment is given to the support of the School of Physics, University of Sydney, through its Denison Foundation and the International Science Linkages program through the ISL DEST grant. The silicon samples were fabricated in the framework of the EU-FP6 funded ePIXnet Nanostructuring Platform for Photonic Integration (www.nanophotonics.eu).

References

1. J. D. Joannopoulos, R. D. Meade, and J. N. Winn, *Photonic Crystals: Molding the Flow of Light*, Princeton University Press, Princeton, NJ, 1995.
2. T. Baba, "Slow light in photonic crystals," *Nature Photonics* **2**, 465–473 (2008).
3. P. Halevi and F. Ramos-Mendieta, "Tunable photonic crystals with semiconducting constituents," *Phys. Rev. Lett.* **85**, 1875–1878 (2000).
4. B. Wild, R. Ferrini, R. Houdre, M. Mulot, S. Anand, and C. J. M. Smith, "Temperature tuning of the optical properties of planar photonic crystal microcavities," *Appl. Phys. Lett.* **84**, 846–848 (2004).
5. T. Tanabe, M. Notomi, S. Mitsugi, A. Shinya, and E. Kuramochi, "Fast bistable all-optical switch and memory on a silicon photonic crystal on-chip," *Opt. Lett.* **30**, 2575–2577 (2005).
6. K. Busch and S. John, "Liquid-crystal photonic-band-gap materials: The tunable electromagnetic vacuum," *Phys. Rev. Lett.* **83**, 967–970 (1999).
7. C. Schuller, F. Klopf, J. P. Reithmaier, M. Kamp, and A. Forchel, "Tunable photonic crystals fabricated in III-V semiconductor slab waveguides using infiltrated liquid crystals," *Appl. Phys. Lett.* **82**, 2767–2769 (2003).
8. K. Yoshino, Y. Shimoda, Y. Kawagishi, K. Nakayama, and M. Ozaki, "Temperature tuning of the stop band in transmission spectra of liquid-crystal infiltrated synthetic opal as tunable photonic crystal," *Appl. Phys. Lett.* **75**, 932–934 (1999).
9. S. W. Leonard, J. P. Mondia, H. M. van Driel, O. Toader, S. John, K. Busch, A. Birner, U. Gosele, and V. Lehmann, "Tunable two-dimensional photonic crystals using liquid-crystal infiltration," *Phys. Rev. B* **61**, R2389–R2392 (2000).
10. G. Mertens, T. Roder, R. Schweins, K. Huber, and H. S. Kitzerow, "Shift of the photonic band gap in two photonic crystal/liquid crystal composites," *Appl. Phys. Lett.* **80**, 1885–1887 (2002).
11. G. Mertens, T. Roder, H. Matthias, H. Marsmann, H. S. R. Kitzerow, S. L. Schweizer, C. Jamois, R. B. Wehrspohn, and M. Neubert, "Two- and three-dimensional photonic crystals made of macroporous silicon and liquid crystals," *Appl. Phys. Lett.* **83**, 3036–3038 (2003).
12. J. Martz, R. Ferrini, F. Nuesch, L. Zuppiroli, B. Wild, L. A. Dunbar, R. Houdre, M. Mulot, and S. Anand, "Liquid crystal infiltration of InP-based planar photonic crystals," *J. Appl. Phys.* **99**, 1031051–1031054 (2006).
13. H. Takeda and K. Yoshino, "Disappearances of uncoupled modes in two-dimensional photonic crystals due to anisotropies of liquid crystals," *Phys. Rev. E* **67**, 0566121–0566125 (2003).
14. H. Takeda and K. Yoshino, "Coupling of the TE and TM modes of electromagnetic waves in two-dimensional photonic crystals with surface defects of liquid crystals," *Phys. Rev. E* **68**, 0466021–0466025 (2003).
15. H. Takeda and K. Yoshino, "Tunable refraction effects in two-dimensional photonic crystals utilizing liquid crystals," *Phys. Rev. E* **67**, 0566071–0566075 (2003).
16. S. M. Weiss, H. M. Ouyang, J. D. Zhang, and P. M. Fauchet, "Electrical and thermal modulation of silicon photonic bandgap microcavities containing liquid crystals," *Opt. Express* **13**, 1090–1097 (2005).
17. C. Schuller, J. P. Reithmaier, J. Zimmermann, M. Kamp, A. Forchel, and S. Anand, "Polarization-dependent optical properties of planar photonic crystals infiltrated with liquid crystals," *Appl. Phys. Lett.* **87**, 1211051–1211053 (2005).
18. G. Mertens, R. B. Wehrspohn, H. S. Kitzerow, S. Matthias, C. Jamois, and U. Gosele, "Tunable defect mode in a three-dimensional photonic crystal," *Appl. Phys. Lett.* **87**, 2411081–2411083 (2005).
19. D. Kang, J. E. Maclennan, N. A. Clark, A. A. Zakhidov, and R. H. Baughman, "Electro-optic behavior of liquid-crystal-filled silica opal photonic crystals: Effect of liquid-crystal alignment," *Phys. Rev. Lett.* **86**, 4052–4055 (2001).

20. Y. Shimoda, M. Ozaki, and K. Yoshino, "Electric field tuning of a stop band in a reflection spectrum of synthetic opal infiltrated with nematic liquid crystal," *Appl. Phys. Lett.* **79**, 3627–3629 (2001).
21. M. Ozaki, Y. Shimoda, M. Kasano, and K. Yoshino, "Electric field tuning of the stop band in a liquid-crystal-infiltrated polymer inverse opal," *Adv. Mater.* **14**, 514–518 (2002).
22. G. Pucker, A. Mezzetti, M. Crivellari, P. Bellutti, A. Lui, N. Daldosso, and L. Pavesi, "Silicon-based near-infrared tunable filters filled with positive or negative dielectric anisotropic liquid crystals," *J. Appl. Phys.* **95**, 767–769 (2004).
23. R. Ozaki, T. Matsui, M. Ozaki, and K. Yoshino, "Electrically color-tunable defect mode lasing in one-dimensional photonic-band-gap system containing liquid crystal," *Appl. Phys. Lett.* **82**, 3593–3595 (2003).
24. R. Ozaki, Y. Matsuhisa, M. Ozaki, and K. Yoshino, "Electrically tunable lasing based on defect mode in one-dimensional photonic crystal with conducting polymer and liquid crystal defect layer," *Appl. Phys. Lett.* **84**, 1844–1846 (2004).
25. B. Maune, M. Loncar, J. Witzens, M. Hochberg, T. Baehr-Jones, D. Psaltis, A. Scherer, and Y. M. Qiu, "Liquid-crystal electric tuning of a photonic crystal laser," *Appl. Phys. Lett.* **85**, 360–362 (2004).
26. M. Haurylau, S. P. Anderson, K. L. Marshall, and P. M. Fauchet, "Electrical modulation of silicon-based two-dimensional photonic bandgap structures," *Appl. Phys. Lett.* **88**, 0611031–0611033 (2006).
27. M. Haurylau, S. P. Anderson, K. L. Marshall, and P. M. Fauchet, "Electrically tunable silicon 2-D photonic bandgap structures," *IEEE J. Sel. Top. Quantum Electron.* **12**, 1527–1533 (2006).
28. S. M. Weiss, M. Haurylau, and P. M. Fauchet, "Tunable photonic bandgap structures for optical interconnects," *Opt. Mater.* **27**, 740–744 (2005).
29. S. Kubo, Z. Z. Gu, K. Takahashi, Y. Ohko, O. Sato, and A. Fujishima, "Control of the optical band structure of liquid crystal infiltrated inverse opal by a photoinduced nematic-isotropic phase transition," *J. Am. Chem. Soc.* **124**, 10950–10951 (2002).
30. S. Kubo, Z. Z. Gu, K. Takahashi, A. Fujishima, H. Segawa, and O. Sato, "Tunable photonic band gap crystals based on a liquid crystal-infiltrated inverse opal structure," *J. Am. Chem. Soc.* **126**, 8314–8319 (2004).
31. M. Moritsugu, S. N. Kim, T. Ogata, T. Nonaka, S. Kurihara, S. Kubo, H. Segawa, and O. Sato, "Photochemical switching behavior of azofunctionalized polymer liquid crystal/SiO_2 composite photonic crystal," *Appl. Phys. Lett.* **89**, 1531311–1531313 (2006).
32. P. El-Kallassi, R. Ferrini, L. Zuppiroli, N. Le Thomas, R. Houdre, A. Berrier, S. Anand, and A. Talneau, "Optical tuning of planar photonic crystals infiltrated with organic molecules," *J. Opt. Soc. Am. B-Opt. Phys.* **24**, 2165–2171 (2007).
33. B. Maune, J. Witzens, T. B. Jones, M. Kolodrubetz, H. Atwater, A. Scherer, R. Hagen, and Y. M. Qiu, "Optically triggered Q-switched photonic crystal laser," *Opt. Express* **13**, 4699–4707 (2005).
34. T. Tanabe, M. Notomi, S. Mitsugi, A. Shinya, and E. Kuramochi, "All-optical switches on a silicon chip realized using photonic crystal nanocavities," *Appl. Phys. Lett.* **87**, 1511121–1511123 (2005).
35. T. Yasuda, Y. Tsuji, and M. Koshiba, "Tunable light propagation in photonic crystal coupler filled with liquid crystal," *IEEE Photonics Technol. Lett.* **17**, 55–57 (2005).
36. M. Loncar, A. Scherer, and Y. M. Qiu, "Photonic crystal laser sources for chemical detection," *Appl. Phys. Lett.* **82**, 4648–4650 (2003).
37. A. Sharkawy, D. Pustai, S. Y. Shi, and D. W. Prather, "Modulating dispersion properties of low index photonic crystal structures using microfluidics," *Opt. Express* **13**, 2814–2827 (2005).
38. L. Diehl, B. G. Lee, P. Behroozi, M. Loncar, M. A. Belkin, F. Capasso, T. Aellen, D. Hofstetter, M. Beck, and J. Faist, "Microfluidic tuning of distributed feedback quantum cascade lasers," *Opt. Express* **14**, 11660–11667 (2006).

39. S.-H. Kim, J.-H. Choi, S.-K. Lee, S.-H. Kim, S.-M. Yang, Y.-H. Lee, C. Seassal, P. Regreny, and P. Viktorovitch, "Optofluidic integration of a photonic crystal nanolaser," *Opt. Express* **16**, 6515–6527 (2008).
40. R. van der Heijden, C. F. Carlstrom, J. A. P. Snijders, R. W. van der Heijden, F. Karouta, R. Notzel, H. W. M. Salemink, et al., "InP-based two-dimensional photonic crystals filled with polymers," *Appl. Phys. Lett.* **88**, 1611121–1611123 (2006).
41. K. Yoshino, S. Tatsuhara, Y. Kawagishi, M. Ozaki, A. A. Zakhidov, and Z. V. Vardeny, "Amplified spontaneous emission and lasing in conducting polymers and fluorescent dyes in opals as photonic crystals," *Appl. Phys. Lett.* **74**, 2590–2592 (1999).
42. J. H. Park, W. S. Choi, H. Y. Koo, and D-Y. Kim, "Colloidal photonic crystals with graded refractive index distribution" *Adv. Mater.* 17, 879–885 (2005).
43. S. Tay, J. Thomas, B. Momeni, M. Askari, A. Adibi, P. J. Hotchkiss, S. C. Jones, S. R. Marder, R. A. Norwood, and N. Peyghambarian, "Planar photonic crystals infiltrated with nanoparticle/polymer composites," *Appl. Phys. Lett.* **91**, 2211091–2211093 (2007).
44. Z. Wu, Z. Mi, P. Bhattacharya, T. Zhu, and J. Xu, "Enhanced spontaneous emission at 1.55 mu m from colloidal PbSe quantum dots in a Si photonic crystal microcavity," *Appl. Phys. Lett.* **90**, 1711051–1711053 (2007).
45. M. Haurylau, J. Zhang, S. M. Weiss, P. M. Fauchet, D. V. Martyshkin, V. I. Rupasov, and S. G. Krivoshlykov, "Nonlinear optical response of photonic bandgap structures containing PbSe quantum dots," *J. Photochem. Photobiol. A-Chem.* **183**, 329–333 (2006).
46. P. Lodahl, A. F. van Driel, I. S. Nikolaev, A. Irman, K. Overgaag, D. L. Vanmaekelbergh, and W. L. Vos, "Controlling the dynamics of spontaneous emission from quantum dots by photonic crystals," *Nature* **430**, 654–657 (2004).
47. K. Yoshino, S. B. Lee, S. Tatsuhara, Y. Kawagishi, M. Ozaki, and A. A. Zakhidov, "Observation of inhibited spontaneous emission and stimulated emission of rhodamine 6G in polymer replica of synthetic opal," *Appl. Phys. Lett.* **73**, 3506–3508 (1998).
48. A. F. Koenderink, L. Bechger, H. P. Schriemer, A. Lagendijk, and W. L. Vos, "Broadband fivefold reduction of vacuum fluctuations probed by dyes in photonic crystals," *Phys. Rev. Lett.* 88, 1439031–1439034 (2002).
49. M. Gersborg-Hansen and A. Kristensen, "Optofluidic third order distributed feedback dye laser," *Appl. Phys. Lett.* 89, 1035181–1035183 (2006).
50. S. Balslev and A. Kristensen, "Microfluidic single-mode laser using high-order Bragg grating and antiguiding segments," *Opt. Express* **13**, 344–351 (2005).
51. Z. Li, Z. Zhang, A. Scherer, and D. Psaltis, "Mechanically tunable optofluidic distributed feedback dye laser," *Opt. Express* **14**, 10494–10499 (2006).
52. M. Gersborg-Hansen and A. Kristensen, "Tunability of optofluidic distributed feedback dye lasers," *Opt. Express* **15**, 137–142 (2007).
53. K. Busch, S. Lölkes, R. B. Wehrspohn, and H. Föll, *Photonic Crystals-Advances in Design, Fabrication, and Characterization*, Wiley-VCH, Berlin, Germany, 2004.
54. C. Monat, P. Domachuk, C. Grillet, M. Collins, B. J. Eggleton, M. Cronin-Golomb, S. Mutzenich, T. Mahmud, G. Rosengarten and A. Mitchell, "Optofluidics: a novel generation of reconfigurable and adaptive compact architectures," *Microfluid. Nanofluid.* **4**, 81–95 (2008).
55. H. Takeda and K. Yoshino, "Tunable light propagation in Y-shaped waveguides in two-dimensional photonic crystals utilizing liquid crystals as linear defects," *Phys. Rev.* B **67**, 0731061–0731063 (2003).
56. S. F. Mingaleev, M. Schillinger, D. Hermann, and K. Busch, "Tunable photonic crystal circuits: concepts and designs based on single-pore infiltration," *Opt. Lett.* **29**, 2858–2860 (2004).
57. T. F. Krauss, R. M. DeLaRue, and S. Brand, "Two-dimensional photonic-bandgap structures operating at near infrared wavelengths," *Nature* **383**, 699–702 (1996).

58. E. Chow, S. Y. Lin, S. G. Johnson, P. R. Villeneuve, J. D. Joannopoulos, J. R. Wendt, G. A. Vawter, W. Zubrzycki, H. Hou, and A. Alleman, "Three-dimensional control of light in a two-dimensional photonic crystal slab," *Nature* **407**, 983–986 (2000).
59. M. Notomi, A. Shinya, S. Mitsugi, E. Kuramochi, and H. Y. Ryu, "Waveguides, resonators and their coupled elements in photonic crystal slabs," *Opt. Express* **12**, 1551–1561 (2004).
60. H. Kurt, and D. S. Citrin, "Reconfigurable multimode photonic-crystal waveguides," *Opt. Express* **16**, 11995–12001 (2008).
61. F. Intonti, S. Vignolini, V. Turck, M. Colocci, P. Bettotti, L. Pavesi, S. L. Schweizer, R. Wehrspohn, and D. Wiersma, "Rewritable photonic circuits," *Appl. Phys. Lett.* **89**, 2111171–2111173 (2006).
62. C. L. C. Smith, D. K. C. Wu, M. W. Lee, C. Monat, S. Tomljenovic-Hanic, C. Grillet, B. J. Eggleton, et al., "Microfluidic photonic crystal double heterostructures," *Appl. Phys. Lett.* **91**, 1211031–1211033 (2007).
63. C. L. C. Smith, U. Bog, S. Tomljenovic-Hanic, M. W. Lee, D. K. C. Wu, L. O'Faolain, C. Monat, C. Grillet, T. F. Krauss, C. Karnutsch, R. C. McPhedran, and B. J. Eggleton, "Reconfigurable microfluidic photonic crystal slab cavities," *Opt. Express* **16**(20), 15887–15896 (2008).
64. P. El-Kallassi, S. Balog, R. Houdré, L. Balet, L. Li, M. Francardi, A. Gerardino, A. Fiore, R. Ferrini, and L. Zuppiroli, "Local infiltration of planar photonic crystals with UV-curable polymers," *J. Opt. Soc. Am. B-Opt. Phys.* **25**, 1562–1567 (2008).
65. D. Erickson, T. Rockwood, T. Emery, A. Scherer, and D. Psaltis, "Nanofluidic tuning of photonic crystal circuits," *Opt. Lett.* **31**, 59–61 (2006).
66. M. L. Adams, M. Loncar, A. Scherer, and Y. M. Qiu, "Microfluidic integration of porous photonic crystal nanolasers for chemical sensing," *IEEE J. Sel. Areas Commun.* **23**, 1348–1354 (2005).
67. S. John, "Strong Localization of Photons in Certain Disordered Dielectric Superlattices," *Phys. Rev. Lett.* **58**, 2486–2489 (1987).
68. E. Yablonovitch, T. J. Gmitter, R. D. Meade, A. M. Rappe, K. D. Brommer, and J. D. Joannopoulos, "Donor and acceptor modes in photonic band-structure," *Phys. Rev. Lett.* **67**, 3380–3383 (1991).
69. O. Painter, R. K. Lee, A. Scherer, A. Yariv, J. D. O'Brien, P. D. Dapkus, and I. Kim, "Two-dimensional photonic band-gap defect mode laser," *Science* **284**, 1819–1821 (1999).
70. H. G. Park, S. H. Kim, S. H. Kwon, Y. G. Ju, J. K. Yang, J. H. Baek, S. B. Kim, and Y. H. Lee, "Electrically driven single-cell photonic crystal laser," *Science* **305**, 1444–1447 (2004).
71. Y. Akahane, T. Asano, B. S. Song, and S. Noda, "Investigation of high-Q channel drop filters using donor-type defects in two-dimensional photonic crystal slabs," *Appl. Phys. Lett.* **83**, 1512–1514 (2003).
72. D. O'Brien, A. Gomez-Iglesias, M. D. Settle, A. Michaeli, M. Salib, and T. F. Krauss, "Tunable optical delay using photonic crystal heterostructure nanocavities," *Phys. Rev. B* **76**, 1151101–1151104 (2007).
73. Y. Yoshie, A. Scherer, J. Hendrickson, G. Khitrova, H. M. Gibbs, G. Rupper, C. Ell, O. B. Shchekin, and D. G. Deppe, "Vacuum Rabi splitting with a single quantum dot in a photonic crystal nanocavity," *Nature* **432**, 200–203 (2004).
74. D. Englund, D. Fattal, E. Waks, G. Solomon, B. Zhang, T. Nakaoka, Y. Arakawa, Y. Yamamoto, and J. Vuckovic, "Controlling the spontaneous emission rate of single quantum dots in a two-dimensional photonic crystal", *Phys. Rev. Lett.* **95**, 0139041–0139044 (2005).
75. K. J. Vahala, "Optical microcavities," *Nature* **424**, 839–846 (2003).
76. Y. Akahane, T. Asano, B. S. Song, and S. Noda, "High-*Q* photonic nanocavity in a two-dimensional photonic crystal", *Nature* **425**, 944–947 (2003).
77. D. Psaltis, S. R. Quake, and C. H. Yang, "Developing optofluidic technology through the fusion of microfluidics and optics," *Nature* **442**, 381–386 (2006).
78. C. Monat, P. Domachuk, and B. J. Eggleton, "Integrated optofluidics: A new river of light," *Nat. Photonics* **1**, 106–114 (2007).

79. X. D. Fan, I. M. White, S. I. Shopoua, H. Y. Zhu, J. D. Suter, and Y. Z. Sun, "Sensitive optical biosensors for unlabeled targets: A review," *Anal. Chim. Acta* **620**, 8–26 (2008).
80. I. M. White and X. D. Fan, "On the performance quantification of resonant refractive index sensors," *Opt. Express* **16**, 1020–1028 (2008).
81. E. Chow, A. Grot, L. W. Mirkarimi, M. Sigalas, and G. Girolami, "Ultracompact biochemical sensor built with two-dimensional photonic crystal microcavity," *Opt. Lett.* **29**, 1093–1095 (2004).
82. S. Kita, K. Nozaki, and T. Baba, "Refractive index sensing utilizing a cw photonic crystal nanolaser and its array configuration," *Opt. Express* **16**, 8174–8180 (2008).
83. S. H. Kwon, T. Sunner, M. Kamp, and A. Forchel, "Optimization of photonic crystal cavity for chemical sensing," *Opt. Express* **16**, 11709–11717 (2008).
84. S. Mandal and D. Erickson, "Nanoscale optofluidic sensor arrays," *Opt. Express* **16**, 1623–1631 (2008).
85. M. Lee and P. M. Fauchet, "Two-dimensional silicon photonic crystal based biosensing platform for protein detection," *Opt. Express* **15**, 4530–4535 (2007).
86. S. Chakravarty, J. Topol'ancik, P. Bhattacharya, S. Chakrabarti, Y. Kang, and M. E. Meyerhoff, "Ion detection with photonic crystal microcavities," *Opt. Lett.* **30**, 2578–2580 (2005).
87. M. R. Lee and P. M. Fauchet, "Nanoscale microcavity sensor for single particle detection," *Opt. Lett.* **32**, 3284–3286 (2007).
88. N. Skivesen, A. Tetu, M. Kristensen, J. Kjems, L. H. Frandsen, and P. I. Borel, "Photonic-crystal waveguide biosensor," *Opt. Express* **15**, 3169–3176 (2007).
89. S. S. Xiao and N. A. Mortensen, "Proposal of highly sensitive optofluidic sensors based on dispersive photonic crystal waveguides," *J. Opt. A-Pure Appl. Opt.* **9**, S463–S467 (2007).
90. M. E. V. Pedersen, L. S. Rishoj, H. Steffensen, S. H. Xiao, and N. A. Mortensen, "Slow-light enhanced optical detection in liquid-infiltrated photonic crystals," *Opt. Quantum Electron.* **39**, 903–911 (2007).
91. N. A. Mortensen, S. S. Xiao, and J. Pedersen, "Liquid-infiltrated photonic crystals: enhanced light-matter interactions for lab-on-a-chip applications," *Microfluid. Nanofluid.* **4**, 117–127 (2008).
92. J. Pedersen and N. A. Mortensen, "Enhanced circular dichroism via slow light in dispersive structured media," *Appl. Phys. Lett.* **91**, 2135011–2135013 (2007).
93. N. A. Mortensen and S. S. Xiao, "Slow-light enhancement of Beer-Lambert-Bouguer absorption," *Appl. Phys. Lett.* **90**, 1411081–1411083 (2007).
94. J. Topol'ancik, P. Bhattacharya, J. Sabarinathan, and P. C. Yu, "Fluid detection with photonic crystal-based multichannel waveguides," *Appl. Phys. Lett.* **82**, 1143–1145 (2003).
95. T. Hasek, H. Kurt, D. S. Citrin, and M. Koch, "Photonic crystals for fluid sensing in the subterahertz range," *Appl. Phys. Lett.* **89**, 1735081–1735083 (2006).
96. H. Kurt and D. S. Citrin, "Coupled-resonator optical waveguides for biochemical sensing of nanoliter volumes of analyte in the terahertz region," *Appl. Phys. Lett.* **87**, 2411191–2411193 (2005).
97. H. Kurt and D. S. Citrin, "Photonic crystals for biochemical sensing in the terahertz region," *Appl. Phys. Lett.* **87**, 0411081–0411083 (2005).
98. B. Cunningham, P. Li, B. Lin, and J. Pepper, "Colorimetric resonant reflection as a direct biochemical assay technique," *Sens. Actuator B-Chem.* **81**, 316–328 (2002).
99. C. J. Choi and B. T. Cunningham, "Single-step fabrication and characterization of photonic crystal biosensors with polymer microfluidic channels," *Lab Chip* **6**, 1373–1380 (2006).
100. C. J. Choi and B. T. Cunningham, "A 96-well microplate incorporating a replica molded microfluidic network integrated with photonic crystal biosensors for high throughput kinetic biomolecular interaction analysis," *Lab Chip* **7**, 550–556 (2007).
101. P. S. Nunes, N. A. Mortensen, J. P. Kutter, and K. B. Mogensen, "Photonic crystal resonator integrated in a microfluidic system," *Opt. Lett.* **33**, 1623–1625 (2008).

102. S. Tomljenovic-Hanic, C. M. de Sterke, and M. J. Steel, "Design of high-Q cavities in photonic crystal slab heterostructures by air-holes infiltration," *Opt. Express* **14**, 12451–12456 (2006).
103. B. S. Song, S. Noda, T. Asano and Y. Akahane, "Ultra-high-*Q* photonic double-heterostructure nanocavity," *Nat. Mater.* **4**, 207–210 (2005).
104. S. Noda, M. Fujita, and T. Asano, "Spontaneous-emission control by photonic crystals and nanocavities," *Nat. Photonics* **1**, 449–458 (2007).
105. E. Kuramochi, M. Notomi, S. Mitsugi, A. Shinya, and T. Tanabe, "Ultrahigh-*Q* photonic crystal nanocavities realized by the local width modulation of a line defect," *Appl. Phyis. Lett.* **88**, 041112 (2006).
106. E. Istrate and E. H. Sargent, "Photonic crystal heterostructures and interfaces," *Rev. Modern Phys.* **78**, 455–481 (2006).
107. S. Tomljenovic-Hanic, M. J. Steel, C. M. de Sterke and D. J. Moss, "High-*Q* cavities in photosensitive photonic crystals," *Opt. Lett.* **32**, 542–544 (2007).
108. G. Wu, J. Wang, J. Shen, T. Yang, Q. Zhang, B. Zhou, Z. Deng, F. Bin, D. Zhou, and F. Zhang, "Properties of sol-gel derived scratch-resistant nano-porous silica films by a mixed atmosphere treatment," *J. Non-Cryst. Solids* **275**, 169–174 (2000).
109. V. A. Mandelshtam and H. S. Taylor, "Harmonic inversion of time signals," *J. Chem. Phys.* **107**, 6756–6769 (1997).
110. P. E. Barclay, K. Srinivasan, M. Borselli, and O. Painter, "Probing the dispersive and spatial properties of photonic crystal waveguides via highly efficient coupling from fiber tapers," *Appl. Phys. Lett.* **85**, 4–6 (2004).
111. C. Grillet, C. Smith, D. Freeman, S. Madden, B. Luther-Davies, E. C. Mägi, D. J. Moss, and B. J. Eggleton, "Efficient coupling o chalcogenide glass photonic crystal waveguide via silica optical fiber nanowires," *Opt. Express* **14**, 1070–1078 (2006).
112. G. P. Harmon, "Optical Polymers: Fibers and Waveguides" in *Optical Polymers Fibers and Waveguides*, J. P. Harmon and G. K. Noren, eds. American Chemical Society, Washington, D.C, U.S, 2001, pp. 1–23.
113. P.-G. de Gennes, F. Brochard-Wyart, D. Quéré, and A. Reisinger, *Capillarity and Wetting Phenomena: Drops, Bubbles, Pearls, Waves.* Springer, New York, US, 2004.
114. T. M. Squires and S. R. Quake, "Microfluidics: Fluid physics at the nanoliter scale," *Reviews of Modern Physics* **77**, 977–950 (2005).
115. J. C. Knight, G. Cheung, F. Jacques, and T. A. Birks, "Phase-matched excitation of whispering-gallery-mode resonances by a fiber taper," *Opt. Lett.* **22**, 1129–1131 (1997).
116. K. Srinivasan, P. E. Barclay, M. Borselli, and O. Painter, "Optical-fiber based measurement of an ultra-small volume high-Q photonic crystal microcavity," *Phys. Rev. B* **70**, 081306(R) (2004).
117. K. Srinivasan, P. E. Barclay, M. Borselli, and O. Painter, "An optical-fiber based probe for photonic crystal microcavities," *IEEE J. Sel. Areas Commun.* **23**, 1321–1329 (2005).
118. I. Hwang, S. Kim, J. Yang, S. Kim, S. Lee, and Y. Lee, "Curved-microfiber photon coupling for photonic crystal light emitter," *Appl. Phys. Lett.* **87**, 131107 (2005).
119. W. Kuang, C. Kim, A. Stapleton, and J. D. O'Brien, "Grating-assisted coupling of optical fibers and photonic crystal waveguides," *Opt. Lett.* **27**, 1604–1607 (2002).
120. P. E. Barclay, K. Srinivasan, M. Borselli, and O. Painter, "Experimental demonstration of evanescent coupling from optical fiber tapers to photonic crystal waveguides," *Electron. Lett.* **39**, 842 (2003).
121. P. E. Barclay, K. Srinivasan, M. Borselli, and O. Painter, "Efficient input and output optical fiber coupling to a photonic crystal waveguide," *Opt. Lett.* **29**, 697–699 (2004).
122. J. C. Knight, G. Cheung, F. Jacques, and T. A. Birks, "Phase-matched excitation of whispering-gallery-mode resonances by a fiber taper," *Opt. Lett.* **22**, 1129–1131 (1997).

123. Y. K. Lizé, E. C. Mägi, V. G. Ta'eed, J. A. Bolger, P. Steinvurzel, and B. J. Eggleton, Microstructured optical fiber photonic wires with subwavelength core diameter," *Opt. Express* **12**, 3209–3217 (2004).
124. T. A. Birks and Y. W. Li, "The shape of fiber tapers," IEEE Journal of Lightwave Technology **10**, 432–438 (1992).
125. M. W. Lee, C. Grillet, C. G. Poulton, C. Monat, C. L. Smith, E. Mägi, D. Freeman, S. Madden, B. Luther-Davies, and B. J. Eggleton, "Characterizing photonic crystal waveguides with an expanded k-space evanescent coupling technique," *Opt. Express* **16**, 13800–13808 (2008).
126. J. Li, T. P. White, L. O'Faolain, A. Gomez-Iglesias, and T. F. Krauss, "Systematic design of flat band slow light in photonic crystal waveguides," *Opt. Express* **16**, 6227–6232 (2008).
127. J. C. Knight, G. Cheung, F. Jacques, and T. A. Birks, "Phase-matched excitation of whispering-gallery-mode resonances by a fiber taper," *Opt. Lett.* **22**, 1129–1131 (1997).
128. X. Letartre, C. Seassal, C. Grillet, P. Rojo-Romeo, P. Viktorovitch, M. L. V. d'Yerville, D. Cassagne, and C. Jouanin, "Group velocity and propagation losses measurement in a single-line photonic-crystal waveguide on InP membranes," *Appl. Phys. Lett.* **79**, 2312–2314 (2001).
129. K. Srinivasan and O. Painter, "Fourier space design of high-Q cavities in standard and compressed hexagonal lattice photonic crystals," *Opt. Express* **11**, 579–593 (2003).
130. U. Bog, C. L. C. Smith, M. W. Lee, S. Tomljenovic-Hanic, C. Grillet, C. Monat, L. O'Faolain, et al., "High-Q microfluidic cavities in silicon-based 2D photonic crystal structures," *Opt. Lett.* **33** (2008).
131. M. Loncar, A. Scherer, and Y. M. Qiu, "Photonic crystal laser sources for chemical detection," *App. Phys. Lett.* **82**(26), 4648–4650 (2003).
132. N. A. Mortensen, S. Xiao, and J. Pedersen, "Liquid-infiltrated photonic crystals," *Microfluid. Nanofluid.* **4**, 117–127 (2008).
133. S. H. Kwon, T. Sunner, M. Kamp, and A. Forchel, "Optimization of photonic crystal cavity for chemical sensing," *Opt. Express* **16**, 11709–11717 (2008).
134. E. Chow, A. Grot, L. W. Mirkarimi, M. Sigalas, and G. Girolami, "Ultracompact biological sensor built with two-dimensional photonic crystal microcavity," *Opt. Lett.* **29**(10), 1093–1095 (2004).
135. E. Verpoorte, "Chip vision-optics for microchips," *Lab Chip* **3**, 42N–52N (2003).
136. V. G. Kozlov, G. Parthasarathy, P. E. Burrows, V. B. Khalfin, J. Wang, S. Y. Chou, S. R. Forrest, "Structures for organic diode lasers and optical properties of organic semiconductors under intense optical and electrical excitations," *IEEE J. Quantum Electron.* **36**(1), 18–26 (2000).
137. T. Spehr, A. Siebert, T. Fuhrmann-Lieker, J. Salbeck, T. Rabe, T. Riedl, H.-H. Johannes, W. Kowalsky, J. Wang, T. Weimann, P. Hinze, "Organic solid-state ultraviolet-laser based on spiro-terphenyl," *Appl. Phys. Lett.* **87**, 161103 (2005).
138. G.A. Turnbull, A. Carleton, G.F. Barlow, A. Tahraouhi, T.F. Krauss, K.A. Shore, I.D.W. Samuel, "Influence of grating characteristics on the operation of circular-grating distributed-feedback polymer lasers," *J. Appl. Phys.* **98**, 023105 (2005).
139. C. Karnutsch, V. Haug, C. Gärtner, U. Lemmer, T. Farrell, B. Nehls, U. Scherf, J. Wang, T. Weimann, G. Heliotis, C. Pflumm, J.C. deMello, D.D.C. Bradley, "Low threshold blue conjugated polymer lasers with first- and second-order distributed feedback," *Appl. Phys. Lett.* **89**, 201108 (2006).
140. I. D. W. Samuel and G. A. Turnbull, "Organic semiconductor lasers," *Chem. Rev.* **107**, 1272–1295 (2007).
141. C. Karnutsch, C. Pflumm, G. Heliotis, J. C. deMello, D. D. C. Bradley, J. Wang, T. Weimann, V. Haug, C. Gärtner, and U. Lemmer, "Improved organic semiconductor lasers based on a mixed-order distributed feedback resonator design," *Appl. Phys. Lett.* **90**, 131104 (2007).
142. U. Scherf, S. Riechel, U. Lemmer, R. F. Mahrt, "Conjugated polymers: lasing and stimulated emission," *Curr. Opin. Solid State Mat. Sci.* **5**, 143–154 (2001).

143. R. Xia, G. Heliotis, D. D. C. Bradley, "Semiconducting polyfluorenes as materials for solid-state polymer lasers across the visible spectrum," *Synth. Met.* **140**, 117–120 (2004).
144. T. Kobayashi, J. B. Savatier, G. Jordan, W. J. Blau, Y. Suzuki, and T. Kaino, "Near-infrared laser emission from luminescent plastic waveguides," *Appl. Phys. Lett.* **85**(2), 185–187 (2004).
145. K. Yamashita, T. Kuro, K. Oe, and H. Yanagi, "Low threshold amplified spontaneous emission from near-infrared dye-doped polymeric waveguide," *Appl. Phys. Lett.* **88**, 241110 (2006).
146. C. Karnutsch, V. Haug, C. Gärtner, U. Lemmer, T. Farrell, B. Nehls, U. Scherf, J. Wang, T. Weimann, G. Heliotis, C. Pflumm, J. C. deMello, and D. D. C. Bradley, "Low threshold blue conjugated polymer DFB lasers," *Conf. on Lasers and Optoelectronics (CLEO)*, CFJ3, Long Beach, CA, (2006).
147. C. Karnutsch, M. Stroisch, M. Punke, U. Lemmer, J. Wang, and T. Weimann, "Laser diode-pumped organic semiconductor lasers utilizing two-dimensional photonic crystal resonators," *IEEE Photonics Technol. Lett.* **19**(10), 741–743 (2007).
148. T. Riedl, T. Rabe, H.-H. Johannes, W. Kowalsky, T. Weimann, J. Wang, P. Hinze, B. Nehls, T. Farrell, and U. Scherf, "Tunable organic thin-film laser pumped by an inorganic violet diode laser," *Appl. Phys. Lett.* **88**, 241116 (2006).
149. A. E. Vasdekis, G. Tsiminis, J.-C. Ribierre, L. O' Faolain, T. F. Krauss, G. A. Turnbull, I. D. W. Samuel, "Diode pumped distributed Bragg reflector lasers based on a dye-to-polymer energy transfer blend," *Opt. Express* **14**(20), 9211–9216 (2006).
150. Y. Yang, G. A. Turnbull, and I. D. W. Samuel, "Hybrid optoelectronics: A polymer laser pumped by a nitride light-emitting diode," *Appl. Phys. Lett.* **92**, 163306–163303 (2008).

CHAPTER 19

Micro and Nano Optofluidic Flow Manipulation

G. Logan Liu

Department of Electrical and Computer Engineering, University of Illinois at Urbana-Champaign

Luke P. Lee

Department of Bioengineering, University of California—Berkeley

19-1 Introduction to Optofluidic Flow Manipulation

Flow manipulation especially at microfluidic level is the key technical issue in dynamic fluidic devices. Microfluidic manipulation is geometrically constrained to submillimeter scale and has unique features different from the macroscale flow. The aspects such as surface tension and fluidic resistance dominate the microfluidic flow behavior as typically the Reynolds number, the number to characterize turbulent flow is low. Due to its unique properties including minute volume, fast speed, and predictable flow pattern, microfluidics has been widely used in chemical and biological fluidic processing related to chemical synthesis, emulsification, gene and protein analysis, and cell and tissue culture. The complexity of microfluidic circuits is increasing with more and more integrated functions and so is the flow manipulation in such circuits. All these microfluidic applications require precise manipulations of the speed and direction of the fluidic flow. Mechanical fluid actuation still remains the dominant flow-control methods in microfluidics. Mechanical fluidic manipulation has the advantages in reliability and stability; however, it also has apparent disadvantages such as macroscale pump- and valve-control units. The microfluidic systems with mechanical fluidic control usually occupy large space, which compromised their portability and limit field applications. Although microscale valves and pumps have been developed and a relatively large-scale integrated microfluidic device with 10s of mechanical fluidic manipulation units was demonstrated, the realization of very large-scale microfluidic networks is intimidated by the cumbersome flow manipulation.

Optofluidics is a new emerging concept for microfluidic flow manipulation using light. With the rapid development of optical and photonic technologies in the past half century, various light-control methods have been invented. The increasing demands in optical communication and holography industry for large-scale dynamic light manipulation have significantly propelled the technology progress in this respect. Among all the light manipulation technologies, digital micromirror device (DMD), liquid crystal spatial light modulation (SMD), and vertical-cavity surface-emitting laser (VCSEL) arrays are the mostly ready ones that are able to generate large-area dynamic light illumination patterns and can be integrated in a portable system. In optofluidic flow control, light is either provided as the energy source for the dynamic fluidic flow or as the cue signal to direct the movement of fluidic flow. The dynamic light patterning and scanning technology makes the versatile optofluidic flow control available for very large-scale microfluidic networks and promises the insurmountable advantage in terms of flexibility over conventional mechanical fluidic control.

The idea of the direct conversion from optical energy into hydrodynamic energy is stimulating, although the photovoltaic driven electrical engine can be used to actuate fluidic flow mechanically. The discussion of the optofluidic actuation shall be limited to the microfluidic devices, where surface tension and flow resistance play far more important roles. In this chapter two major optofluidic actuation principles will be reviewed. One is optically changing the surface tension via photochemistry to move microscale droplet and another one is the rapid microscale photothermal evaporation and recondensation process with the assistance from nanoplasmonic structures. Besides optofluidic actuation another aspect discussed in this chapter is the optofluidic particle manipulation in liquid. Primarily, the photothermophoresis molecular trapping and optoelectronic tweezer will be discussed.

19-2 Optical Manipulation of Liquid Surface Tension

Surface tension is the driving force to attract the surface of a portion of liquid to another surface or another portion of liquid. Surface tension is caused by the intermolecular attraction forces between the liquid molecules. Inside bulk liquid each molecule is pulled equally in all directions by neighboring liquid molecules, resulting in a net force of zero. At the surface of the liquid, all the molecules are also subject to an inward force from the molecular attraction inside the liquid which is balanced only by the liquid's resistance to compression. Due to the interaction force between the adjacent molecules, there is a driving force at the liquid surface to diminish the surface area resembling a stretched elastic membrane. The liquid has to minimize its number of boundary molecules to minimize its energy state and therefore minimize its surface area. The liquid will keep changing its surface profile until it reaches the lowest surface area possible.

Surface tension is mathematically defined as the force along a line of unit length and the force is parallel to the surface. In thermodynamics, the surface tension is defined as work done per unit area, that is, the potential energy needed for the unit surface area increase of a mass of liquid. A free droplet of liquid naturally assumes a spherical shape which has the minimum surface area and energy for a given volume. If no force acts normal to a tensioned liquid surface from the supporting surface, the surface will remain flat. However, the pressure difference on either side of the surface area will result in a normal force. The liquid surface must be curved in order for the surface tension forces to cancel the force due to pressure. When all the forces are balanced, the resulting equation is known as the Young-Laplace equation [1] $\Delta p = \gamma(1/R_x + 1/R_y)$, where Δp is the pressure difference, γ is surface tension, and R_x and R_y are radii of curvature in the orthogonal axes tangential to the liquid surface. The shape of the liquid surface can be determined by this equation.

For a liquid droplet sitting on a flat surface, the pressure will result in certain geometry of the liquid droplet on the surface. Surface tension in this case is not only the surface tension of the liquid, but the surface tension between the liquid interface with the supporting flat surface and air. The shape of the liquid must be in the geometry to make all surface tension forces balanced. Where the two surfaces, for example, the liquid surface and supporting substrate surface meet, they form a contact angle, which is the angle between the tangential line of the liquid surface and the solid surface. The diagrams in Fig. 19-1 show the contact angle of a liquid droplet on a surface in the case of hydrophilic and hydrophobic surfaces, respectively. Tension forces exist at the liquid-air interface, the liquid-solid interface, and the solid-air interface. The contact angle has to be the value satisfying the surface tension balance equation $\gamma_{ls} - \gamma_{sa} = -\gamma_{ls} \cos\theta$, where γ_{ls}, γ_{sa}, and γ_{la} are the surface tensions for the liquid-solid interface, the solid-air interface, and the liquid-air interface, respectively. The liquid-air surface tension is larger than the differential surface tension of the liquid-solid interface and the solid-air interface. For highly hydrophobic surfaces, the contact angle is larger than 90° and the liquid-air surface tension component along the solid surface is pointing outward

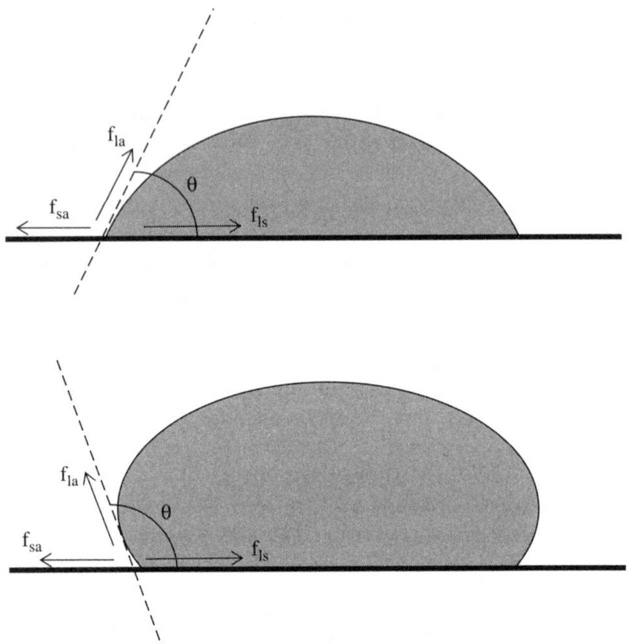

Figure 19-1 Schematic diagram of a liquid-solid interface. The solid surface can be hydrophilic (upper image) or hydrophobic. The contact angle and surface tension forces are denoted.

from the liquid body. For highly hydrophilic surface, the contact angle is smaller than 90° and the liquid-air surface tension component along the solid surface is pointing inward the liquid body. The change of the solid surface property will result in the change of contact angle and the liquid-wetting area on the solid surface. Therefore, the liquid flow may be actuated by dynamically changing the surface tension of the liquid-solid interface.

19-2-1 Photochemical Control of Surface Tension

A gradient in surface tension can induce a net mass transport of liquids. Microfluidic flow controlled by surface tension at the liquid-solid interface was discussed in the previous section. The gradient of surface tension force can be harnessed to move liquid droplets on the surface even uphill against the gravitational force. Such nonmechanical flow arising from the action of a surface tension gradient can be created by several approaches, including thermal [2], chemical [3], and electrochemical [4] methods. In most cases the surface tension property of the supporting substrate was patterned with a fixed chemical layer, the microfluidic flow or droplet movement has to follow a fixed pattern. Dynamical control of the microfluidic flow pattern utilizing surface tension forces thus requires a dynamic patterning of surface chemistry.

Ichimura et al. have shown an optofluidic liquid droplet actuation on a solid surface by patterning the surface with photo responsive chemical molecules [5]. When a liquid droplet was placed on a substrate surface modified with a layer of photoresponsive chemical molecules having photochromic azobenzene units, asymmetrical photoirradiation can cause a gradient in surface free energy due to the photoisomerization of surface azobenzenes, leading to the directional motion of the droplet. The macroscopic motion of liquids on a flat solid surface was manipulated reversibly by photoirradiation of a photoisomerizable monolayer covering the surface. The direction and velocity of the motion are tunable by varying the direction and steepness of the gradient in light intensity.

Because the surface free energies of flat solid substrates are determined by atomic level constitutions of their outermost surfaces [6,7], alteration of chemical structures of the outermost monomolecular layers by light can be used to trigger and manipulate various interfacial phenomena, including wettability [8,9], liquid crystal alignment [10], and dispersibility [11]. Thus, if a gradient in surface energy is generated photochemically as a result of spatially controlled changes of chemical structures of an outermost surface, the motion of a liquid can be guided by spatially controlled photoirradiation of the photoresponsive substrate surface.

The photoresponsive chemical molecule used here is a crown conformer of *O*-carboxymethylated calix[4]resorcinarene (CRA-CM, Fig. 19-2) bearing four *p*-octylazobenzene residues at one of the rims of the cyclic skeleton [12] to assemble a photoresponsive monolayer. The photoresponsive self-assembled monolayer was prepared simply by immersing an aminosilylated silica plate in a dilute solution of CRA-CM, yielding a robust monolayer with dense packing [12]. As a result of the flat-laid adsorption of CRA-CM molecules on a silica surface [13], the octylazobenzene units in their trans state are stretched out to be exposed to the air. The outermost surface of a UV-exposed CRA-CM monolayer is likely terminated by the polar *cis*-azogroups, leading to an increase in surface free energy. Photoirradiation of the cis-rich surface with blue (436 nm) light causes the cis isomer to reverse into the trans isomer.

They placed a droplet of olive oil on the photoresponsive surface of the CRA-CM–modified plate and then used spatially controlled irradiation to generate a gradient in the level of photoisomerization. Figure 19-3 shows the directional motion of a droplet on a cis-rich surface upon asymmetrical irradiation with blue light. The surface energy gradient between the advancing and receding edges of the droplet was constantly maintained by moving the light beam, which

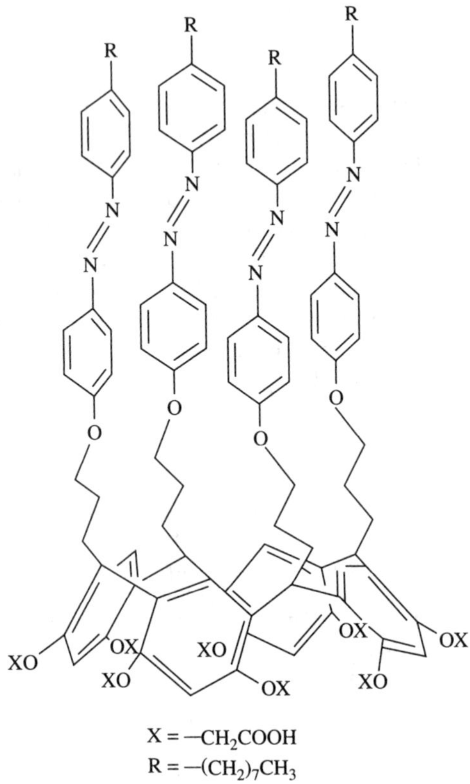

FIGURE 19-2 A photoresponsive chemical molecule that can be coated on the substrate surface to change the surface tension force at the liquid-solid interface with light irradiation. (*From K. Ichimura, S-K. Oh, and M. Nakagawa, "Light-driven motion of liquids on a photoresponsive surface,"* Science, *288, (2000) 1624–1626. Reprinted with permission from AAAS.*)

continued moving the droplet. To stop the movement of the droplet, the photoresponsive surface is irradiated with a homogeneous blue light (Fig. 19-3c). The velocity of the droplet relies on the intensity and gradient of the light. A typical speed of 35 mm/s was reported. The surface-assisted liquid motion was also workable for other surface chemicals including 1-methylnaphthalene and 1,1,2,2-tetrachloroethane, and even for nematic liquid crystals including NPC-02 (a binary mixture of 4-propyl-49-ethoxy- and 4-propyl-49-butoxyphenylcyclohexanes) and 5CB (4-pentyl-49-cyanobiphenyl).

Alternating irradiation of a CRA-CM monolayer with homogeneous UV and blue light (1.0 mW/cm^2, 100 s) can lead to reversible in situ symmetrical spreading and dewetting of droplets of the liquids as shown in Fig. 19-4. The spreading is delayed markedly when compared with the photoisomerization processes. The delay likely

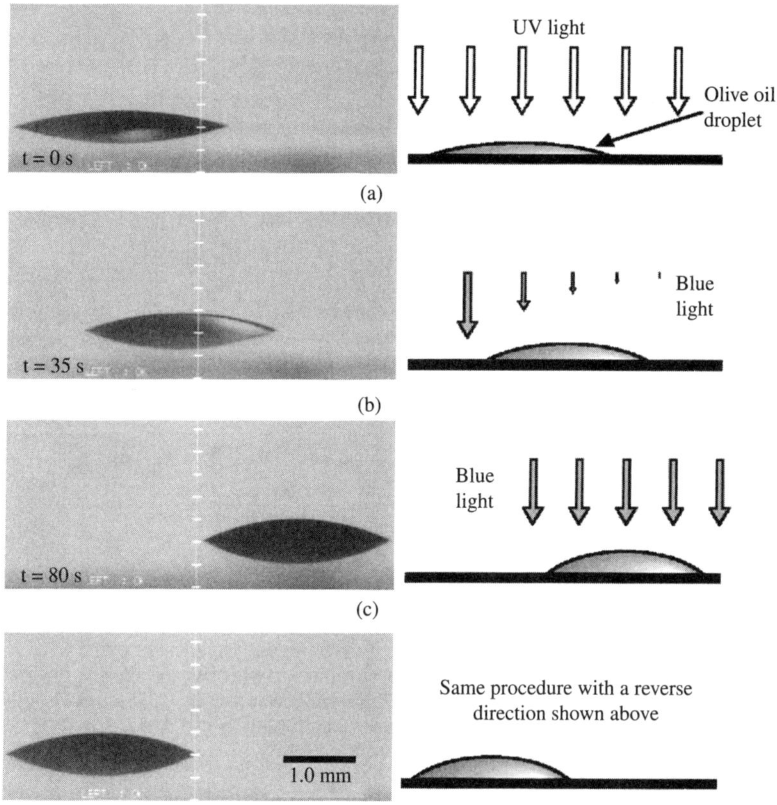

FIGURE 19-3 Lateral photographs of light-driven motion of an olive oil droplet on a silica plate modified with CRA-CM. The olive oil droplet on a cis-rich surface moved in a direction of higher surface energy by asymmetrical irradiation with 436-nm light perpendicular to the surface. (*From K. Ichimura, S-K. Oh, and M. Nakagawa, "Light-driven motion of liquids on a photoresponsive surface,"* Science, *288, (2000) 1624–1626. Reprinted with permission from AAAS.*)

arises from dynamic processes involving the reorientation of photoisomerized azobenzenes so as to minimize an interfacial energy [14,15]. This effect should be one of the factors affecting the velocity of a surface chemistry–assisted light-driven motion of a liquid droplet (Fig. 19-3) in addition to the steepness of the gradient in surface energy, the droplet volume, and the surface tension and viscosity of the droplet.

Because photochemical events can be controlled precisely in time and space, the surface-assisted, light-driven motion of liquids can lead to improved understanding of interface phenomena, including the spreading kinetics and the role of surface tension gradients.

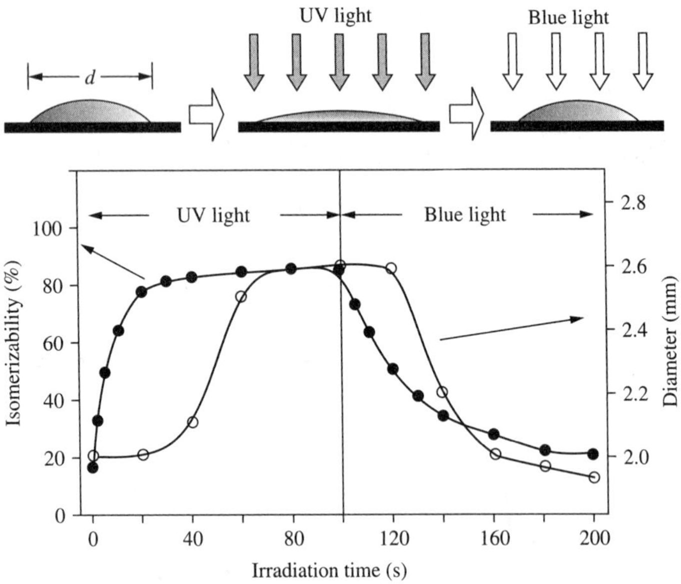

FIGURE 19-4 Correlation between the level of cis isomer (solid circles) and the diameter (open circles) of an NPC-02 droplet placed on a silica plate modified with CRA-CM upon homogeneous irradiation with UV and blue light (1.0 mW/cm^2). (*From K. Ichimura, S-K. Oh, and M. Nakagawa, "Light-driven motion of liquids on a photoresponsive surface,"* Science, *288, (2000) 1624–1626. Reprinted with permission from AAAS.*)

19-2-2 Optoelectronic Liquid Surface Wetting

Besides the surface chemistry, electrical potential can also affect the surface tension at the liquid-solid interface, which is known as electrowetting effect [16,17]. The electrowetting effect is the change in solid electrolyte contact angle due to an applied voltage potential between the solid and the electrolyte as shown in Fig. 19-5. The electrowetting

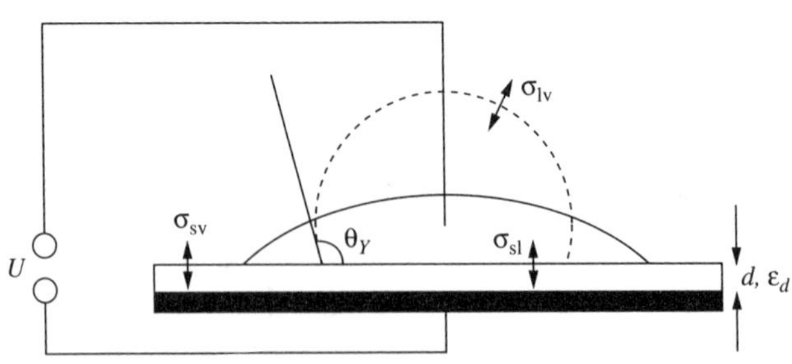

FIGURE 19-5 Generic electrowetting setup. Partially wetting liquid droplet at zero voltage (dotted line) and at high voltage (solid). (*Source: F. Mugele and J.-C. Baret, "Electrowetting: from basics to applications,"* J. Phs.: Condens. Matter, *17, (2005) R705–774.*)

force result from the applied electric field will change the total force along the liquid-solid interface and thus the liquid droplet has to change its contact angle to satisfy the surface tension balance equation again. The fringing field at the corners of the electrolyte droplet also tends to pull the droplet down onto the electrode, lowering the macroscopic contact angle and increasing the droplet contact area.

The electrowetting behavior can be described using the thermodynamic model and the surface tension at the liquid-solid interface will change due to the interfacial charge induced by the applied electrical field [18]. Upon applying a voltage U, an electric double layer builds up spontaneously at the solid-liquid interface consisting of charges on the metal surface on the one hand and of a cloud of oppositely charged counter-ions on the liquid side of the interface. For simplicity we can make the assumption that the counter-ions are all located at the upper surface of the insulating dielectric thin film and have fixed distance d from the electrode underneath the dielectric film. The contact angle of the liquid droplet with the electrowetting effect is defined as

$$Cos\theta = \frac{\gamma_{sa} - \gamma_{la}}{\gamma_{la}} + \frac{\varepsilon_0 - \gamma_l}{2d\gamma_{la}} U^2$$

here γ_{ls}, γ_{sa}, and γ_{la} are the surface tensions for the liquid-solid interface, solid-air interface, and liquid-air interface, respectively and ε_l is the dielectric constant of liquid. The surface tension at the liquid-solid surface due to electrowetting effect is defined as

$$\gamma'_{la} = \gamma_{la} - \frac{\varepsilon_0 \varepsilon_l}{2d} U^2$$

When the electrical field is only applied around a subportion of a liquid drop, the contact angle is not uniform everywhere and the center of mass of the liquid droplet will deviate from the geometrical center. The unbalancing of the gravitational force will then lead to the contact-line movement of the liquid droplet. Electrowetting-induced motion is analogous to the motion of liquid droplets on chemically patterned substrates. The driving force per unit length Δy of the contact line is determined by the energy gain upon displacing the contact line by Δx

$$\Delta E/\Delta y = -\varepsilon_0 \varepsilon_l U^2 \Delta x/d$$

The force f per unit length Δy is given by $f = -\Delta E/\Delta x = \varepsilon_0 \varepsilon_l U^2 dy/d$ and this force is perpendicular to the contact line. The net force on a droplet can be calculated by integrating the force vector per unit length along the contact line. If bulk viscous effects dominate in the liquid droplet, the hydrodynamic pressure gradient arises and drives fluid flow within the droplet. The liquid motion can only be achieved above some threshold applied voltage when the angle on the leading edge of the droplet exceeds the local contact angle.

Multiple fluidic functions, such as liquid injection, transportation, mixing, and separation can be integrated on a single microfluidic chip using eletrowetting controls [19]. Manipulating multiple droplets simultaneously requires a two-dimensional array of electrodes to control the local surface tension, however, the integration of a large number of electrodes presents a challenge for device fabrications. In order to solve this challenge, Chiou et al. report a novel mechanism "optoelectrowetting (OEW)" for light actuation of liquid droplets [20]. They integrate a photoconductive material underneath the electrowetting electrodes. In the OEW microfluidic chip, a microliter droplet can be transported to any location on the chip by translating the light spots illuminating on the droplet. This OEW fluidic control method eliminates the wiring bottleneck of conventional electrowetting devices. By controlling the light illumination spot size, nanoliter or smaller droplets may be manipulated.

Figure 19-6 shows the configuration of the optoelectrowetting device. The photoconductive material, that is, amorphous silicon is

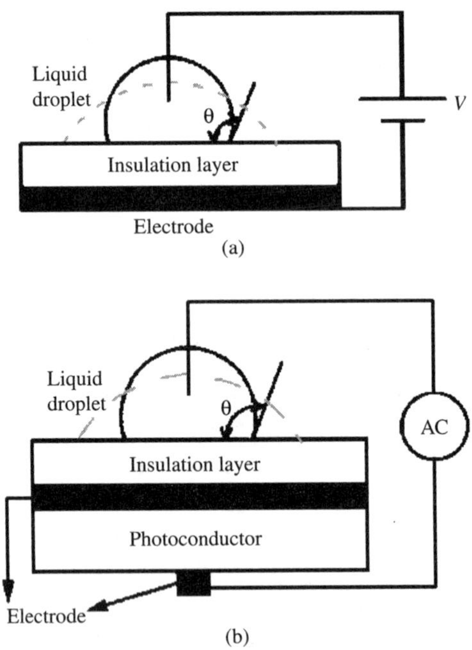

FIGURE 19-6 (a) Conventional electrowetting under dc bias. (b) Optoelectrowetting with an integrated photoconductor under ac bias. (*Reprinted from P.-Y. Chiou, H. Moon, H. Toshiyoshi, C.-J. Kim, and M. C. Wu, "Light actuation of liquid by optoelectrowetting,"* Sensors and Actuators A, *104, (2003) 222–228, with permission from Elsevier.*)

integrated under the electrodes of conventional electrowetting circuit. As the electrical circuit model, the electrical impedance of the liquid, thin-film insulator, photoconductor, and bottom electrode are serially connected and ac voltage is applied across the liquid and the bottom electrode. Since the contact angle of the droplet on the OEW surface is determined by the voltage drop across the insulating layer, the droplet contact angle may be changed with the impedance change of the photoconductor. The frequency of the ac voltage is adjusted such that the impedance of the photoconductor dominates in the absence of light and there is very little voltage across the insulating layer, in which case the contact angle remains the original state. While the light illumination on the photoconductor will significantly increase its conductivity due to electron-hole pair generation. As a result, most of the voltage drop is across the insulating layer and the contact angle is therefore reduced by light illumination.

The liquid droplet is sandwiched between a top hydrophobic surface and a bottom OEW surface. The topside is a transparent conductive indium-tin-oxide (ITO) glass coated with 20 nm of Teflon. The OEW structure is realized by integrating a two-dimensional array of electrowetting electrodes on a photoconductive material, amorphous silicon deposited by plasma-enhanced chemical vapor deposition (PECVD). The electrodes are then covered by a thin layer of SiO_2 and Teflon coating. The Al electrode below the photoconductor is connected to a power supply. The principle of moving a liquid droplet on an OEW surface by light actuation is as follows: An ac voltage is applied between the top ITO electrode and the bottom Al electrode. Shining an optical beam on one edge of the liquid droplet decreases the contact angle and creates a pressure difference between two ends of the droplet. The liquid droplet then follows the movement of the optical beam. The light actuation scheme described here requires ac bias. Liquid droplet does not move under dc voltage bias because the majority of the voltage drops is across the insulating layer. The contact angle on both ends of the droplet decreases and no pressure difference is created.

Figure 19-7 shows the optofluidic liquid droplet manipulation on an OEW device. The chip area is 1 cm × 1 cm and a water droplet with a diameter of 2 mm is sandwiched between a Teflon-coated ITO glass and an OEW surface with a gap spacing of 0.5 mm. A 4-mW laser at 532-nm wavelength is used to drag the liquid droplet. The droplet can be moved across the 1 cm × 1 cm surface by the laser beam and the movement speed can be as fast as 7 mm/s, which is limited by the scanning speed of the laser. The four snap shots in the figure clearly show the transport of the liquid droplets. The laser illuminated spot at the leading edge of the droplet is also visible in the figure.

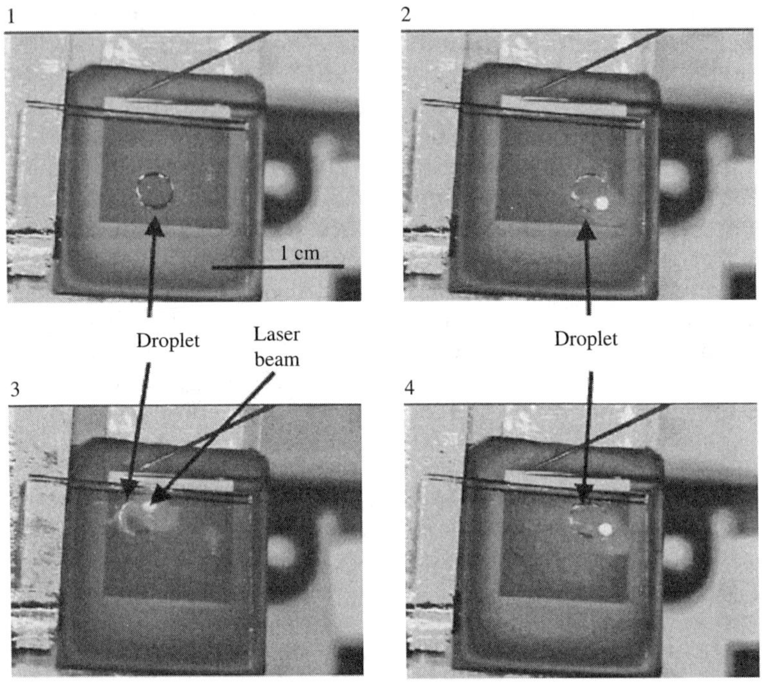

FIGURE 19-7 Images of liquid transport across a 1 cm × 1 cm OEW area actuated by an optical beam. (*Reprinted from P.-Y. Chiou, H. Moon, H. Toshiyoshi, C.-J. Kim, and M. C. Wu, "Light actuation of liquid by optoelectrowetting,"* Sensors and Actuators A, *104, (2003) 222–228, with permission from Elsevier.*)

19-3 Photothermal Fluidic Actuations

In this section, the optofluidic manipulation by photothermal heating will be discussed. The temperature gradient can induce the liquid flow in the microfluidic flow cell when heated nonuniformly. For a thin liquid layer sandwiched between two plates like the microfluidic flow cell, the initial movement is the upwelling of warmer liquid from the heated bottom layer. When applying heat from below, the temperature at the top layer will show temperature fluctuations. With increase in temperature, surface tension decreases and a lateral flow of liquid at the surface will take place from warmer areas to cooler areas. It is known that liquids flow from places of lower surface tension to places of higher surface tension. This is called the Marangoni effect [21]. In order to preserve a horizontal liquid surface, liquid from the cooler places on the surface have to go down into the liquid. Therefore, the vertical convection flow can be actuated in the liquid by heating it from the bottom. Liquid droplet can be moved in microfluidic flow cells by thermocapillar pumping. Thermocapillary pumping is referred as fluid motion induced by temperature-dependent variations of surface tension within the liquid as well as different contact angles. The surface tension of a droplet can be locally manipulated by local heating to achieve two different contact angles and curvatures. Since the curvature is related to a specific pressure, a pressure difference within the droplet occurs and that makes the droplet move.

When the temperature is above the boiling point of the liquid, evaporation will occur. Evaporation is a type of phase transition and slow vaporization of a liquid. It is the reaction by which molecules in a liquid state such as water spontaneously become gaseous such as water vapor. Condensation, the opposite of evaporation, is the change of the physical state of aggregation of matter from gaseous phase into liquid phase. Condensation commonly occurs when a vapor is cooled to its dew point. The condensed vapor is called a condensate, the laboratory or the industrial equipment used for condensation is called a condenser. Water vapor will only condense onto another surface when the temperature of that surface is cooler than the temperature of the water vapor. When water vapor condenses into liquid water, the hydrogen bonds form again and release latent heat, which increases the sensible heat and causes the air temperature to rise. Sensible heat is removed from the air and the temperature drops when evaporation is occurring, and latent heat is converted to sensible heat and the temperature rises when condensation occurs.

The water vapor evaporated from the surface of liquid droplet or wave front quickly condenses in the cooler air to form new liquid condensates on the surface in front of the droplet contact line. The condensates close to the droplet contact line will not only change the surface tension locally but rejoin the droplet and move the droplet forward. The local temperature rising and liquid evaporation can be controlled by local photothermal heating of the liquid. This effect can be enhanced by integrating nanoscale photothermal enhancing nanoparticles in microfluidic flow. The details will be elaborated in the following sections.

19-3-1 Fluidic Actuation via Photothermal Nanoparticles

Photothermal metallic nanoparticles were used as the localized heat source in liquid. Liu et al. have presented optofluidic manipulations based on a direct optical-to-hydrodynamic energy conversion using suspended photothermal nanoparticles near the liquid-air interface [21]. With a focused laser spot, liquid flows are driven and guided in microfluidic channels to transport biomolecules and living cells at controlled speeds and directions.

They demonstrated a new mechanism of optofluidic effect using photothermal nanoparticles as shown in Fig. 19-8. The substrate surface on which the microfluidic flow is conducted is coated with hydrophobic chemical molecules. Without any actuation the liquid on the hydrophobic surface remains stationary. Proper concentrations of photothermal nanoparticles are suspended in the liquid and the local

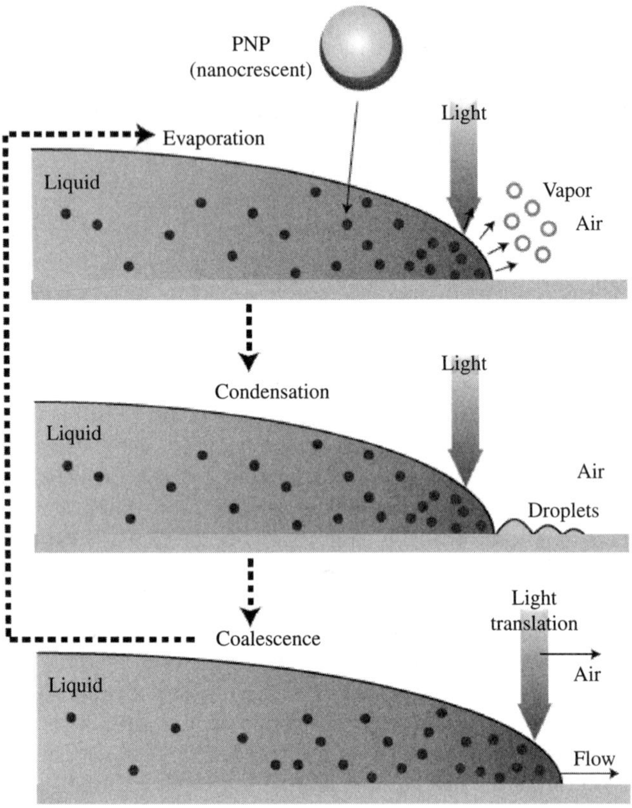

Figure 19-8 The principle of the optically controlled advance of the liquid-air interface. First, the focused-light illumination on the photothermal nanoparticles increases the local temperature of the liquid and leads to water evaporation at the liquid-air interface. Second, the vapor in the relatively cold air condenses into droplets in front of the liquid-air interface. Third, the droplets coalesce with the original bulk liquid body and the liquid-air interface advances. The processes are repeated as the light is translated, so the optofluidic flow can be continuous. (*Reprinted by permission from Macmillan Publishers Ltd: G. L. Liu, J. Kim, Y. Lu, and L. P. Lee, "Optofluidic control via photothermal nanoparticles."* Nature Mater. *5, (2005) 27–32.*)

particle concentration near the liquid-air interface is higher due to the "coffee-ring" effect [23]. When a focused light illuminates the photothermal nanoparticles near the liquid-air interface, heat is generated and immediately transferred from the photothermal nanoparticles to the surrounding liquid, that is, water. Beyond the boiling point, the heated liquid rapidly evaporates from the interface and produces vapor. The original liquid contact line is pinned and liquid lost in evaporation is replenished from the interior region, so the liquid does not retreat from the contact line. Since the heat source is in the nanoscale dimension, the air temperature remains relatively static. The vapor in the colder air condenses almost immediately after the evaporation and droplets form right in front of the original liquid-air interface. The droplets then coalesce with each other and grow into larger ones that eventually merge with the original liquid body and extend its contact line. The surface wetting by the coalesced droplets also assists the advance of the liquid-air interface. The photothermal nanoparticles are drawn toward the new contact line because of the liquid motion and convection. The above processes can occur repeatedly and concurrently, and the liquid flow can be continuous if the light illumination is translated along with the advancing liquid-air interface.

Using the suspended photothermal nanoparticles, optofluidic liquid guiding in polydimethylsiloxane (PDMS) microfluidic chips are demonstrated. PDMS devices fabricated by soft lithography have been widely used in chemical, biomolecular, and cellular analysis. The application of optofluidic liquid manipulation in PDMS devices benefits a very large research and industrial community. Especially, the optically controlled fluidic flow in predefined microchannels is laminar and unidirectional. It shows a much higher flow speed as the vapor and droplets are bound within the channel and contribute to the liquid advance only along the channel direction and the minimized vertical convection in microchannels favors the heat concentration at the liquid-air interface. The configuration of the microfluidic device is straightforward and simply formed by directly placing a PDMS slab with recessed grooves on the hydrophobic glass slide as shown in Fig. 19-9.

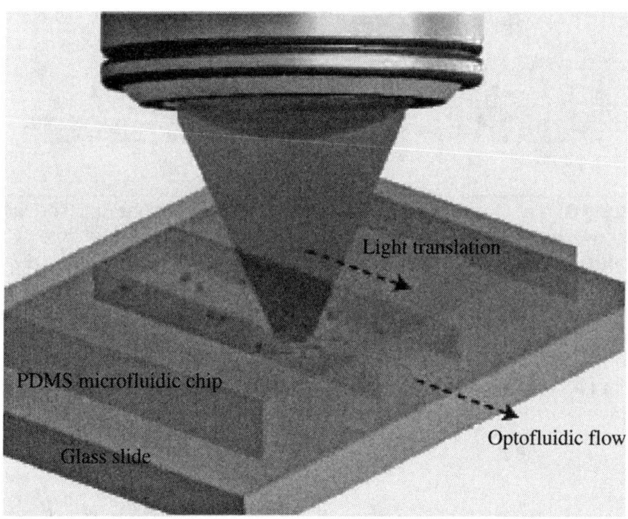

FIGURE 19-9 Illustration of the experimental system configuration for optofluidic control in straight microfluidic channels. (*Reprinted by permission from Macmillan Publishers Ltd: G. L. Liu, J. Kim, Y. Lu, and L. P. Lee, "Optofluidic control via photothermal nanoparticles."* Nature Mater. *5, (2005) 27–32.*)

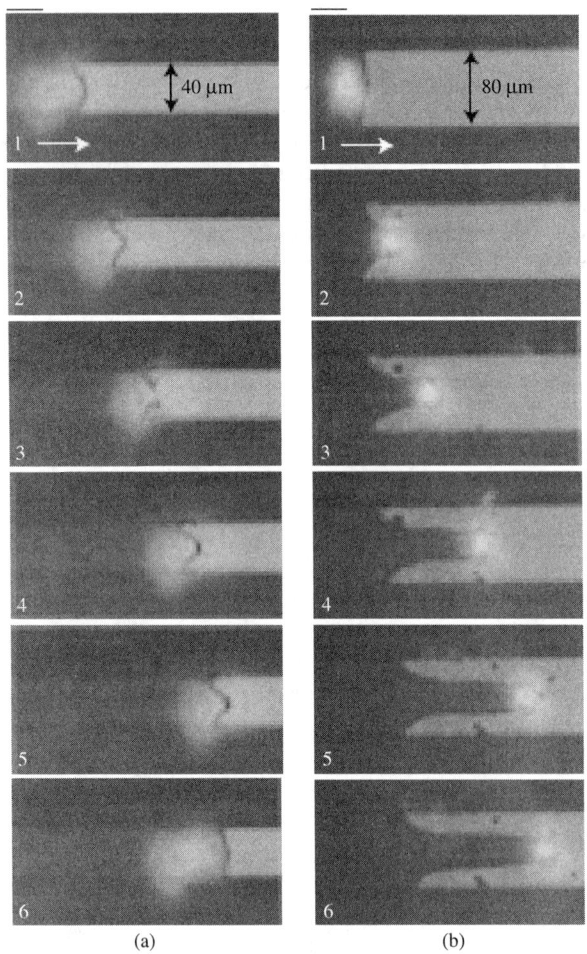

FIGURE 19-10 (a) Optofluidic control in a 40-μm-wide channel. The video prints show that the flow of the 0.5-nM photothermal nanoparticle suspended 1X PBS buffer solution follows the optical guiding of a 10-mW, 785-nm laser spot at a speed of ~50 μm/s (frames 1–5) and stops otherwise (frame 6). (b) Optofluidic control in an 80-μm-wide channel. The 1-nM photothermal nanoparticle suspended solution is guided by a 10-mW, 785-nm laser spot at a speed of ~50 μm/s. (*Reprinted by permission from Macmillan Publishers Ltd: G. L. Liu, J. Kim, Y. Lu, and L. P. Lee, "Optofluidic control via photothermal nanoparticles."* Nature Mater. *5, (2005) 27–32.*)

Figure 19-10 shows that the photothermal nanoparticle-suspended liquid in a 40-μm-wide and 5-μm-high channel was driven and guided by the translation of a focused 785-nm laser spot. The liquid remains stationary in the hydrophobic channel without the light guide due to the balanced surface energy, and no thermocapillary flow is seen when the light spot illuminates the interior of the liquid. The liquid flow stops immediately after the light translation stops, and liquid motion in the microchannel is under complete control without any valve or pump. For a channel width (80 μm) much larger than the focused light spot (10 μm), the optofluidic flow can also be realized. An extruding liquid flow is generated as only a portion of the liquid-air interface is illuminated by the light spot (Fig. 19-10b).

Previously, the positioning of liquid flow at channel junctions requires complicated valve systems [24]. The light illumination spot can be shaped into various geometries and used to actuate different flow patterns. The optofluidic liquid flows in parallel microfluidic channels, at channel junctions or converging mixing channels are controlled with excellent directionality by precise light control. It is to be believed that photothermal nanoparticle-assisted optofluidic control can be realized in a microfluidic "maze" with many junctions in various shapes. Besides small soluble or suspended molecules can be transported in microfluidic devices with optofluidic control, but living cells can also be transported. Since the photothermal heating through the nanoparticles is an extremely localized effect, most of the molecules and cells can be transported intact.

The maximum speed shown is around 1 mm/s, which is limited by the rate of droplet formation, growth, and coalescence. Light illumination power, microchannel dimension, and nanoparticle concentration are three major tunable factors to determine the rate of droplet formation and coalescence. The photothermal nanoparticle-assisted optofluidic flow speed can be further increased by adopting narrower microchannels, more accurate light control, and nanoparticles with higher photothermal conversion efficiency.

19-3-2 Fluidic Actuation via Photothermal Nanocarpet

Other than being suspended in liquid, photothermal nanostructures can be immobilized as a carpet on the substrate surface in microfluidic devices. The photothermal heating on the nanocarpet will induce a new liquid mass–transfer effect. Boyd et al. reported interphase mass-transfer in microfluidics through the bubble when a small amount of heat is added close to a liquid-vapor interface of a captive gas bubble and the flow rate can be controlled with only a slight change in the temperature of the fluid [25]. This method is referred to as bubble-assisted interphase mass-transfer (BAIM). BAIM can be realized with an all-optical technique by illuminating an array of nanoscale plasmonic metal structures with a stationary low-power laser. In optofluidic actuation assisted by the plasmonic nanocarpet, the interphase mass-transfer occurs at both ambient temperature and pressure and without the need for active cooling for complete recovery of the vapor. The fluid from a warmer portion of the interface is vaporized and then condensed on a cooler portion [26].

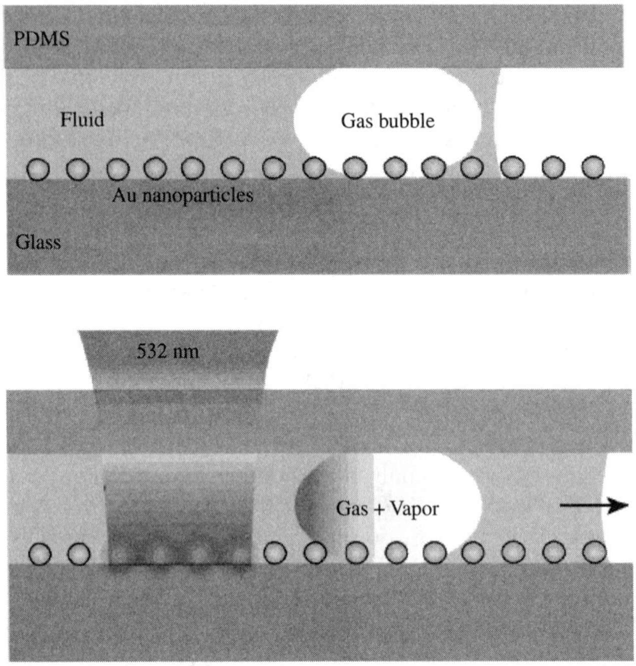

Figure 19-11 Schematic of the microchannel assembly (side view). An array of nanoparticles is placed on a glass slide, which serves as the base of the channel (top). A laser near the resonant frequency of the nanoparticle array is focused on the substrate, heating the nanoparticles. The heat from the nanoparticles is transferred to the surrounding fluid resulting in evaporation into the gas bubble. The vapor is subsequently condensed on the opposite side of the bubble causing an increase in the volume of the fluid to the right of the bubble and a corresponding movement to the right of the position of the free surface of the fluid column, the far right interface. (*Reprinted with permission from D. A. Boyd, J. R. Adleman, D. G. Goodwin, and D. Psaltis, "Chemical separations by bubble-assisted interphase mass-transfer,"* Anal. Chem., *80, (2008) 2452–2456. Copyright 2008 American Chemical Society.*)

A schematic of the plasmonic nanocarpet–integrated microchannel system is illustrated in Fig. 19-11; the system consists of a microfluidic channel, which has a quasi-ordered array of gold nanoparticles fabricated on the surface and a captive gas bubble. A laser with a frequency near the plasmon resonant frequency of the nanoparticles array is focused on the particles near the edge of a gas bubble causing them to be heated. The heat from the nanoparticles is transferred to the surrounding fluid causing evaporation from the surface near the laser and subsequent condensation on the far surface of the bubble. The fabrication process of such a system is as following. The channels are made in PDMS and sealed to a glass substrate coated with an array of Au nanoparticles which was created by block copolymer lithography and metal evaporation. The average particle diameter was 14.5 nm with an average spacing of 46 nm. The width of the channels in the demonstrated experiments was 30 μm, and the height was 5 μm. The power of the illuminating laser at the sample is around 14 mW, and the diameter of the beam spot was approximately 10 μm.

Air bubbles were formed in the liquid by trapping air in the partially filled channel. To do so, the laser spot is placed near the free surface of the liquid causing evaporation and recondensation on the channel walls 10 to 30 μm away. The droplets grew together to form a continuous liquid plug, trapping an air bubble with a width of 10 to 20 μm between the original free surface and the plug. This process is illustrated in Fig. 19-12. Gas bubbles could also be injected into the channel. Placing the laser several micrometers behind the captive air bubble allowed steady mass-transfer across the bubble, increasing the volume of fluid on the opposite side. This process is illustrated in Fig. 19-13. This "pumping" action can be continued indefinitely, as liquid from the supply reservoir will replace the vapor that passes through the bubble. The maximum volume of fluid pumped in these experiments was a few nanoliters and was only limited by the length of the channels. The bubble remains stationary throughout this process. The measurements show a rapid drop-off in the pumping rates as the beam is moved beyond a few micrometers from the initial spot. This demonstrates that it is optimal to apply heat only near the interface and thereby achieving interphase mass transport without heating the entire volume of the fluid.

19-4 Optofluidic Particle Manipulation

Optofluidic flow actuation and direction have been discussed in details. Another important technical question is the manipulation of microscale and nanoscale particles in liquid by optical control. The particles can be molecules, cells, or any small objects in liquid. The ability to manipulate biological cells and micrometre-scale particles is critical in many biological and colloidal science applications. Conventional manipulation techniques include optical tweezers [27–32], electrophoresis [33,34], dielectrophoresis (DEP) [35], travelling-wave DEP [36,37], magnetic tweezers [38,39], acoustic traps [40], and hydrodynamic flows [41–43]. The optofluidic particle manipulation to be discussed here is different from the direct optical manipulation of the particle itself such as optical tweezer. It is defined here as the particle manipulation either induced by optofluidic flow actuation or induced by electrical field addressed optically. The optofluidic particle manipulation, as will be seen, demonstrates high spatial resolution and flexibility in manipulating molecules and cells and shows great promises in the biomedical applications.

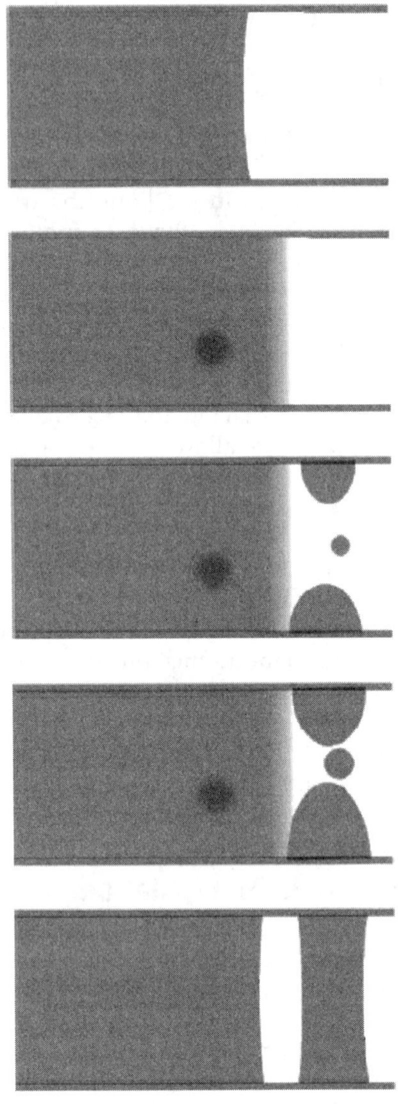

Figure 19-12 Schematic of the process for forming a captive air bubble used in this experiment (view from above). Starting with a still channel (top), the laser is placed near the free surface of the liquid. Local laser heating of the nanoparticles causes accelerated evaporation of the free surface (right), and vapor recondenses on the channel walls 10 to 30 μm from the surface. The droplets on the walls tend to grow together into a continuous liquid slug, trapping an air bubble with a size of 10 to 20 μm (bottom). (*Reprinted with permission from D. A. Boyd, J. R. Adleman, D. G. Goodwin, and D. Psaltis, "Chemical separations by bubble-assisted interphase mass-transfer,"* Anal. Chem., *80, (2008) 2452–2456. Copyright 2008 American Chemical Society.*)

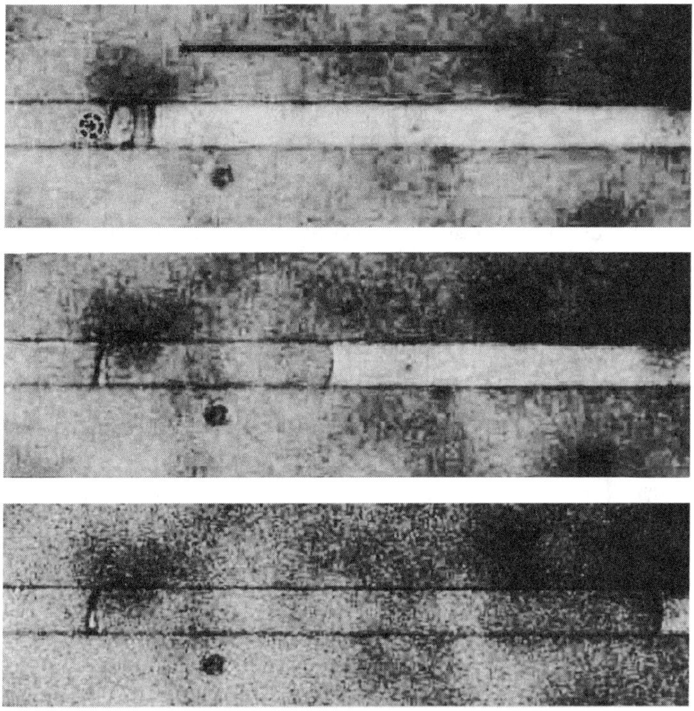

FIGURE 19-13 Images of BAIM process taken during pumping in a 40-μm channel. The scale bar, top, is 200 μm. The position of the laser spot, which is not visible, is represented as dashed outline in the top image. (*Reprinted with permission from D. A. Boyd, J. R. Adleman, D. G. Goodwin, and D. Psaltis, "Chemical separations by bubble-assisted interphase mass-transfer,"* Anal. Chem., *80, (2008) 2452–2456. Copyright 2008 American Chemical Society.*)

19-4-1 Photothermophoretic Molecular Trapping

As described in the previous section, temperature gradient in liquid will induce convection flows. Thermophoresis refers to the effect that particles and molecules are moved by fluidic flow induced by thermal gradients in liquid [44,45]. Thermophoresis is also called thermodiffusion or the Ludwig–Soret [44,45] effect and describes the particle movement due to a temperature gradient, typically from hot to cold. Photothermophoresis is the thermophoresis effect created by photothermal heating of liquid. In bulk-scale liquid particles and molecules are repelled along temperature gradients and depleted from a heated spot by a weak thermophoresis process [46–52]. Interestingly Braun et al. reported that with convection, this depletion in a flat liquid microchamber can be turned into an accumulation [51]. This is an unexpected result since convection normally mixes a solution toward equal concentration.

Braun et al. heats a DNA buffer solution modestly in a 25-μm-thin PDMS chamber with an infrared laser with 0.13 mW [51]. The laser light is focused through a 25-μm-thick chamber using a 32× microscope objective lens with a numerical aperture of 0.4 and the Gaussian focus depth is determined to be 680 μm. By locally heating water with the infrared laser as shown in Fig. 19-14, the DNA molecules in the liquid can be repelled from the heating spot at the beginning. The DNA molecules are labelled with temperature-sensitive fluorescence dye. The local temperature difference is measured by fluorescence intensity imaging to be 2.3 K and 73% of DNA molecules are depleted from the heating center due to thermophoretic repulsion (Fig. 19-14b). The temperature distribution is checked against a 2D numerical simulation. It is shown in grey equithermal

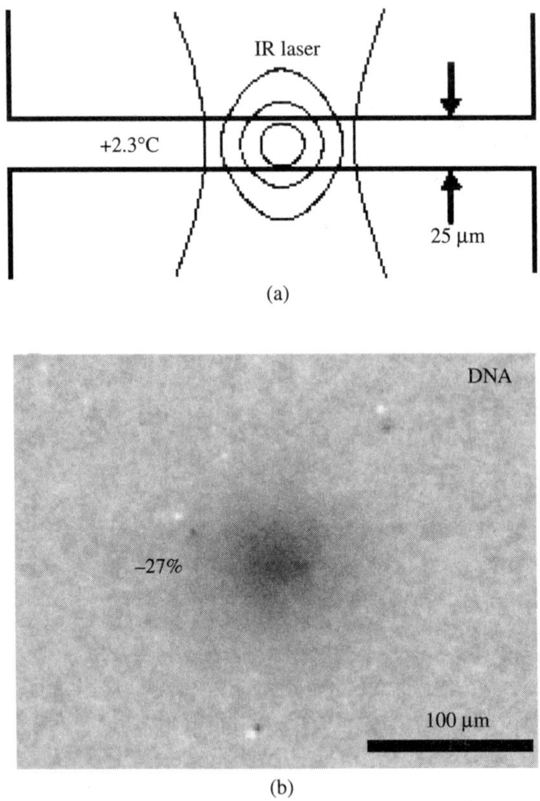

FIGURE 19-14 Thermophoresis of DNA. (a) The temperature in a thin chamber is raised by 2.3 K with infrared heating. (b) DNA (5.6 K base pair) is repelled along the gradients by thermophoresis. A concentration drop of 27% is imaged. (*Reprinted with permission from D. Braun and A. Libchaber, "Trapping of DNA by thermophoretic depletion and convection,"* Phys. Rev. Lett., *89, (2002) 188103. Copyright 2002 by American Physical Society.*)

lines with a spacing of 0.5 K. The gradient is mostly parallel to the chamber walls due to the low-conducting PDMS chamber material (Fig. 19-14a). Therefore, the optical averaging across the chamber yields a small error.

Trapping can be found after doubling the chamber thickness to 50 μm, using cooling glass cover slips and increasing the heating power to 10 mW (Fig. 19-15). Thermal convection is then greatly enhanced and gives rise to an accumulation of DNA at the lower surface of the chamber near the heating spot. The repulsion by thermophoresis is now much stronger (Fig. 19-15c vs. Fig. 19-14b). Shortly after repulsion, the DNA is accumulated in a ring geometry at the bottom of the chamber by a synergistic combination of fousr phenomena (Fig. 19-15d): (1) DNA is repelled from the heated center by lateral thermophoresis; (2) convection breaks the symmetry and transports the repelled DNA downward, while upward convection occurs in the depleted center; (3) axial thermophoresis pins the DNA with high temperature gradients toward the cooling glass most efficiently in the slow-flow area near the central stagnation point; and (4) the DNA brought by

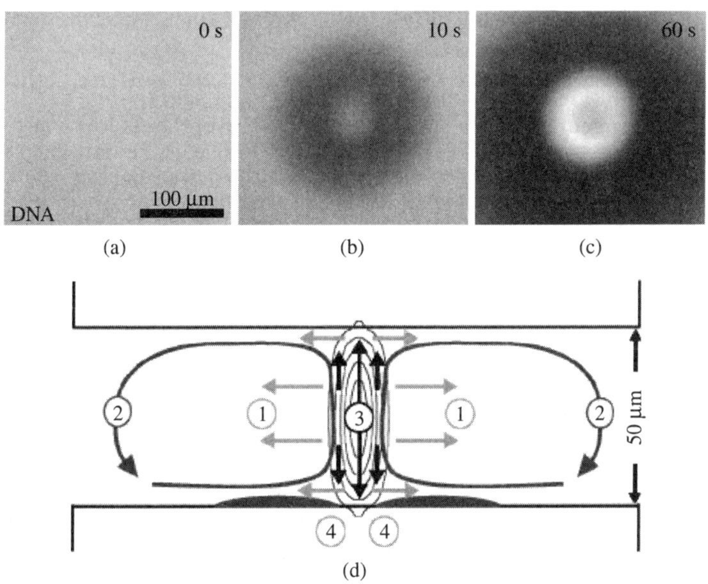

FIGURE 19-15 Mechanism of thermophoretic trapping of DNA. (a) Image of stained DNA before heating. (b) Thermophoresis from central heating first repels DNA laterally. (c) Thermophoresis and toroidal convection trap DNA in the center toward the lower chamber wall in a ring geometry. The concentration enhancement is 13-fold. (d) The mechanism of thermophoretic trapping in the center is an interplay of lateral thermophoresis (1), (4) and axial thermophoresis (3) with convection (2). (*Reprinted with permission from D. Braun and A. Libchaber, "Trapping of DNA by thermophoretic depletion and convection,"* Phys. Rev. Lett., *89, (2002) 188103. Copyright 2002 by American Physical Society.*)

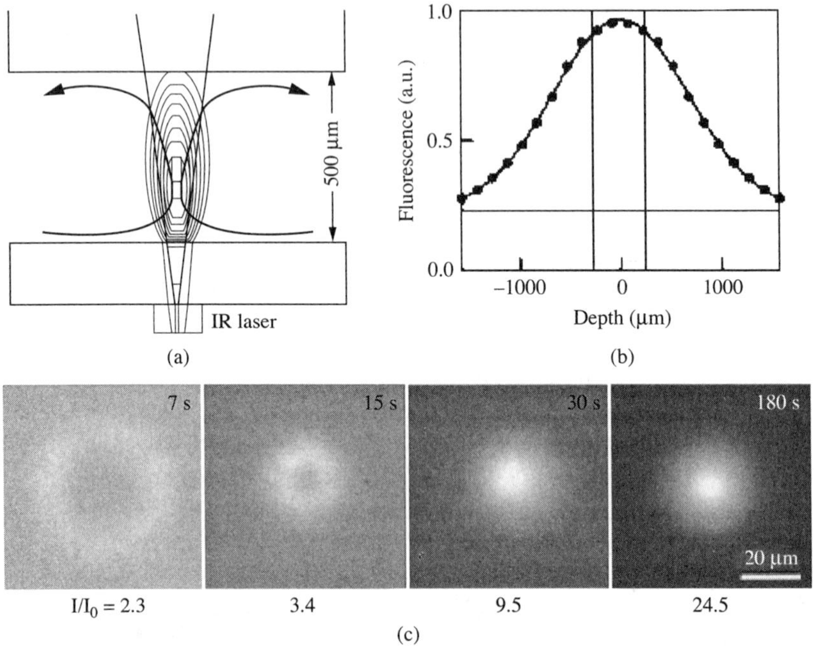

FIGURE 19-16 Strong thermophoretic trapping of DNA. (a) Infrared heating is applied from a divergent beam in a 500-μm-thick chamber. (b) The fluorescence focus detects across the whole chamber, given by vertical lines. (c) Strong trapping of DNA is achieved and initially ring-shaped as imaged with fluorescence of a DNA stain. (*Reprinted with permission from D. Braun and A. Libchaber, "Trapping of DNA by thermophoretic depletion and convection,"* Phys. Rev. Lett., *89, (2002) 188103. Copyright 2002 by American Physical Society.*)

convection is laterally repelled by the edges of the Gaussian heating spot into a ring of accumulation. To summarize, the trapping is the result of the combination of convection and thermophoresis which are both induced by temperature gradients.

Even stronger trapping of DNA molecules can be realized by increasing the chamber thickness to 500 μm (Fig. 19-16). Infrared heating can be applied from below directly out of a single-mode fiber to increase the focal depth of the light illumination. Under these conditions DNA is trapped to a point geometry and the axial temperature gradients are dominating the final equilibrium.

More recently Duhr et al. oriented the fluid flow along the thermal gradient made by optical heating with an infrared laser focus and manipulate biological molecules [52]. The combination of thermophoresis and fluid flow results in strong trapping of small biomolecules (Fig. 19-17). They directly oppose a fluid flow with a thermal gradient in a microfluidic channel with 10 μm × 10 μm cross section. The channel is surrounded by PDMS silicone. They oppose the flow of DNA containing water with a locally enhanced temperature gradient,

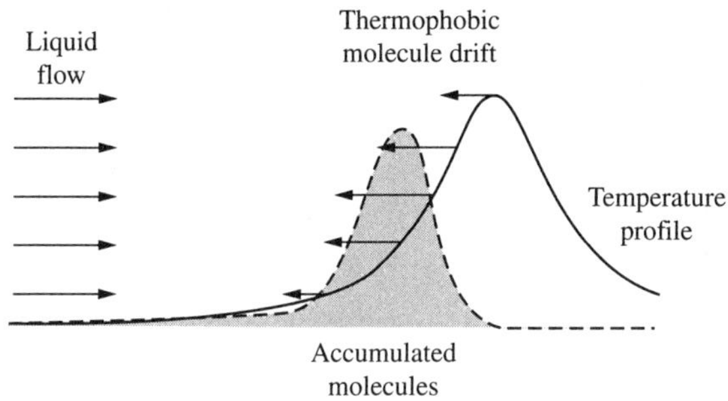

FIGURE 19-17 Principle of thermophoretic flow trap. A warm spot repels molecules by thermophoresis. Counteracting fluid flow leads to accumulation of molecules upstream of the warm spot. (*Reprinted with permission from S. Duhr and D. Braun, "Optothermal molecule trapping by opposing fluid flow with thermophoretic drift,"* Phys. Rev. Lett., *97, (2006) 038103. Copyright 2002 by American Physical Society.*)

created with a focused infrared laser. DNA molecules are flowing with an effective velocity of 0.55 μm/s against the warm spot. First, DNA is depleted at the warm spot, followed by accumulation in front of the depletion as the water drift from the left brings in DNA. Convection has proven insignificant in these thin-fluid films [52] with convection velocities on the order of nanometers per second which leads to a fast and strong accumulation of DNA molecules.

It is also demonstrated that DNA molecules in a two-dimensional 20-μm-high water sheet can be accumulated by optically imposed circular heat rings, which move concentrically toward a spot (Fig. 19-18). With a radial drift velocity of 2.3 μm/s and an average temperature difference of 20 K, short 1000 base pair DNA was accumulated sixfold in a spot of 10-μm radius within 80 s by radially moving a circular temperature ring. This result illustrates how optically triggered thermophoretic traps can accumulate molecules in a variety of geometries. Opposing thermophoretic flow against a fluid flow creates a molecule trap. Fluorescence measurements in a microfluidic channel agree well with the one-dimensional superposition of both flows. Within 15 min, 10,000 base pair DNA can be accumulated 16-fold by a warm microscopic spot with a temperature difference of 16 K.

19-4-2 Optofluidic Dielectrophoretic Manipulation

Optical tweezer is a very important tool in biological research to manipulate micro- and nanoscale objects in liquid [54]. Alternatively electrophoresis and dielectrophoresis effects can be utilized to manipulate microparticles in liquid. Dielectrophoresis refers the motion of an electrically neutral particle resulting from the interaction between the applied electric field and the induced dipole. It has been used widely in the manipulation of micro- or submicroparticles and biological cells [55,56]. Recently the advantages of optical tweezing and electric- or

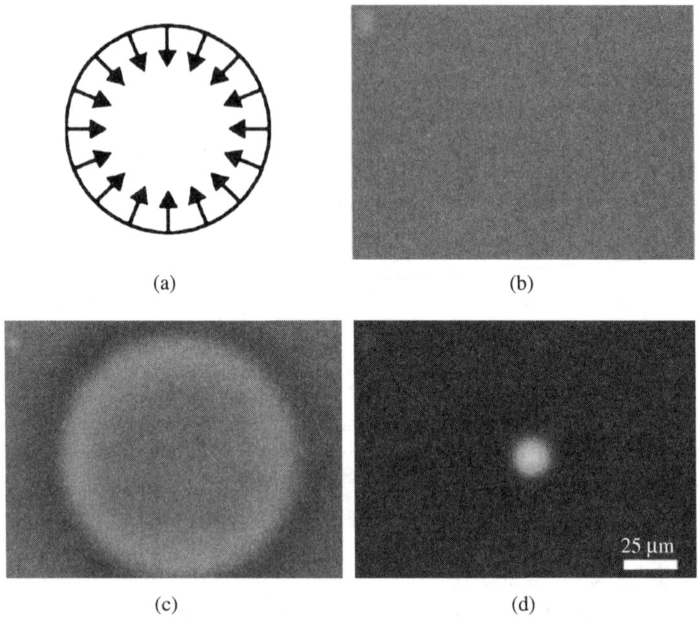

Figure 19-18 Principle of thermophoretic flow trap. A warm spot repels molecules by thermophoresis. Counteracting fluid flow leads to accumulation of molecules upstream of the warm spot. (*Reprinted with Permission from S. Duhr and D. Braun, "Optothermal molecule trapping by opposing fluid flow with thermophoretic drift ,"* Phys. Rev. Lett., *97, (2006) 038103. Copyright 2002 by American Physical Society.*)

dielectricphoretic trapping are combined in optofluidic manipulations. Ozkan et al. proposed a light-induced electrophoresis mechanism to optically address polymer beads by using dc electric bias [57]. The electrically charged particles are attracted to the electrode with opposite polarity. Chiou et al. presented a light-induced dielectrophoresis mechanism that would allow the optical addressing of electrically neutral microparticles with microwatt optical energy [58], which is much lower than the 1 ~ 100 mW optical energy used by optical tweezer.

The schematic structure of the optoelectronic tweezer is shown in Fig. 19-19. The liquid solution containing the particles is sandwiched between two surfaces separated by a gap spacing of 100 μm. The top surface is a commercial ITO glass. The bottom surface is a glass substrate coated with three pattern-less layers: a 200-nm-thick aluminum layer, a 2-μm-thick photoconductive (amorphous silicon) layer, and a 20-nm-thick silicon nitride layer. An ac bias is applied between the top (ITO) and the bottom (aluminum) electrodes. In the dark state, most of the voltage drops across the photoconductor due to its high electrical impedance. This results in a very weak electric field in the liquid layer. When the laser beam is focused on the photoconductive

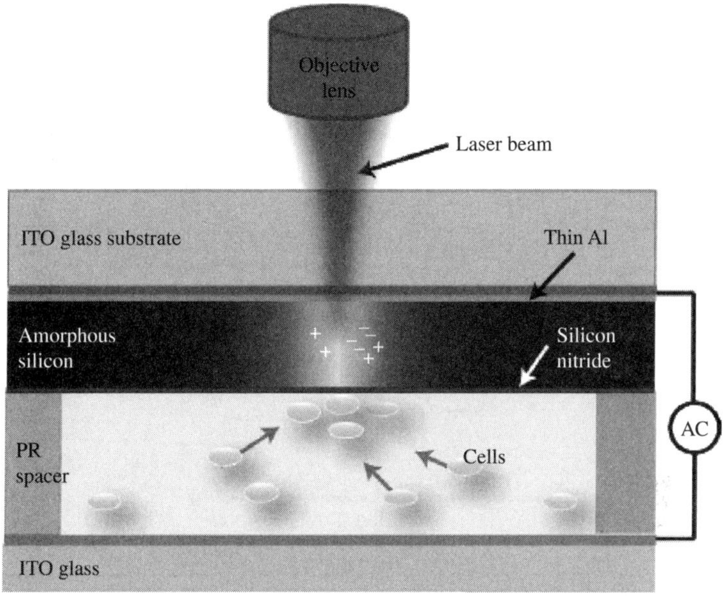

FIGURE 19-19 Schematic structure of optoelectronic tweezers. A virtual electrode is created at the laser-illuminated area. (*Source: P. Y. Chiou, W. Wong, J. C. Liao, and M. C. Wu, "Cell addressing and trapping using novel optoelectronic tweezers,"* IEEE 17th Annual International Conference on Micro Electro Mechanical Systems (MEMS '04), *(2004) 21–24.*)

layer, the local photoconductivity at the site under light illumination is greatly increased due to the photo-generated electron-hole pairs. A light-defined microelectrode is turned on locally and creates a highly nonuniform field in the liquid layer. The laser spot creates a light-defined electrode and a highly nonuniform electric field in the liquid layer. The particles inside the liquid is polarized by the nonuniform field and pushed away from the illuminated site by the negative DEP force. Since light is used to switch the ac voltage drop between the photoconductive layer and the liquid layer, rather than to directly trap the particles, the required optical power is orders of magnitudes lower than that of conventional optical tweezers.

The photoconductor layer and the focused-light address provide a high spatial confinement of the electrical field. Figure 19-20 shows the simulated electric field distribution in the liquid layer for a 17-μm spot generated by a focused laser beam and a bias voltage of 10 V. The conductivity of the liquid is 1 mS/m. The photoconductivity of the amorphous silicon layer is assumed to have a Gaussian distribution, following the profile of the incident light, with a peak conductivity of 10 mS/m at the center. Since the DEP force is proportional to the gradient of E^2, the electric field distribution shows that the OET can

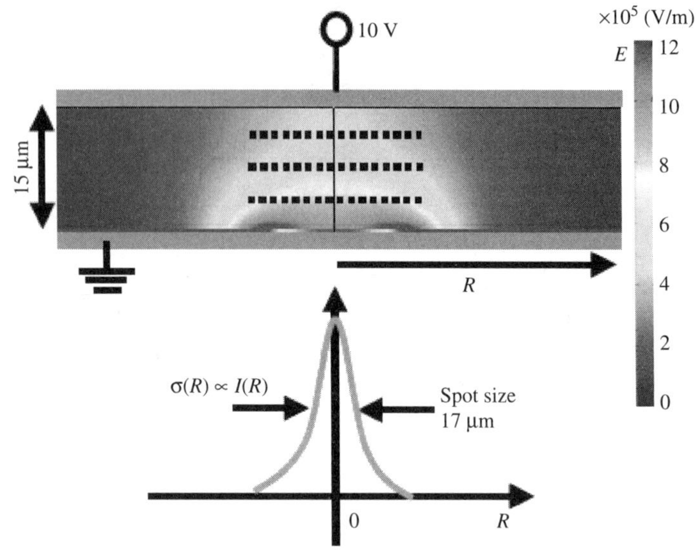

FIGURE 19-20 Electric field distribution in the liquid layer when the photoconductor is illuminated by a focused laser with 17-μm spot size. *(Source: P. Y. Chiou, W. Wong, J. C. Liao, and M. C. Wu, "Cell addressing and trapping using novel optoelectronic tweezers,"* IEEE 17th Annual International Conference on Micro Electro Mechanical Systems (MEMS '04), *(2004) 21–24.*)

generate strong DEP force within a radius of ~20 μm in the lateral direction. There is also a vertical gradient that will attract microparticles toward the photoconductor surface. Both the lateral and the vertical gradients are strongest near the edge of the laser spot, similar to those generated by a physical electrode.

OET allows a focused optical beam to create a virtual electrode on a photoconductive surface, producing a highly nonuniform electric field. This enables optical addressing of dielectrophoresis forces with a spatial resolution of ~ 1 μm. By scanning the laser beam, the trapped cells can be moved to any position on a 2D surface. Experimentally, Chiou et al. have successfully demonstrated the concentration and transport of multiple *Escherichia coli* cells using a HeNe laser with a focused spot size of 17 μm and an optical power as low as 8 μW is sufficient to trap the *E. coli* cells [59].

The schematic structure of the OET device for cell manipulation is the same as shown in Fig. 19-19. The top wafer is flipped over and placed on top of the bottom wafer with a 15-μm-thick photoresist spacer. The top and the bottom surfaces are not physically bonded together. Instead, since both surfaces are hydrophilic, they are pulled toward each other by the surface tension of the liquid until stopped by the photoresist spacer. An ac bias is applied between the top and the bottom ITO electrodes. When the laser beam is focused

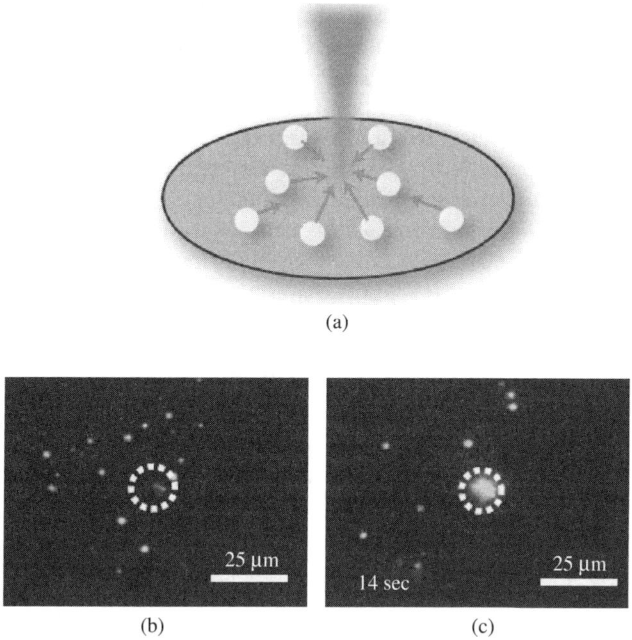

FIGURE 19-21 (a) Schematic diagram of cell concentrator. (b) Images of fluorescent *E. coli* cells before OET is turn on. (c) The same image after the OET is turned on for 14 s. The *E. coli* cells are "focused" by the OET to the laser spot. (*Source: P. Y. Chiou, W. Wong, J. C. Liao, and M. C. Wu, "Cell addressing and trapping using novel optoelectronic tweezers,"* IEEE 17th Annual International Conference on Micro Electro Mechanical Systems (MEMS '04), *(2004) 21–24.*)

on a fixed spot, the OET attracts cells within the trapping area toward the center of the focus. It functions as a cell concentrator. Figure 19-21 shows the captured video images of the *E. coli* cells before and after the OET is turned on. In the experiment, we use the *E. coli* cells that can express green fluorescent protein (GFP) for the convenience of observation under fluorescent microscope. The liquid has a conductivity of 1 mS/m. A 100-kHz, 10-V peak-to-peak ac electric bias was applied between the top and the bottom ITO electrodes. The *E. coli* cells experience positive DEP force under these conditions. Before the laser is turned on, the electric field in the liquid is very weak and the *E. coli* cells are randomly distributed (Fig. 19-21). When the laser is turned on, the cells within a 20-μm radius start to move toward the laser beam, and eventually are trapped at the focal spot, as shown in Fig. 19-21c. Due to the electric field gradient in the vertical direction, the cells are trapped right on top of the photosensitive surface.

Larger OET trapping array can be created by illuminating an optical image on the photoconductive surface. Chiou et al. created an optical image in OET by combining a light-emitting diode and a digital micromirror spatial light modulator (Texas Instruments, 1024 × 768 pixels, 13.68 mm × 13.68 mm pixel size) [60]. The pattern is imaged onto the photoconductive surface through a 10× objective. The resulting pixel size of the virtual electrode is 1.52 μm. The illumination source is a red light–emitting diode (625-nm wavelength) with a 1-mW output power (measured after the objective lens), which is sufficient to actuate 40,000 pixels. Tight focusing is not required for OET, and the optical manipulation area can be magnified by choosing an appropriate objective lens. Using a 10× objective, the manipulation area (1.3 mm × 1.0 mm) is 500 times larger than that of optical tweezers. A demonstration of the high-resolution capabilities of OET is the creation of 15,000 DEP traps across an area of 1.3 mm × 1.0 mm (Fig. 19-22). The particles are trapped

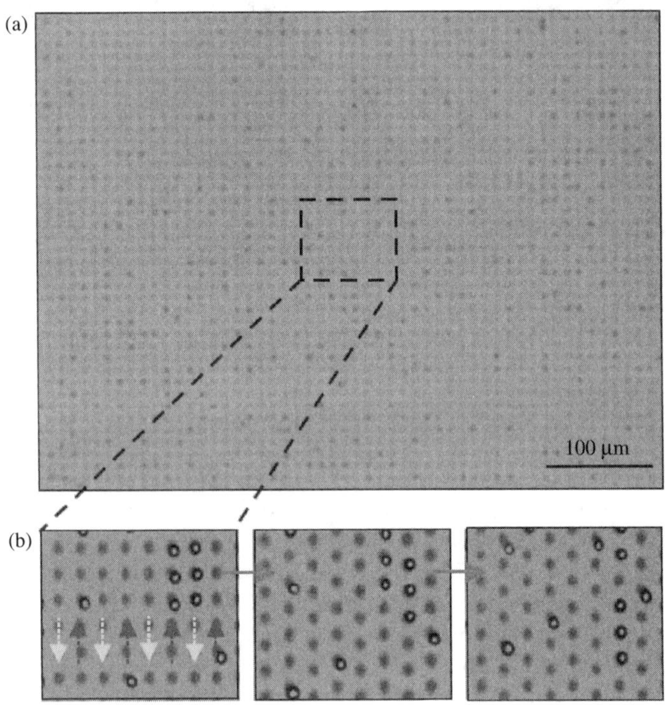

FIGURE 19-22 Massively parallel manipulation of single particles. (a) 15,000 particle traps are created across a 1.3 mm × 1.0 mm area. The 4.5-mm-diameter polystyrene beads experiencing negative DEP forces are trapped in the darker circular areas. Each trap has a diameter of 4.5 mm, which is adjusted to fit a single particle. (b) Parallel transportation of single particles. Three snapshots show the particle motion in part of the manipulation area. The trapped particles in two adjacent columns move in opposite directions, as indicated by the arrows. (*Reprinted by permission from Macmillan Publishers Ltd: P. Y. Chiou, A. T. Ohta, and M. C. Wu: "Massively parallel manipulation of single cells and microparticles using optical images."* Nature, *436, (2005) 370–372.*)

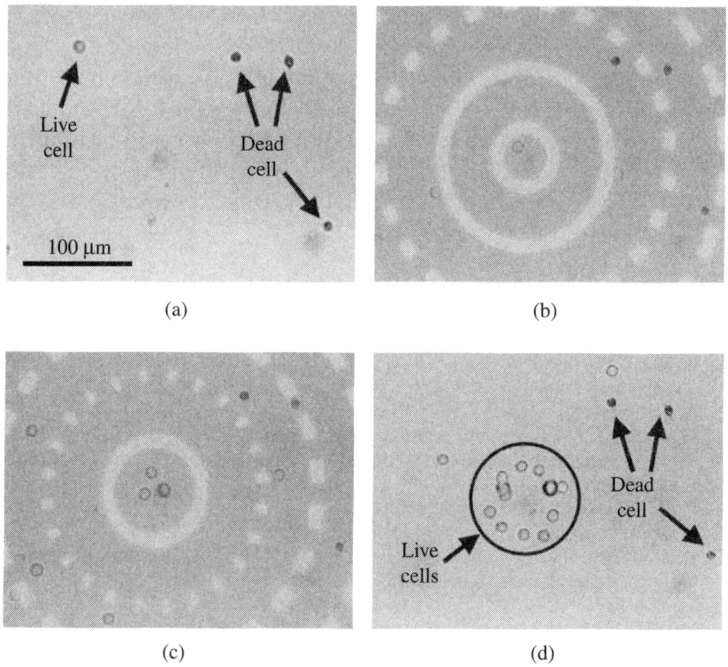

FIGURE 19-23 Selective collection of live cells from a mixture of live and dead cells. (a) Randomly positioned cells before OET. (b and c) Cell sorting. The live cells experience positive OET, trapping them in the bright areas, and pulling the live cells into the pattern's centre. The dead cells (stained with Trypan blue dye) leak out through the dark gaps and are not collected. The optical pattern has a yellowish colour, while weak background scattered light results in a pinkish hue in the nonpatterned areas. (d) Sorted cells. (*Reprinted by permission from Macmillan Publishers Ltd: P. Y. Chiou, A. T. Ohta, and M. C. Wu: "Massively parallel manipulation of single cells and microparticles using optical images."* Nature, *436, (2005) 370–372.*)

in the darker circular areas by the induced negative DEP forces, which push the beads into the nonilluminated regions, where the electric field is weaker. The size of each trap is optimized to capture a single 4.5-mm-diameter polystyrene bead. By exploiting the dielectric differences between different particles or cells, the selective concentration of live human B cells from a mixture of live and dead cells is demonstrated using OET in Fig. 19-23.

19-5 Conclusion

As the merge product of two rapidly emerging fields—optics and microfluidics—optofluidics opens a new path in physical and biological sciences. Light is always the catalyst in the development of new technology in human history. The idea of manipulating liquid and small objects in the liquid using light intrigues many scientists and engineers. This chapter provides a brief review of the latest development of optofluidic liquid actuation and particle manipulation technology. It is reasonable to believe in the near future more and more interesting research results and new discovery will be reported and an all-optical microfluidic processing device will be invented to facilitate the advancement of other fields in physical sciences and biological sciences.

References

1. P. G. de Gennes, F. Brochard-Wyart, and D. Quéré. *Capillary and Wetting Phenomena—Drops, Bubbles, Pearls, Waves*, 2002, Springer, New York.
2. A. M. Cazabat, F. Heslot, S. M. Troian, and P. Carles, "Fingering instability of thin spreading films driven by temperature gradients." *Nature* **346**, (1990) 824–826.
3. M. K. Chaudhury and G. M. Whitesides, "How to make water run uphill." *Science* **256**, (1992)1539–1541.
4. B. S. Gallardo et al., "Electrochemical principles for active control of liquids on submillimeter scales." *Science* **283**, (1999) 57–60.
5. K. Ichimura, S-K. Oh, and M. Nakagawa, "Light-driven motion of liquids on a photoresponsive surface," *Science,* **288,** (2000) 1624–1626.
6. G. M. Whitesides and P. E. Laibinis, "Wet chemical approaches to the characterization of organic surfaces: self-assembled monolayers, wetting, and the physical-organic chemistry of the solid-liquid interface." *Langmuir* **6**, (1990) 87–96.
7. C. D. Bain and G. M. Whitesides, "Modeling organic surfaces with self-assembled monolayers." *Angew. Chem. Int. Ed. Engl.* **28**, (1989) 506–512.
8. L. M. Siewierski, W. J. Brittain, S. Petrash, and M. D. Foster, "Photoresponsive monolayers containing in-chain azobenzene." *Langmuir* **12**, (1996) 5838–5844.
9. G. Moller, M. Harke, H. Motschmann, and D. Prescher, "Controlling microdroplet formation by light." *Langmuir* **14**, (1998) 4955–4957.
10. K. Ichimura, Y. Suzuki, T. Seki, A. Hosoki, and K. Aoki, "Reversible change in alignment mode of nematic liquid crystals regulated photochemically by command surfaces modified with an azobenzene monolayer." *Langmuir* **4**, (1988) 1214–1216.
11. M. Ueda, H.-B. Kim, and K. Ichimura, "Photocontrolled aggregation of colloidal silica." *J. Mater. Chem.* **4**, (1994) 883–889.
12. S.-K. Oh, M. Nakagawa, and K. Ichimura, "Self-assembled monolayers derived from calix4.resorcinarenes exhibiting excellent desorption-resistance and their applicability to surface energy photocontrol." *Chem. Lett.* **4**, (1999) 349–350.
13. E. Kurita et al., "Macrocyclic amphiphiles. Part 2—Multi-point adsorptivity of the crown conformer of calix4.resorcinarenes and their derivatives on surfaces of amorphous polar substrates." *J. Mater. Chem.* **8**, (1998) 397–403.
14. T. Seki, H. Sekizawa, S. Morino, and K. Ichimura, "Inherent and cooperative photomechanical motions in monolayers of an azobenzene containing polymer at the air–water interface." *J. Phys. Chem.* **B 102**, (1998) 5313–5321.
15. I. Panaiotov, S. Taneva, A. Bois, and F. Rondelez, "Photoinduced dilatational motion in monolayers of poly(methyl methacrylate) having benzospiropyran side groups." *Macromolecules* **24**, (1991) 4250–4254.
16. G. Lippmann, "Relations Entre Les Phénomènes Électriques et Capillaires." *Ann. Chim. Phys.* **5**, (1875) 494–549.
17. G. Beni and S. Hackwood, "Electro-wetting displays." *Appl. Phys. Lett.* **38**, (1981) 207–209.
18. F. Mugele and J.-C. Baret, "Electrowetting: from basics to applications." *J. Phys. Condens. Matter* **17**, (2005) R705–R774.

19. S. K. Cho, S. K. Fan, H. Moon, and C.-J. Kim, "Towards digital microfluidic circuits: creating, transporting, cutting, and merging liquid droplets by electrowetting-based actuation, technical digest." *Proceedings of the 15th IEEE International Conference on Micro Electro Mechanical Systems, MEMS,* 2002, pp. 32–35.
20. P.-Y. Chiou, H. Moon, H. Toshiyoshi, C.-J. Kim, and M. C. Wu, "Light actuation of liquid by optoelectrowetting." *Sensors and Actuators* **A 104**, (2003) 222–228.
21. D. Edwards, H. Brenner, and D. Wasan, *Interfacial Transport Processes and Rheology,* 1991, Bufferworth-Heinemann, Boston.
22. G. L. Liu, J. Kim, Y. Lu, and L. P. Lee, "Optofluidic control via photothermal nanoparticles." *Nature Mater.* **5**, (2005) 27–32.
23. R. D. Deegan, et al. "Capillary flow as the cause of ring stains from dried liquid drops." *Nature* **389**, (1997) 827–829.
24. M. A. Unger, H. P. Chou, T. Thorsen, A. Scherer, and S. R. Quake, "Monolithic microfabricated valves and pumps by multilayer soft lithography." *Science* **288**, (2002)113–116.
25. D. A. Boyd, J. R. Adleman, D. G. Goodwin, and D. Psaltis, "Chemical separations by bubble-assisted interphase mass-transfer." *Anal. Chem.* **80**, (2008) 2452–2456.
26. M. S. Plesset and A. Prosperetti, "Flow of vapour in a liquid enclosure." *J. Fluid Mech.* **78**, (1976) 433–444
27. D. G. Grier, "A revolution in optical manipulation." *Nature* **424**, (2003) 810–816.
28. A. Ashkin, J. M. Dziedzic, and T. Yamane, "Optical trapping and manipulation of single cells using infrared laser beams." *Nature* **330**, (1987) 769–771.
29. M. P. MacDonald, G. C. Spalding, and K. Dholakia, "Microfluidic sorting in an optical lattice." *Nature* **426**, (2003) 421–424.
30. J. E. Curtis, B. A. Koss, and D. G. Grier, "Dynamic holographic optical tweezers." *Opt. Commun.* **207**, (2002) 169–175.
31. D. McGloin, G. C. Spalding, H. Melville, W. Sibbett, and K. Dholakia, "Three-dimensional arrays of optical bottle beams." *Opt. Commun.* **225**, (2003) 215–222.
32. V. Garces-Chavez, K. Dholakia, and G. C. Spalding, "Extended-area optically induced organization of microparticles on a surface." *Appl. Phys. Lett.* **86**, (2005) 031106.
33. L. Kremser, D. Blaas, and E. Kenndler, "Capillary electrophoresis of biological particles: Viruses, bacteria, and eukaryotic cells." *Electrophoresis* **25**, (2004) 2282–2291.
34. C. R. Cabrera and P. Yager, "Continuous concentration of bacteria in a microfluidic flow cell using electrokinetic techniques." *Electrophoresis* **22**, (2001) 355–362.
35. M. P. Hughes, "Strategies for dielectrophoretic separation in laboratory-on-a-chip systems." *Electrophoresis* **23**, (2002) 2569–2582.
36. R. Pethig, M. S. Talary, and R. S. Lee, "Enhancing traveling-wave dielectrophoresis with signal superposition." *IEEE Eng. Med. Biol. Mag.* **22**, (2003) 43–50.
37. H. Morgan, N. G. Green, M. P. Hughes, W. Monaghan, and T. C. Tan, "Large-area travelling-wave dielectrophoresis particle separator." *J. Micromech. Microeng.* **7**, (1997) 65–70.
38. J. Yan, D. Skoko, and J. F. Marko, "Near-field-magnetic-tweezer manipulation of single DNA molecules." *Phys. Rev.* **E 70**, (2004) 011905.
39. H. Lee, A. M. Purdon, and R. M. Westervelt, "Manipulation of biological cells using a microelectromagnet matrix." *Appl. Phys. Lett.* **85**, (2004)1063–1065.
40. H. M. Hertz, "Standing-wave acoustic trap for non-intrusive positioning of microparticles." *J. Appl. Phys.* **78**, (1995) 4845–4849.
41. J. O. Kessler, "Hydrodynamic focusing of motile algal cells." *Nature* **313**, (1985) 218–220.
42. N. Sundararajan, M. S. Pio, L. P. Lee, and A. A. Berlin, "Three-dimensional hydrodynamic focusing in polydimethylsiloxane (PDMS) microchannels." *J. Microelectromech. Syst.* **13**, (2004) 559–567.

43. G. B. Lee, B. H. Hwei, and G. R. Huang, "Micromachined pre-focused M×N flow switches for continuous multi-sample injection." *J. Micromech. Microeng.* **11**, (2001) 654–661.
44. C. Ludwig, "Diffusion zwischen ungleich erwärmten Orten gleich Zusammengesetz Losungen," *Sitzber. Akad. Wiss. Wien, Math.-Nat. Wiss. Kl.* **20**, (1856) 539.
45. C. Soret, "Sur l'état d'équilibre que prend, du point de vue de sa concentration, une dissolution saline primitivement homogène, dont deux parties sont portées à des températures différentes," *Arch. Sci. Phys. Nat. Geneve* **2**, (1879) 48–61.
46. J. Rauch and W. Kohler, "Diffusion and thermal diffusion of semidilute to concentrated solutions of polystyrene in toluene in the vicinity of the glass transition." *Phys. Rev. Lett.* **88**, (2002)185901.
47. R. Piazza, S. Iacopini, and B. Triulzi, "Thermophoresis as a probe of particle–solvent interactions: The case of protein solutions." *Phys. Chem. Chem. Phys.* **6**, (2004) 1616–1622.
48. M. Giglio and A. Vendramini, "Thermal-diffusion measurements near a consolute critical point." *Phys. Rev. Lett.* **34**, (1975) 561–564.
49. M. Giglio and A. Vendramini, "Soret-type motion of macromolecules in solution." *Phys. Rev. Lett.* **38**, (1977) 26–30.
50. R. Piazza and A. Guarino, "Soret effect in interacting micellar solutions." *Phys. Rev. Lett.* **88**, (2002) 208302.
51. D. Braun and A. Libchaber, "Trapping of DNA by thermophoretic depletion and convection." *Phys. Rev. Lett.* **89**, (2002) 188103.
52. S. Duhr, S. Arduini, and D. Braun, "Thermophoresis of DNA determined by microfluidic fluorescence." *Eur. Phys. J.* **E 15**, (2004) 277–286.
53. S. Duhr and D. Braun, "Optothermal molecule trapping by opposing fluid flow with thermophoretic drift." *Phys. Rev. Lett.* 97, (2006) 038103.
54. A. Ashkin and J. M. Dziedzic, "Optical levitation by radiation pressure." *Appl. Phys. Lett.* **19**, (1974) 283–285.
55. N. G. Green and H. Morgan, "Dielectrophoresis of sub-micrometre latex spheres. Part I: experimental results." *J. Phys. Chem.* **B 103**, (1999) 41–50.
56. X. B. Wang, J. Yang, Y. Huang, J. Vykoukal, and F. F. Becker, "Cell separation by dielectrophoretic field-flow-fractionation." *Anal. Chem.* **72**, (2000) 832–839.
57. M. Ozkan, S. Bhatia, S. C. Esener, "Optical addressing of polymer beads in microdevices," *Sensors and Materials* **14**, (2002) 189–197.
58. P. Y. Chiou, Z. Chang, and M. C. Wu, "A novel optoelectronic tweezer using light induced dielectrophoresis," *Proceedings IEEE/LEOS International Conference on Optical MEMS and Their Applications (OMEMS '03)*, 2003, 8–9.
59. P. Y. Chiou, W. Wong, J. C. Liao, and M. C. Wu, "Cell addressing and trapping using novel optoelectronic tweezers," *IEEE 17th Annual International Conference on Micro Electro Mechanical Systems (MEMS '04)*, 2004, 21–24.
60. P. Y. Chiou, A. T. Ohta, and M. C. Wu, "Massively parallel manipulation of single cells and microparticles using optical images." *Nature* **436,** (2005) 370–372.

Index

Note: Page numbers referencing figures are followed by an "*f*"; page numbers referencing tables are followed by a "*t*".

P

T